Econometrics of Financial High-Frequency Data

Nikolaus Hautsch

Econometrics of Financial High-Frequency Data

Springer

Professor Dr. Nikolaus Hautsch
Institute for Statistics and Econometrics
School of Business and Economics
Humboldt-Universität zu Berlin
Spandauer Str. 1
10178 Berlin
Germany
nikolaus.hautsch@wiwi.hu-berlin.de

ISBN 978-3-642-21924-5 e-ISBN 978-3-642-21925-2
DOI 10.1007/978-3-642-21925-2
Springer Heidelberg Dordrecht London New York

Library of Congress Control Number: 2011938812

Cover design: eStudio Calamar S.L.

Printed on acid-free paper

Springer is part of Springer Science+Business Media (www.springer.com)

To Christiane, Elias and Emilia

Preface

This book is an extended revision of *Modelling Irregularly Spaced Financial Data – Theory and Practice of Dynamic Duration Models* (Hautsch 2004) which has been written as a doctoral dissertation at the Department of Economics at the University of Konstanz. Six years later, when I started thinking about a second edition of this book accounting for recent developments in the area of high-frequency finance, I realized that an extension of the scope of the book and the inclusion of more topics and material are inevitable. Given the developments in high-frequency finance, the number of new approaches and the current challenges induced by technological progress in market structures as well as in the trading industry, I decided to change the title of the book, to revise and restructure existing material and to include additional topics resulting in a new monography.

Compared to Hautsch (2004), the list of topics has been extended, among others, by various types of univariate and multivariate multiplicative error models, autoregressive count data approaches, dynamic specifications for integer-valued variables as well as models for quote dynamics. Moreover, different approaches to quantify intraday volatility are discussed involving realized volatility measures, trade-based volatility concepts, and intensity-based measures. A further focus lies on the modeling of liquidity and order book dynamics. Finally, institutional settings, market structures, issues of data preparation, preprocessing, and implementation pitfalls as well as illustrations of the empirical properties of high-frequency data are discussed more extensively and thoroughly using updated data from trading at the New York Stock Exchange, NASDAQ and the Deutsche Börse.

The book is intended for researchers interested in methods, approaches and applications in the area of high-frequency econometrics. Moreover, it is written for students and scholars covering this subject, for instance, in a course on financial econometrics, financial statistics, or empirical finance. Students using the book should have a basic knowledge in mathematical statistics, time series analysis, and econometric estimation theory. Finally, the book addresses the needs of financial practitioners who require statistical methods to model and predict high-frequency market processes as well as intraday volatility and liquidity dynamics.

Needless to say that a book focusing on a rapidly developing and growing field, such as high-frequency financial econometrics, is never complete and entirely up-to-date. Moreover, it is impossible to cover *all* specific topics, approaches, and applications. Furthermore, it is obvious that each topic could be addressed in more depth both from a methodological (and mathematical) viewpoint and from an applied side. In fact, some of the topics, such, for instance, the concept of realized volatility, are only touched without going into deep mathematical details. Therefore, I tried to find a compromise between elaborateness, compactness, and topical broadness.

I wish to thank Axel Groß-Klußmann, Gustav Haitz, Ruihong Huang, Peter Malec, and Stefanie Schulz for helpful comments, proof reading, and editing work. Moreover, I am grateful to many colleagues, coworkers and coauthors for inspiring discussions and joint work building the basis for many aspects and topics covered in this book. In this context, I wish to express my special gratitude to Luc Bauwens, Wolfgang Härdle, Anthony Hall, Lada Kyj, Roel Oomen, Mark Podolskij, and Melanie Schienle.

Last but not least, I would like to thank my family. I am exceptionally indebted to my wonderful wife Christiane. Without her love, support, and encouragement, this book could not have been written. I also wish to thank my children Elias and Emilia for providing refreshing and valuable distraction from research and writing a book.

Berlin
May 2011

Nikolaus Hautsch

Contents

Chapter 1
Introduction

1.1 Motivation

The availability of financial data recorded on high-frequency level has inspired a research area which over the last decade emerged to a major area in econometrics and statistics. The growing popularity of high-frequency econometrics is triggered by technological progress in trading systems and trade recording as well as an increasing importance of intraday trading, optimal trade execution, order placement and liquidity dynamics. Technological progress and the growing dominance of electronic trading allows to record market activity on high frequency and with high precision leading to advanced and comprehensive data sets. The informational limiting case is reached when *all* market events, e.g., in form of order messages, are recorded. Such recording schemes result in data bases which are even more detailed than transaction data and allow to reproduce the entire order flow as well as the underlying order book.

A major reason for the academic as well as practical interest in high-frequency finance is that market structures and the process of trading are subject to ongoing changes. This process is induced by technological progress, further developments in trading systems, increasing competition between exchanges and a strong increase in intraday trading activity. The introduction of electronic trading platforms have automatized and speeded up trade execution as well as trade reporting and allow investors to automatize trading strategies, order routing and real-time order management. Particularly the launch of alternative trading systems, e.g., in form of electronic communication networks (ECNs), challenges traditional and established exchanges, induces mergers and take-offers and increases the competition for liquidity. As a consequence, trading costs declined and the speed of trading and order submission substantially increased over the last decade. In blue chip assets, traded, for instance, at the NASDAQ, we can nowadays easily observe more than 100,000 transactions per day. Moreover, competition between exchanges and ECNs creates a greater variety in trading forms, rules and institutional settings challenging economic theory as well as econometric modelling.

N. Hautsch, *Econometrics of Financial High-Frequency Data*,
DOI 10.1007/978-3-642-21925-2_1, © Springer-Verlag Berlin Heidelberg 2012

Transaction and limit order book data, as provided by various exchanges and trading platforms, create strong academic interest as they allow to analyze the impact of institutional settings on the trading process, price discovery as well as the market outcome and enable to study market dynamics and traders' behavior on the lowest possible aggregation level. Monitoring asset prices as well as liquidity supply and demand on a maximally high observation frequency opens up possibilities to construct more efficient estimators and predictors for price volatility, time-varying correlation structures as well as liquidity risks. Modelling limit order books on high frequencies provide insights into the interplay between liquidity supply and demand, execution risks, traders' order submission strategies and the market impact of order placements. Addressing these issues requires the development and application of econometric models which are tailor-made for specific data and research tasks.

Methods and models for high-frequency data are also of growing importance in financial industry. Important tasks in financial practice are high-frequency predictions of trading volume, volatility, market depth, bid-ask spreads and trading costs to optimize order placement and order execution with minimal price impact and transaction costs. Moreover, econometric and statistical techniques are required to quantify dynamic interdependencies between order flow, volatility and liquidity as well as between markets and assets. Finally, from a regulation viewpoint, (il)liquidity risks, intraday price risks and the consequences of automated high-frequency trading for market outcomes are not fully understood yet and require ongoing empirical investigations.

High-frequency data embody distinct properties which challenge econometric and statistical modelling. One major feature of transaction data is the irregular spacing in time. The question of how this salient property should be treated in an econometric model is not obvious. Indeed, the time between market events carries information and is a valuable economic variable serving as a measure of trading activity and affects price and volume behavior. Accounting for the timing of market events requires to consider the data statistically as *point processes*. Point processes characterize the random occurrence of single events along the time axis in dependence of observable characteristics and of the process history. The importance of point process models in financial econometrics has been discussed for the first time by the 2003 Nobel laureate Robert F. Engle on the 51th European Meeting of the Econometric Society in Istanbul, 1996. His paper, which is published under Engle (2000), can be regarded as the starting point for a fast growing body of research in high-frequency financial econometrics.

A further important property of financial high-frequency data is the discreteness of prices, quotes, bid-ask spreads or, e.g., trade counts in fixed intervals. Moreover, most high-frequency variables are positive-valued, positively autocorrelated, strongly persistent and follow distinct intraday periodicities. Finally, trading and order book processes are inherently high-dimensional and reveal complex multivariate dynamic structures. To capture these properties, new types of econometric models are developed combining features of (multivariate) time series models, microeconometric, e.g., categorical, approaches, point process models as well as factor specifications.

The objective of this book is to provide a state-of-the art overview of the most important approaches in high-frequency econometrics. A major aim is to discuss implementation details including insights into properties of high-frequency data and institutional settings and to present applications to volatility and liquidity estimation, order book modelling and market microstructure analysis.

An important task of successful high-frequency modelling is to appropriately capture the dynamics in the data. In this context, *autoregressive conditional mean* models play a dominant role in the literature. The underlying principle is to model the conditional mean as an autoregressive process which is updated based on observation driven or parameter driven innovations. If the conditional mean function multiplicatively interacts with positive-valued error terms, the class of *multiplicative error models* (MEMs) – named according to Engle (2002) – is obtained. These specifications are popular to model the dynamics of continuous positive-valued random variables, such as trade-to-trade durations, trading volumes or market depth. They have been originally introduced by Engle and Russell (1997, 1998) in form of the *autoregressive conditional duration* (ACD) model to capture the dynamics of trade-to-trade durations. Alternatively, autoregressive specifications of the (conditional) mean of count data distributions yield a class of autoregressive count data models, as, e.g., the *Autoregressive Conditional Poisson* (ACP) model. As illustrated in this book, this class of models is readily modified to capture non-linearities in dynamics, long range dependence, explanatory variables as well as intraday seasonalities and can be extended to multivariate settings.

A further methodological focus lies on dynamic models for the (stochastic) intensity function. The latter is a central concept in the theory of point processes and is defined as the instantaneous rate of occurrence given the process history and observable factors. The intensity function describes a point process in continuous time and thus allows to account for events and information arrival occurring at *any* point in time, as in the case of time-varying covariates or induced by the arrival of other point processes. As illustrated in this book, dynamic intensity processes can be either specified as so-called *self-exciting* intensity processes, where the intensity is driven by functions of the backward recurrence time to all previous points, or as time series specifications where the intensity function follows a dynamic structure that is updated at each occurrence of a new point. For instance, multivariate intensity processes can be used to model order arrival processes in continuous time or to study multivariate instantaneous price change intensities.

Moreover, to reduce the dimensionality of multivariate trading processes or order book dynamics, factor-based approaches play an increasingly important role. For instance, latent dynamic factor models are applied to capture commonalities in market processes stemming from a common underlying unobservable factor. Alternatively, multi-factor models are used to model order book dynamics and time-varying order book curves.

These types of models, applications thereof and corresponding empirical evidence will be discussed in more detail in the sequel of this book.

1.2 Structure of the Book

Chapter 2 provides insights into trading rules, trading forms and institutional settings. We discuss different forms of trading, most importantly quote-driven markets vs. order-driven markets, and present typical order precedence and pricing rules. As an illustration, we show the institutional settings of selected international exchanges. Moreover, the chapter gives a compact overview on major branches in market microstructure theory. We briefly explain the main principles underlying sequential trade and strategic trade models, inventory-based approaches as well as models for limit order markets.

Chapter 3 discusses the different types of high-frequency data and illustrates preparation and processing steps as well as potential pitfalls and problems which should be taken into account. In this context, the concept of so-called financial durations arising from aggregations based on specific trading events is introduced. Here, we consider the (multivariate) point process describing the complete trading process of a financial asset over a given time span. By systematically selecting individual points of this process, different types of financial duration processes, such as trade-to-trade durations or price change durations are generated.

The major part of the chapter focuses on the presentation and discussion of the major empirical properties of different types of high-frequency variables, such as returns, trading intensities, volatilities, trading volumes, bid-ask spreads and market depth based on different aggregation levels. The illustrations are based on blue chip stocks traded at the New York Stock Exchange (NYSE), NASDAQ and XETRA.

Chapter 4 introduces to the theory of point processes and provides methodological background. The main focus lies on the intensity function and the integrated intensity function as the key concepts to describe point processes in a continuous-time framework. We briefly review their statistical properties and show how to perform intensity-based inference. By concentrating on a non-dynamic framework, we illustrate the basic statistical concepts and discuss different ways to model point processes. In this context, we review different possibilities to classify point process models. This leads to a distinction between *proportional intensity (PI) models* and *accelerated failure time (AFT) models* as well as a classification in intensity models, duration models and count data approaches. These different classes of point processes build the basis for dynamic extensions considered in the remainder of the book.

In Chap. 5, we present the most common type of dynamic duration model, the *autoregressive conditional duration (ACD) model* proposed by Engle and Russell (1998) which is equivalent to a multiplicative error model (MEM) for durations. It allows to model autocorrelated durations in a discrete-time framework and combines elements of GARCH specifications with features of duration models. We discuss the statistical properties of the basic (linear) ACD model, illustrate how to estimate it using (quasi) maximum likelihood and how to account for explanatory variables and intraday periodicities. Moreover, as the most important alternative to a linear specification, we present a logarithmic ACD model, as proposed by Bauwens and Giot (2000), which does not require explicit non-negativity restrictions of the parameters and allows to apply quasi maximum likelihood techniques as well.

A major part of the chapter is devoted to specification tests for ACD models. We discuss various tests against distributional and dynamical misspecification including Portmanteau tests and independence tests for ACD residuals, nonparametric distribution tests as well as Lagrange multiplier tests and conditional moment tests. The latter are designed to test for misspecifications of the functional form and thus violations of the conditional mean restriction implied by the ADC model.

Empirical evidence, however, shows that in most applications, linear multiplicative error models are too restrictive to sufficiently fit the data. This has given birth to a plethora of papers extending the linear MEM/ACD model in various directions. In Chap. 6, we show generalizations of basic MEMs which can be presented in terms of a generalized polynomial random coefficient model according to Carrasco and Chen (2002). Moreover, we illustrate several other types of MEM/ACD models including (smooth) transition ACD models, Markov Switching ACD specifications and ACD models accommodating for long range dependence in the data. Particular emphasis is put on mixture ACD models, where the conditional mean function is driven by a dynamic latent factor, and component models combining intradaily dynamics with daily dynamics.

Chapter 7 focuses on multivariate extensions of multiplicative error models. We discuss statistical properties and inference for the basic multivariate model and illustrate applications. Particular focus lies on the presentation of stochastic vector MEMs. The underlying principle is to augment a VMEM process by a common latent factor which jointly affects the individual components. The latent process follows its own (parameter driven) dynamic and serves as a driving force for commonalities and common autoregressive structures in the multivariate process. The individual components are subject to own (observation driven) dynamics capturing idiosyncratic effects. As a result, the model combines both parameter driven as well as observation driven dynamics and is useful to describe the complex dynamics of multivariate trading processes. As the dynamic latent factor has to be integrated out, the model cannot be estimated using standard maximum likelihood and requires simulation-based techniques. In this context, we discuss simulated maximum likelihood estimation employing efficient importance sampling techniques.

Chapter 8 presents different intraday volatility estimators. We discuss realized measures to estimate the intraday quadratic variation of prices and review selected estimators, such as the maximum likelihood estimator by Aït-Sahalia et al. (2005), the realized kernel estimator by Barndorff-Nielsen et al. (2008a) as well as the pre-averaging estimator by Jacod et al. (2009). In applications, we illustrate how to use these estimators to estimate intraday variances over windows ranging between 1 h and 5 min. As high-frequency variances are closely related to spot variances and particularly affected by asset price jumps, we also discuss basic ideas behind jump-robust and spot variance estimators. As an alternative to spot variances on ultra-high frequencies, we illustrate estimators of trade-to-trade return variances based on high-frequency GARCH models as proposed by Engle (2000) and Ghysels and

Jasiak (1998). Finally, we present the idea of price change durations and illustrate how to construct intensity-based volatility estimators as alternatives to volatility estimators based on calendar time aggregations.

In Chap. 9, we focus on the estimation and prediction of market liquidity. Here, we discuss different dimensions of liquidity and present simple bid-ask spread and price impact measures. As alternative liquidity concepts, we study the use of volume durations to capture the time and volume dimension of liquidity and excess volume durations as measures of one-sided trading intensity. However, since the *traded* volume indicates only the liquidity *demand*, the pending volume in order book queues, reflects the *supply* of liquidity. To model the latter, we present dynamic factor approaches to estimate and predict time-varying order curves. The major principle is to reduce the high dimensionality of order curves by parsimoniously describing their shapes by a low number of parametric or non-parametric factor specifications. Then, time variations are captured by multivariate time series models of the factor loadings. We illustrate the usefulness of such approaches to model order book dynamics.

Chapter 10 presents dynamic semiparametric proportional hazard (PH) models as an alternative and direct counterpart to the class of ACD models considered in Chap. 5. In the semiparametric PH model, the hazard function is specified based on the product between some non-specified baseline hazard characterizing the underlying distribution of the durations and a function of covariates. The advantage of this model is that no explicit distributional assumptions are needed since the baseline hazard is estimated semiparametrically. The so-called *autoregressive conditional proportional hazard (ACPH) model* proposed by Gerhard and Hautsch (2007) exploits the close relationship between ordered response models and approaches for categorized durations. The main idea is to categorize the durations and formulate the semiparametric PH model in terms of a specific type of ordered response model which is augmented by an observation driven dynamic. Such an approach allows for a consistent estimation of the dynamic parameters without requiring explicit distributional assumptions for the baseline hazard. Discrete points of the baseline survivor function can be estimated simultaneously with the dynamic parameters. It is illustrated that the approach allows to account for censoring structures, induced, e.g., by trading halts or non-trading periods and therefore is useful to estimate price change intensities based on aggregation levels covering longer time spans.

Chapter 11 addresses univariate autoregressive intensity models. We consider dynamic parameterizations of the intensity function which allow for a modelling of point processes in continuous time. The focus is on two general types of models: *autoregressive conditional intensity* (ACI) models and *self-exciting intensity models*. The first class of models is proposed by Russell (1999) and is based on an autoregressive structure of the intensity function that is updated at each occurrence of a new point. In the latter class of models, the intensity is driven by a function of the backward recurrence time to all previous points. Hawkes (1971) introduces a linear self-exciting process based on an exponentially decaying backward recurrence function. For both types of models we discuss their theoretical properties as well as estimation issues.

Multivariate autoregressive intensity models are introduced in Chap. 12. We present multivariate extensions of ACI and Hawkes models. Moreover, as a generalized framework for the modelling of point processes, *stochastic conditional intensity (SCI) processes* as proposed by Bauwens and Hautsch (2006) are discussed. The main idea is to assume that the conditional intensity, given the (observable) history of the process, is not deterministic but stochastic and itself follows a dynamic process. Accordingly, the SCI model embodies characteristics of a doubly stochastic Poisson process (see, e.g., Grandell 1976, or Cox and Isham 1980). A multivariate SCI model is obtained by assuming the latent dynamic factor serving as a common component jointly driving all individual processes. Then, similar to the principle underlying stochastic vector MEMs in Chap. 7, idiosyncratic effects are captured by component-specific dynamics. We illustrate probabilistic properties of the model, statistical inference as well as applications to the analysis of multivariate price intensities.

Chapter 13 presents models for autoregressive discrete-valued processes and quote dynamics. A major focus lies on the discussion of *autoregressive conditional Poisson* (ACP) models where the conditional mean of the Poisson distribution is dynamically parameterized. Extensions thereof, such as the Negative Binomial distribution and Double Poisson distribution lead to Autoregressive Conditional Negative Binomial and Autoregressive Conditional Double Poisson models, respectively. As shown in this chapter, these approaches have the advantage of being straightforwardly extended to a multivariate framework. Moreover, we review the classical approach by Hasbrouck (1991) to model (mid-)quote and price dynamics in terms of a vector autoregressive framework. While this approach is easily implemented and extended in various ways, it has the disadvantage of not explicitly accounting for the discreteness of prices and quotes on transaction level. Addressing the latter issue leads to the class of dynamic models for integer-valued variables allowing to model discrete-valued transaction price changes, bid-ask spreads, indicators of the trading direction (buy vs. sell) or trade sizes occurring only in round lot sizes. In this context, we present autoregressive conditional multinomial models and integer count hurdle models decomposing integer-valued random variables into their directional components (negative, zero, positive) as well as their magnitudes. Moreover, the chapter presents structural approaches to capture the joint dynamics of bid and ask quotes.

The individual chapters include extensive empirical illustrations of the presented frameworks and models. We illustrate the models' usefulness and potential in various applications focusing on the modelling of trading processes, the estimation and prediction of volatility and liquidity as well as the modelling of limit order books.

Note that it is not necessary to read all chapters in a strict order. For readers who are mainly interested in methodological issues, it is recommended to first read Chap. 4 and then to focus on the methodological Chaps. 5–7 and 10–13 or the application-orientated Chaps. 8 and 9 which can be read separately. Readers who are interested in major empirical properties of high-frequency data, data preparation issues as well as institutional and economic background are referred to the Chaps. 2 and 3. Finally, the appendix contains a review of the most important distributions relevant for high-frequency data.

References

Aït-Sahalia Y, Mykland P, Zhang L (2005) How often to sample a continuous-time process in the presence of market microstructure noise. Rev Financ Stud pp. 351–416

Barndorff-Nielsen O, Hansen P, Lunde A, Shephard N (2008a) Designing realized kernels to measure the ex-post variation of equity prices in the presence of noise. Econometrica 76:1481–1536

Bauwens L, Giot P (2000) The logarithmic ACD model: an application to the bid/ask quote process of two NYSE stocks. Annales d'Economie et de Statistique 60:117–149

Bauwens L, Hautsch N (2006) Stochastic conditional intensity processes. J Financ Econom 4:450–493

Carrasco M, Chen X (2002) Mixing and moment properties of various GARCH and stochastic volatility models. Econom Theory 18(1):17–39

Cox DR, Isham V (1980) Point processes. Chapman and Hall, London

Engle RF (2000) The econometrics of ultra-high-frequency data. Econometrica 68(1):1–22

Engle RF (2002) New frontiers for ARCH models. J Appl Econom 17:425–446

Engle RF, Russell JR (1998) Autoregressive conditional duration: a new model for irregularly spaced transaction data. Econometrica 66:1127–1162

Gerhard F, Hautsch N (2007) A dynamic semiparametric proportional hazard model. Stud Nonlinear Dynamics Econometrics 11, http://www.bepress.com/snde/vol11/iss2/art1

Ghysels E, Jasiak J (1998) GARCH for irregularly spaced financial data: The ACD-GARCH model. Stud Nonlinear Dynamics Econometrics 2:133–149

Grandell J (1976) Doubly stochastic poisson processes. Springer, Berlin

Hasbrouck J (1991) Measuring the information content of stock trades. J Finance 46:179–207

Hautsch N (2004) Modelling irregularly spaced financial data – theory and practice of dynamic duration models vol. 539 of Lecture Notes in Economics and Mathematical Systems. Springer, Berlin

Hawkes AG (1971) Spectra of some self-exciting and mutually exciting point processes. Biometrika 58:83–90

Jacod J, Li Y, Mykland P, Podolskij M, Vetter M (2009) Microstructure noise in the continuous case: the pre-averaging approach. Stochastic Process Appl 119:2249–2276

Russell JR (1999) Econometric modeling of multivariate irregularly-spaced high-frequency data. Working Paper, University of Chicago

Chapter 2
Microstructure Foundations

This chapter gives an overview of institutional and theoretical market microstructure foundations. Section 2.1 introduces to the institutional framework of trading on modern financial markets. We discuss different forms of trading, types of traders as well as types of orders. Moreover, we present fundamental types of market structures, most importantly quote-driven markets vs. order-driven markets, and provide insights into common order precedence and pricing rules. In the last section, market structures and trading rules on some selected international exchanges are discussed. This section is understood as an overview of the most important institutional aspects of financial market microstructures and mainly follows Harris (2003).

Section 2.2 provides a compact overview on the fundamental strings of theoretical market microstructure literature. We review classical approaches of asymmetric information based market microstructure theory, such as sequential trade models and strategic trade models, as well as inventory-based approaches and models for limit order markets. The purpose of this section is not to provide an in-depth derivation and discussion of the individual models. Rather we compactly illustrate the basic principles underlying the different approaches and review their major predictions for market microstructure relationships and trading dynamics. Readers interested in more details are referred, e.g., to Hasbrouck (2007).

2.1 The Institutional Framework of Trading

2.1.1 Types of Traders and Forms of Trading

Traders either trade on their own account, arrange trades for others or have others arranging trades for them. Accordingly we distinguish between *proprietary traders* trading on their own accounts and *brokers* arranging trades as agents for their clients. This results either in *proprietary trading* or *agency or brokerage trading*. *Dealers*

N. Hautsch, *Econometrics of Financial High-Frequency Data*,
DOI 10.1007/978-3-642-21925-2_2,

are traders who stand ready to trade with other traders (or clients) when they want to trade. Correspondingly, they serve as liquidity suppliers taking the opposite side of the market if required. They operate as market makers, specialists or floor traders. Often traders simultaneously serve as brokers and dealers and are known as *broker-dealers*.

The buy side of a market is typically driven by investors (e.g., individuals, funds, money managers, endowments), borrowers (e.g., individuals or companies), hedgers (e.g., banks) or speculators. The market's sell side consists of dealers serving, (e.g., as market makers or floor traders), brokers (e.g., retail brokers or institutional brokers), or broker-dealers representing well-known companies, like Goldman Sachs or Merrill Lynch.

Exchanges are the platform where traders arrange their trades. On most exchanges only member brokers are allowed to trade. Non-members can only trade by instructing a member to trade for them. We distinguish between *floor trading*, where traders meet on exchange floors to arrange trades and *electronic trading* where traders trade via electronic systems. *Order-driven systems* are, most generally, platforms where matches between buy and sell orders are arranged according to certain trading rules. Here, trades can be processed via computers, clerks or member brokers. For instance, classical floor-based oral auctions are also order-driven systems where trades are arranged by personally exchanging information. Conversely, in an order-driven computerized system, the order matching is performed by a computer. Often, brokerages have their own (typically order-driven) trading platforms to arrange trades for their clients. Important examples are the *electronic communication networks* (ECNs), such as Island ECN or Archipelago, which are electronic order-driven systems that are not regulated as exchanges and are owned by brokerages and dealers. These trading platforms are best-known as *alternative trading systems* and are competitors to regulated exchanges. Finally, trades can be also arranged *over the counter* without involving an exchange. An example is the corporate bond market, where most trading is arranged over the counter.

2.1.2 Types of Orders

An order represents an instruction of a trader who cannot personally negotiate his trades and therefore determines what to trade, when to trade and how much to trade. A *bid* (*offer*) reflects a trader's willingness to buy (sell) and contains the respective price and quantity the trader will accept. Correspondingly, bid and ask (offer) prices are the prices at which the trader is willing to trade. The highest (lowest) bid (ask) price available is called *best ask (bid) price* or *ask (bid) quote*. A *market quotation* gives the best bid and offer in a market and is called *Best Bid and Offer* (BBO). The best bid and offer across consolidated markets for National Market System (NMS) stocks is called the *National Best Bid and Offer* (NBBO). The difference between the best ask and best bid is called the *bid-ask* spread.

A *market order* is an order that trades immediately at the best price currently available in the market. The corresponding price at which an order is executed is called *transaction price*. Market order traders "pay" the bid-ask spread as long as the order is filled with the offered quantity at the best ask or bid price. If the size of the market order is larger than the quantity offered at the best ask or bid, the trader must move prices and thus has to pay an extra premium ("*price concession*"). Then, buyers (sellers) have to bid prices up (down) in order to find a counter-party who is willing to take the other side a large trade. The resulting price movements are called (instantaneous) *market impact* or *price impact* and naturally increase with the order size and are the dominant part of the trading costs (on top of the bid-ask spread). These trading costs induced by a potential market impact and the execution price uncertainty are often referred to as the price traders have to pay to obtain priority in the market, i.e., the "price of immediacy". In some markets traders can negotiate prices and receive prices which are better than the currently available best ask and bid. In this case, the trader receives price improvement due to a counter-party who is willing to step in front of the currently best quote.

A *limit order* is a trade instruction to trade at a price which is no worse than the so-called *limit price* specified by the trader. As the corresponding limit price is not necessarily offered on the other side of the market, a limit order faces execution risk. If no one is willing to take the opposite side at the required limit price, the order is not executed and is placed in the *limit order book* where all non-executed limit orders are queued according to price and time priority. Correspondingly, the larger the distance between the limit order and the best quote, the worse is the order's position in the queue and the lower is its execution probability in given time. Hence, in order to increase the execution probability and to reduce the time until execution, the limit order trader has to bid more aggressively with a limit price closer to the market. Limit orders at the best bid or offer are called *at the market*. Accordingly, limit orders with prices worse (better) than the current best quotes are called *behind (in) the market*. A limit order with a limit price at or above (below) the best ask (bid) price in case of a buy (sell) order is executed immediately and, if necessary, filled until the limit price level is reached. Such an order is called a *marketable limit order* corresponding to a market order where the trader limits the potential price impact (by correspondingly setting the limit price). If the limit price is worse than the current best price, the order has to "walk up (down)" the book in case of a buy (sell) order. Finally, a *market-to-limit order* is a market order, which is executed at the best ask/bid quote in the order book. Any unfilled part of a market-to-limit order automatically enters the order book.

Besides the risk of execution uncertainty, limit order traders face *adverse selection risk* corresponding to the risk that the order is executed ("picked up") and then the markets moves against their new position causing a loss. This happens, for instance, if the market moves downwards, picks up a standing buy (bid) limit order and continues declining. Adverse selection risk can only be reduced by posting far away from the market which, however, increases the execution uncertainty.

A *stop order* is an order which automatically buys (sells) a given quantity after the price rises (falls) to the "stop price". In case of a sell, the stop order prevents further losses in case of further falling prices and thus is called *stop loss order*. Hence, in contrast to a limit order, a stop order is activated as soon as the price passes the specified stop price. In contrast, a limit order is only executed as long as the price is *better* than the corresponding limit price. Combinations of both trade instructions result in a *stop limit order* with the stop price determining when the limit order becomes active and the limit price indicating the limit until which the order is executed.

A so-called *market-if-touched order* is a market order which becomes only valid if the price reaches a certain *touch price*. While stop orders buy (sell) after the price rises (falls) to the stop price, a corresponding market-if-touched order sells (buys) after the price falls (rises) to the touch price. While loss orders enforce price movements by trading in line with the market trend and thus generate momentum, market-if-touched orders do exactly the opposite. They buy when prices drop and sell when prices rise and thus generate contrarian trading patterns.

Limit orders are typically combined with specific attributes specifying for instance, for how long an order is valid and under which conditions it might be canceled. Examples are *fill-or-kill order* whose portions that cannot filled immediately are canceled or *all-or-none* orders which are only executed if they can be filled at once. For other types of orders, see, e.g., Harris (2003).

Finally, on a growing number of markets, orders can be partly or even completely hidden. A common type of order is the so-called *iceberg order* (or *reserve order*) where a certain proportion of the order can be non-displayed and thus is non-visible for other market participants. Such orders are used to hide trading strategies and to reduce adverse selection risk. The minimum display size is fixed by trading rules and differs across exchanges. Some markets, such as the NASDAQ allow to post *hidden orders*, where the order is entirely hidden. As shown by Hautsch and Huang (2011) based on NASDAQ data, the possibility to post completely hidden orders in the spread, might create substantial market activity as traders "search" for hidden volume by submitting "fleeting limit orders" to test for potential execution.

2.1.3 Market Structures

The market structure determines who can trade, what can be traded, when can be traded, and how it can be traded. We generally distinguish between *continuous trading*, where traders can trade whenever the market is open and *call markets* where all traders trade simultaneously in a call auction when the market is called. Call markets are often used to open a trading session and to settle the price before the market switches to continuous trading. Likewise they are also used to end a trading session, to have a mid-day call auction interrupting continuous trading or to re-start it after a trading halt. In some markets, call markets are also used to trade less liquid assets.

The most important characteristic of a market is the form of its execution system. We distinguish between the three major types of execution systems leading to *quote-driven dealer markets*, *order-driven markets* and *brokered markets*.

2.1.3.1 Quote-Driven Dealer Markets

In a quote-driven market (dealer market), trades are only executed by dealers. The dealers quote the ask and bid prices and supply liquidity by standing ready on the opposite side of the market. They often trade among themselves. In a *pure* quote-driven market, traders (or brokers acting on behalf of traders) cannot trade themselves (even if they have matching positions) but must execute their trades by dealers which earn the bid-ask spread. In some (though not pure) dealer markets, however, traders can trade directly without interacting with a dealer. A prominent example is the NASDAQ Stock Market as described in Sect. 2.1.5.2.

Dealers are often specialized in serving specific clients which are trustworthy and creditworthy and tend to refuse trading with counter-parties outside their clientele or which might be better informed. Traders who do not have credit relationships with dealers have to trade via brokers stepping in and guaranteeing credit worthiness. *Interdealer brokers* are brokers arranging trades among dealers if dealers prefer keeping anonymity and not informing their rivals about their quotes.

Quote-driven dealer markets are popular forms of trading for bonds, currencies and stocks. Examples are the NASDAQ Stock Market or the London Stock Exchange.

2.1.3.2 Order-Driven Markets

In an order-driven market, traders trade directly with each other. As there are no dealers serving as intermediaries, trading occurs according to specific trading rules. *Order precedence rules* determine which buyers trade with which sellers and *trade pricing rules* determine the resulting transaction prices. Liquidity provision in an order-driven market is ensured by traders taking the opposite side of the market or by dealers serving as traders. In some order-driven markets, most of the liquidity is still provided by dealers. However, a main characteristic of an order-driven market is that dealers cannot choose their clients but have to trade with anyone accepting the offer.

Order-driven markets are mostly *auction markets*. An auction is a formalized process, the so-called *price discovery process* by which buyers seek the lowest available prices and sellers seek the highest available prices. In a *single-price auction* all trades at the same price following a call are simultaneously arranged. Conversely, in *continuous two-sided auctions*, traders continuously arrange their trades on both sides of the market based on prices varying over time. In markets with *oral auctions* traders trade face-to-face on a trading floor. As they negotiate their trades by crying out their bids and offers these markets are called *open outcry markets*. Order-driven markets which use an electronic execution system employ *rule-based order matching*.

Only order-driven markets which are not organized as auctions are *crossing networks* where trading takes place at prices which are determined on other markets. These prices are called *crossing prices*. Crossing networks are call markets where submitted orders are matched according to order precedence rules. Such market structures are sometimes used to organize after-hours trading based on closing prices from continuous trading (e.g., at the New York Stock Exchange). Alternatively, they are used to organize intraday calls where the crossing prices correspond to a price in a continuous market at a random time within a certain time interval following the call.

As in an order-driven market, in contrast to quote-driven dealer markets, trades are arranged according to precedence rules and traders cannot choose with whom they trade, traders typically trade with counter-parties with whom they do not have credit relationships. To ensure proper order settlement, these markets require elaborate clearing mechanisms.

2.1.3.3 Brokered Markets

In *brokered markets*, brokers initiate the matches between buyers and sellers. The broker's role is to find liquidity provided by a counter-party. We distinguish between *concealed traders* and *latent traders*. Concealed traders are traders who intend to trade but do not make public offers to hide their strategies. However, they trade if brokers offer them suitable matches. A latent trader has no concrete trade intention but only trades if she is confronted with an attractive trading opportunity. Successful brokers in a brokered market have to find both concealed traders and latent traders.

2.1.4 Order Precedence and Pricing Rules

The order precedence rules in an oral auction are *price priority* and *time precedence*. According to the price priority rule, traders who offer the most competitive prices have priority. Traders are not allowed to accept bids and offers at inferior prices. The time precedence rule gives priority to traders whose quotes *first* improve the prevailing best ask and bid prices. Time precedence is retained as long as the quotes are maintained or accepted by the counter-party. In oral auctions, according to traders "a quote is good only as long as the breath is warm" requiring to repeating offers to maintain precedence. However, time precedence is only helpful as long as the minimum price increment, the so-called *minimum tick size*, is not too small. Otherwise, time precedence gives only little privilege as competitors can easily improve the best quotes. In oral auctions, price priority is self-fulfilling as traders try to trade to best possible prices which encourages seeking for the best offer. This, however, is not guaranteed for time precedence as for traders it makes no difference with whom to trade as long as quotes are equal.

Most exchanges and electronic trading platforms use *rule-based order-matching systems*. Here, trading is typically anonymous, decentralized and traders trade by electronically submitting, canceling or amending (already existing) orders. Order matching is arranged by a typically automatized system using a sequence of procedures. All systems use price priority as primary order precedence rule. As second order precedence rule, mostly time precedence is used. Some markets use *display precedence* giving priority to displayed orders over undisclosed orders or *size precedence* where orders with either large or small sizes are privileged. *Pro rata matching* yields an allocation where each order is filled in proportion to its size.

In *single-price auctions*, the market ranks the orders and starts by matching the highest-ranking buy and sell orders to each other. This match results in a trade as long as the buyer pays at least as much as the seller demands. If the two orders have the same size, both will execute completely. Otherwise, the unfilled part of an order will be matched with the next highest-ranking orders. This process continues until all possible trades are matched. The price of the last possible match is the *market-clearing price*. In a single price auction, all trades take place at the market clearing price.

The market-clearing price of a single price auction corresponds to the price where the supply schedule equals the demand schedule. The supply schedule lists the total offered volume according to price priority and thus is a non-decreasing curve. Likewise, the demand schedule lists the total sell volume and is a non-increasing curve. Correspondingly, at prices below the clearing price there is excess demand in the market while there is excess supply for prices above the clearing price. By choosing the market-clearing price to match all orders, a maximum amount of possible volume is traded. Moreover, it is easily shown that traders' overall benefits from participating in this auction are maximized.

If the buy order in the last feasible trade bids at a higher price than the sell order, the resulting market-clearing price can be either of these two prices or can be between them. In this case, the market-clearing price does not provide an exact match and some excess demand or supply might be left. In case of excess supply (demand), all buyers (sellers) at the market-clearing price fill their orders, while the secondary precedence rules determines which sell (buy) orders to be filled.

The pricing rule in an oral auction and in a continuous rule-based order matching system is called *discriminatory pricing rule*. According to this rule, every trade takes place at the price proposed by the trader whose ask or bid is accepted. Consequently, the entire offer or bid side of the market is matched in a discriminatory fashion with progressively inferior prices. This is similar to the way how a large order is simultaneously matched with all trades pending on the opposite side yielding best prices for the first piece and progressively inferior prices for the other pieces as the order walks up or down the ask or bid schedule and exhausts liquidity.

In continuous auction markets, this process is maintained by an order book which is automatically updated whenever a new order arrives. New orders enter the book according to order precedence rules. If they are marketable, they are matched against the standing orders. Then, under discriminatory pricing, the limit price of any standing order determines the price for each match. Accordingly, large orders trade their individually matched parts at different prices as they walk up or down the book. In contrast, under uniform pricing, the individual orders trade at the same price corresponding to the worst price under discriminatory pricing. Hence, large impatient traders prefer discriminatory pricing while liquidity suppliers prefer uniform pricing.

Finally, refined and modified rules are applied to special type of orders. For instance, in order-driven markets allowing for hidden orders, special rules regulate order precedence for displayed and non-displayed parts of an order. A common rule is that, for a given limit price, visible order volume has priority over non-visible volume (even if the non-displayed volume has been posted before). For more specific details, see the trading rules, e.g., on NASDAQ.

2.1.5 *Trading Forms at Selected International Exchanges*

In this section, we illustrate the current trading forms at some selected international exchanges. We concentrate on those exchanges where most of the data used in the remainder of the book come from. These are the two most important exchanges in the U.S., the New York Stock Exchange (NYSE) and NASDAQ as well as the electronic trading systems of the Frankfurt stock exchange (XETRA) and the Australian Stock Exchange (ASX) as representatives of typical electronic limit order book markets commonly used for equities.

2.1.5.1 The New York Stock Exchange (NYSE)

Trading at the NYSE is based on a hybrid system, i.e., the trading mechanism combines elements from quote-driven, order-driven and brokered markets. Essentially, it is an order-driven market, however, there are still specialists who have to provide liquidity. In fact, NYSE trading combines an open outcry system, a dealer market and an electronic limit order book.

The NYSE started in 1792 as an order-driven open outcry floor market and was historically the dominant trading platform for U.S. equities. Later, components of a dealer market have been integrated in the open outcry system. As a result, for each stock, one market maker (specialist) has to manage the trading and quote process and has to guarantee the provision of liquidity, when necessary, by taking the other side of the market. The specialist also has to maintain the underlying limit order book. Besides from orders posted by the specialist, the limit order book also receives orders directly from the exchange's routing system. These are orders which do not require to be handled by a broker and are sent directly to the specialists's workstation. The specialist sets bid and ask quotes representing his own interest, those of a floor broker or the best ask and bid implied by the limit order book. In his function to serve as an agent for the book, she is considered as a single floor trader. According to present rules, the specialist must display limit orders (from the book) that are better than the prevailing ones within 30 s. Hence, during short time spans, orders from the book might not have price priority over prices offered by floor brokers. Moreover, the book might not have time priority over floor traders. The dealer market starts with a single-price call auction as opening procedure.

Regular trading at the NYSE starts at 9:30 and ends at 16:00. The closing procedure has some similarities to an auction and requires the specialist to balance supply and demand and to establish a (preferably smooth) transition path to the closing prices. Accordingly, trading "at the close" is governed by relatively complex rules.

While in the (open outcry) dealer market trades are only executed by the specialists, NYSE offers also automatic (electronic) execution. The NYSE Direct+ system, which was introduced in 2000, is an electronic trading system which runs parallel to the dealer market. Induced by the rising competition with ECNs, the NYSE established several mergers. One important merger was the merger between the NYSE and the ECN "Archipelago" yielding "NYSE Arca".

2.1.5.2 NASDAQ

The name NASDAQ is an acronym for "National Association of Securities Dealers Automated Quotations" and was founded in 1971 by the National Association of Securities Dealers (NASD). Historically, it was primarily organized as a dealer market linking geographically non-centralized dealers via an electronic system. In the early 1990s NASDAQ gave little protection to customer limit orders. For instance, competitive customer quotes were not necessarily immediately displayed but were used by the dealers to trade on their own account. Such a proceeding was prohibited by the so-called "Manning Rules" adopted in 1994/1995. In 1994, the SEC started investigations on coordinated quote setting behavior of NASDAQ dealers resulting in unnecessarily discrete price grids and large spreads. These investigations resulted in civil lawsuits against NASDAQ dealers and more explicit order execution rules set by the SEC (which were not only limited to NASDAQ). The two most important components of these rules are the "display rule" requiring the display of customer orders with prices better than the currently prevailing quotes and the "quote rule" requiring a market maker to make publicly available any (potentially superior) prices that she quotes in inter-dealer markets. As a result of these reforms, NASDAQ spreads declined significantly.

Subsequently, NASDAQ established an electronic system ("SuperMontage") which organizes trade execution, reporting, confirmation, and interdealer communication. The underlying principles resemble mostly those of an electronic limit order book, which, however, is maintained only by the dealers and not by the customers. NASDAQ's position further increased by numerous alliances and take-overs. For instance, NASDAQ purchased the American Stock Exchange (AMEX) in 2003, the

ECN's INET (formerly Island/Instinet) and Brut in 2004/2005 and the Philadelphia Stock Exchange in 2007. As a result, NASDQA is currently the biggest electronic stock market in the U.S.

2.1.5.3 XETRA

The German Exchange Electronic Trading (XETRA) is the electronic trading system of the Deutsche Börse AG for cash market trading in equities and a variety of other instruments including Exchange Traded Funds, mutual funds, bonds, warrants, certificates, among others. It has been introduced in November 1997 as an electronic supplement of the classical floor trading at the Frankfurter Wertpapierbrse. XETRA is a double continuous auction system with an opening and closing call auction at the beginning and at the end of the trading day, respectively, and a mid-day call auction. During the normal trading period, trading is based on an automatic order matching procedure. Limit orders enter the queues of the order book according to strict price-time priority. Auctions consider all order sizes for price determination, whereas continuous trading is based upon round lots only. XETRA trading is completely anonymous and does not reveal the identity of the traders. A trader can act as agent trader or as proprietary trader. Some traders might act as market makers on behalf of XETRA and are obliged to guarantee liquidity on both sides of the market as well as to adjust supply and demand imbalances. Normal trading currently takes place from 09:00 to 17:00.

2.1.5.4 Australian Stock Exchange

The Australian Stock Exchange (ASX) is a continuous double auction electronic market and as such is an example for an electronic limit order book trading system similar to those operating, for instance, in Paris, Hong Kong and Sao Paulo. The continuous auction trading period is preceded and followed by an opening call auction. Normal trading takes place continuously between 10:09 and 16:00 Sydney time on Monday to Friday. Limit orders are placed in the buy and sell queues according to a strict time-price priority order. Any buy (sell) order entered that has a price that is greater (less) than existing queued sell (buy) orders, will be executed immediately. The order will be automatically matched to the extent of the volume that is available at the specified limit price. All orders and trades are always visible to the public. Order prices are always visible, however orders may be entered with an undisclosed (hidden) volume if the total value of the order exceeds AUD 200,000. The identity of the broker who entered an order is not public information, but is available to all other brokers. A comprehensive description of the trading rules of the Stock Exchange Automated Trading System (SEATS) on the ASX can be found in the SEATS Reference Manual available at www.asxonline.com.

2.2 A Review of Market Microstructure Theory

In this section, we give a compact overview of the fundamental approaches and directions in theoretical market microstructure literature. More in-depth treatments of this material are given, for example, by O'Hara (1995), the surveys by Madhavan (2000) and Biais et al. (2005) or the monograph by Hasbrouck (2007).

As stated by Madhavan (2000), market microstructure theory is concerned with "the process by which investors' latent demands are ultimately translated into prices and volumes." Consequently, central topics in market microstructure theory deal with price formation, price discovery, inventory, liquidity, transaction costs as well as information diffusion and dissemination in markets. Traditional microstructure theory provides two major directions to explain price setting behavior: asymmetric information based models and inventory models. The former branch models market dynamics and adjustment processes of prices using insights from the theory of asymmetric information and adverse selection. As discussed in Sect. 2.2.1, two main approaches are *sequential trade models* and *strategic trade models*. The branch of inventory models, as discussed in Sect. 2.2.2, investigates the uncertainty in order flow and the inventory risk and optimization problem of liquidity suppliers under possible risk aversion. We summarize the major theoretical implications of the individual approaches in Sect. 2.2.3. Finally, more recent work addresses trading behavior and equilibria on limit order book markets. We briefly review this literature in Sect. 2.2.4.

2.2.1 Asymmetric Information Based Models

2.2.1.1 Sequential Trade Models

In *sequential trade models*, randomly selected traders sequentially arrive at the market. The framework is based on the assumption of the existence of differently informed traders. Accordingly, there are so-called "informed traders", who trade due to private information on the fundamental value of the asset and "liquidity traders", who trade due to exogenous reasons, like portfolio adjustments or liquidity aspects. The assumption of heterogeneous groups of traders provides the basis for a plethora of asymmetric information based models. Seminal papers in this direction are Copeland and Galai (1983) and Glosten and Milgrom (1985).

In the Glosten and Milgrom (1985) model, securities have a payoff which is either high or low with given probability and is revealed after market closure. The population of traders consist of informed traders knowing the true asset payoff and uninformed traders who buy or sell randomly with equal probability. Informed traders buy (sell) if the true asset value is high (low). The proportion of informed traders in the market is given. Dealers are uninformed and infer on the asset's true value based on the trade history. In particular, observing a buy (sell) request of a trader, the dealer computes the conditionally expected value of the asset given a trade is a buy or sell. Then, she sets the ask (bid) quote such that the expected gain from an uninformed buyer (seller) are balanced by the loss to an informed buyer (seller). After the next trade, the dealer updates her beliefs on the asset's true value using her initial beliefs as priors. This results into updating recursions on the probabilities for the asset's true values. The resulting bid-ask spread is a function of the asset's potential values (high vs. low), their corresponding probabilities, and the relative proportion of informed traders. Fundamental implications of this sequential trade model is that trade prices follow a martingale, order flow is correlated (buys tend to follow buys, sells tend to follow sells), bid-ask spreads decline over time as the dealer's uncertainty is reduced and individual trades have price impact.

The Glosten and Milgrom (1985) model has been extended and modified in various directions. Easley and O'Hara (1992) allow for event uncertainty by assuming the random occurrence of a trade event at the beginning of each day. In case of no information event, informed traders refrain from trading and only uninformed traders (randomly) trade in the market. Easley and O'Hara (1992) assume that uninformed traders do not necessarily always buy or sell but can also refrain from trading. Consequently, also the occurrence of no trade (i.e., a slow-down of the trading process) carries information. Therefore, besides bid-ask spreads also the time between trades is informative. Variations of this framework are Easley and O'Hara (1987) where different order sizes are possible and Easley and O'Hara (1991) allowing for different types of orders.

Easley et al. (1997) and Easley et al. (2002) extend the framework of Easley and O'Hara (1992) to allow for Poisson arrival of the events determining the asset's true value. Then, traders do not sequentially arrive in discrete time but arrive randomly in continuous time. This arrival process is governed by further Poisson processes with different intensities for informed and uninformed traders. As a result, the numbers of daily buys and sells are jointly distributed based on a mixture of Poisson distributions. Then, based on the information arrival intensity as well as the arrival intensities for informed and uninformed traders, the probability of informed trading (PIN), i.e., the probability that a randomly chosen trader is informed, can be computed. Easley et al. (2008) extend this approach to a dynamic framework and estimate time-varying PINs.

2.2.1.2 Strategic Trade Models

In a sequential trade model, a trader participates in a market only once. Therefore, she does not take into account the impact of her trade decision on the subsequent behavior of others. As a consequence, informed traders trade largest possible quantities as they do not have to account for possible adverse price effects in future trades. This situation is completely different in a *strategic trade model*, where a trader repeatedly participates in the market and therefore has to behave strategically. A seminal paper in this area is Kyle (1985). In the Kyle model, the security's value is stochastic but is known by an informed trader. Uninformed traders ("noise traders") trade independently of the asset's true value and submit a stochastic order flow. The market maker receives the demand of both the uninformed and informed traders and has to set a price such that all trades are cleared. However, as the informed trader might trade aggressively, the market maker has to protect herself against being on the wrong side of the market by setting the price as a linearly increasing function of the net order flow (i.e., the total net volume requested by informed and uninformed traders). This, however is anticipated by the informed trader who computes her profits given her conjecture on the market maker's price setting rule and her demand. Note that in contrast to a sequential trade model, the informed trader's profit is not necessarily positive as a high demand from liquidity traders might drive up the price set by the market maker. The informed trader's optimization problem is to choose her demand such that her expected profit is maximized. This yields a linear demand function in the asset's true value.

When the market maker conjectures the informed trader's underlying optimization problem, she can compute the trader's linear demand function in dependence of the parameters of her own price setting rule. This yields an inverse relationship between the slopes of the trader's demand and the market maker's price setting rule. The slope of the market maker's price setting rule determines the price impact of net order flow and is commonly referred to as "Kyle's lambda".

Kyle (1985) makes this quantity operational by exploiting properties of bivariate normal random variables. In such a framework, it can be computed as a function of the covariance between the total asset demand and the true value of the asset as well as the variance of noise trading. An implication of the Kyle model is that the informed trader's expected profit is increasing in the divergence between the asset's true value and the market maker's unconditional price (irrespective of the order flow) and in the variance of noise trading. The latter effect is interesting as it implies that the informed trader's profit is higher when there is more liquidity in the market.

Kyle's model has been extended in various directions. For instance, Admati and Pfleiderer (1988) and Foster and Viswanathan (1990) allow for uninformed traders who behave strategically themselves. Foster and Viswanathan (1996), among others, allow for multi-period models.

2.2.2 Inventory Models

Inventory models consider the inventory problem of a dealer who is facing buyers and sellers arriving asynchronously. This string of the literature originates from Garman (1976) who models the arrival processes of buyers and sellers as Poisson processes. The arrival intensities depend on the price they pay or receive, respectively. Hence, as long as the intensities are equal, the dealer is on average buying and selling at the same rate. The dealer makes profits by setting a spread. Then, the larger is the bid-ask spread, the higher are the profits per trade but the lower is the trade arrival rate. Garman (1976) characterizes the inventory problem of the market maker who has to ensure that her holdings of the security and cash do not drop below a given level. If ask and bid quotes are set such that the resulting arrival intensities of buyers and sellers are equal, holding of stock follow a zero-drift random walk while cash holding follow a positive-drift random walk (as long as the spread is strictly positive). This causes the market maker to go bankrupt with probability one as a zero-drift random walk hits any finite level with probability one. Hence, as long as the market maker keeps ask and bid quotes constant, she must expect to be ruined within short time. Amihud and Mendelson (1980) present a similar framework where the market maker's inventory is constrained to lie between upper and lower bounds. They show that the market maker updates her quotes whenever the inventory approaches these boundaries to drive up or down, respectively, the arrival rates of buyers and sellers. As a result, bid and ask quotes are monotonically decreasing in the inventory levels and quotes are not necessarily set symmetrically around the asset's true value.

Dealer's price setting can be also analyzed in a framework where the dealer is risk averse and sets ask and bid quotes to appropriately balance her portfolio. This is, e.g., studied by Stoll (1978), Ho and Stoll (1981), Stoll (1989) and Huang and Stoll (1997), among others.

2.2.3 Major Implications for Trading Variables

The main theoretical findings on the properties and determinants of key microstructure variables and relationships thereof are summarized as follows:

1. *Transaction volumes:* In the Easley and O'Hara (1987) model, traders are allowed to trade either small or large quantities, but are not allowed to refrain from trading. Thus, large quantities indicate the existence of information. Blume et al. (1994) investigate the informational role of volume when traders receive information signals of different quality in each period. The authors analyze how the statistical properties of volume relate to the behavior of market prices and show that traders can infer from the volume about the quality and quantity of information in the market. An important result is that the volume provides additional information that cannot be deduced from price statistics. As a consequence, volume and volatility are correlated.

2. *Bid-ask spreads:* In the Glosten and Milgrom (1985) model, the market maker determines the spread in a way that it compensates for the risk due to adverse selection. The higher the probability that she transacts at a loss due to trading with market participants with superior information, the higher the bid-ask spread. Moreover, bid-ask spreads are positively correlated with market maker's inventory risks and risk aversion. In the Easley and O'Hara (1992) approach, the market maker uses no-trade-intervals to infer the existence of new information. Consequently, lagged durations and the size of spreads are negatively correlated.

3. *Trade durations:* Diamond and Verrecchia (1987) propose a rational expectation model with short selling constraints. They assert that the absence of a trade

is associated with the occurrence of "bad" news. Then, the absence of a trade is informative and is correlated with price volatility. In this framework, time matters only because of the imposed short selling restrictions. In Easley and O'Hara (1992), however, informed traders enter the market whenever there are information signals while non-informed traders might also refrain from trading. As a consequence, short trade-to-trade durations indicate the existence of information. Admati and Pfleiderer (1988) provide an explanation for temporal clustering of durations. In their setting, liquidity traders prefer to minimize their transaction costs and to trade if other traders are in the market. In equilibrium, it is optimal for informed traders to behave similarly. As a consequence, trading is clustered, and trade durations are positively autocorrelated.

2.2.4 Models for Limit Order Book Markets

A seminal paper to model limit order markets is Glosten (1994). In this model, all market participants have access to an electronic screen. Posting limit orders is done costlessly and the execution of a trade against the book occurs in a "discriminatory" fashion. That is, each limit order transacts at its limit price. Investors are rational and risk averse and maximize a quasi-concave utility function of their cash and share position as well as personal preferences. The trading behavior of market order traders depends on their marginal valuation functions and the prevailing terms of trade, i.e., the list of bid and ask quotes available, which influence the changes in investors' cash and share positions. It is assumed that an investor chooses the trade quantity such that her marginal valuation equals the marginal price corresponding to the price paid for the last share in a transaction. There is informed trading if an investor's marginal valuation is associated with the future payoff. Then, incoming market orders reveal information about the unknown "full information value" of the traded security. Due to the anonymity of the electronic market, the underlying marginal valuation implied by an arriving market order can be assessed by the liquidity suppliers only through the observed limit price and the traded quantity given the terms of trades offered by the book.

Glosten assumes that there is a large number of uninformed, risk-neutral and profit-maximizing limit order submitters who set limit prices and quantities on the basis of their "upper tail expectation". The latter corresponds to the conditional expectation of the asset's full information liquidation value given that the next arrival's marginal valuation is greater than or equal to the traded quantity. In the presence of private information, liquidity suppliers protect themselves against adverse selection by setting the limit price at least equal to the upper tail expectation given a market order trading at the corresponding price. It is shown that such a strategy leads to a Nash equilibrium which is characterized by a zero-profit condition for prices at which positive quantities are offered.

Glosten's model is extended in several directions by Chakravarty and Holden (1995), Handa and Schwartz (1996), Seppi (1997), Kavajecz (1999), Viswanathan and Wang (2002) and Parlour and Seppi (2003). However, while static equilibrium models provide insights into the structure of the limit order book, they do not allow to analyze (dynamic) interactions between the order flow and the state of the limit order book. For this reason, Parlour (1998) proposes a dynamic game theoretical equilibrium model where traders have different valuations for the asset and choose between submitting a market order, a limit order or refraining from trading. Since the expected future order flow is affected by their own order submission strategies, the execution probabilities of limit orders are endogenous. This leads to systematic patterns in traders' order submission strategies even when there is no asymmetric information in the market. The basic underlying mechanism is a "crowding out" effect whereby market orders and limit orders on the individual market sides crowd out one another when the ask or bid queue is changed. In particular, the probability of the arrival of a buy (sell) trade after observing a buy (sell) trade is higher than after observing a sell (buy) trade. This results from a buy transaction reducing the depth on the ask side which in turn increases the execution probability for limit sell orders. Hence, for a potential seller, the attractiveness of limit orders relative to market orders rises inducing a crowding out of market sell orders in favor of limit sell orders. Handa et al. 2003 extend this approach by introducing an adverse selection component due to the presence of privately informed traders.

An alternative dynamic game theoretical equilibrium model has been proposed by Foucault (1999) in order to study the cross-sectional behavior of the mix between market orders and limit orders and the implied trading costs. He analyzes the influence of the risks of being picked off and of non-execution on traders' order submission strategy and derives testable implications regarding the relationship between the proportion of limit orders and market orders in the order flow, the fill rate (i.e., the percentage of executed limit orders), the trading costs and the volatility of the asset price. Handa et al. (2003) extend the approach by Foucault (1999) by introducing private information in his model. While in Foucault's model trading occurs because of differences in traders' valuation for the security, Handa et al. introduce an adverse selection component due to the presence of privately informed traders. As a result, the size of the spread is a function of the differences in valuation among investors and of adverse selection. Further extensions of these frameworks are, among others, Foucault et al. (2005) and Goettler et al. (2005, 2009). Recent literature focuses on the theoretical analysis on automated trading and smart order routing in electronic trading platforms. See, e.g., Foucault and Menkveld (2008), Hendershott et al. (2011) or Biais et al. (2010), among others.

References

Admati A, Pfleiderer P (1988) A theory of intraday patterns: volume and price variability. Rev Financ Stud 1:3–40

Amihud Y, Mendelson H (1980) Dealership markets: market making with inventory. J Finan Econ 8:31–53

Biais BR, Glosten L, Spatt C (2005) Market microstructure: a survey of microfoundations, empirical results, and policy implications. J Finan Markets 8:217–264
Biais BR, Hombert J, Weill P-O (2010) Trading and liquidity with limited cognition. NBER Working Paper No. 16628
Blume L, Easley D, O'Hara M (1994) Market statistics and technical analysis. J Finance 49(1):153–181
Chakravarty S, Holden C (1995) An integrated model of market and limit orders. J Financial Intermediation 4:213–241
Copeland TE, Galai D (1983) Information effects and the bid-ask spread. J Finance 38:1457–1469
Diamond DW, Verrecchia RE (1987) Constraints on short-selling and asset price adjustment to private information. J Finan Econ 18:277–311
Easley D, Engle RF, O'Hara M, Wu L (2008) Time-varying arrival rates of informed and uninformed traders. J Financ Econom 6:171–207
Easley D, Hvidkjaer S, O'Hara M (2002) Is information risk a determinant of asset returns? J Finance 57:2185–2221
Easley D, Kiefer NM, O'Hara M (1997) The information content of the trading process. J Empir Financ 4:159–186
Easley D, O'Hara M (1987) Price, trade size, and information in securities markets. J Finan Econ 19:69–90
Easley D, O'Hara M (1991) Order form and information in securities markets. J Finance 46:905–927
Easley D, O'Hara M (1992) Time and process of security price adjustment. J Finance 47:577–605
Foster DF, Viswanathan S (1990) A theory of the interday variations in volume, variance, and trading costs in securities markets. Rev Financ Stud 3:593–624
Foster DF, Viswanathan S (1996) Strategic trading when agents forecast the forecasts of others. J Finance 51:1437–1478
Foucault T (1999) Order flow composition and trading costs in a dynamic limit order market. J Finan Markets 2:99–134
Foucault T, Kadan O, Kandel E (2005) Limit order book as a market for liquidity. Rev Financ Stud 18:1171–1217
Foucault T, Menkveld AJ (2008) Competition for order flow and smart order routing systems. J Finance 63:119–158
Garman M (1976) Market Microstructure. J Finan Econ 3:257–275
Glosten LR (1994) Is the electronic open limit order book inevitable. J Finance 49:1127–1161
Glosten LR, Milgrom PR (1985) Bid, ask and transaction prices in a specialist market with heterogeneously informed traders. J Finan Econ 14:71–100
Goettler RL, Parlour CA, Rajan U (2005) Equilibrium in a dynamic limit order market. J Finance 60:2149–2192
Goettler RL, Parlour CA, Rajan U (2009) Informed traders and limit order markets. J Finan Econ 93:67–87
Handa P, Schwartz R (1996) Limit order trading. J Finance 51:1835–1861
Handa P, Schwartz R, Tiwari A (2003) Quote setting and price formation in an order driven market. J Finan Markets 6:461–489
Harris L (2003) Trading & exchanges – market microstructure for practitioners. Oxford University Press, Oxford
Hasbrouck J (2007) Empirical market microstructure. Oxford University Press, Oxford
Hautsch N, Huang R (2011) On the dark side of the market: identifying and analyzing hidden order placements. Working Paper, Humboldt-Universität zu Berlin
Hendershott T, Jones CM, Menkveld AJ (2011) Does algorithmic trading improve liquidity? J Finance 66:1–33
Ho T, Stoll H (1981) Optimal dealer pricing under transactions and return uncertainty. J Finan Econ 9:47–73
Huang RD, Stoll HR (1997) The components of the bid/ask-spread: a general approach. Rev Financ Stud 10:995–1034

Kavajecz K (1999) The specialist's quoted depth and the limit order book. J Finance 52:747–771
Kyle AS (1985) Continuous auctions and insider trading. Econometrica 53(6):1315–1335
Madhavan A (2000) Market microstructure: a survey. J Finan Markets 3(3):205–258
O'Hara M (1995) Market microstructure. Basil Blackwell, Oxford
Parlour CA (1998) Price dynamics in limit order markets. Rev Financ Stud 11(4):789–816
Parlour CA, Seppi DJ (2003) Liquidity-based competition for order flow. Rev Financ Stud 16:301–343
Seppi DJ (1997) Liquidity provision with limit orders and strategic specialist. Rev Financ Stud 1(1):103–150
Stoll HR (1978) The supply of dealer services in securities markets. J Finance 33:1133–1151
Stoll HR (1989) Inferring the components of the bid-ask spread: theory and empirical tests. J Finance 44(1):115–134
Viswanathan S, Wang JJD (2002) Market architecture: limit-order books versus dealership markets. J Financ Markets 5:127–167

Chapter 3
Empirical Properties of High-Frequency Data

In this chapter, we present financial high-frequency data and their empirical properties. We discuss data preparation issues and show the statistical properties of various high-frequency variables based on blue chip assets traded at the NYSE, NASDAQ and XETRA. Section 3.1 focuses on peculiar problems which have to be taken into account when transaction data sets are prepared. Section 3.2 discusses the concept of so-called financial durations arising from aggregations based on trading events. Section 3.3 illustrates the statistical features of different types of financial durations including trade durations, price (change) durations and volume durations. In Sect. 3.4, we discuss the properties of further trading characteristics such as high-frequency returns, trading volumes, bid-ask spreads and market depth. Section 3.5 presents the empirical features of time aggregated data. Finally, Sect. 3.6 gives a compact summary of the major empirical features of high-frequency data.

3.1 Handling High-Frequency Data

3.1.1 Databases and Trading Variables

As illustrated in Chap. 2, the most dominant trading forms for equities are floor trading, limit order book trading or combinations thereof yielding hybrid forms of trading. Typical datasets arising from floor trading contain information on *trades* and *quotes* whereas data from electronic trading often contains information on the process of *order arrivals* as well as (at least partly) of the *order book*. Typically, the data is recorded whenever a trade, quote or – in the informational limiting cases – a limit order occurs. This data is called *transaction data*, *(ultra-)high frequency data* or sometimes *tick data*.[1]

[1]Strictly speaking, the terminology "tick data" refers to settings where the data is only recorded whenever the price changes (by at least one tick). The literature is not always stringent with these terminologies.

N. Hautsch, *Econometrics of Financial High-Frequency Data*,
DOI 10.1007/978-3-642-21925-2_3, © Springer-Verlag Berlin Heidelberg 2012

Regarding the detailedness of high-frequency information, we can distinguish between five major levels:

1. *Trade data.* The transaction level is associated with information on individual trades consisting of
   (a) the time stamp of trades,
   (b) the price at which a trade was executed,
   (c) the traded volume (in number of shares).
2. *Trade and quote data.* Information on trades *and* quotes provides the most common form of transaction data containing
   (a) the time stamp of trades and best ask/bid quote updates,
   (b) the underlying best ask/bid quotes,
   (c) the price at which a trade was executed,
   (d) the traded volume (in number of shares),
   (e) the trade direction (up to identification rules as described below),
   (f) the indicative depth associated with best ask and bid quotes.

   The most common database of this type is the Trade & Quote (TAQ) database released by the NYSE which is illustrated in more detail below.
3. *Fixed level order book data.* If the underlying trading system is a fully computerized system, often also (at least partial) information on the depth behind the market is available. This type of data contains the same information as above but provides also information on limit order activities behind the market. Based on such data it is possible to reconstruct the limit order book up to a fixed level.
4. *Messages on all limit order activities.* Such data provide full information on any limit order activities, including time stamps, (limit) prices, sizes and specific attributes of limit order submissions, executions, cancellations and amendments. It allows to fully re-produce the trading flow and to re-construct the limit order book at any point in time during continuous trading and allows for an exact identification of buyer-initiated or seller-initiated trades. Sometimes such data contains also information on hidden orders or iceberg orders. For instance, TotalView-ITCH data from NASDAQ trading (see for an illustration below) provides information on execution against hidden orders. See, e.g., Hautsch and Huang (2011) for more details.
5. *Data on order book snap-shots.* Some data sets provide snap-shots of the limit order book at equi-distant time intervals avoiding the need for order book re-constructions. However, as they are recorded on an equi-distant grid, the matching with the corresponding underlying trading process is difficult. Therefore, this data is only useful to study limit order book dynamics but is of limited use to analyze interactions between the book and the trading process.

Even based on full-information limit order book data, a complete reconstruction of the limit order book is a difficult task. Two major problems have to be addressed in this context: Firstly, a complete and correct re-construction of the limit order book requires accounting also for order book activities outside the continuous trading hours including opening auctions, pre-trading and late-trading periods. Secondly,

Table 3.1 TAQ data record on trades for Microsoft on June 1, 2009

SYMBOL	DATE	TIME	EX	PRICE	SIZE	COND	CORR	G127
MSFT	2009-06-01	36601	Z	21.1900	200		0	0
MSFT	2009-06-01	36601	Z	21.1900	1000		0	0
MSFT	2009-06-01	36601	Z	21.1900	100		0	0
MSFT	2009-06-01	36601	B	21.1900	400	@F	0	0
MSFT	2009-06-01	36601	B	21.1900	400	@F	0	0
MSFT	2009-06-01	36602	D	21.1912	470		0	0
MSFT	2009-06-01	36602	Z	21.1900	200		0	0
MSFT	2009-06-01	36602	Q	21.1900	900		0	0
MSFT	2009-06-01	36602	Q	21.1900	100	@F	0	0
MSFT	2009-06-01	36602	Q	21.1900	100	@F	0	0
MSFT	2009-06-01	36602	Q	21.1900	300		0	0
MSFT	2009-06-01	36602	Q	21.1900	100		0	0
MSFT	2009-06-01	36602	D	21.1900	100	@F	0	0
MSFT	2009-06-01	36602	D	21.1900	100	@F	0	0

SYMBOL stock symbol, *DATE* trade date, *TIME* trade time, *EX* exchange on which the trade occurred, *PRICE* transaction price, *SIZE* trade size, *COND* sale condition, *CORR* correction indicator of correctness of a trade, *G127* indicating G trades (trades of NYSE members on their own behalf) and rule 127 transactions (block trades)

as briefly discussed in Chap. 2, most modern electronic exchanges allow traders to submit iceberg orders or hidden orders. The trading rules associated with partial (or complete) display differ across exchanges. For instance, some exchanges, such as the NASDAQ, even allow to post completely hidden orders in the bid-ask spread (and thus providing execution priority) while this is not possible on other trading platforms. As long as information on hidden orders is not available, limit order books can be only incompletely constructed. For more details on iceberg orders and hidden orders, see, e.g., Bessembinder et al. (2009), Frey and Sandas (2009) or Hautsch and Huang (2011).

The quality as well as the format of the data strongly depends on the underlying institutional settings and the recording system. See, Chap. 2 or, e.g., Harris (2003) for more details on institutional frameworks. Though to the growing importance of electronic trading, the quality and detailedness of transaction data has increased during recent years, rigorous data handling and processing is still an important task and essential prerequisite for empirical studies.

To illustrate possible forms of raw high-frequency data, Tables 3.1 and 3.2 show extracts of raw files from the "Trade and Quote" (TAQ) database released by the NYSE. The TAQ database is one of the most popular and widely used transaction datasets and contains detailed information on the intraday trade and quote process at the NYSE, NASDAQ and numerous local exchanges in the U.S. The TAQ database consists of two parts: the trade database and the quote database. The trade database contains transaction prices, trading volumes, the exact time stamp (to the second) and attribute information on the validity of the transaction. The quote database consists of time stamped (best) bid and ask quotes, the volume for which the particular quote is valid (market depth), as well as additional information on the validity of the quotes. As the NYSE features a hybrid trading mechanism (see

Table 3.2 TAQ data record on quotes for Microsoft on June 1, 2009

SYMBOL	DATE	EX	TIME	BID	BID SZ	OFFER	OFF SZ	MODE
MSFT	2009-06-01	Z	36001	21.1100	43	21.1300	38	12
MSFT	2009-06-01	T	36001	21.1100	97	21.1200	6	12
MSFT	2009-06-01	T	36001	21.1100	92	21.1200	6	12
MSFT	2009-06-01	T	36001	21.1100	82	21.1200	6	12
MSFT	2009-06-01	I	36001	21.1100	9	21.1200	5	12
MSFT	2009-06-01	T	36001	21.1100	72	21.1200	6	12
MSFT	2009-06-01	B	36001	21.1100	30	21.1300	22	12
MSFT	2009-06-01	D	36001	21.1000	8	21.2100	2	12
MSFT	2009-06-01	B	36001	21.1100	31	21.1300	22	12
MSFT	2009-06-01	B	36002	21.1100	30	21.1300	22	12
MSFT	2009-06-01	B	36002	21.1100	21	21.1300	22	12
MSFT	2009-06-01	T	36002	21.1100	72	21.1200	5	12
MSFT	2009-06-01	T	36002	21.1100	78	21.1200	5	12
MSFT	2009-06-01	I	36002	21.1100	9	21.1300	33	12

SYMBOL stock symbol, *DATE* quote date, *TIME* quote time, *EX* exchange on which the trade occurred, *BID* bid price, *BID SZ* bid size in number of round lots (100 shares), *OFFER* offer (ask) price, *OFF SZ* offer size in number of round lots (100 shares), *MODE* quote condition

Chap. 2), the quotes reported in the quote database can be quotes that are posted by the specialist, limit orders from market participants posted in the limit order book, or limit orders submitted by traders in the trading crowd.

Table 3.3 shows raw data from the TotalView-ITCH data feed offered by NASDAQ which is more detailed than the TAQ database and also provides information on incoming limit orders. The "event classification" allows for a quite precise reconstruction of all limit order book activities and thus the resulting order book. Here, only messages on the events "A" and "P" yield information on prices and types of limit orders. Corresponding information on other event types can be retrieved by tracing the limit order according to its order ID. This data can be used to fully re-construct the limit order book and to partly identify the location of hidden volume.[2] For more details on the properties and the use of ITCH data, see Hautsch and Huang (2011).

3.1.2 *Matching Trades and Quotes*

Many exchanges, such as, e.g., the NYSE, NASDAQ, EURONEXT or XETRA, record trades and quotes separately which raises the problem of appropriately matching the two data files. This step is necessary whenever trade characteristics, like trade prices and trade sizes, have to be linked to the underlying quotes prevailing

[2]Automatic and efficient limit order book reconstruction can be performed by a limit order book system reconstructor ("LOBSTER") which is developed at Humboldt-Universität zu Berlin and can be accessed on http://lobster.wiwi.hu-berlin.de.

Table 3.3 TotalView-ITCH data record on market and limit orders for Microsoft trading at NASDAQ on September 1, 2009

TIME	ORDER ID	EVENT	PRICE	SIZE	SIDE
40900995	135132726	E		500	
40900995	135133117	E		100	
40900996	135126512	D		100	
40900996	135135501	A	2428	100	−1
40900996	135125636	D		200	
40900996	132601833	P	2427	500	−1
40900996	132601833	P	2427	250	−1
40900996	132601833	P	2427	144	−1
40900997	135135542	A	2427	100	1
40900997	135135544	A	2428	200	−1
40900997	135135580	A	2426	200	1
40900998	135135501	D		100	
40900998	135135591	A	2432	100	−1
40900999	135135631	A	2428	4000	−1

TIME milliseconds from midnight, *ORDER ID* unique ID for each limit order, *EVENT*: *E* execution of an order, *A* posting a new limit order, *D* (partial or total) deletion of limit orders, *P* execution against a hidden order, *PRICE* limit price, *SIZE* order size, *SIDE*: −1 – sell side, 1 – buy side

in the market. This matching process induces an identification problem as long as the corresponding time stamps are not exact. Particularly in systems where the trading process is recorded manually, the time stamps are not necessarily reliable and thus comparable. Even if the recording system is exact, latency and technological limitations circumvent a perfect matching of trades and corresponding quotes. For NYSE data of the early nineties, Lee and Ready (1991) show that the problem of potential mismatching can be reduced by the so-called "five-seconds rule". Accordingly, a trade is linked to the quote posted at least 5 s before the corresponding transaction. This is due to the fact that quotes are posted more quickly than trades can be recorded. Lee and Ready (1991) illustrate that this rule leads to the lowest rates of mismatching. However, while this rule was sensible for transaction data during the nineties and early 2000s, it is not applicable anymore to more recent data. In fact, the speed and precision of order processing has been increased substantially reducing the average 5-s delay of trade records. For instance, using NYSE data and estimating the adverse selection cost component in bid-ask spreads, Henker and Wang (2006) show that the time delay is rather 1 s than 5 s. This result is in line with most recent studies which use the most recent quote as the relevant one at each trade arrival.

However, given the variety of trading forms and systems as well as specific trading rules on the individual markets, the application of universal matching rules is rather inappropriate. If, for instance, reliable data on market depth at (and eventually behind) the best quotes are available, data-driven matching methods could be more sensible. Using data on market orders, limit orders and market depth associated with the three best levels from EURONEXT Amsterdam, Hautsch and Huang (2009)

propose an automatized algorithm yielding a specific matching for each trade. It consists of three steps, where the first step searches for a perfect match between the trade and the corresponding order book update while the following steps search for an approximate match whenever a perfect match is not possible:
Step 1: Perfect matching. Consider a specified time window, e.g., $[-10, 10]$ seconds around the time stamp of the corresponding trade. Then, pick every order book record in this time window and perform the following analysis:

1. if the trade price equals the current best bid (ask) and the difference in best bid (ask) order book volumes between the current record and the previous one equals the trade size, or,
2. if the trade price equals the previous best bid (ask), the difference in best bid (ask) order book volumes between the current record and the previous one equals the trade size, and the best bid (ask) decreased (increased) since the last order book update,

then, match the most previous order book record with the current trade and record the corresponding delay time. Case (1) is associated with a trade which absorbs parts of the pending depth on the first level. Accordingly, case (2) is associated with a trade which completely removes the first depth level and thus moves the best ask/bid quote. If for none of the order book records a match can be achieved in the given time window, the trade remains unmatched and we move to Step 2.
Step 2: Imperfect matching. Pick any unmatched trade record's time stamp and consider a time window of size which is twice the average delay time computed in Step 1. Moreover, if

1. the trade price equals to the best bid (ask) and the best bid (ask) size is less than the previous one, or,
2. the best bid (ask) decreases (increases) between two consecutive records,

then, match the trade with the corresponding order book entry. This step accounts for the possibility that trades might be executed against hidden liquidity. If for none of the order book records a match can be achieved in the given time window, the trade remains unmatched and we move to Step 3.
Step 3: Round time matching. Pick any unmatched trade and match it with the order book record that is closest to the trade's time stamp plus the average delay time.

Obviously, this procedure has to be adapted to specific trading rules at individual exchanges.

3.1.3 Data Cleaning

After matching trades and quotes, obvious data errors should be filtered out. Typical data errors are due to (a) a wrong recording or (b) a delayed recording of trade or quote information. Delayed records arise from trades which are recorded too late or from subsequent corrections of mis-recorded trades which are lined into the

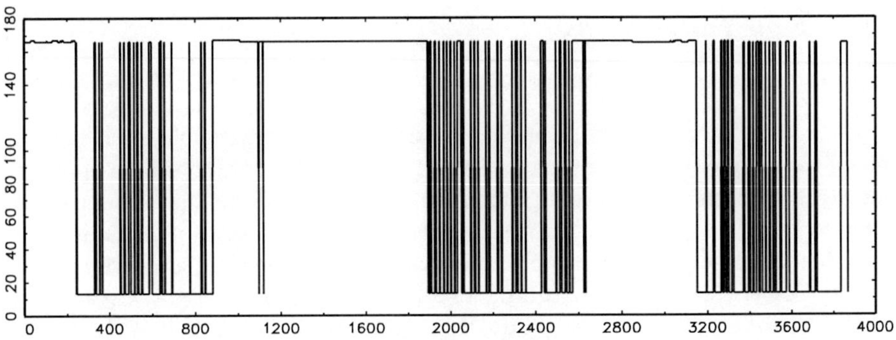

Fig. 3.1 Plot of Apple trade prices, NASDAQ, 11/10/2007, 13:11 to 13:15

trading system with a time delay. As an illustration of mis-recorded trades, Fig. 3.1 shows the price evolution of the Apple stock on October 11, 2007, over the course of 4 min.[3] While the underlying price level is around 165, we observe a massive number of price jumps due to the imputation of mis-recorded prices.

Such recording errors are most easily identified if transaction prices or quotes show severe jumps between consecutive observations which are reverted immediately thereafter. To remove such types of errors, a set of filters, similar to those shown below, is commonly applied:

1. Delete observations which are directly indicated to be incorrect, delayed or subsequently corrected.
2. Delete entries outside the regular trading hours.
3. Delete entries with a quote or transaction price equal to zero or being negative.
4. Delete all entries with negative spreads.
5. Delete entries whenever the price is outside the interval [bid $- 2 \times$ spread ; ask $+ 2 \times$ spread].
6. Delete all entries with the spread being greater or equal than 50 times the median spread of that day.
7. Delete all entries with the price being greater or equal than 5 times the median mid-quote of that day.
8. Delete all entries with the mid-quote being greater or equal than 10 times the mean absolute deviation from the local median mid-quote.
9. Delete all entries with the price being greater or equal than 10 times the mean absolute deviation from the local median mid-quote.

Obviously, the choice of window sizes in rules (v) to (ix) is somewhat arbitrary and dependent on the overall quality of the data. However, the parameters here are rather typical and in accordance with, e.g., Brownlees and Gallo (2006), Barndorff-Nielsen et al. (2008b) or Hautsch et al. (2011), among others.

[3]This example was kindly provided by Roel Oomen.

3.1.4 Split-Transactions

After having the data synchronized and cleaned, so-called "split-transactions" have to be taken into account. Split-transactions arise when a marketable order on one side of the market is matched against several standing limit orders on the opposite side. Such observations occur in electronic trading systems when the volume of an order exceeds the capacities of the first level of the opposing queue of the limit order book. In these cases, the orders are automatically matched against several opposing order book entries. Often, these sub-trades are recorded individually. Consequently, the recorded time between the particular "sub-transactions" is extremely small[4] and the corresponding transaction prices are equal or show an increasing (or decreasing, respectively) sequence. Depending on the research objective, it is sometimes justified to aggregate these sub-trades to one single transaction. However, as argued by Veredas et al. (2008), the occurrence of such observations might also be due to the fact that the limit orders of many traders are set for being executed at round prices, and thus, trades executed at the same time do not necessarily belong to the same trader. Moreover, in very actively traded stocks, the occurrence of different transactions within a couple of milliseconds is not unlikely. In these cases, a simple aggregation of observations with zero trade-to-trade durations would lead to mismatching. Grammig and Wellner (2002) propose identifying a trade as a split-transaction when the durations between the sub-transactions are smaller than 1 s, and the sequence of the prices (associated with a split-transaction on the bid (ask) side of the order book) are non-increasing (non-decreasing). Then, the volume of the particular sub-trades is aggregated and the price is computed as the (volume weighted) average of the prices of the sub-transactions.

3.1.5 Identification of Buyer- and Seller-Initiated Trades

Often, it is not possible to directly identify whether a trade is seller- or buyer-initiated. In such a case, the initiation of trades has to be indirectly inferred from the price and quote process. The most commonly used methods of inferring the trade direction are the quote method, the tick test as well as hybrid methods combining both methods (see, e.g., Finucane 2000). The quote method is based on the comparison of the transaction price and the mid-quote. Whenever the price is above (below) the mid-quote, the trade is classified as a buy (sell). Trades which are executed directly at or above (below) the prevailing best ask (bid) are most easily identified. However, the closer transaction prices are located to current mid-quotes, the higher the risk of misclassification is.

[4]Often it is a matter of measurement accuracy that determines whether sub-transactions have exactly the same time stamp or differ only by hundredths of a second.

If trades are executed at the mid-quote (and no other information is available), only the sequence of previous prices can be used to identify the current trade direction. According to the so-called *tick test*, a trade is classified as a buy (sell) if the current trade occurs at a higher (lower) price than the previous trade. If the price change between consecutive transactions is zero, the trade classification is based on the last price that differs from the current price. However, if information on the underlying market depth is available, the comparison of transaction volume and corresponding changes of market depth on one side of the market provides additional information which increases the precision of the identification algorithm.

3.2 Aggregation by Trading Events: Financial Durations

Transaction data is often also used in an aggregated way. Though aggregation schemes naturally induce a loss of information, there are three major reasons for the use of specific sampling schemes. Firstly, as discussed below, data aggregation allows to construct economically as well as practically interesting and relevant variables. Secondly, aggregation schemes allow to reduce the impact of market microstructure effects whenever the latter are of less interest and might cause noise in a given context. A typical example is the use of aggregated (squared) returns in realized volatility measures to estimate daily quadratic price variations. In such a context, a well-known phenomenon is the trade-off between on the one hand using a maximum amount of information increasing estimators' efficiency and on the other hand the impact of market microstructure effects causing biases in volatility estimates.[5] Thirdly, aggregation schemes allow to reduce the amount of data which is helpful whenever long sample periods or large cross-sections of assets are studied.

In general, we can distinguish between two major types of sampling and aggregation schemes: (i) *Event aggregation*, i.e., aggregations of the process according to specific trading events. This type of sampling scheme will be discussed in more detail in this section. (ii) *Time aggregation*, i.e., aggregations of the process according to calendar time which will be discussed in Sect. 3.5.

Consider in the following a (multivariate) point process associated with the complete order arrival process of a financial asset over a given time span. By selecting points of this process according to certain trading events, different types of so-called *financial point processes* are generated. The selection of individual points is commonly referred to as a "thinning" of the point process.

3.2.1 *Trade and Order Arrival Durations*

Sampling the process whenever a trade occurs is commonly referred to as *transaction time sampling* or *business time sampling* and is often used as a sampling

[5]For more details, see Chap. 8.

scheme underlying realized volatility measures. For a discussion, see e.g., Hansen and Lunde (2006), Andersen et al. (2010) or Hautsch and Podolskij (2010). The time between subsequent transactions is called *trade duration*, is the most common type of financial duration and is a natural measure of the trading intensity. Since a trade reflects demand for liquidity, a trade duration is naturally associated with the intensity of liquidity demand. Correspondingly, *buy* or *sell trade durations* are defined as the time between consecutive buys or sells, respectively, and measure the demand for liquidity on the individual sides of the market.

In an order driven market, *limit order (arrival) durations*, defined as the time between consecutive arrivals of limit orders, reflect the activity in the limit order book and thus on the supply side of liquidity. In studying limit order book dynamics, it is of particular interest to distinguish between different types of limit order activities reflecting, for instance, traders' order aggressiveness.

3.2.2 Price and Volume Durations

Price (change) durations are generated by selecting points according to their price information. Let p_i, a_i and b_i be the process of transaction prices, best ask quotes and best bid quotes, respectively. Define in the following i' with $i' < i$ as the index of the *most recently* selected point of the point process. Then, a series of price durations is generated by thinning the process according to the following rule:

Retain point i, $i > 1$, if $|p_i - p_{i'}| \geq dp$.

The variable dp gives the size of the underlying cumulative absolute price change and is chosen exogenously. The first point typically corresponds to the first point ($i = 1$) of the original point process. In order to avoid biases caused by a bouncing of transaction prices between ask and bid quotes ("bid-ask bounce"), an alternative is to generate price durations not on the basis of transaction prices p_i, but based on midquotes $mq_i := (a_i + b_i)/2$. As discussed in more detail in Chap. 8, price durations are closely related to volatility measures. Sampling whenever prices (or mid-quotes) change by a tick (corresponding to the smallest possible price movement), i.e., $dp = 1$, is commonly referred to as *tick time sampling*.

In technical analysis, turning points of local price movements are of particular interest since they are associated with optimal times to buy or to sell. The time between such local extrema, i.e., the so-called *directional change duration*, provides information on the speed of mean reversion in price processes. A directional change duration is generated according to the following procedure:

Retain point i, $i > 1$, if

1. $p_i - p_i' \geq (\leq)\, dp$
   and if there exists a point, indexed by i'' with $i'' > i'$, for which
2. $p_i \geq (\leq)\, p_j$ with $j = i + 1, \ldots, i'' - 1$ and $p_i - p_{i''} \geq (\leq)\, dp$.

Here, dp gives the *minimum* price difference between consecutive local extreme values.

Volume durations are defined as the time until a certain aggregated volume is traded on the market. It is generated formally by retaining point i, $i > 1$ if $\sum_{j=i'+1}^{i} v_j \geq dv$, where dv represents the chosen amount of the cumulated volume and v_i the transaction volume associated with trade i. Volume durations capture not only the speed of trading but also the size of trades. Consequently, they naturally reflect the intensity of liquidity demand.

An indicator for the presence of information on the market is the time it takes to trade a given amount of excess (or net) buy or sell volume. Consequently, so-called *excess volume duration* measure the intensity of (one-sided) demand for liquidity and are formally created by retaining point i, $i > 1$, if $|\sum_{j=i'+1}^{i} y_j^b v_j| \geq dv$, where y_j^b is an indicator variable that takes the value 1 if a trade is buyer-initiated and -1 if a transaction is seller-initiated. The threshold value dv is fixed exogenously and determines the level of one-sided volume under risk.

While excess volume durations reflect market-side specific imbalances in liquidity *demand*, the same idea can be applied to liquidity *supply* in limit order book markets. Correspondingly, we can quantify the time it takes until a given imbalance between ask and bid depth is realized. We refer this to an *excess depth duration*.

3.3 Properties of Financial Durations

During the remainder of this chapter we analyze the empirical properties of high-frequency data using the stocks JP Morgan (JPM), traded at NYSE, Microsoft (MSFT), traded at NASDAQ, and Deutsche Telekom (DTEK), traded in the German XETRA system. JP Morgan and Microsoft data are extracted from the TAQ database for June 2009. For Deutsche Telekom we employ trade data as well as 1-s snapshots of the (displayed) limit order book during September 2010. Throughout this chapter, overnight effects are omitted. Hence, aggregated data cover only observations within a trading day.

The present section discusses major empirical properties of financial durations. Figure 3.2 shows the evolution of monthly averages of the number of trades per day for JP Morgan traded at the NYSE from 2001 to 2009. We observe a clear increase in trading intensities since 2001, particularly during the financial crisis in 2008. Though trading activity declined after the crisis, we still observe a trading frequency in 2009 which is more than two times as high as in 2001.

Figure 3.3 shows the distribution of the time between trades over the universe of S&P 1500 stocks between 2006 and 2009. We observe that nearly 600 out of 1,500 assets trade more frequently than every 10 s. These assets are mostly constituents of the S&P 500 index. However, even beyond the S&P 500 assets, we observe around 1,000 assets trading more frequently than every 20 seconds on average.

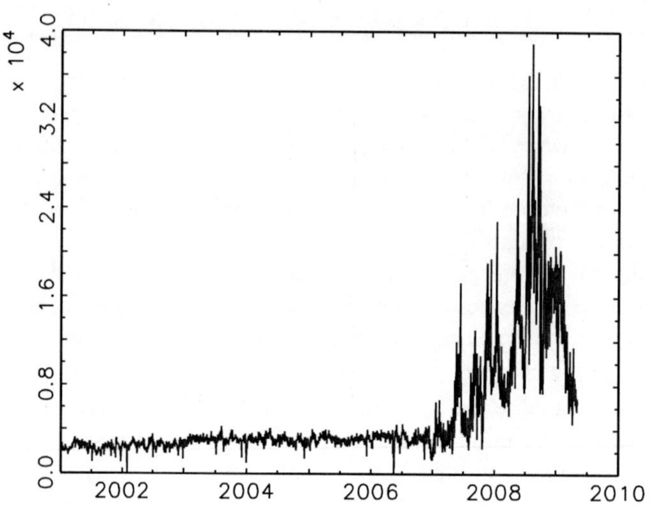

Fig. 3.2 Monthly averages of the number of trades per day for JP Morgan, NYSE, 2001–2009

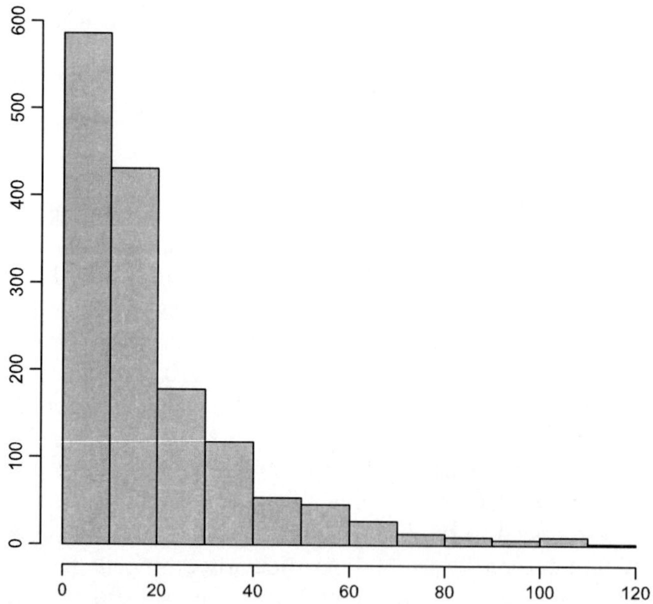

Fig. 3.3 Histogram of the time between trades (horizontal axis; in seconds) over the S&P 1500 universe, 2006–2009

Figure 3.4 shows histograms of trade durations, midquote (change) durations, volume durations, excess volume durations and excess depth durations for JPM, MSFT and DTEK. We observe that more than 50% of all JPM trade durations are less or equal than 1 s whereas trade durations longer than 10 s happen only very infrequently. This amounts to an average trade duration of 2.9 s. A similar picture arises for MSFT with an average trade duration of 4.5 s. DTEK trades occur slightly less frequent (with an average trade duration of 9.5 s) but reveal a more fat-tailed distribution with higher probabilities for longer durations ($\geq$10 s) and very short durations ($\leq$1 s). For all stocks, a striking feature is the non-trivial proportion of zero durations caused by split-trades (see Sect. 3.1.4) or truly simultaneous (or close-to-simultaneous) trade occurrences.

The second panel shows price durations associated with midquote changes of 10 basis points of the average price level for JPM (corresponding to 3.5 cents) and 3 ticks for MSFT and DTEK (corresponding to 1.5 cents and 0.15 cents, respectively). These price movements last on average 47.6, 64.1 and 66.7 s for JPM, MSFT and DTEK, respectively, and reveal a clearly more dispersed unconditional distribution as for trade durations. This dispersion becomes higher if the magnitude of the underlying price changes increases (third panel). The distributions of both trade and price durations reveal overdispersion, i.e., the standard deviation exceeds the mean. We observe dispersion ratios between 1.6 and 1.9 for trade durations and price durations yielding clear evidence against an exponential distribution.

The fourth panel of Fig. 3.4 shows the distribution of the time it takes to trade 10 times the average (single) trade size. These waiting times last on average 40.5, 67.4 and 118.3 s for JPM, MSFT and DTEK, respectively, with dispersion ratios around approximately 1.1. Hence, we observe a lower proportion of extremely long or extremely small durations inducing a more symmetric distribution. Finally, excess volume durations (for JPM and MSFT) and excess depth durations (for DTEK), shown in the bottom panel, give the time it takes until a certain imbalance in market side specific trading volume and depth arises on the market. Also here, we observe substantial variations in waiting times associated with clearly overdispersed distributions. Figure 3.5 shows time series plots of trade durations and price durations. We observe that financial durations are clustered in time with long (short) durations following on long (short) durations. This suggests positive serial dependence in duration series and is confirmed by Fig. 3.6 depicting the corresponding autocorrelation functions (ACF) of the individual financial duration series.

Note that these are autocorrelations of (irregularly spaced) time intervals. Accordingly, the (calendar) time distance to a lagged observation is time-varying. We observe highly significant positive autocorrelations revealing a strong persistence of the process. This is particularly true for trade durations having ACFs which decay very slowly. Actually, though not explicitly documented here, corresponding tests show evidence for long range dependence. For more details, see Chap. 6.

Price durations have higher autocorrelations but reveal slightly less persistence. Nevertheless, price durations based on relatively small price changes still reveal significant long range dependence. Recalling the close link between price durations and price volatility,[6] these results show that high-frequency volatility is obviously

[6]See Chap. 8 for more details on this relationship.

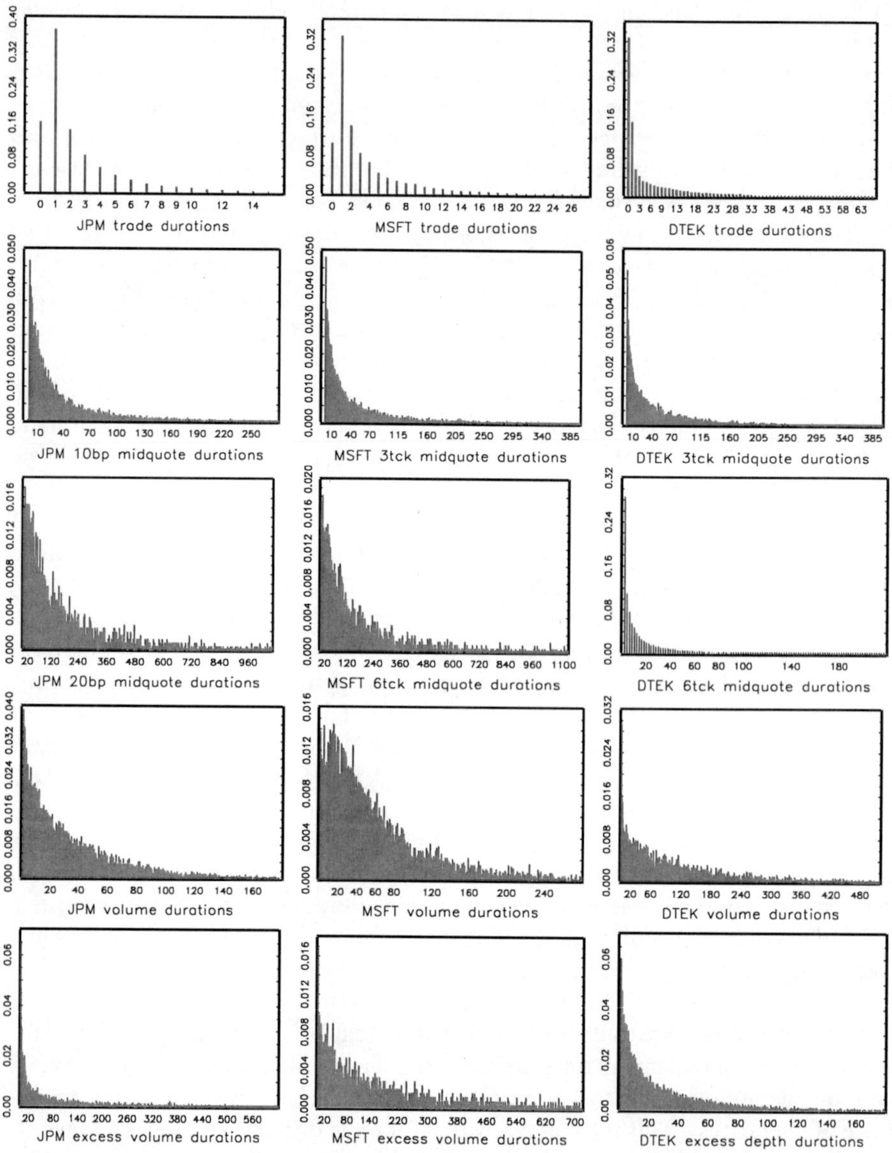

Fig. 3.4 Histograms of trade durations, midquote durations, volume durations, excess volume durations and excess depth durations for JP Morgan (NYSE), Microsoft (NASDAQ) and Deutsche Telekom (XETRA). Aggregation levels for price durations in basis points of the average price level for JP Morgan and in minimum tick size for Microsoft and Deutsche Telekom. The aggregation level for (excess) volume durations is 10 times of the average trade size. Aggregation level for excess depth durations: 10% of average order book depth up to the tenth level. Sample period: June 2009 for JP Morgan and Microsoft and September 2010 for Deutsche Telekom

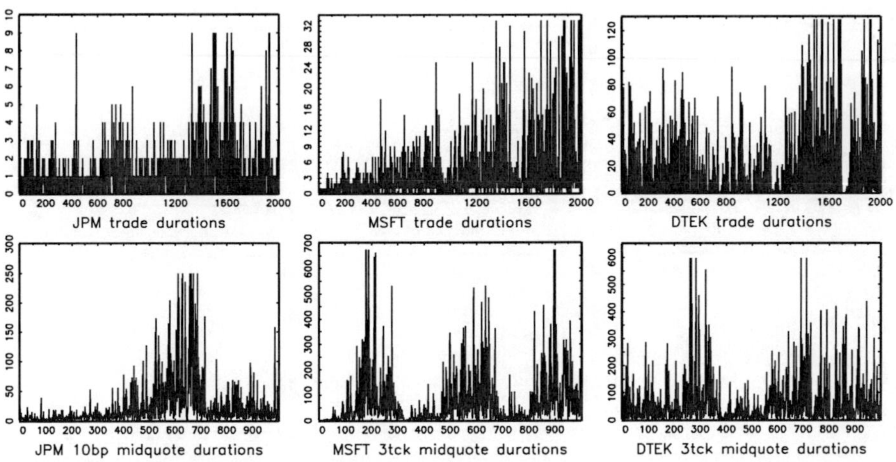

Fig. 3.5 Time series of trade durations and price durations for JP Morgan (NYSE), Microsoft (NASDAQ) and Deutsche Telekom (XETRA). Aggregation levels for price durations in basis points of the average price level for JP Morgan and in minimum tick size for Microsoft and Deutsche Telekom. Sample period: First 2,000 observations for trade durations and 1,000 observations for price durations in June 2009 for JP Morgan and Microsoft and in September 2010 for Deutsche Telekom

very persistent. This persistent declines when the underlying aggregation level (i.e., the size of underlying price movements) raises and first passage times become longer. Then, as shown in the third panel, the ACFs decay quickly and become even negative for JPM and MSFT after approximately 25 lags. This negative dependence is driven by intraday periodicities since higher lags of comparably long price durations might easily go beyond the current trading day and are driven by price activity of the day before.

The ACFs of volume durations start on a comparably high level. For example, for JPM volume durations, the first order autocorrelation is around 0.65. This high autocorrelation is caused by the fact that trade sizes are strongly clustered themselves (see Sect. 3.4 for more details).[7] Finally, also excess volume duration series are strongly autocorrelated but are less persistent than volume durations. In contrast, excess depth durations are very persistent. The time it takes to build up a certain excess liquidity supply reveals long memory. Hence, (excess) liquidity shocks are quite long-lived.

Figure 3.7 shows the intraday seasonality patterns based on cubic spline regressions.[8] For JPM and MSFT we find a distinct inverse U-shaped pattern with lowest durations in the morning and before closure and significantly longer spells

[7]Note that these effects are *not* caused by split-transactions since such effects have been taken into account already.

[8]For more details on the estimation of seasonality effects, see Sect. 5.4.

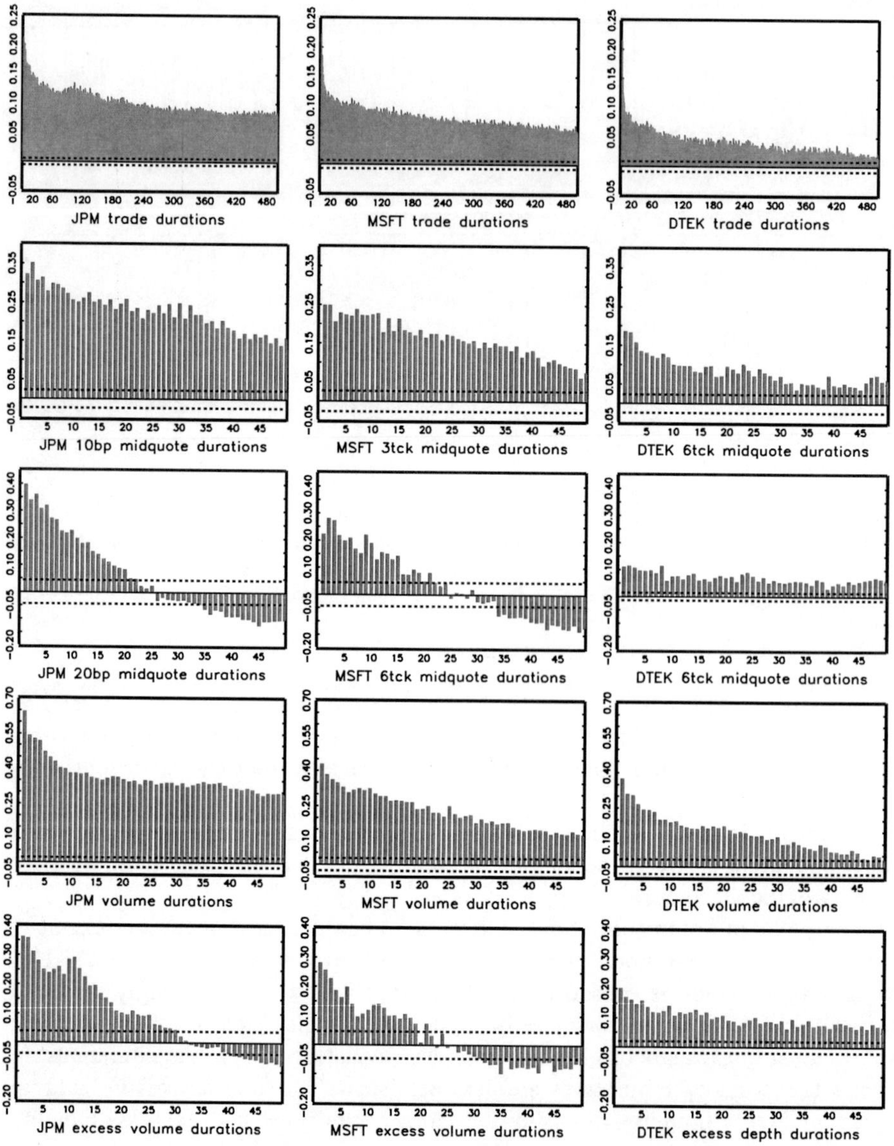

Fig. 3.6 Autocorrelation functions of trade durations, midquote change durations, volume durations, excess volume durations and excess depth durations for JP Morgan (NYSE), Microsoft (NASDAQ) and Deutsche Telekom (XETRA). Data description see Fig. 3.4. *Dotted lines*: approximately 99% confidence interval. The x-axis denotes the lags in terms of durations

around lunch time. High trading activities after market opening are driven by the dissemination of information occurring over night. As soon as information from other markets is processed, market activity declines and reaches its minimum around

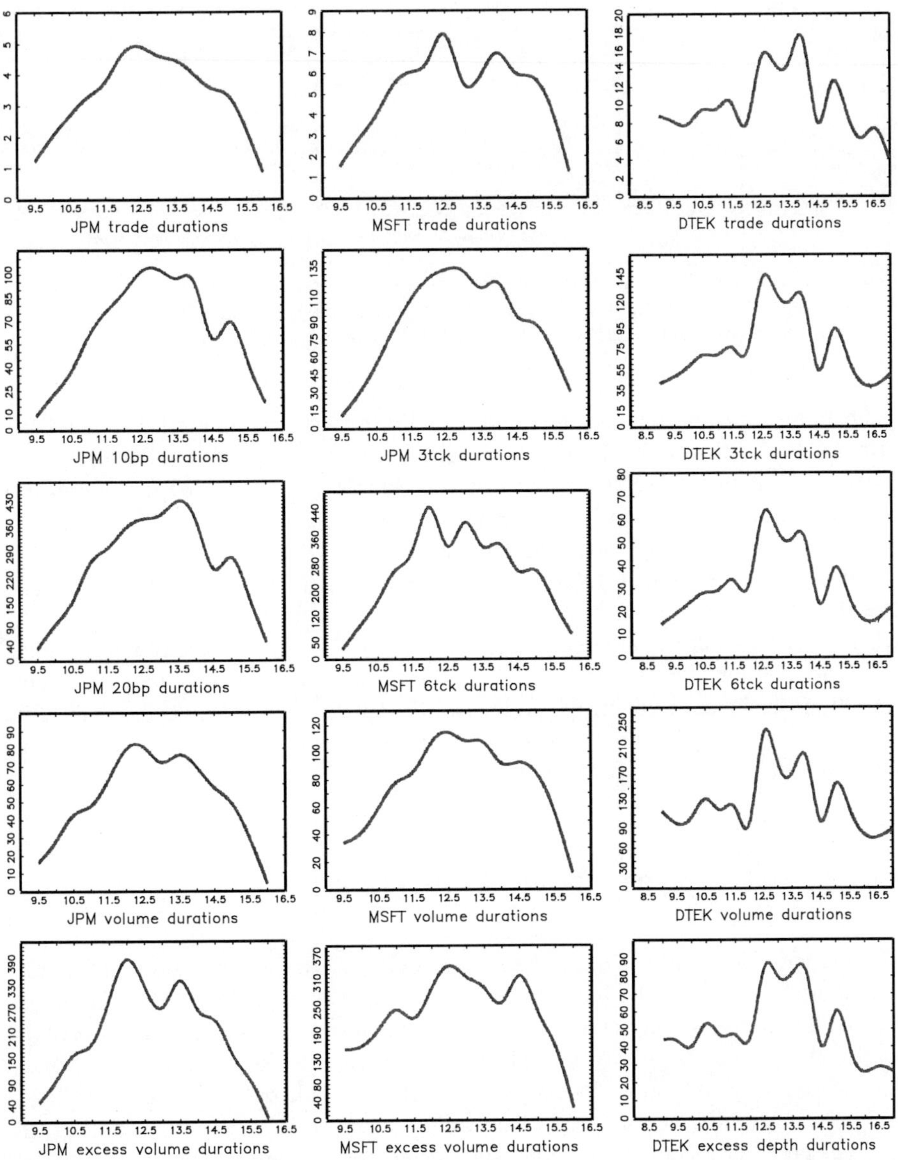

Fig. 3.7 Cubic spline functions (30 min nodes) of trade durations, midquote change durations, volume durations, excess volume durations and excess depth durations for JP Morgan (NYSE), Microsoft (NASDAQ) and Deutsche Telekom (XETRA). Data description see Fig. 3.4. The x-axis denotes local calendar time

lunch time. Then, approaching market closure, trading activity steadily increases again as many traders tend to close or to re-balance their positions before continuous trading stops. XETRA trading reveals a slightly different pattern with the mid-day

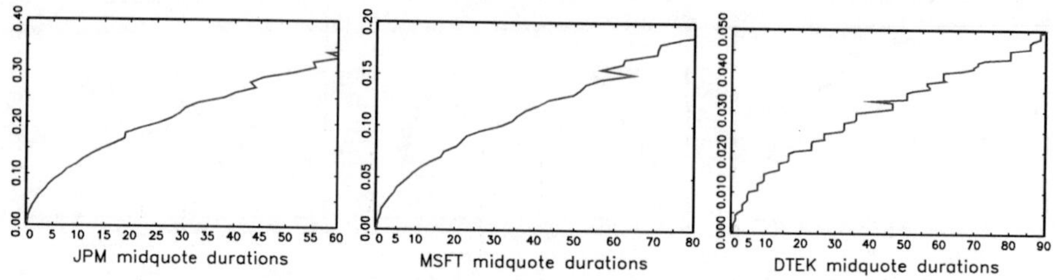

Fig. 3.8 Relationship between the size of cumulative absolute price changes dp (y-axis) and the average length of the resulting price duration (in minutes) (x-axis). Based on JP Morgan (NYSE), Microsoft (NASDAQ) and Deutsche Telekom (XETRA). Data description see Fig. 3.4

spike being more pronounced and a sharp decline of durations (and thus an increase of market activities) around 14:30. The latter "dip" is likely to be induced by the opening of the major U.S. exchanges (CBOT, NYSE and NASDAQ) and the need to process new information. We observe that this significantly increases temporary trading activity, volatility and market imbalances.

Finally, Fig. 3.8 plots the relationship between the size of the underlying cumulative absolute price change dp and the average length of the resulting price duration, where overnight periods are discarded. We observe slightly concave functions with relatively similar shapes. Note that the scaling on the y-axis is quite different reflecting that overall volatility differs from stock to stock.[9] While for DTEK it takes on average approximately 40 min to move the price by 0.03, this movement just takes around 2 min in JPM trading.

3.4 Properties of Trading Characteristics

In this section, we discuss the statistical properties of the most important trading characteristics, such as the bid-ask spread, the trade size, the order book depth as well as the trade-to-trade return. These characteristics are observed on transaction level and thus are irregularly spaced in time. Figure 3.9 shows the unconditional distributions of trade sizes, bid-ask spreads, trade-to-trade price changes as well as trade-to-trade midquote changes. We observe that JPM trade sizes are quite discrete and clustered in round lots. To model such outcomes, a count data distribution (see Chap. 13) seems to be most appropriate. Conversely, for MSFT and DTEK, the distribution of trade sizes is significantly more dispersed. Nevertheless, we still observe a concentration of probability mass at round numbers. This is particularly striking for DETK trading where trade sizes of multiples of 5 lots are particularly preferred. This reflects the well-known phenomenon of traders' preference for round numbers.

[9]The discontinuities are caused by finite-sample properties as for high aggregation levels the number of underlying observations naturally shrinks.

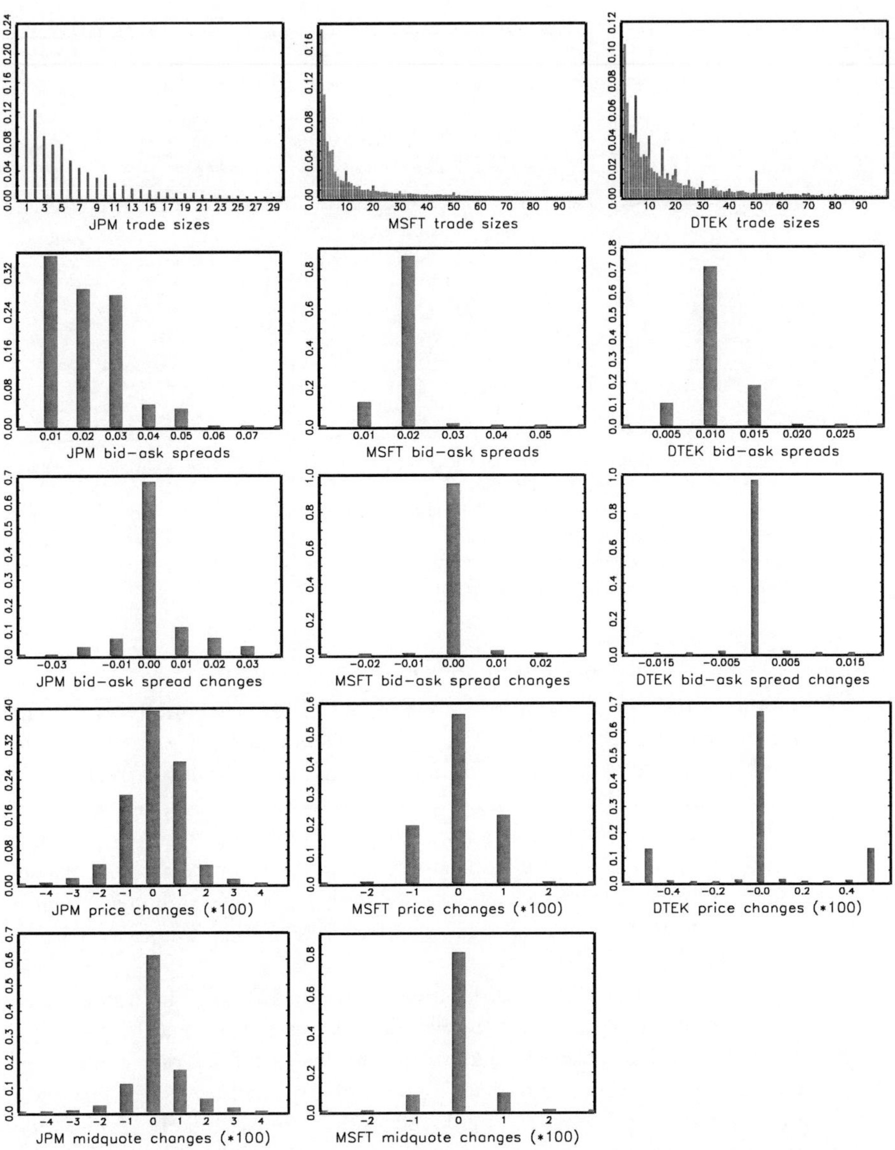

Fig. 3.9 Histograms for trade sizes (in 100 share lots), bid-ask spreads, trade-to-trade spread changes, trade-to-trade price changes and trade-to-trade midquote changes for JP Morgan (NYSE), Microsoft (NASDAQ) and Deutsche Telekom (XETRA). Sample period: June 2009 for JP Morgan and Microsoft and September 2010 for Deutsche Telekom

The distribution of bid-ask spread realizations is very discrete. In 90% of all cases, JPM spreads take the values 0.01, 0.02 and 0.03. In case of JPM and DTEK, the distribution is even more extreme with a clear concentration of spread

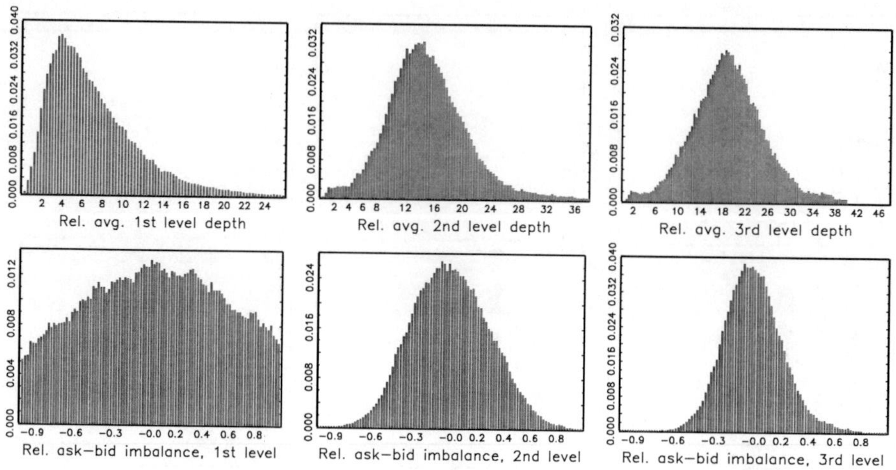

Fig. 3.10 *Top panel*: Histograms of first, second and third level depth (ask and bid average) for Deutsche Telekom (XETRA) as multiples of the average trade size. *Bottom panel*: Percentage difference between cumulated ask and bid depth, weighted by the price difference to the opposite side of the market. Sample period: September 2010

realizations at one or two ticks. This discreteness is naturally also reflected in the distribution of trade-to-trade spread changes which are essentially one-point distributions in case of MSFT and DTEK. This reflects that for these very active stocks, bid-ask spreads are close to be minimal and are mostly constant over time.

The last two panels in Fig. 3.9 show the distributions of trade-to-trade price and mid-quote changes. Probability mass clearly concentrates at zero indicating that transaction prices are mostly constant from trade to trade.[10] This distribution is even more concentrated in case of mid-quote price changes which reflects that trade-to-trade quote changes are less likely than transaction price changes.[11]

Figure 3.10 plots distributions of order book depth for DTEK. The upper panel depicts the histograms of depth (averaged over the ask and bid side) at the first best, second best and third best observed quote as multiples of the average trade size. First level depth is right-skewed indicating only small probabilities for observing a comparably thin market. On average, first level depth is approximately seven times the average trade size with a standard deviation of 5. Note that the second and third best *observed* quote is not necessarily the second and third best *theoretically possible* quote as there might be empty price grids in the book. However, since DTEK belongs to the most actively traded XETRA stocks during the observation

[10]The asymmetries in the distributions are induced by the underlying sample period where, e.g., for JPM, upward price movements are slightly less likely than downward movements.

[11]We do not record trade-to-trade midquote changes for Deutsche Telekom since for this stock, we only employ 1-s limit order book snapshots.

period, the probability to observe gaps in the order book close to the market is relatively small. Therefore, the distributions shown in Fig. 3.10 are very similar to depth distributions shown for fixed price grids around the best ask and bid quote. We observe that these distributions are rather symmetric with averages of 15 and 19 for the second and third level, respectively. Hence, on average, the depth *behind* the market is on average significantly higher than the depth at the market.

In the bottom panel of Fig. 3.10, we show the difference between the cumulated ask and bid depth, weighted by the spread to the opposite side of the market, relative to the sum of cumulated ask and bid weighted depth. In particular, for $k \in \{1, 2, 3\}$ we compute

$$\upsilon_i(k) = \frac{\sum_{j=1}^{k} v_i^{a,j}(a_i^j - b_i^1) - v_i^{b,j}(a_i^1 - b_i^j)}{\sum_{j=1}^{k} v_i^{a,j}(a_i^j - b_i^1) + v_i^{b,j}(a_i^1 - b_i^j)},$$

where $v_i^{a,j}$ and $v_i^{b,j}$ denote the ask and bid depth at the jth level, respectively and a_i^j and b_i^j are the corresponding ask and bid quotes, respectively. Hence, $\upsilon_i(k) = 0$ indicates a completely balanced market whereas values of 1 (-1) reflect that all depth is cumulated on the ask (bid) side of the market. Accordingly, $\upsilon_i(1)$ just equals the relative ask-bid depth difference. For this quantity, we observe a fat-tailed distribution assigning significant probability mass to extreme values of completely imbalanced order books. Conversely for $k > 1$, the distributions of $\upsilon_i(k)$ are clearly less fat-tailed with shapes very similar to that of a normal distribution. Hence, if we cumulate depth over several price levels, the probability for extreme market imbalances is clearly smaller. Figure 3.11 shows time series plots of trade sizes and bid-ask spreads. We observe that trade sizes themselves are clearly clustered over time. Hence, large trade sizes tend to follow large trade sizes suggesting positive autocorrelations. The time series plots of bid-ask spreads naturally reflect the discreteness of quotes shown in Fig. 3.9. Moreover, even spread realizations tend to be clustered over time as well. This is most evident for JPM spreads but is also visible for the very discrete series of MSFT and DTEK.

The notion of clustering in trade sizes and spreads is confirmed by Fig. 3.12 depicting trade-to-trade autocorrelations of the corresponding trading characteristics. JPM trade sizes reveal a slowly decaying autocorrelation function similar to that observed for trade durations. The fact that serial correlations in trade sizes are clearly lower and less persistent for MSFT and DTEK, indicates that trade-to-trade dynamics vary over the cross-section of stocks and are obviously driven by specific underlying institutional structures. Nonetheless, it is remarkable that trade sizes are significantly autocorrelated up to at least 100 lags. The second panel shows that also bid-ask spreads are strongly clustered over time. Again, there is substantial variation in the ACF shapes across the different stocks. While for JPM and DTEK, bid-ask spread dynamics are very persistent and reveal long range dependence, MSFT autocorrelations are lower and decay quite fast. Overall these results indicate that bid-ask spreads – and as such important components of transaction costs – are clearly predictable.

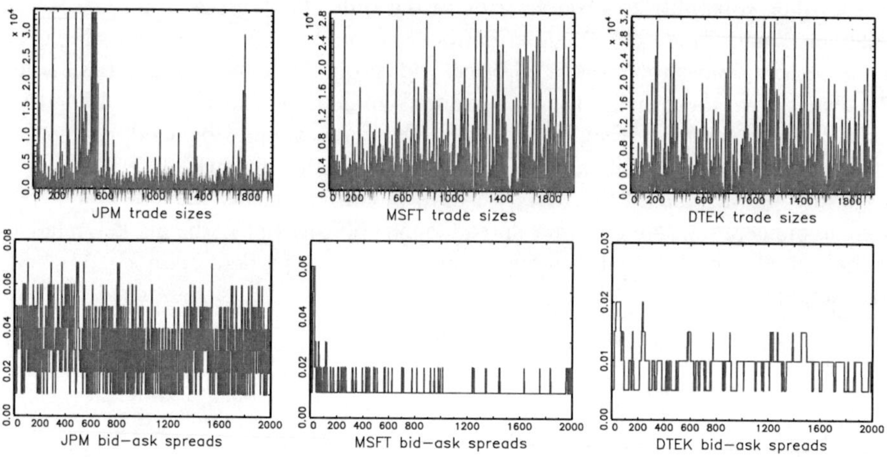

Fig. 3.11 Time series of trade sizes (in 100 share lots) and bid-ask spreads for JP Morgan (NYSE), Microsoft (NASDAQ) and Deutsche Telekom (XETRA). Sample period: First 2,000 observations in June 2009 for JP Morgan and Microsoft and in September 2010 for Deutsche Telekom

The third panel depicts the ACFs of absolute trade-to-trade price changes. The latter can be seen as proxies for high-frequency (or trade based) volatility (for more details, see Chap. 8). It is shown that trade volatility is strongly serially dependent and persistent. The shape of the ACFs are quite similar to those of trade durations and trade sizes. Note that the clustering of absolute trade-to-trade transaction price changes is partly driven by a bouncing of trade prices between ask and bid quotes (see also the discussion below). Indeed, trade-to-trade movements of absolute *midquote* changes (bottom panel) reveal slightly weaker (but still persistent) dependencies. Using the mid-quote as a proxy for the underlying (unobservable) "efficient" price of the asset, these plots reveal substantial volatility clustering even on the transaction level.

Figure 3.13 shows the evolution of trade prices and corresponding quotes over an arbitrary 1-min interval for JPM and MSFT trading. The pictures illustrate the irregular spacing of trades, the discrete movements of prices and quotes as well as the up-ward and down-ward bouncing of trade prices between ask and bid quotes (as well as within the spread). As confirmed by Fig. 3.14, this bid-ask bouncing of trade prices causes a highly significant and negative first order autocorrelation in trade-to-trade (signed) price changes. This feature is well-known in the market microstructure literature and is formally discussed by Roll (1984). Roll illustrates that in the most simple case where (unobservable) "efficient prices" follow a random walk and transactions can occur only on ask and bid quotes (with the spread being constant and symmetric around the efficient price), resulting trade price changes follow an MA(1) process with negative coefficient. For a deeper discussion, see also Hasbrouck (2007). The second panel of Fig. 3.14 shows the ACFs of trade-to-trade midquote changes. Though there is still some evidence for an MA(1) process,

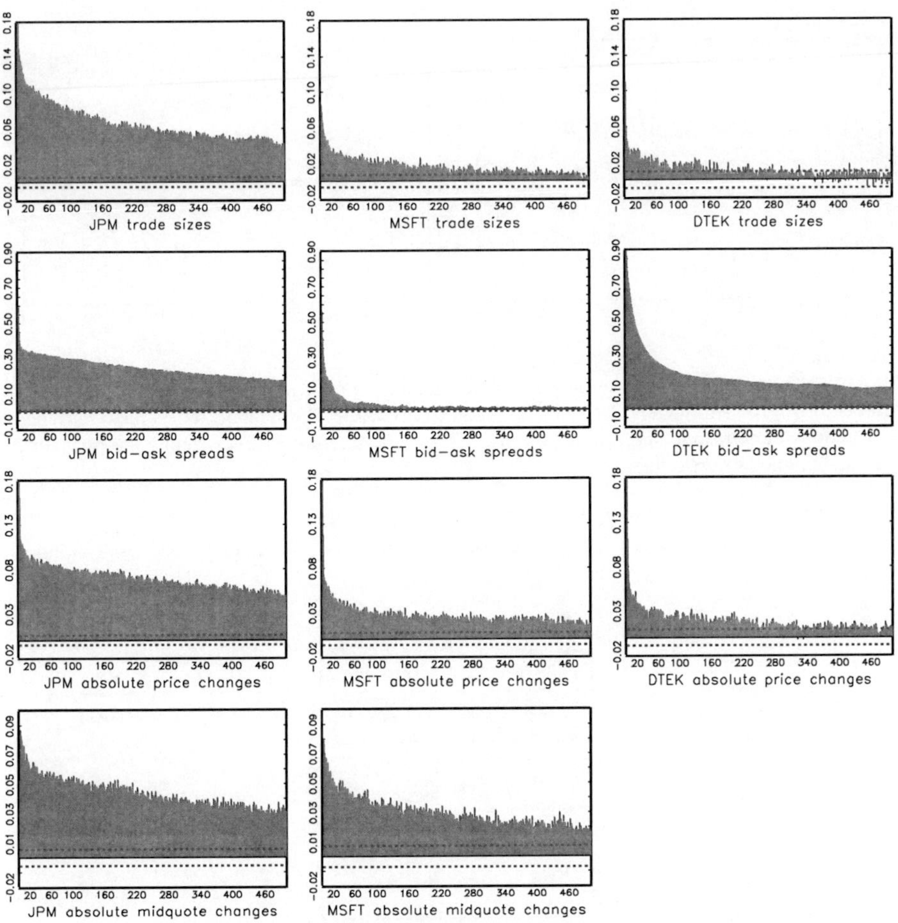

Fig. 3.12 Autocorrelation functions of trade sizes, bid-ask spreads, absolute trade-to-trade price changes and absolute trade-to-trade midquote changes for JP Morgan (NYSE), Microsoft (NASDAQ) and Deutsche Telekom (XETRA). Sample period: June 2009 for JP Morgan and Microsoft and September 2010 for Deutsche Telekom

the dependence is clearly weaker as for trade price changes. The remaining negative autocorrelation provides evidence for reversal effects in quote changes, i.e., changes in quotes tend to be reversed thereafter. This effect is confirmed by the ACFs of ask quote changes shown in the bottom panel.[12]

Figure 3.15 depicts the intraday seasonality patterns of trading characteristics. We observe distinct seasonality shapes for trading volumes of JPM and MSFT with high trade sizes after opening and before market closure. This pattern is well

[12]The ACFs for bid quote changes look very similar and are not shown here.

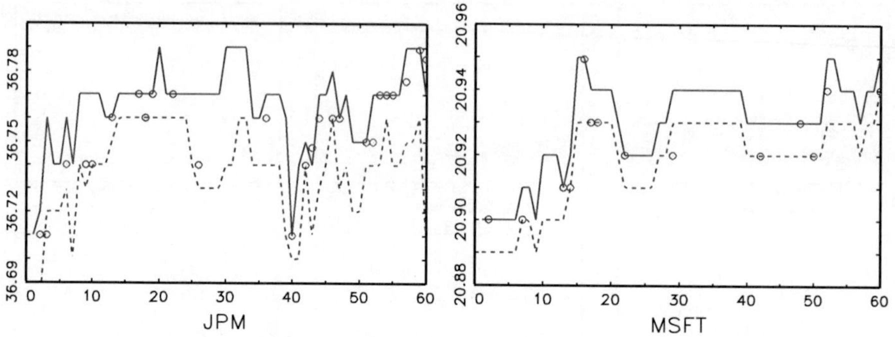

Fig. 3.13 Time series of transaction prices and ask and bid quotes over a one-minute interval at June 1, 2009 for JP Morgan (NYSE) and Microsoft (NASDAQ). The symbols reflect the occurrence of a trade with corresponding transaction price. The *solid* and *dotted lines* denote the prevailing ask and bid quotes, respectively

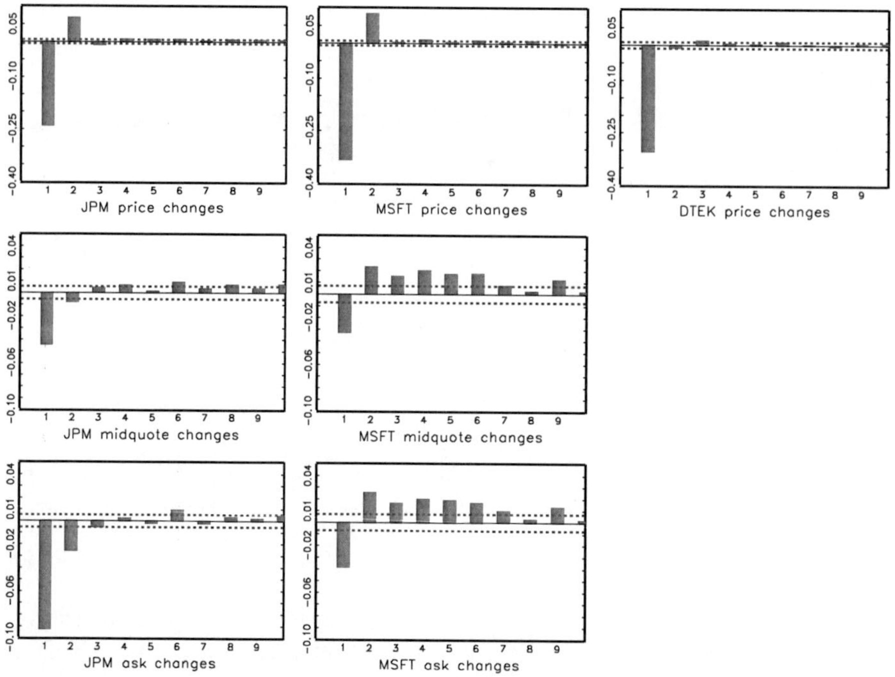

Fig. 3.14 Autocorrelation functions of trade-to-trade price changes, trade-to-trade midquote changes and trade-to-trade ask changes for JP Morgan (NYSE), Microsoft (NASDAQ) and Deutsche Telekom (XETRA). Sample period: June 2009 for JP Morgan and Microsoft and September 2010 for Deutsche Telekom

in accordance with the seasonalities found for financial durations in Fig. 3.7 and indicates that high trading activities at the beginning and before the end of a trading session are not only reflected in the speed of trading but also in trade sizes. However,

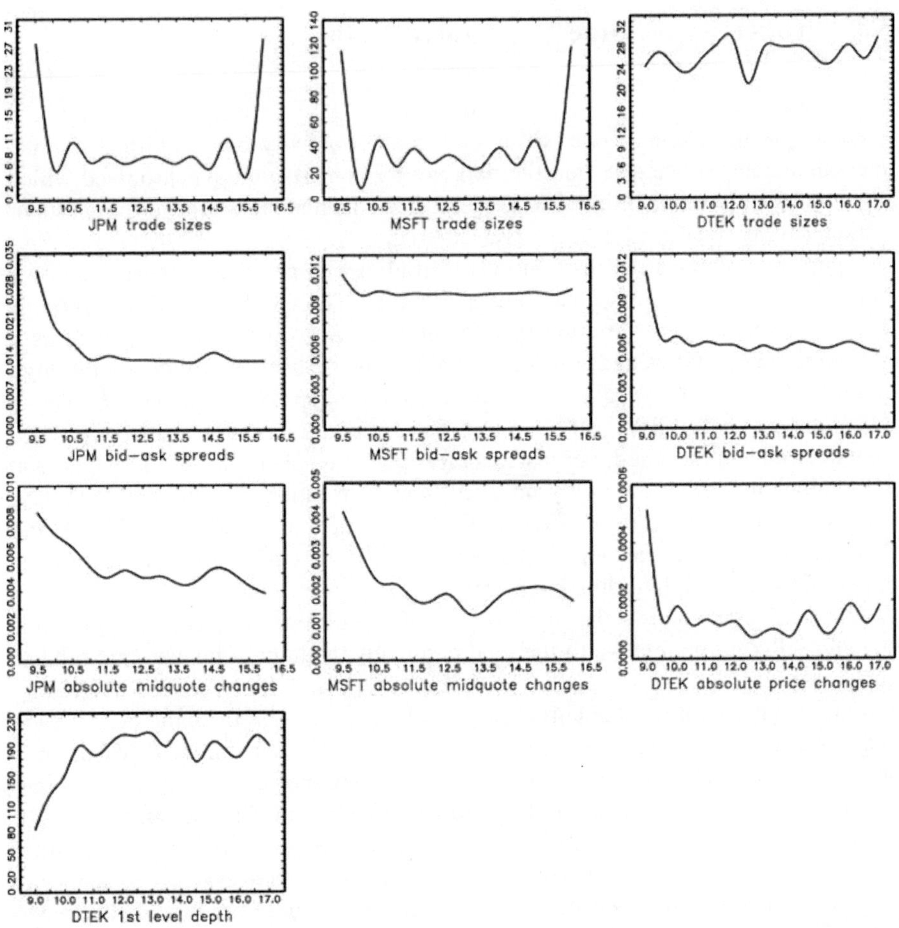

Fig. 3.15 Cubic spline functions (30 min nodes) for trade sizes (in 100 share lots), bid-ask spreads, absolute trade-to-trade price changes, absolute trade-to-trade midquote changes and first level depth (in 100 share lots) for JP Morgan (NYSE), Microsoft (NASDAQ) and Deutsche Telekom (XETRA). Sample period: June 2009 for JP Morgan and Microsoft and September 2010 for Deutsche Telekom

while this pattern is quite pronounced for JPM and MSFT, for DTEK trading no clear patterns are observable. Hence, intraday periodicities differ across markets and depend on institutional settings and time zones.

For bid-ask spreads and absolute midquote changes, the intraday seasonality pattern is quite pronounced with high spreads and high volatility after opening which then remain quite constant during the trading day. Finally, also market depth reveals clear intraday periodicities. Accordingly, depth is lowest after opening, successively increases during the morning and remains on a widely constant level during the day. Hence, it takes some time after market opening until sufficient order book depth is built up.

3.5 Properties of Time Aggregated Data

As discussed in Sect. 3.3, an alternative way to aggregate high-frequency data is to sample in calendar time. Evaluating market activity over equi-distant time intervals has the advantage that the data are by construction synchronized which eases multi-variate modelling. Moreover, it is advantageous whenever forecasts of market activity over fixed time intervals are required.

Figure 3.16 depicts the distributions of trading characteristics aggregated over 2 min. The first panel shows the number of trades. Though the underlying variable is a count variable, its variation is sufficiently large to justify its treatment as a continuous (positive-valued) random variable. The distributions indicate the high liquidity of the underlying assets with on average approximately 41, 26 and 13 trades per 2 min for JPM, MSFT and DTEK, respectively. The distributions are right-skewed and fat-tailed. For instance, for JPM, the occurrence of more than 100 transactions in 2 min is not very unlikely. The distributions of cumulative trading volumes are even more right-skewed which is particularly evident for MSFT and DTEK. Also here, significant probability mass in the right tail reflects the occurrence of periods of very high trading intensity.

The third panel shows the distributions of the difference between cumulative buy and sell volume relative to the total cumulative volume. This variable reflects imbalances in liquidity demand. The spikes at -1 and 1 are caused by trading periods where the entire cumulative trading volume is on one side of the market. Due to the concentration of probability mass at single points, the resulting distribution is obviously not purely continuous but is rather a mixture of continuous and discrete components. The right picture in the third panel shows the distribution of *changes* in the DTEK relative ask-bid (first level) depth imbalance (relative to the total prevailing depth) evaluated at 2 min intervals. As the relative ask-bid imbalance is by construction bounded between -1 and 1, changes thereof are bounded between -2 and 2. Accordingly, values of -2 or 2 indicate that over a 2-min interval one-sided depth in the book has been entirely shifted to the other side of the market. In our dataset, such a situation, however, never occurred. Nevertheless, values of higher than $|1.5|$ indicate that quite substantial shifts of order book depth from one side of the market to the other side within 2 min are possible. Finally, the two bottom panels of Fig. 3.16 show the distributions of 2-min quote changes (here representatively only for the ask side) and transaction price changes. The discreteness of quote and price changes is still clearly visible even over 2-min periods.

Figure 3.17 gives the histograms of 2-min percentage changes of the first, second and third level depth (averaged over ask and bid sides). It turns out that the overall order book depth does not change very dramatically over 2 min with zero values occurring with probability around 50%. Conversely, as illustrated above, the *relative* allocation over the two sides of the market can vary quite significantly.

Figures 3.18 and 3.19 display the corresponding distributions over 30 s and 10 s aggregates, respectively. The ranges of realizations for trade counts naturally shrink. Likewise the distributions become even more right skewed and traders' preferences

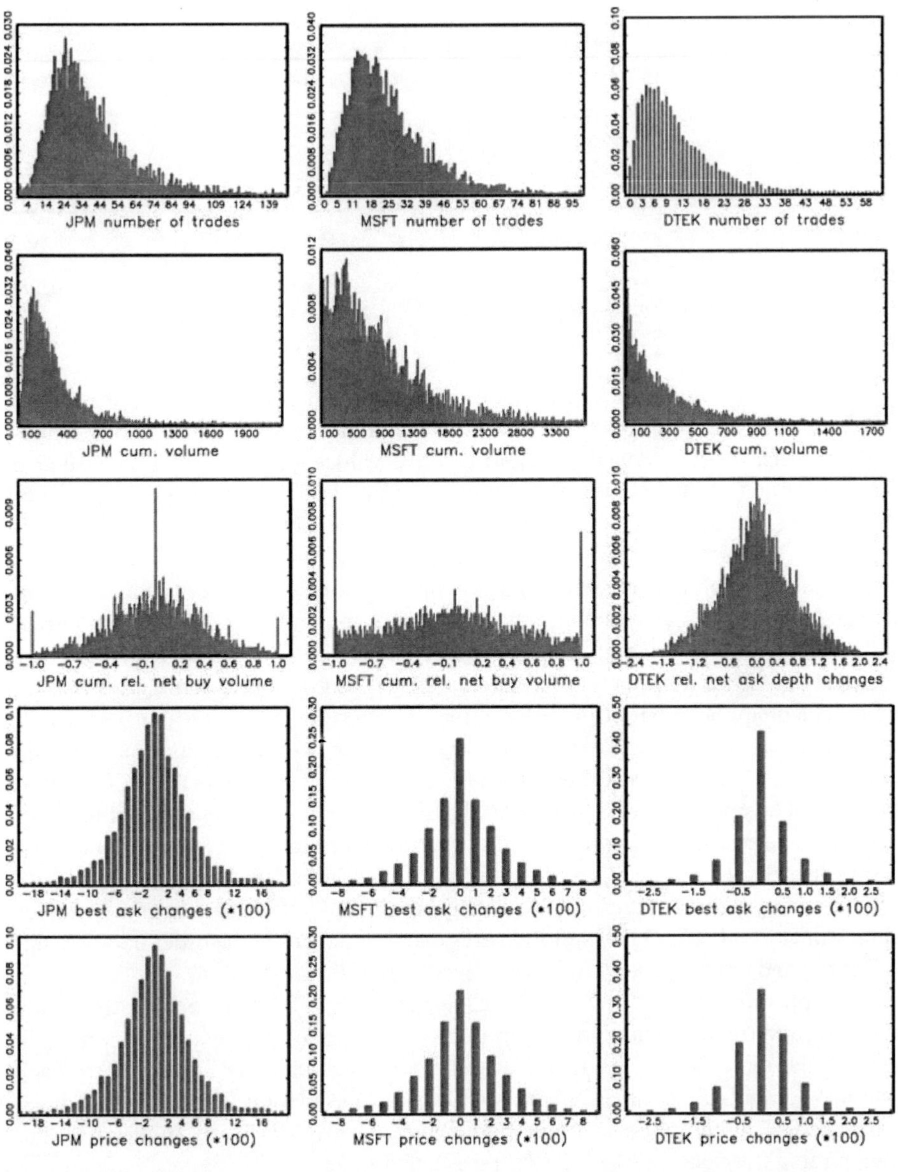

Fig. 3.16 Histograms for 2-min number of trades, cumulated trading volume (in 100 share lots), cumulated net buy volume (in %), relative net ask depth changes, best ask changes and transaction price changes for JP Morgan (NYSE), Microsoft (NASDAQ) and Deutsche Telekom (XETRA). Sample period: June 2009 for JP Morgan and Microsoft and September 2010 for Deutsche Telekom

for round lot sizes (see Sect. 3.4) become visible again if the aggregation level declines. This is most evident for 10-s DTEK volume revealing a concentration of probability mass at round lot sizes. Moreover, also the distribution of cumulative

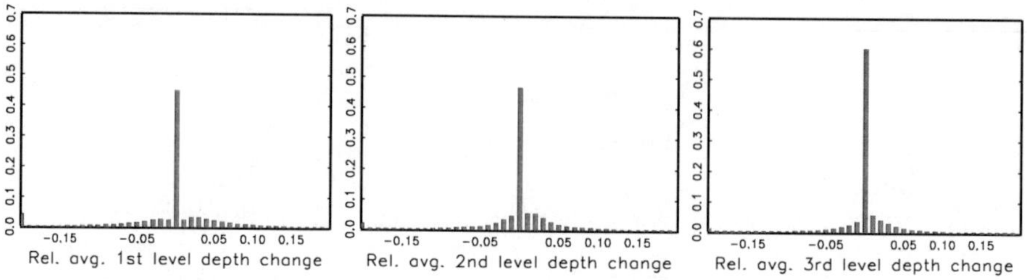

Fig. 3.17 Top panel: Histograms of 2-min relative changes of first, second and third level depth (ask and bid average) for Deutsche Telekom (XETRA). Sample period: September 2010

relative net buy volume becomes much more discrete since the probability for the occurrence of one-sided volume rises with shrinking time intervals. Correspondingly, the distributions reveal a clustering at the values −1, 0, and 1. Likewise also the distribution of quote and price changes become less dispersed and converge to the distributions observed on transaction level (see Fig. 3.9).

Figure 3.20 plots the ACFs of the corresponding 2-min aggregates. Trade counts and cumulated volumes reveal a strong serial dependence indicating that liquidity demand over short intervals is highly predictable. In contrast to the ACFs reported for financial durations (see Fig. 3.6), the persistence is lower with ACFs decaying relatively fast. The third panel reveals that also signed (relative) cumulative trading volume is predictable with first order autocorrelations between 0.1 and 0.15. Likewise, also bid-ask spreads are still autocorrelated over 2-min intervals though we see clear differences across the different assets. The bottom panel plots the ACFs of first level depth (averaged over the ask and bid side) and the relative net ask depth change for DTEK over 2-min intervals. A first order autocorrelation of approximately 0.4 shows that depth is clearly predictable. Relative changes in the excess ask depth show a strong reversal pattern with a highly significant first order autocorrelation of approximately −0.46. This result indicates that imbalances in the order book are not persistent and are very likely to be re-moved within the next 2 min.

The top three panels of Fig. 3.21 give the ACFs of quote log returns, trade price log returns and midquote log returns computed over 2 min. With very few exceptions, there is no significant evidence for predictability in high-frequency returns. The bid-ask bounce effect as shown in Fig. 3.14 on the transaction level is not visible anymore on a 2-min frequency which is obviously induced by the high trading frequency of the underlying assets. In fact, bid-ask bounces can still be significant over longer time intervals if the underlying trading frequency is lower. The bottom panel in Fig. 3.21 reports the ACFs of squared midquote log returns showing that 2-min price volatility is clearly clustered.

Figures 3.22–3.25 present the corresponding autocorrelation functions for 30 and 10 s. The plots illustrate how the dynamics of the individual variables change when the sampling frequency increases and ultimately converge to transaction level. It turns out that the persistence in trade counts and cumulative volumes clearly increases and the processes tend to reflect long range dependence. Similar

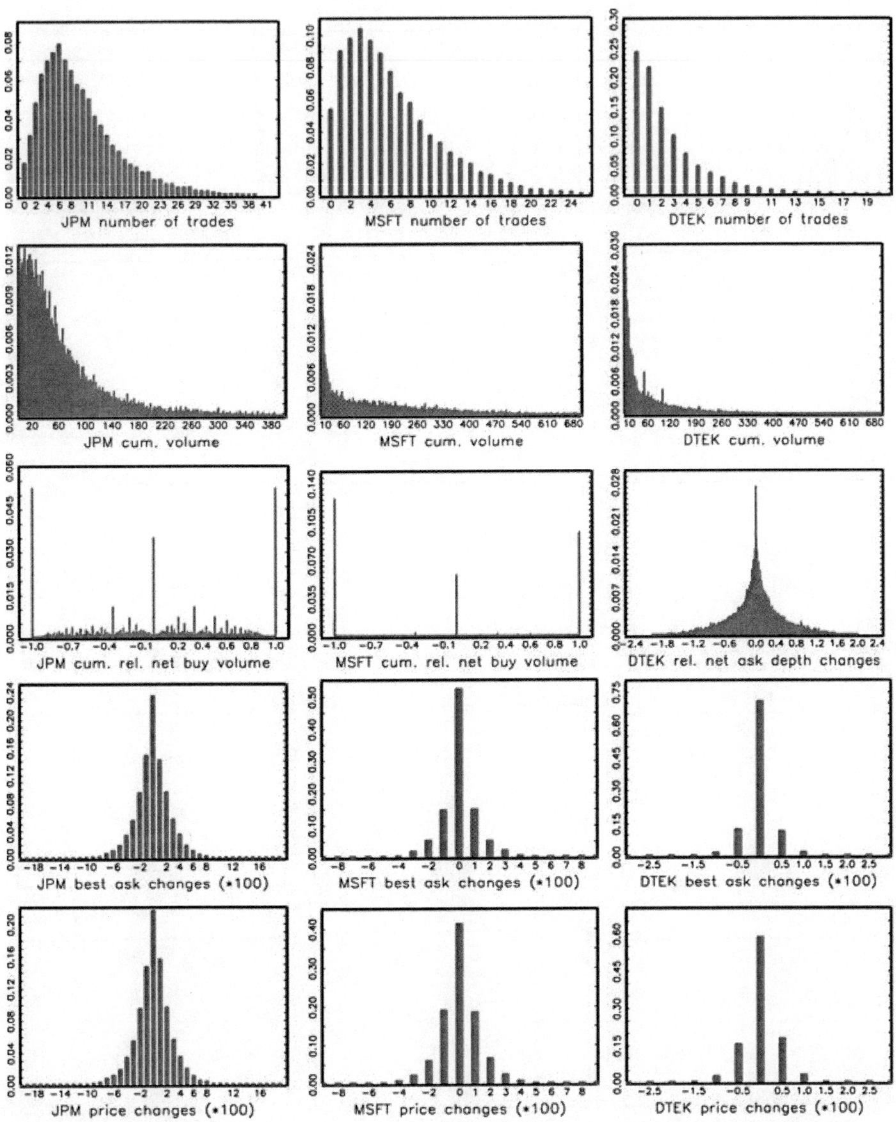

Fig. 3.18 Histograms for 30-s number of trades, cumulated trading volume, cumulated net buy volume, relative net ask depth changes, best ask changes and transaction price changes for JP Morgan (NYSE), Microsoft (NASDAQ) and Deutsche Telekom (XETRA). Sample period: June 2009 for JP Morgan and Microsoft and September 2010 for Deutsche Telekom

effects are observed for spread and depth dynamics. It is also illustrated how return dynamics of quotes and prices become more pronounced if we approach transaction level. Besides slightly significant (though very small) autocorrelations, the bid-ask bounce effect becomes most dominant. The intraday seasonality of

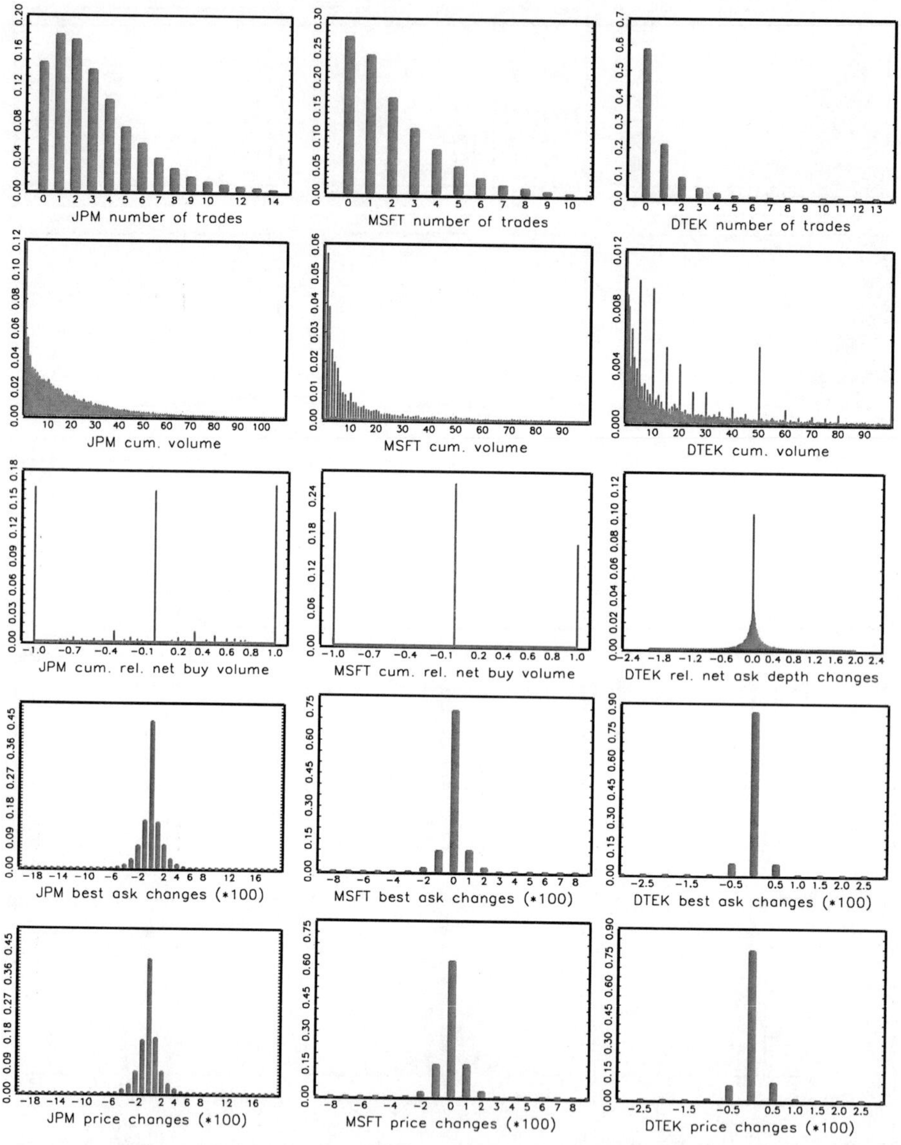

Fig. 3.19 Histograms for 10-s number of trades, cumulated trading volume, cumulated net buy volume, relative net ask depth changes, best ask changes and transaction price changes for JP Morgan (NYSE), Microsoft (NASDAQ) and Deutsche Telekom (XETRA). Sample period: June 2009 for JP Morgan and Microsoft and September 2010 for Deutsche Telekom

2-min aggregated data shown in Fig. 3.26 confirm the findings above: All trading activity variables reveal a distinct U-shaped pattern with activities being highest after opening and before closure.

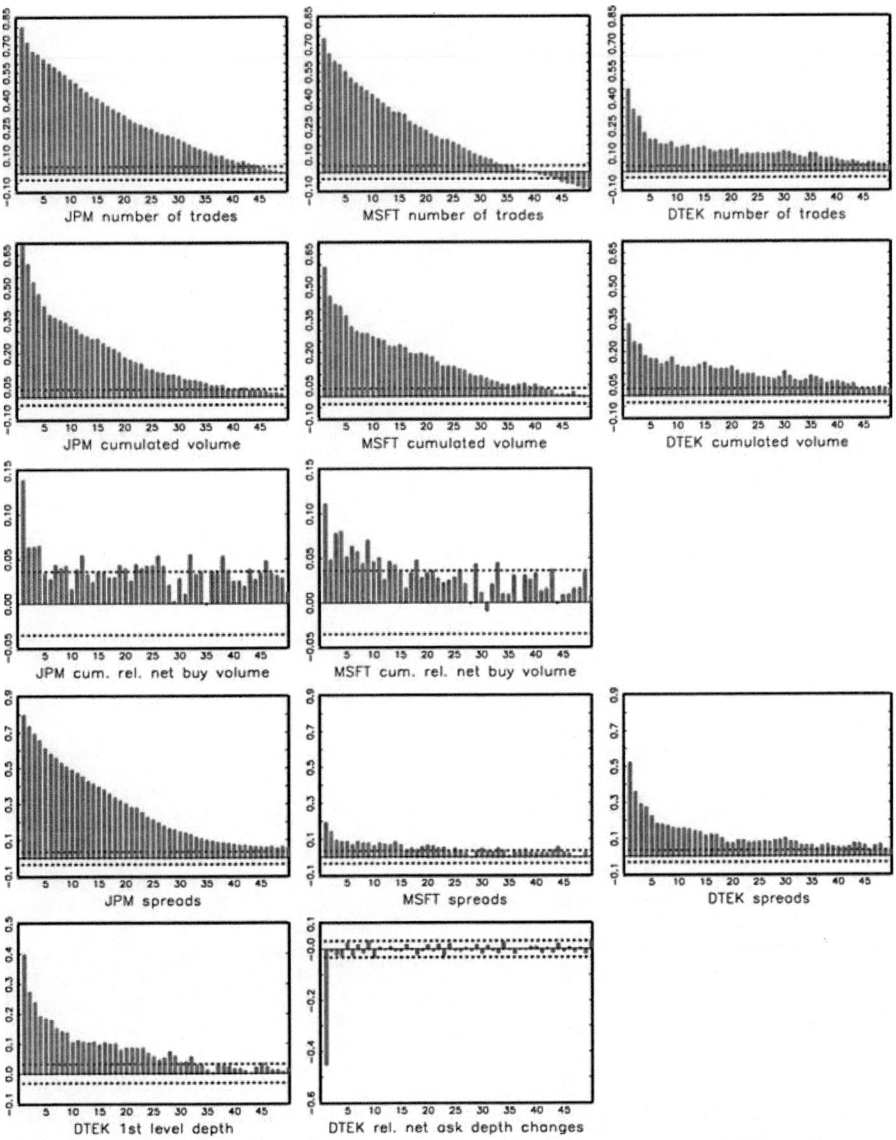

Fig. 3.20 Autocorrelation functions for 2-min number of trades, cumulated trading volume, cumulated net buy volume, bid-ask spreads, first level depth and relative net ask depth changes for JP Morgan (NYSE), Microsoft (NASDAQ) and Deutsche Telekom (XETRA). Sample period: June 2009 for JP Morgan and Microsoft and September 2010 for Deutsche Telekom

Figures 3.27–3.29 plot the cross-autocorrelations between the different trading variables for 2 min, 30 and 10 s aggregates. For JPM and MSFT, we observe strong temporal cross-dependencies between trade counts, cumulative trading volumes

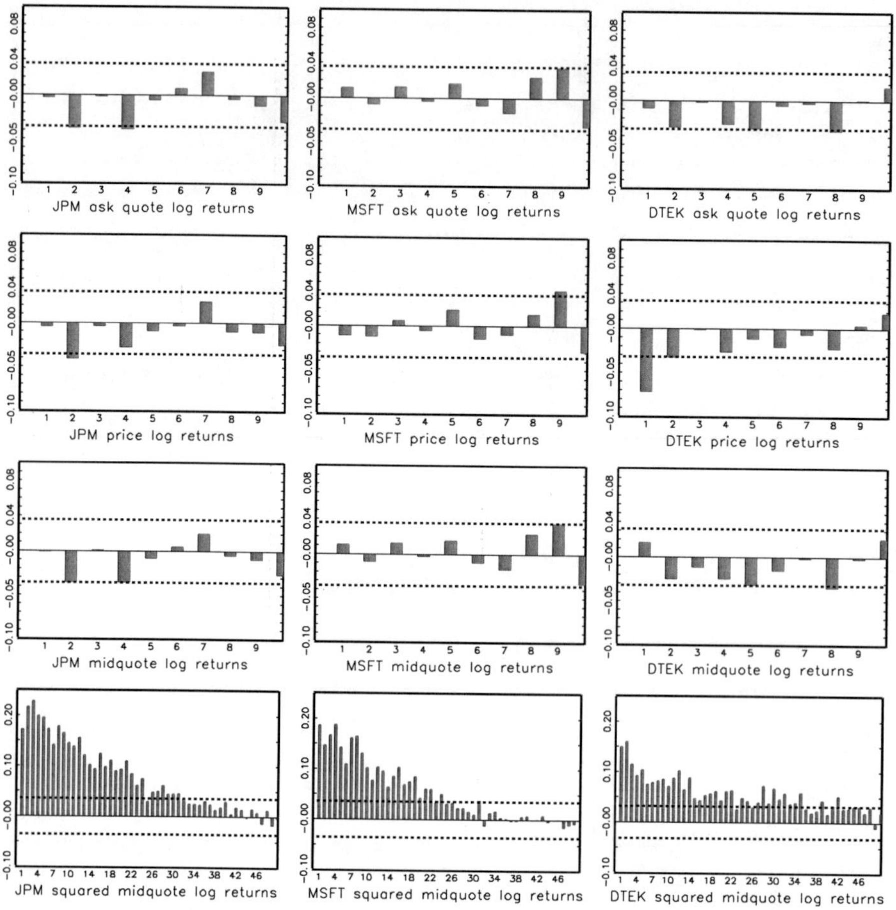

Fig. 3.21 Autocorrelation functions for 2-min ask quote log returns, price log returns, midquote log returns and squared midquote log returns for JP Morgan (NYSE), Microsoft (NASDAQ) and Deutsche Telekom (XETRA). Sample period: June 2009 for JP Morgan and Microsoft and September 2010 for Deutsche Telekom

and absolute returns, where causalities work in all directions. In contrast, DTEK characteristics reveal only very little cross-dependencies. These findings suggest that causalities between different trading variables are obviously quite dependent on the underlying stock and the exchange. Overall, the results show that many trading variables are predictable not only based on their own history but also based on other variables. Besides variables reflecting the liquidity demand (such as trading intensities and volumes) and volatility this is also evident for liquidity supply variables such as bid-ask spreads and market depth.

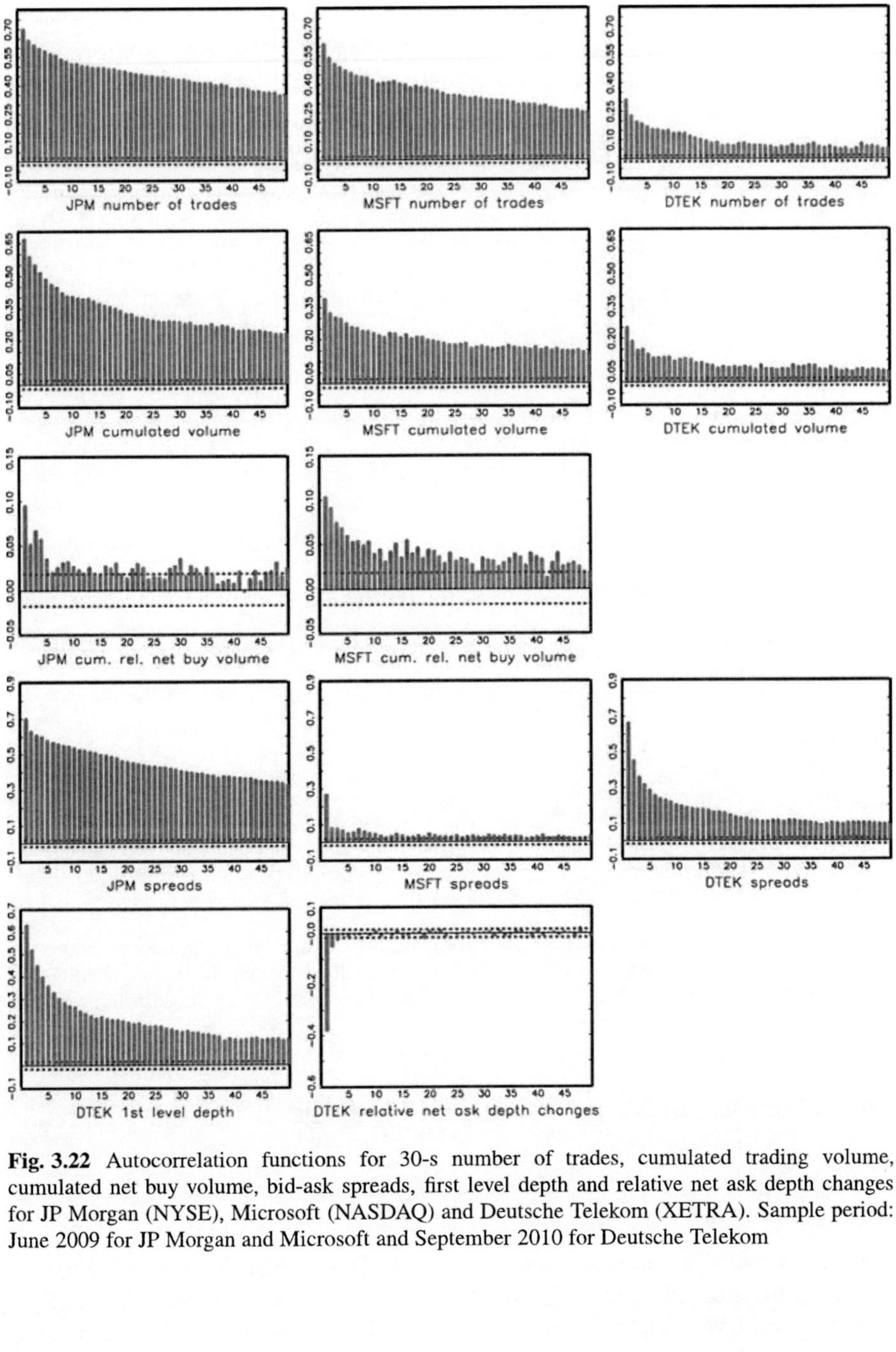

Fig. 3.22 Autocorrelation functions for 30-s number of trades, cumulated trading volume, cumulated net buy volume, bid-ask spreads, first level depth and relative net ask depth changes for JP Morgan (NYSE), Microsoft (NASDAQ) and Deutsche Telekom (XETRA). Sample period: June 2009 for JP Morgan and Microsoft and September 2010 for Deutsche Telekom

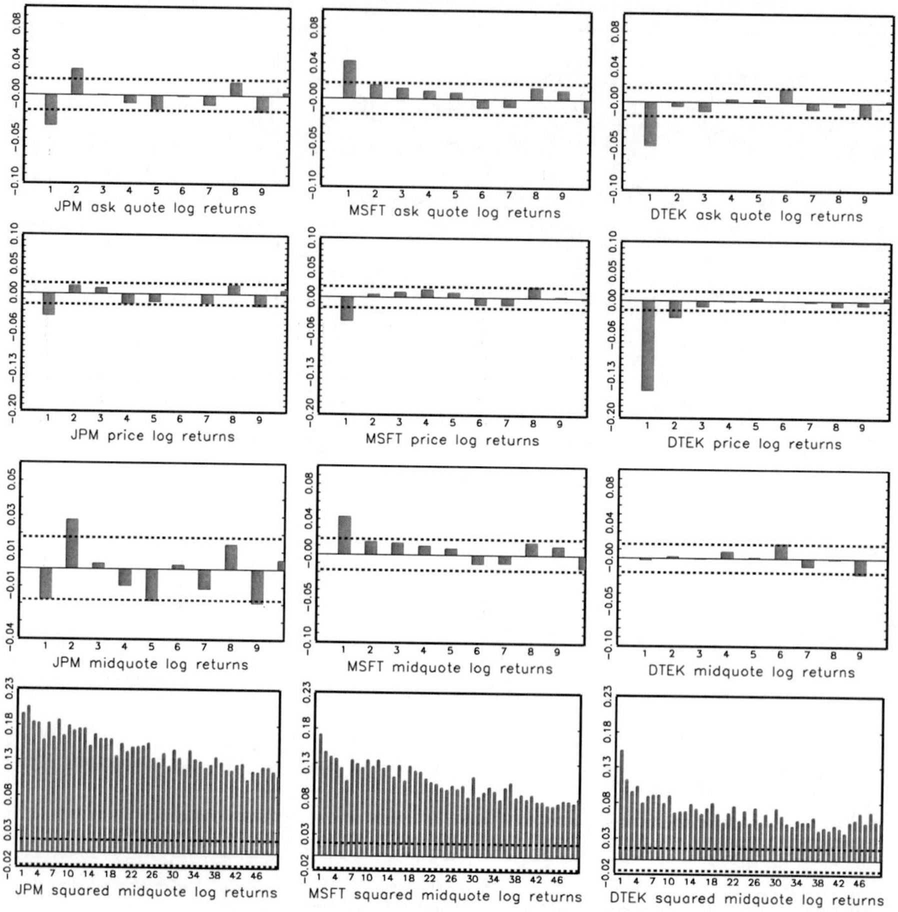

Fig. 3.23 Autocorrelation functions for 30-s ask quote log returns, price log returns, midquote log returns and squared midquote log returns for JP Morgan (NYSE), Microsoft (NASDAQ) and Deutsche Telekom (XETRA). Sample period: June 2009 for JP Morgan and Microsoft and September 2010 for Deutsche Telekom

3.6 Summary of Major Empirical Findings

Summarizing the major empirical features of financial high-frequency data results in the following main findings:

1. Virtually all high-frequency trading characteristics (apart from returns themselves) are strongly serially correlated. This holds for characteristics observed on transaction level, data which are aggregated over time (resulting in equi-distant observations) and aggregated based on trading events (resulting in irregularly spaced financial durations). To capture this feature, appropriate dynamic models are needed which are defined either in calendar time or in business time.

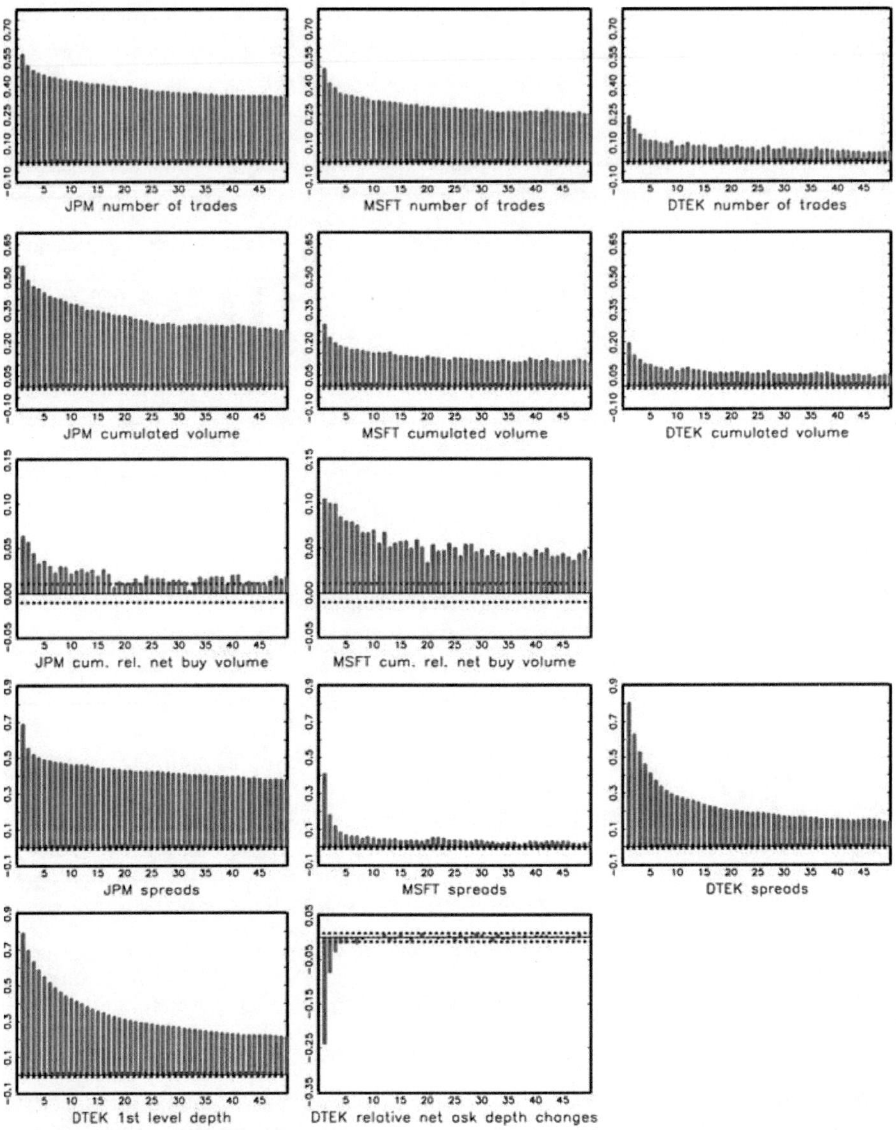

Fig. 3.24 Autocorrelation functions for 10-s number of trades, cumulated trading volume, cumulated net buy volume, bid-ask spreads, first level depth and relative net ask depth changes for JP Morgan (NYSE), Microsoft (NASDAQ) and Deutsche Telekom (XETRA). Sample period: June 2009 for JP Morgan and Microsoft and September 2010 for Deutsche Telekom

2. Many high-frequency characteristics are very persistent over time and reveal long range dependence. This calls for models allowing not only for ARMA-type dynamics but also long memory behavior. This is particularly evident for the dynamics of trading intensities, volumes, spreads and market depth.

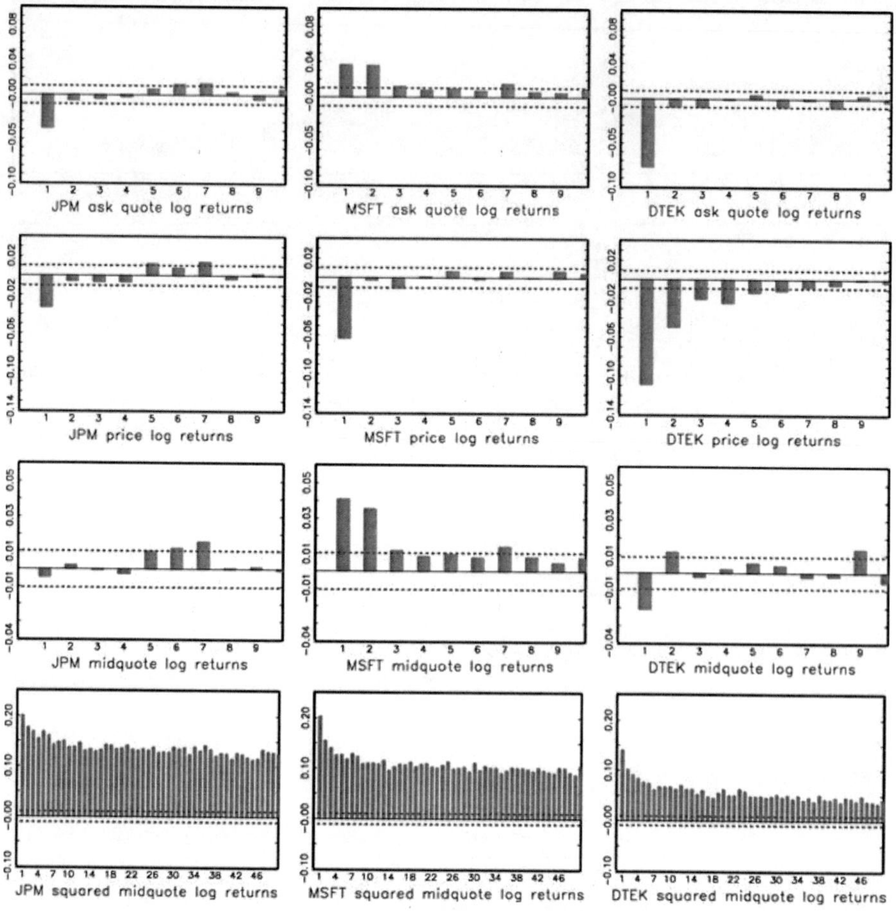

Fig. 3.25 Autocorrelation functions for 10-s ask quote log returns, price log returns, midquote log returns and squared midquote log returns for JP Morgan (NYSE), Microsoft (NASDAQ) and Deutsche Telekom (XETRA). Sample period: June 2009 for JP Morgan and Microsoft and September 2010 for Deutsche Telekom

3. Most high-frequency variables take only positive values calling for specific models for positive-valued variables. This is particularly true for all volatility-related variables as well as characteristics capturing different dimensions of liquidity.
4. Nearly all high-frequency variables are subject to strong intraday periodicities. A common feature is the typical U-shaped intraday seasonality pattern associated with high market activities after opening and before closure and less activity over lunch time. Additional periodicities might occur due to the opening of markets in other time zones.
5. Some high-frequency variables are quite discrete. This is mostly true for trade-to-trade price, quote or spread changes but might also occur if trading is only

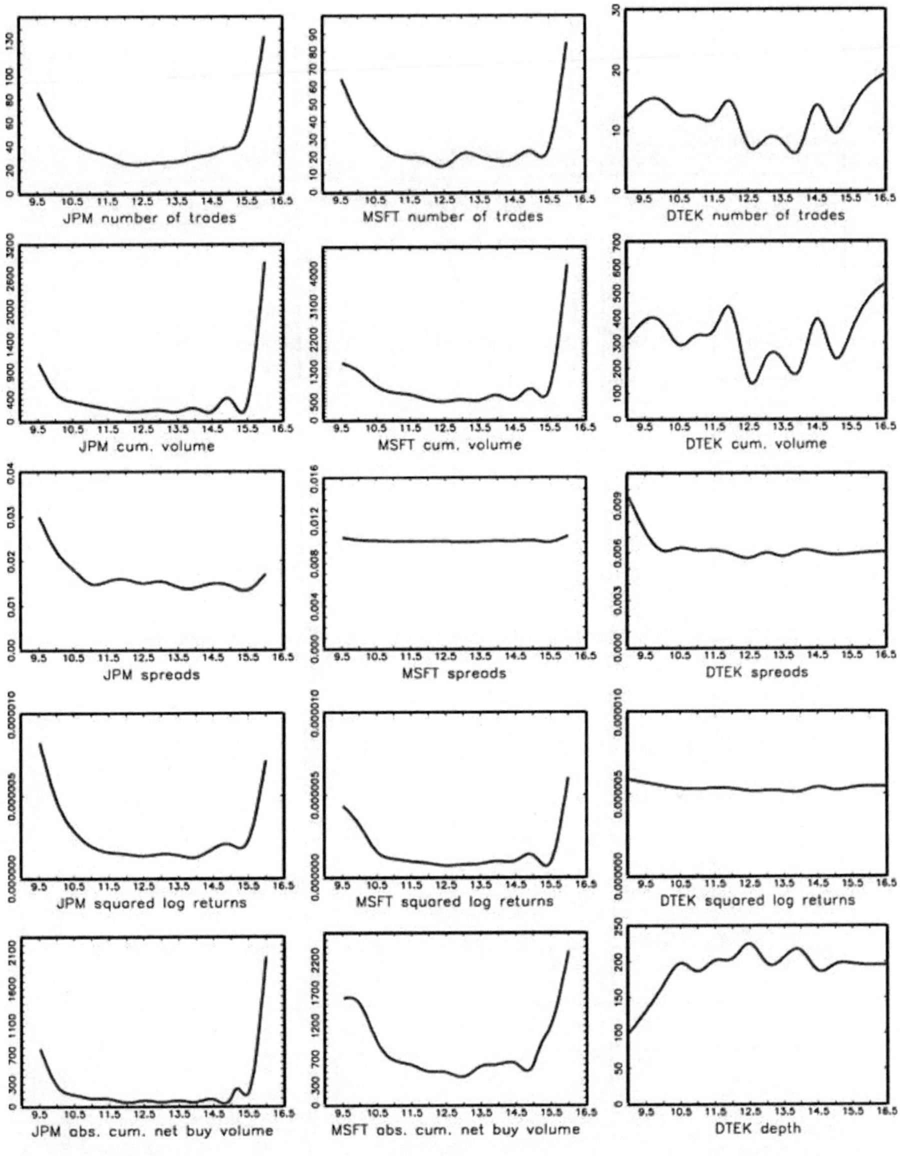

Fig. 3.26 Cubic spline functions (30 min nodes) for 2-min number of trades, cumulated trading volume (in 100 share lots), bid-ask spreads, squared log returns and absolute cumulated net buy volume (in 100 share lots) for JP Morgan (NYSE), Microsoft (NASDAQ) and Deutsche Telekom (XETRA). Sample period: June 2009 for JP Morgan and Microsoft and September 2010 for Deutsche Telekom

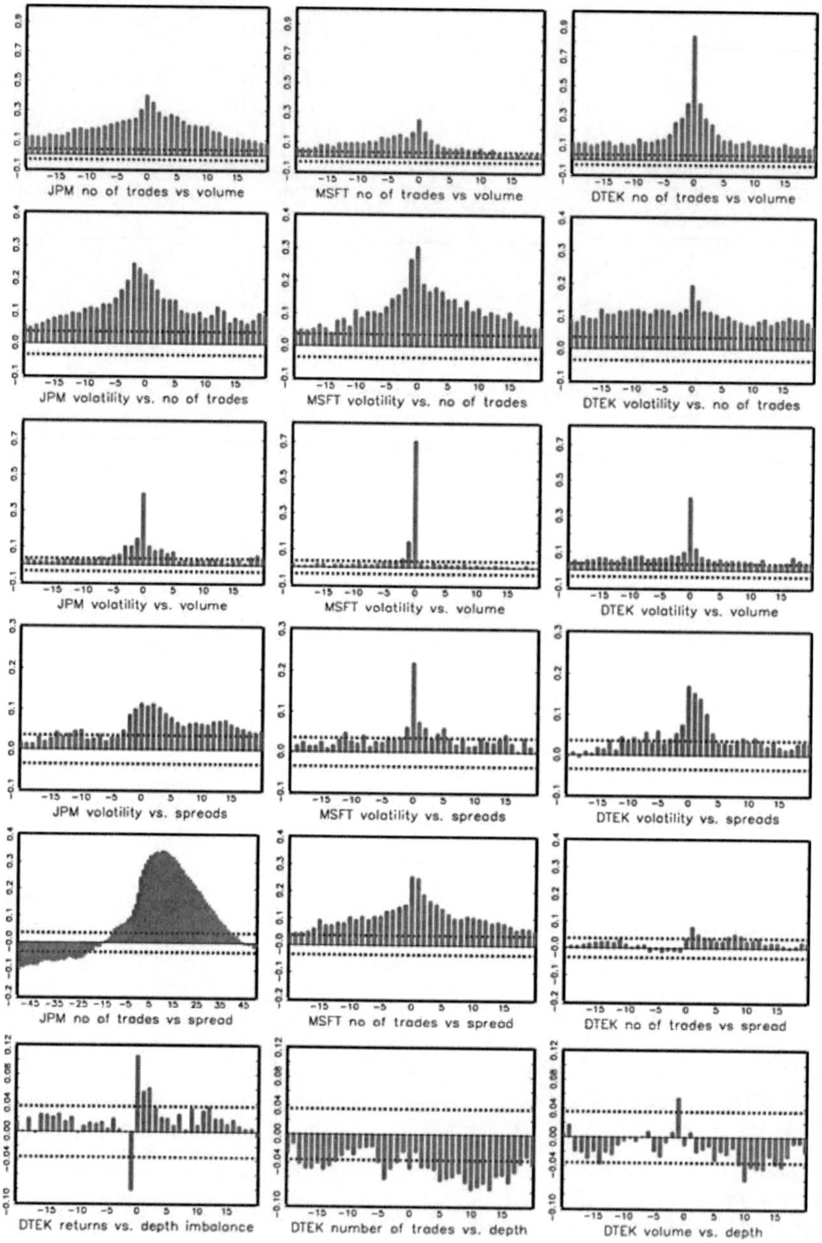

Fig. 3.27 Cross-autocorrelation functions for 2-min number of trades vs. cumulated trading volume, squared log returns vs. number of trades, squared log returns vs. cumulated trading volume, squared log returns vs. bid-ask spreads, number of trades vs. bid-ask spreads, returns vs. ask-bid depth imbalance, number of trades vs. depth, and cumulated trading volume vs. depth for JP Morgan (NYSE), Microsoft (NASDAQ) and Deutsche Telekom (XETRA). Sample period: June 2009 for JP Morgan and Microsoft and September 2010 for Deutsche Telekom

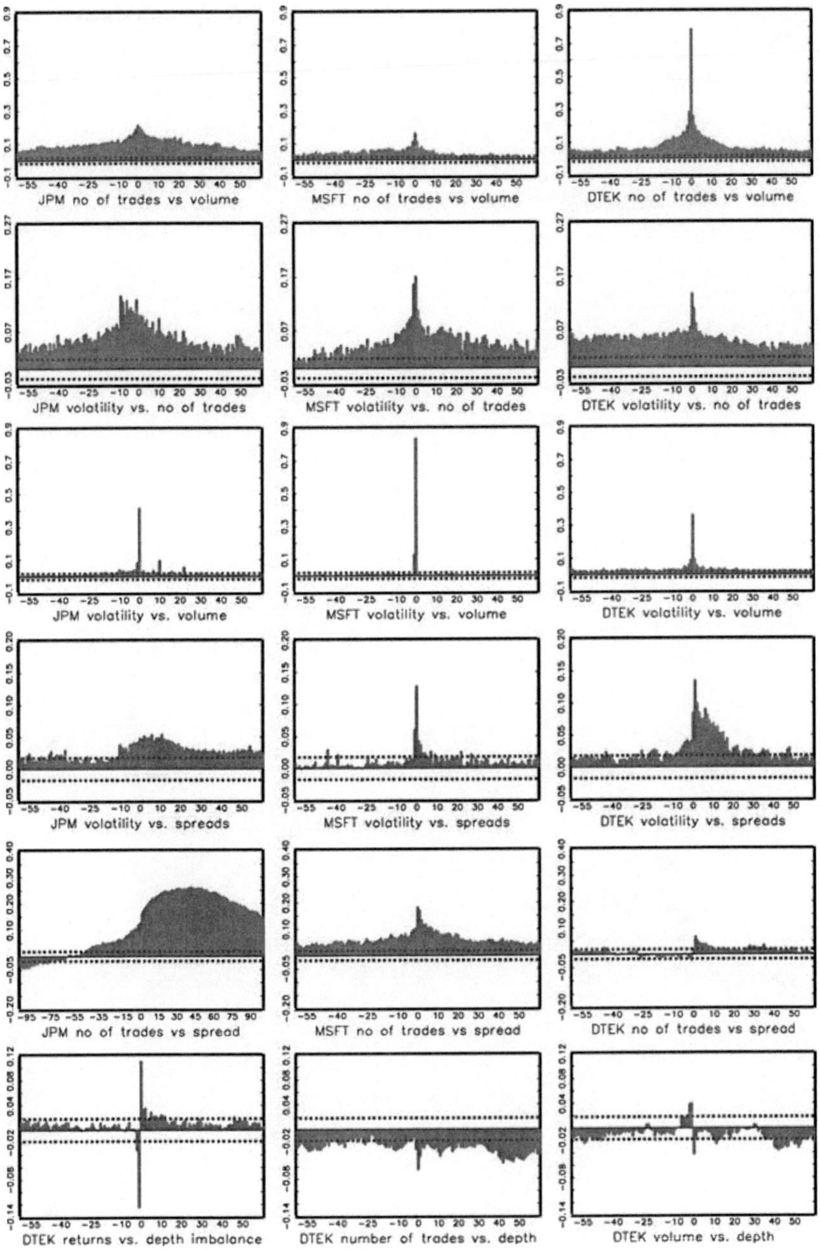

Fig. 3.28 Cross-autocorrelation functions for 30-s number of trades vs. cumulated trading volume, squared log returns vs. number of trades, squared log returns vs. cumulated trading volume, squared log returns vs. bid-ask spreads, number of trades vs. bid-ask spreads, returns vs. ask-bid depth imbalance, number of trades vs. depth, and cumulated trading volume vs. depth for JP Morgan (NYSE), Microsoft (NASDAQ) and Deutsche Telekom (XETRA). Sample period: June 2009 for JP Morgan and Microsoft and September 2010 for Deutsche Telekom

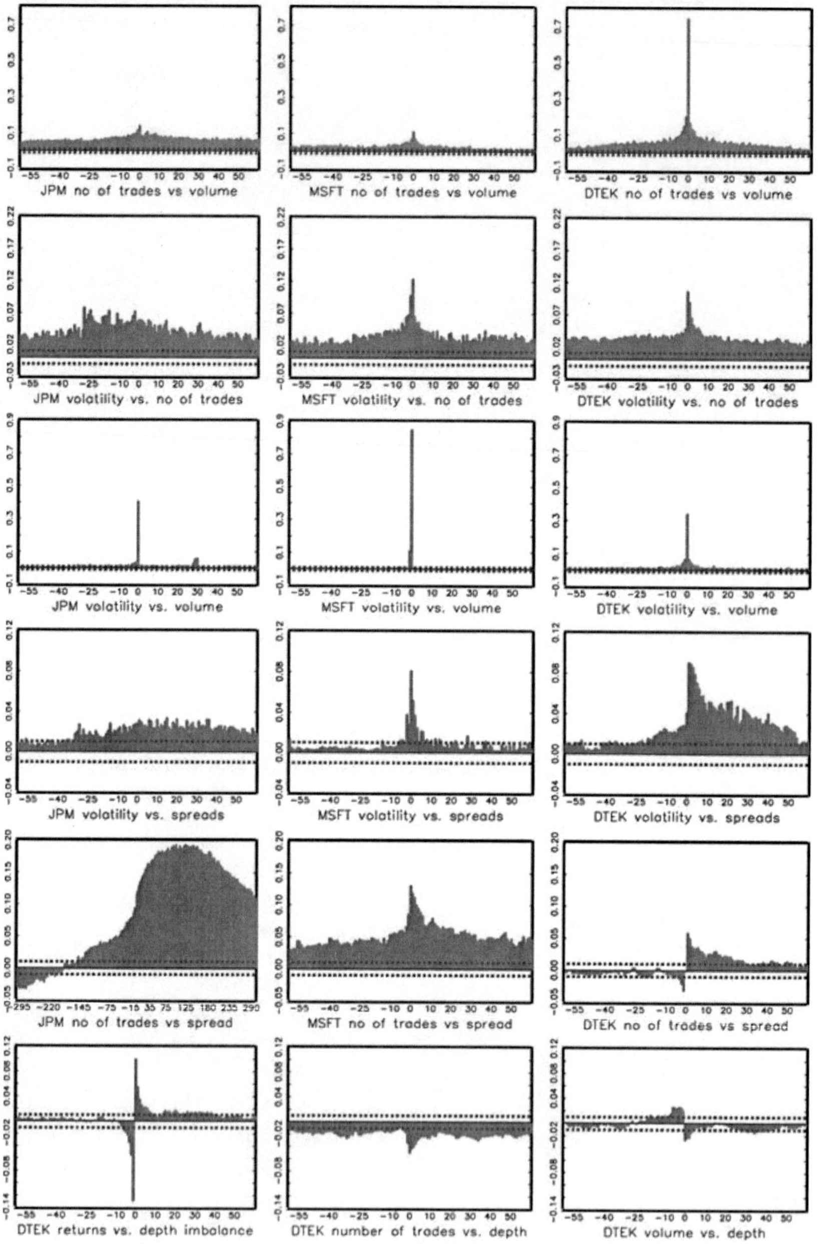

Fig. 3.29 Cross-autocorrelation functions for 10-s number of trades vs. cumulated trading volume, squared log returns vs. number of trades, squared log returns vs. cumulated trading volume, squared log returns vs. bid-ask spreads, number of trades vs. bid-ask spreads, returns vs. ask-bid depth imbalance, number of trades vs. depth, and cumulated trading volume vs. depth for JP Morgan (NYSE), Microsoft (NASDAQ) and Deutsche Telekom (XETRA). Sample period: June 2009 for JP Morgan and Microsoft and September 2010 for Deutsche Telekom

possible in round lot sizes. This calls for dynamic approaches for discrete-valued random variables.

6. Some distributions of high-frequency variables contain mixtures of discrete and continuous components. A typical example is the high proportion of zero outcomes in cumulative trading volumes measured in calendar time. Moreover, mixtures of discrete and continuous components are also observed in distributions of trade sizes reflecting traders' preference for round numbers.
7. Trading processes are inherently high-dimensional calling for multivariate dynamic models either defined in discrete calendar time or transaction time in case of time-synchronized data or defined in continuous time if variables occur asynchronously over time.

Econometric frameworks and models to capture these specific properties are discussed in the following chapters.

References

Andersen T, Dobrev D, Schaumburg E (2010) Jump-robust volatility estimation using nearest neighbor truncation Federal Reserve Bank of New York Staff Report, No. 465

Bessembinder H, Panayides M, Venkataraman K (2009) Hidden liquidity: an analysis of order exposure strategies in electronic stock markets. J Finan Econ 94:361–383

Barndorff-Nielsen O, Hansen P, Lunde A, Shephard N (2008b) Realised kernels in practice: trades and quotes. Econom J 4:1–32

Brownlees C, Gallo G (2006) Financial econometric analysis of ultra-high frequency: data handling concerns. Comput Stat Data Anal 51:2232–2245

Finucane TJ (2000) A direct test of methods for inferring trade direction from intra-day data. J Financ QuantAnal 35:553–676

Frey S, Sandas P (2009) The impact of iceberg orders in limit order books. Working Paper 09-06, Centre for Financial Research, Cologne

Grammig J, Wellner M (2002) Modeling the interdependence of volatility and inter-transaction duration process. J Econom 106:369–400

Harris L (2003) Trading & exchanges – market microstructure for practitioners. Oxford University Press, Oxford

Hasbrouck J (2007) Empirical market microstructure. Oxford University Press, Oxford

Hautsch N, Huang R (2009) The market impact of a limit order. Discussion Paper 2009/23, Collaborative Research Center 649 "Economic Risk", Humboldt-Universität zu Berlin

Hautsch N, Huang R (2011) On the dark side of the market: identifying and analyzing hidden order placements. Working Paper, Humboldt-Universität zu Berlin

Hautsch N, Kyj L, Oomen R (2011) A blocking and regularization approach to high-dimensional realized covariance estimation. J Appl Econom, in press

Hautsch N, Podolskij M (2010) Pre-averaging based estimation of quadratic variation in the presence of noise and jumps: theory, implementation, and empirical evidence. Discussion Paper 2010/38, Collaborative Research Center 649 "Economic Risk", Humboldt-Universität zu Berlin

Henker T, Wang J-X (2006) On the importance of timing specifications in market microstructure research. J Finan Markets 9:162–179

Hansen PR, Lunde A (2006) Realized variance and market microstructure noise. J Bus Econ Stat 24(2):127–161

Lee CMC, Ready MJ (1991) Inferring trade direction from intraday data. J Finance 46:733–746

Roll R (1984) A simple implicit measure of the effective bid-ask spread in an efficient market. J Finance 39:1127–1139

Veredas D, Rodriguez-Poo J, Espasa A (2008) Semiparametric estimation for financial durations. In: Bauwens WPL, Veredas D (eds) High frequency financial econometrics. Physica-Verlag, Heidelberg, pp 225–251

Chapter 4
Financial Point Processes

This chapter provides the methodological background for the specification and estimation of financial point processes. We give a brief introduction to the fundamental statistical concepts and the basic ways to model point processes. For ease of introduction, we restrict our attention to non-dynamic point processes. In Sect. 4.1, we discuss the most important theoretical concepts in point process theory. Here, the focus lies on the idea of the intensity function as a major concept in the theory of point processes. In Sect. 4.2, different ways to model point processes are discussed. Section 4.3 is concerned with the treatment of censoring mechanisms and time-varying covariates. Section 4.4 gives an outlook on different ways to dynamically extend basic point process models.

4.1 Basic Concepts of Point Processes

In this section, we discuss central concepts of point process theory. The mathematical level of this chapter is chosen in a way such that concepts and relationships are understandable without requiring deep foundations in stochastics. In this sense, it provides a compromise between mathematical exactness and accessibility. Mathematically more rigorous textbook treatments of point process theory are given by Karr (1991), Snyder and Miller (1991) or Daley and Vere-Jones (2005). Classical textbooks on duration and failure time analysis are Kalbfleisch and Prentice (1980) and Lancaster (1997). A recent review of the literature on financial point processes is also given by Bauwens and Hautsch (2009).

4.1.1 Fundamental Definitions

4.1.1.1 Point Processes

Let t denote physical (calendar) time and let $\{t_i\}_{i \in \{1,2,\dots\}}$ be a random sequence of increasing event arrival times $0 \leq t_i \leq t_{i+1}$. Then, the sequence $\{t_i\}$ is called a

N. Hautsch, *Econometrics of Financial High-Frequency Data*,
DOI 10.1007/978-3-642-21925-2_4,

point process on $[0, \infty)$ (on the line).[1] Here, we restrict ourselves to point processes defined in time. If $t_i < t_{i+1} \ \forall \ i$, the process is called a *simple point process* excluding the possibility of the simultaneous occurrence of events. During the remainder of the book, we only focus on simple point processes. Furthermore, denote n as the number of observed events and t_n as the last observed point of the process. Then, $\{t_i\}_{i=1}^n$ is the complete observable process. Let $\{W_i\}_{i \in \{1,\ldots,n\}}$ denote a sequence of *marks* with $W_i \in \{1, 2, \ldots, K\}$ denoting the type of the ith event. Then, the double sequence $\{t_i, W_i\}_{i \in \{1,2,\ldots,n\}}$ is called a (simple) *marked point process* or, alternatively, a (simple) K-dimensional point process. Marks specifically indicate different types of events, as, e.g., the arrival of buys, sells or certain limit orders in financial trading processes. The event-specific arrival times are denoted by $\{t_i^k\}_{i \in \{1,\ldots,n^k\}}$, $k = 1, \ldots, K$, with n^k defining the number of observed k-type events. Considering only the sequence $\{t_i\}$ of a marked point process $\{t_i, W_i\}$ (or, equivalently, superposing the individual k-type arrival times for $k = 1, \ldots, K$) yields the so-called *pooled* process.

4.1.1.2 Counting Processes

The process $N(t)$ with $N(t) := \sum_{i \geq 1} \mathbb{1}_{\{t_i \leq t\}}$ is called a *right-continuous (càdlàg counting process* associated with $\{t_i\}$.[2] $N(t)$ is a right-continuous step function with upward jumps (at each t_i) of magnitude one. Furthermore, the process $\breve{N}(t)$ with $\breve{N}(t) := \sum_{i \geq 1} \mathbb{1}_{\{t_i < t\}}$ is called a *left-continuous counting process* associated with $\{t_i\}$. $\breve{N}(t)$ is a left-continuous step function and counts the number of events that occur *before* t. Often, a point process is characterized by its counting process $N(t)$. If $N(t)$ is a K-variate marked point process, the K sequences of event times are represented by the corresponding k-type counting functions $N^k(t) := \sum_{i \geq 1} \mathbb{1}_{\{t_i \leq t\}} \mathbb{1}_{\{W_i = k\}}$ and $\breve{N}^k(t) := \sum_{i \geq 1} \mathbb{1}_{\{t_i < t\}} \mathbb{1}_{\{W_i = k\}}$.

4.1.1.3 Durations and Backward Recurrence Times

Define x_i as the waiting time between two successive points, defined as

$$x_i := \begin{cases} t_i - t_{i-1}, & i = 2, \ldots, n, \\ t_i, & i = 1, \end{cases}$$

with $t_0 := 0$. Then, $\{x_i\}_{i \in \{1,2,\ldots,n\}}$ is called the *duration process* associated with $\{t_i\}$. Moreover, the process $x(t)$ with $x(t) := t - t_{\breve{N}(t)}$ is called the *backward recurrence*

[1] Point processes can also evolve over space yielding the class of *spatial point processes* or *cluster processes*, see, e.g., Daley and Vere-Jones (2005).

[2] A càdlàg (french: continue à droite, limitée à gauche) function is a function which is right-continuous with left-hand limits.

time at t. The backward recurrence time is the time elapsed since the previous point and is a left-continuous function that grows linearly through time with discrete jumps back to zero *after* each arrival time t_i. Note that $x(t_i) = t_i - t_{i-1} := x_i$.

4.1.1.4 Filtrations and (Time-Varying) Covariates

Throughout this book, we restrict ourselves to point processes which have an evolutionary character in time. That is, at any time t, the event probability of a point process $N(t)$ depends on information which is (at least instantaneously) known prior to t.[3] This information might be the history of $N(t)$ and/or exogenous variables. To formalize such a concept, let $\mathcal{F}_t$ denote the information set up to and including t. The set $\mathcal{F}_t$ might be the so-called *internal history* or *natural filtration* $\mathcal{F}_t := \mathcal{F}_t^N$ with $\mathcal{F}_t^N = \sigma(N^k(s) : 0 \leq s \leq t, k \in \Xi)$, $N^k(s) := \sum_{i\geq 1} \mathbb{1}_{\{t_i \leq s\}} \mathbb{1}_{\{W_i \in \Xi\}}$, where Ξ denotes the σ-field of all subsets of $\{1, \ldots, K\}$.

If the occurrence (but not necessarily the type) of an event is driven by a vector of (weakly exogenous) variables $\{\mathbf{z}_i\}_{i\in\{1,2,\ldots\}}$, which are known and constant prior to an event i, we call them *time-invariant* covariates. Conversely, *time-varying* covariates are characteristics which drive the occurrence of an event and can continuously change between two consecutive events. Correspondingly, they are denoted by the vector $\mathbf{z}(t)$. For the covariates to be weakly exogenous for x_i, the process $\mathbf{z}(t)$ must be càdlàg with $\mathbf{z}_{\breve{N}(t)+1} = \mathbf{z}(t_{i-1})$ for all t with $t_{i-1} < t \leq t_i$.

If $\mathcal{F}_t$ does not only include the internal history of the process but also the processes $\{\mathbf{z}_i\}$ and $\{\mathbf{z}(t)\}$, we have $\mathcal{F}_t^N \subseteq \mathcal{F}_t$.

4.1.2 Compensators and Intensities

A central concept in the theory of point processes is the *intensity function* which is heuristically defined as following:

Definition 4.1. Let $N(t)$ be a simple point process on $[0, \infty)$ that is adapted to some history $\mathcal{F}_t$ and assume that $\lambda(t; \mathcal{F}_t)$ is a positive-valued process with sample paths that are left-continuous and have right-hand limits. Then, the process

$$\lambda(t; \mathcal{F}_t) \approx \lambda(t+) := \lim_{\Delta \downarrow 0} \frac{1}{\Delta} \mathbb{E}\left[N(t+\Delta) - N(t) | \mathcal{F}_t\right], \qquad \lambda(t+) > 0, \ \forall\, t,$$

with $\lambda(t+, \mathcal{F}_t) := \lim_{\Delta\downarrow 0} \lambda(t+\Delta, \mathcal{F}_t)$, is called the $\mathcal{F}_t$-intensity process of the counting process $N(t)$.

This definition is familiar in classical duration literature (see, e.g., Lancaster 1997) and manifests the intensity function as the instantaneous arrival rate of an event

[3] More general types of point processes not fulfilling this property are, e.g., spatial point processes or Neymann-Scott cluster processes. For more details, see, e.g., Daley and Vere-Jones (2005).

in t conditional on the information up to t, $\mathcal{F}_t$. Note that the terminology is not consistent in all papers or textbooks on point processes. Some authors, for example, Brémaud (1981) and Karr (1991), denote $\lambda(t;\mathcal{F}_t)$ as "stochastic intensity function" accounting for the fact that $\lambda(t;\mathcal{F}_t)$ is not necessarily deterministic but might follow a stochastic process. Daley and Vere-Jones (2005) refer $\lambda(t;\mathcal{F}_t)$ to as a "conditional intensity function" as $\lambda(t;\mathcal{F}_t)$ depends on conditioning information. Here, we call $\lambda(t;\mathcal{F}_t)$ just an intensity or, correspondingly, $\mathcal{F}_t$-intensity whenever the type of conditioning information has to be stressed. The point process generated by an $\mathcal{F}_t$-intensity is called $\mathcal{F}_t$-adapted point process.

The above definition provides also the link to the concept of the *hazard function* which is given by

$$h(x_i;\mathcal{F}_{i-1}) := f(x_i;\mathcal{F}_{i-1})/S(x_i;\mathcal{F}_{i-1}) \tag{4.1}$$

$$= \lim_{\Delta\to 0}\frac{1}{\Delta}\Pr[x_i \le X_i < x_i + \Delta | X_i \ge x_i;\mathcal{F}_{i-1}], \tag{4.2}$$

with x_i denoting the (inter-event) duration as represented by the realization of a random variable X_i with conditional probability density function (p.d.f.) $f(x_i;\mathcal{F}_{i-1})$, conditional survivor function $S(x_i;\mathcal{F}_{i-1}) := 1 - F(x_i;\mathcal{F}_{i-1})$, and conditional cumulative distribution function (c.d.f.) $F(x_i;\mathcal{F}_{i-1}) := \mathbb{P}r[X_i \le x_i|\mathcal{F}_{i-1}]$. While the intensity function is defined in continuous time and conditions on a possibly continuously varying information set $\mathcal{F}_t$, the hazard rate is evaluated at the end of each duration x_i and conditions just on the time elapsed since the beginning of the spell and potential covariates (e.g., capturing the process history). It is a central concept in traditional duration or survival analysis, where cross-sectional duration data are analyzed.[4] Formulating the hazard function in terms of the backward recurrence time yields the corresponding continuous time representation

$$h\left(x(t);\mathcal{F}_{\breve{N}(t)}\right) = \lambda(t;x(t),\mathcal{F}_{\breve{N}(t)}) \tag{4.3}$$

as an intensity which just depends on the time since the last event. As there are conflicting definitions and terminologies of hazard and intensity concepts, we distinguish them as follows: In the remainder of the book we call an intensity function a (conditional) hazard function whenever it can be represented in terms of the underlying duration process and can be stated in terms of the ratio of the conditional probability density function and the conditional survivor function of the durations given the information prevailing at the beginning of the spell, see (4.1). With this definition, we rule out that the information set changes over the course of a spell, as, e.g., in the case of time-varying covariates.

In point process theory, the intensity function plays an important role since it is closely linked to a corresponding likelihood function and thus opens up the possibility for likelihood-based inference. This relationship traditionally builds on the evolutionary character of a point process which is mathematically captured by

---

[4]See, e.g., Kalbfleisch and Prentice (1980), Kiefer (1988) or Lancaster (1997).

martingale-based point process theory. Define an $\mathcal{F}_t$-adapted process $N(t)$ with $\mathbb{E}[N(t)] < \infty$ for each t to be a $\mathcal{F}_t$*-martingale* if for $0 \leq s < t < \infty$,

$$\mathbb{E}[N(t)|\mathcal{F}_s] = N(s) \quad \text{a.s. (almost surely)},$$

and to be a $\mathcal{F}_t$*-submartingale* if

$$\mathbb{E}[N(t)|\mathcal{F}_s] \geq N(s) \quad \text{a.s.}$$

Then, according to the Doob–Meyer decomposition,[5] any (bounded) $\mathcal{F}_t$-submartingale $N(t)$ can be decomposed into a unique zero-mean $\mathcal{F}_t$-martingale $M(t)$ and a unique $\mathcal{F}_t$-predictable cumulative process $\tilde{\Lambda}(t)$, i.e.,

$$N(t) = M(t) + \tilde{\Lambda}(t), \tag{4.4}$$

where the process $\tilde{\Lambda}(t)$ is called the *compensator* with the property

$$\tilde{\Lambda}(t) = \int_0^t \lambda(s; \mathcal{F}_s) ds. \tag{4.5}$$

The fact that

$$N(t) - \int_0^t \lambda(t, \mathcal{F}_s) ds = M(t) \tag{4.6}$$

is a martingale process will be exploited in the remainder of the book to construct diagnostic tests for intensity models. Accordingly, we obtain

$$\mathbb{E}[N(t)|\mathcal{F}_s] = \mathbb{E}\left[\int_0^t \lambda(u; \mathcal{F}_u) du \middle| \mathcal{F}_s\right], \quad \text{a.s.},$$

or,

$$\mathbb{E}[N(t) - N(s)|\mathcal{F}_s] = \mathbb{E}\left[\int_s^t \lambda(u; \mathcal{F}_u) du \middle| \mathcal{F}_s\right] := \mathbb{E}\left[\Lambda(s,t) \middle| \mathcal{F}_t\right], \quad \text{a.s.}, \tag{4.7}$$

which gives an alternative implicit definition of the intensity $\lambda(t; \mathcal{F}_t)$. Thus, the expected number of events in the interval $(s, t]$, given $\mathcal{F}_s$, is computed as the conditional expectation of the so-called integrated intensity function defined as

$$\Lambda(s,t) := \int_s^t \lambda(u; \mathcal{F}_u) du \tag{4.8}$$

with $\Lambda(s,t) := \tilde{\Lambda}(t) - \tilde{\Lambda}(s)$.

[5]See, e.g., Daley and Vere-Jones (2005), Theorem A3.4.III.

4.1.3 The Homogeneous Poisson Process

The simplest type of point process is given by the homogeneous Poisson process. The Poisson process of rate λ is defined by the requirements that for all t, as the time interval $\Delta \downarrow 0$,

$$\mathbb{P}r\left[(N(t+\Delta)-N(t))=1\,|\mathcal{F}_t\right]=\lambda\Delta+o(\Delta), \tag{4.9}$$

$$\mathbb{P}r\left[(N(t+\Delta)-N(t))>1\,|\mathcal{F}_t\right]=o(\Delta), \tag{4.10}$$

leading to

$$\mathbb{P}r\left[(N(t+\Delta)-N(t))=0\,|\mathcal{F}_t\right]=1-\lambda\Delta+o(\Delta), \tag{4.11}$$

where $o(\Delta)$ denotes a remainder term with the property $o(\Delta)/\Delta \rightarrow 0$ as $\Delta \rightarrow 0$. One major property of the Poisson process is that the probability for the occurrence of an event in $(t, t+\Delta]$ is independent from $\mathcal{F}_t$, i.e., it does not depend on the number of points observed before (and also exactly at) t. Equations (4.9) and (4.10) are associated with the *intensity representation* of a Poisson process.

Further key properties of the Poisson process are related to the distribution of events in a fixed time interval. Following Lancaster (1997), it is easy to see that the probability for the occurrence of j events *before* $t+\Delta$ is obtained by

$$\begin{aligned}
\mathbb{P}r\left[\breve{N}(t+\Delta)=j\right] &= \mathbb{P}r\left[\breve{N}(t)=j\right]\cdot\mathbb{P}r\left[(N(t+\Delta)-N(t))=0\right]\\
&\quad+\mathbb{P}r\left[\breve{N}(t)=j-1\right]\cdot\mathbb{P}r\left[(N(t+\Delta)-N(t))=1\right]\\
&\quad+\mathbb{P}r\left[\breve{N}(t)=j-2\right]\cdot\mathbb{P}r\left[(N(t+\Delta)-N(t))=2\right]\\
&\quad\vdots\\
&= \mathbb{P}r\left[\breve{N}(t)=j\right]\cdot(1-\lambda\Delta)\\
&\quad+\mathbb{P}r\left[\breve{N}(t)=j-1\right]\cdot\lambda\Delta+o(\Delta).
\end{aligned} \tag{4.12}$$

By rearranging the terms and dividing by Δ, one obtains

$$\begin{aligned}
&\Delta^{-1}\left(\mathbb{P}r\left[\breve{N}(t+\Delta)=j\right]-\mathbb{P}r\left[\breve{N}(t)=j\right]\right)\\
&\quad=-\lambda\mathbb{P}r\left[\breve{N}(t)=j\right]+\lambda\mathbb{P}r\left[\breve{N}(t)=j-1\right]+o(\Delta)\Delta^{-1},
\end{aligned} \tag{4.13}$$

and thus, for $\Delta \downarrow 0$,

$$\frac{d}{dt}\mathbb{P}r\left[\breve{N}(t)=j\right]=-\lambda\mathbb{P}r\left[\breve{N}(t)=j\right]+\lambda\mathbb{P}r\left[\breve{N}(t)=j-1\right]. \tag{4.14}$$

For $j = 0$, (4.14) becomes

$$\frac{d}{dt}\mathbb{P}r\left[\breve{N}(t)=0\right]=-\lambda\mathbb{P}r\left[\breve{N}(t)=0\right], \tag{4.15}$$

because $\mathbb{P}r\left[\breve{N}(t)=j-1\right]=0$ for $j = 0$. By accounting for the initial condition $\mathbb{P}r\left[\breve{N}(0)=0\right]=1$, we obtain

$$\mathbb{P}r\left[\breve{N}(t)=0\right]=\exp(-\lambda t), \tag{4.16}$$

which corresponds to the survivor function of an exponential distribution. Thus, the waiting time until the first event is exponentially distributed. Analogously it is shown that (4.16) is also the probability that no events occur in an interval of length t starting at any arbitrary point on the time axis, like, at the occurrence of the previous event. Therefore, the duration between subsequent points is independent of the length of all other spells. Hence, it can be concluded that the durations are independently exponentially distributed with p.d.f. and c.d.f. given by

$$f(x)=\lambda\exp(-\lambda x),$$
$$F(x)=1-\exp(-\lambda x).$$

Accordingly, the hazard function is given by $h(x) = f(x)/(1 - F(x)) = \lambda$ and thus is identical to the Poisson intensity. These properties are associated with the *duration representation* of a Poisson process.

The third key property of the Poisson process is obtained by solving the differential equation (4.14) successively for $j = 1, 2, \ldots$ given the solution of (4.14) for $j = 0$. It is easily shown that this differential equation has the solution

$$\mathbb{P}r\left[\breve{N}(t)=j\right]=\frac{\exp(-\lambda t)(\lambda t)^j}{j!}, \qquad j=1,2,\ldots. \tag{4.17}$$

Therefore, the number of events during an interval of length t is Poisson distributed with parameter λt. Note that (4.17) does not depend on the starting point on the time axis, thus, the property holds for *any* time interval. Hence, the number of events in the interval $(s, t]$, $N(t) - N(s)$, is Poisson distributed with parameter $\lambda(t - s)$. This relationship is called the *counting representation* of a Poisson process. The mean and variance of the number of points falling in the interval $(s, t]$ is given by $\lambda(t - s)$. This equality is often exploited to test for a Poisson process.

Alternatively, we can express (4.17) as

$$\mathbb{P}r\,[t_i > x] = \mathbb{P}r\,[N(x) < i] = \sum_{j=0}^{i-1} \exp(-\lambda x)\frac{(\lambda x)^j}{j!}. \tag{4.18}$$

Note that the resulting distribution only depends on the length x of the underlying interval but not on its position on the time scale t. Point processes fulfilling this property are called *stationary point processes*.

4.1.4 Generalizations of Poisson Processes

As discussed in the previous subsection, a key property of the homogeneous Poisson process is that the intensity is constant. Thus, the data generating process (DGP) is fully described by the intensity $\lambda(t; \mathcal{F}_t) = \lambda$.

By allowing the intensity rate λ to vary over time, we obtain the general class of *non-homogeneous Poisson processes* or *inhomogeneous* Poisson processes. For example, if the Poisson process depends on càdlàg covariates which are observable at the beginning of the spell, then the DGP is fully described by the $\mathbf{z}_{\breve{N}+1}(t)$-intensity $\lambda(t; \mathbf{z}_{\breve{N}(t)+1})$. One special case is the so-called *non-stationary Poisson process*, where λ is a function of time. It is described by (4.9)–(4.10) but with λ replaced by $\lambda(t)$. A non-stationary Poisson process typically does not imply the independence of consecutive durations.

A further type of inhomogeneous Poisson process is obtained by the class of *doubly stochastic Poisson processes* (or so-called *Cox processes*). In this framework, it is assumed that the intensity is driven by some unobserved stochastic process $\lambda^*(t)$. This leads to a DGP that is characterized by the intensity $\lambda(t; \mathcal{F}_t^*)$, where $\mathcal{F}_t^*$ denotes the history of the unobserved process up to t. Such processes will be discussed in more detail in Chap. 11. For more details concerning specific types of inhomogeneous Poisson processes, see Cox and Isham (1980) or Daley and Vere-Jones (2005).

A general class of point processes is given by *renewal processes* which are characterized by independently identically distributed inter-event durations. In particular, the event arrival times of a renewal process are obtained by summing independent identically distributed non-negative random variables. The independence property is a crucial characteristic of renewal processes and stems from the idea of studying the sequence of intervals between successive replacements of a component which is replaced by a new component in case of failure. In the case where the event arrival times are driven by (partial) sums of i.i.d. standard exponential variables, we obtain the homogeneous Poisson process. Due to the independence of inter-event durations, the process history before t_{i-1} has no impact on the current intensity at $t_{i-1} < t_i$. Therefore, a renewal process is typically used to model cross-sectional duration data.

Deviating from the exponential distribution and allowing for more flexible distributions with density $f(x_i)$ yields hazard functions $h(x_i)$ which are non-constant during a duration spell. In this case, in contrast to Poisson processes, the time since the previous event point does have an influence on the further development of the process. Because of this property, one has to take care of the initial conditions. In general, the *forward recurrence time* x_1 from $t = 0$ to the first subsequent point does not have the density $f(\cdot)$ except for the case when there is a point in $t = 0$. In such a case, the process is called an *ordinary* renewal process, otherwise it is a *modified* renewal process. More details are given in Cox and Isham (1980) or Daley and Vere-Jones (2005). The shape of the hazard function determines the so-called *duration dependence* which is associated with the sign of $dh(s)/ds$. If $dh(s)/ds > (<)\, 0$, the process is said to exhibit positive (negative) duration dependence, i.e., the hazard rate increases (decreases) over the length of the spell.

More general types of point processes are obtained by allowing for serial dependencies in durations $\{x_i\}, i = 1, \ldots, n$. This yields the class of *dynamic point processes* which is more systematically discussed in Sect. 4.4 and in the remainder of the book.

Alternative classes of point processes (on the line) are the classes of *finite point processes*, where the total number of points is finite with probability one and where dependence structures are typically described using combinatorial arguments. More general classes arise by relaxing the stationarity and orderliness of a point process, see, e.g., Daley and Vere-Jones (2005). These type of processes are, however, beyond the scope of this book.

4.1.5 A Random Time Change Argument

A fundamental result in martingale-based point process theory is the random time change theorem by Meyer (1971) which allows to transform a wide class of point processes to a unit-rate Poisson process.

Theorem 4.2. *Let $N(t)$ be a simple point process adapted to a history $\mathcal{F}_t$ with intensity $\lambda(t;\mathcal{F}_t)$ and $\mathcal{F}_t$-compensator $\tilde{\Lambda}(t)$ with $\tilde{\Lambda}(\infty) = \infty$ almost surely. Define for all t, the stopping-time $\tau(t)$ as the solution to*

$$\int_0^{\tau(t)} \lambda(t;\mathcal{F}_s)ds = t. \tag{4.19}$$

Then, the point process

$$\tilde{N}(t) = N(\tau(t)) = N(\tilde{\Lambda}^{-1}(t)) \tag{4.20}$$

is a homogeneous Poisson process with intensity $\lambda = 1$.

Proof. See the proof of Theorem T16 in Brémaud (1981) or Theorem 7.4.I in Daley and Vere-Jones (2005). □

Hence, (4.19) corresponds to a change of the time scale from t to $\tau(t)$ transforming $N(t)$ into a unit rate Poisson process $\tilde{N}(t) = N(\tau(t))$.

In order to exploit this result for the construction of diagnostic tests, we still have to show that the realizations of $\tilde{\Lambda}(t)$ evaluated at t_i, i.e., $\tilde{\Lambda}(t_i) = \int_0^{t_i} \lambda(s;\mathcal{F}_s)ds$ correspond to the arrival times associated with $\tilde{N}(t)$, $\tilde{t}_i$ for all i. This is shown in the following lemma by Bowsher (2007):

Lemma 4.3. *Denote $\tilde{t}_i$ as the arrival times associated with the (transformed) point process $\tilde{N}(t)$ given by* (4.20)*. Then, $\tilde{t}_i = \int_0^{t_i} \lambda(s;\mathcal{F}_s)ds \;\forall\, i$.*

Proof. Assume $\tilde{t}_{i-1} \leq s < \tilde{t}_i$ with $\tilde{N}(s) = N(\tau(s)) = i - 1$. Then, if $s \uparrow \tilde{t}_i$, it follows that $\tau(s) \uparrow \tau(\tilde{t}_i)$. Since $N(\tau(\tilde{t}_i)) = \tilde{N}(\tilde{t}_i) = i$ by (4.20), it follows that $N(\tau(s))$ must jump from $i-1$ to i at $\tau(\tilde{t}_i)$, i.e. $s = \tilde{t}_i$. Since by definition $N(t)$ jumps to i only at t_i, it follows that $t_i = \tau(\tilde{t}_i)$. Then, it follows from (4.19) that $\tilde{t}_i = \int_0^{t_i} \lambda(s;\mathcal{F}_s)ds$. □

Using Lemma 4.3 and Theorem 4.2 yields a theorem building the basis for the construction of diagnostic tests:

Theorem 4.4. *The durations $\tilde{t}_i - \tilde{t}_{i-1}$ associated with the (transformed) point process $\tilde{N}(t)$ according to* (4.20) *are given by*

$$\tilde{t}_i - \tilde{t}_{i-1} = \Lambda(t_{i-1}, t_i) = \int_{t_{i-1}}^{t_i} \lambda(s;\mathcal{F}_s)ds$$

and correspond to i.i.d. unit exponential random variates. Then it follows that

$$\Lambda(t_{i-1}, t_i) \sim \text{ i.i.d. } Exp(1). \tag{4.21}$$

Proof. The proof straightforwardly follows from the fact that $\tilde{t}_i = \int_0^{t_i} \lambda(s;\mathcal{F}_s)ds$ are event times of a unit rate Poisson process. □

Hence, under fairly weak regularity conditions, the integrated intensity function $\Lambda(t_{i-1}, t_i)$ is i.i.d. unit exponentially distributed and establishes a powerful link between the intensity function and the duration until the occurrence of the next point. For example, in a simple case where the intensity function is constant during two points, the duration x_i is given by $x_i = \Lambda(t_{i-1}, t_i)/\lambda(t_i;\mathcal{F}_{t_i})$. Furthermore, $\Lambda(t_{i-1}, t_i)$ can be interpreted as a generalized error (for example, in the spirit of Cox and Snell 1968) and indicates whether the path of the conditional intensity function under-predicts ($\Lambda(t_{i-1}, t_i) > 1$) or over-predicts ($\Lambda(t_{i-1}, t_i) < 1$) the number of events between t_i and t_{i-1}.

The random time change argument also holds in case of a multivariate point process:

Theorem 4.5. *Let $N(t)$ be a K-variate point process adapted to a history $\mathcal{F}_t$ with corresponding intensities $\lambda^k(t;\mathcal{F}_t)$ and $\mathcal{F}_t$-compensators $\tilde{\Lambda}^k(t) = \int_0^t \lambda^k(s;\mathcal{F}_s)ds$ with $\tilde{\Lambda}^k(\infty) = \infty$ almost surely $\forall\, k$. Define for all t, the stopping-time $\tau^k(t)$ as the solution to*

$$\int_0^{\tau^k(t)} \lambda^k(s;\mathcal{F}_s)ds = t. \tag{4.22}$$

Then, the multivariate point process $(\tilde{N}^1(t),\ldots,\tilde{N}^K(t))$ with

$$\tilde{N}^k(t) = N^k(\tau^k(t)) = N^k(\tilde{\Lambda}^k(t)^{-1}), \quad k = 1\ldots,K, \tag{4.23}$$

is a multivariate Poisson process with components each having a unit rate and being independent of each other.

Moreover, the durations $\tilde{t}_i^k - \tilde{t}_{i-1}^k$ associated with the Poisson processes $\tilde{N}^k(t)$, $k = 1\ldots,K$, are given by

$$\tilde{t}_i^k - \tilde{t}_{i-1}^k = \Lambda^k(t_{i-1}^k, t_i^k) = \int_{t_{i-1}^k}^{t_i^k} \lambda^k(s;\mathcal{F}_s)ds$$

and correspond to i.i.d. unit exponential random variates.

Proof. See Daley and Vere-Jones (2005), Meyer (1971) or Brown and Nair (1988) for a more accessible proof. The proof of the exponentiality of $\Lambda^k(t_{i-1}^k, t_i^k)$ follows straightforwardly by adapting Lemma 4.3 to the multivariate case. □

4.1.6 Intensity-Based Inference

Karr (1991) proves the existence of a unique probability measure such that $N(t)$ has the $\mathcal{F}_t$-intensity $\lambda(t;\mathcal{F}_t)$. This theorem is fundamental for intensity-based statistical inference since it implies that a statistical model can be completely specified in terms of the $\mathcal{F}_t$-intensity. A valuable conclusion from Karr's Theorem 5.2 is the establishment of a likelihood function in terms of the intensity. Assume that the intensity process satisfies the condition

$$\int_0^{t_n} \lambda(s;\mathcal{F}_s)ds < \infty.$$

Then, the log-likelihood function associated with the matrix of underlying data $\mathbf{Y}$ and the parameter vector $\boldsymbol{\theta}$ is given by

$$\ln \mathcal{L}(\mathbf{Y}, \boldsymbol{\theta}) = \int_0^{t_n} (1 - \lambda(s; \mathcal{F}_s))ds + \int_{(0,t_n]} \ln \lambda(s; \mathcal{F}_s)dN(s)$$
$$= \int_0^{t_n} (1 - \lambda(s; \mathcal{F}_s))ds + \sum_{i \geq 1} \mathbb{1}_{\{t_i \leq t_n\}} \ln \lambda(t_i), \tag{4.24}$$

where

$$\int_{(0,t]} U(s)dN(s) = \sum_{i \geq 1} \mathbb{1}_{\{t_i \leq t\}} U(t_i) \tag{4.25}$$

defines the stochastic Stieltjes integral[6] of some measurable point process $U(t)$ with respect to $N(t)$.

Note that the relationship above holds under fairly weak conditions. Nevertheless, by imposing more assumptions, we can establish a link between the intensity function $\lambda(t; \mathcal{F}_t)$ and the conditional survivor function $S(x_i; \mathcal{F}_{t_i}) := \mathbb{P}r\,[X_i \geq x_i \,|\mathcal{F}_{t_i}]$ yielding the so called "exponential formula" according to Yashin and Arjas (1988):

$$S(x_i; \mathcal{F}_{t_i}) = \exp\left(-\Lambda(t_{i-1}, t_i)\right). \tag{4.26}$$

When there is no conditioning, this relationship is simply derived by solving the differential equation

$$\frac{d}{dx_i} S(x_i) = -\lambda(t_{i-1} + x_i) S(x_i) \tag{4.27}$$

subject to the initial condition $S(0) = 1$. As illustrated by Yashin and Arjas (1988), this relationship also holds if the conditioning is based on a fixed σ-algebra. However, it does not necessarily hold when the conditioning is "dynamic", i.e., depending on time-dependent random factors (e.g., time-varying covariates). More concretely, Yashin and Arjas (1988) prove that the "exponential formula" (4.26) is only valid if $S(x_i; \mathcal{F}_{t_{i-1}+x_i})$ is absolutely continuous in x_i. Note that his assumption excludes jumps of the conditional survivor function induced by changes of the information set during a spell.

Exploiting the "exponential formula" (4.26) yields another straightforward way showing that $\Lambda(t_{i-1}, t_i)$ is an i.i.d. standard exponential variate (see, for instance, Lancaster 1997). However, this proof requires undesirably restrictive assumptions which are problematic whenever information sets might change continuously, as, e.g., in case of time-varying covariates and multivariate processes. In these situations, the random time change argument as discussed in Sect. 4.1.5 is more universal since it requires less restrictive conditions.

---

[6]See, e.g., Karr (1991), p. 57.

4.1.7 Simulation and Diagnostics

The random time change argument allows us to construct simulation algorithms and diagnostic tests for point processes. A natural simulation algorithm for a univariate $\mathcal{F}_t$-adapted point process $N(t)$ with $\mathcal{F}_t$-intensity λ_t and $\mathcal{F}_t$-compensator $\tilde{\Lambda}(t)$ is obtained by exploiting the fact that $\Lambda(t_{i-1}, t_i) \sim$ i.i.d. $Exp(1)$. Then, using the inverse method we can simulate the process as follows:

(i) Simulate a sequence $U_1, U_2, \ldots, U_n$ of i.i.d. uniform $U(0,1)$ random variables.
(ii) Using the inverse method, generate standard exponential variates by $V_i = -\ln(1-U_i), i = 1, 2, \ldots, n$.
(iii) Set $t_0 = 0$. If $\Lambda(0, t_1) = V_1$ can be expressed as $t_1 = x_1 = \Lambda(0, V_1)^{-1}$ with Λ^{-1} denoting the inverse function of Λ depending only on V_1 (and t_0), set $t_1 = x_1$. Alternatively, solve $\Lambda(0, t_1) = V_1$ implicitly for t_1.
(iv) For $i = 2, \ldots, n$: If $\Lambda(t_{i-1}, t_i) = V_i$ can be expressed as $x_i = \Lambda^{-1}(t_{i-1}, V_i)$, set $t_i = t_{i-1} + x_i$. Alternatively, solve $\Lambda(t_{i-1}, t_i) = V_i$ implicitly for t_i.

Correspondingly, the simulation of a K-dimensional point process is similarly performed but requires to continuously update the (integrated) intensities whenever a new point of the pooled process $N(t)$ occurs:

(i) Simulate K (independent) sequences $U_1^k, U_2^k, \ldots, U_n^k$, $k = 1, \ldots, K$, of i.i.d. uniform $U(0,1)$ random variables.
(ii) Using the inverse method, generate standard exponential variates by $V_i^k = -\ln(1-U_i^k), i = 1, \ldots, n, k = 1, \ldots, K$.
(iii) Set $t_0^k = 0$. If $\Lambda^k(0, t_1^k) = V_1^k$ can be expressed as $t_1^k = \Lambda^k(0, V_1^k)^{-1}$ with $\Lambda^k(\cdot)^{-1}$ denoting the inverse function of Λ^k depending only on V_1^k (and t_0^k), set $t_1^k = x_1^k$. Alternatively, solve $\Lambda^k(0, t_1^k) = V_1^k$ implicitly for t_1^k.
(iv) Set $t_1 = \min\{t_1^1, \ldots, t_1^K\}$ and $W_1 = \arg\min_k\{t_1^1, \ldots, t_1^k, \ldots, t_1^K\}$ for $k = 1, \ldots, K$.
(v) For all $j \neq W_1$: Compute $\Lambda^j(0, t_1)$. If $\Lambda^j(0, t_1^j) = \Lambda^j(0, t_1) + \Lambda^j(t_1, t_1^j) = Y_1^k$ can be expressed as $t_1^j = \Lambda^j(0, V_1^j)^{-1}$, set $t_1^j = x_1^j$.
For $j = W_1$: If $\Lambda^j(t_1, t_2^j) = Y_2^j$ can be expressed as $x_2^j = \Lambda^j(t_1, V_2^j)^{-1}$, set $t_2^j = t_1 + x_2^j$. Alternatively, solve $\Lambda^j(t_1, t_2^j) = Y_2^j$ implicitly for t_2^j.
(vi) Set $t_2 = \min\{t_2^1, \ldots, t_2^K\}$ and $W_2 = \arg\min_k\{t_2^1, \ldots, t_2^k, \ldots, t_2^K\}$ for $k = 1, \ldots, K$.
(vii) Continue this procedure until n_k, $k = 1, \ldots, K$, spells in each dimension are generated.

Alternative simulation techniques are discussed, e.g., by Daley and Vere-Jones (2005).

The random time change argument as discussed in Sect. 4.1.5 provides the possibility to construct different types of diagnostic tests. In particular, we perform diagnostics based on four different types of (generalized) errors in the spirit of Cox and Snell (1968) building on integrated intensities. Assume a K-dimensional point process and define

$$y_i^k := \begin{cases} 1 & \text{if } W_i = k, \\ 0 & \text{otherwise.} \end{cases} \tag{4.28}$$

Then, define

$$\varepsilon_{i,1}^k := \Lambda^k(t_{i-1}^k, t_i^k),$$

$$\varepsilon_{i,2} := \sum_{k=1}^{K} \Lambda^k(t_{i-1}^k, t_i^k) y_i^k,$$

$$\varepsilon_{i,3} := \sum_{k=1}^{K} \Lambda^k(t_{i-1}, t_i) = \Lambda(t_{i-1}, t_i),$$

$$\varepsilon_{i,4} := K(\tilde{t}_i - \tilde{t}_{i-1}),$$

where $\tilde{t}_i$ are the event times associated with the superposition $\tilde{N}_0(t) := \sum_{k=1}^{K} \tilde{N}^k(t)$ and $N^k(t)$, $k = 1, \ldots, K$, are the time-changed point processes according to (4.20). According to Theorems 4.4 and 4.5, all residuals $\varepsilon_{i,j}$, $j = 1 \ldots, 4$, are i.i.d. standard exponentially distributed. The exponentiality of $\varepsilon_{i,1}^k$, $k = 1, \ldots, K$, directly follows from Theorem 4.5. The errors $\varepsilon_{i,2}$ correspond to mixtures of i.i.d. standard exponentially distributed variates and thus must be i.i.d. standard exponentially distributed themselves. The errors $\varepsilon_{i,3}$ correspond to the integrated intensity of the (pooled) market point process and are standard exponentially distributed according to Theorem 4.4. Finally, the errors $\varepsilon_{i,4}$ are computed by sorting the transformed k-type specific event times $\tilde{t}_i^k = \int_0^{t_i^k} \lambda^k(s; \mathcal{F}_s) ds$ for all $k = 1, \ldots, K$ into ascending order yielding the superposed process $\tilde{N}_0(t)$. Since the $\tilde{N}^k(t)$ processes are independent Poisson processes with intensity equal to one, the intensity of $\tilde{N}_0(t)$ is K yielding durations $(\tilde{t}_i - \tilde{t}_{i-1})$ which are exponentially distributed with mean K^{-1}. Thus, scaling by K as in $\varepsilon_{i,4}$ yields unit exponential variates. These types of errors have been proposed by Bowsher (2007) as an "omnibus" test accounting for the multivariate nature of the process. Hence, errors $\varepsilon_{i,j}$, $j = 2, 3, 4$, depend not only on the individual process $N^k(t)$ but also on (potential) cross-process interdependencies. In contrast, $\varepsilon_{i,1}^k$ only depends on the individual k-type series. Obviously, all four types of model residuals coincide in the univariate case ($K = 1$).

Using the residual series, model evaluation can be done by testing the dynamical and distributional properties of $\{\hat{\varepsilon}_{i,1}^k, \hat{\varepsilon}_{i,2}, \hat{\varepsilon}_{i,3}, \hat{\varepsilon}_{i,4}\}$. The dynamical properties are typically evaluated based on Portmanteau statistics or independence tests such as proposed by Brock et al. (1996) and discussed in Chap. 5. The distributional properties can be evaluated based on a test against excess dispersion. Define e_i as the corresponding model induced residual, i.e. $e_i \in \{\hat{\varepsilon}_{i,1}^k, \hat{\varepsilon}_{i,2}, \hat{\varepsilon}_{i,3}, \hat{\varepsilon}_{i,4}\}$. Engle and Russell (1998) propose a specific test on the exponentiality of the series based on the asymptotically standard normally distributed test statistic $\sqrt{n_e/8}\,(\hat{\sigma}_e^2 - 1)$, where n_e denotes the number of residuals and $\hat{\sigma}_e^2$ denotes the empirical variance of the residuals being one under the null hypothesis. Alternatively, the distributional properties of the residuals can be tested based on Pearsons's classical χ^2-goodness-of-fit tests or using graphical methods, such as quantile-quantile plots.

4.2 Four Ways to Model Point Processes

In this section, we briefly discuss four basic ways to model point processes:

(i) models for intensity processes,
(ii) models for hazard processes,
(iii) models for duration processes,
(iv) models for counting processes.

Here, we restrict our focus on the non-dynamic case. In the following chapters of the book, duration and intensity concepts will be reconsidered in more detail and will be extended to a dynamic framework.

4.2.1 Intensity Models

An obvious way to model point processes is to model the intensity $\lambda(t; \mathcal{F}_t)$ directly. A general class of intensity models is given by

$$\lambda(t; \mathcal{F}_t) = \lambda_0(t; g_2(\mathcal{F}_t))g_1(t; \mathcal{F}_t), \tag{4.29}$$

where λ_0 denotes a baseline intensity function which is driven by time t and a function g_2 depending on $\mathcal{F}_t$. The function g_2 might capture the dependence on (time-varying) covariates, past durations or characteristics of other point processes. Likewise, g_1 most generally depends on the same information set but might have another form. It proportionally interacts with λ_0. An example of such a process is a multivariate autoregressive conditional intensity model as discussed in Chap. 12, where g_2 is a function of the backward recurrence time of other point processes and g_1 is a function of the history of the process $\mathcal{F}_{\breve{N}(t)}$ with $\mathcal{F}_{\breve{N}}(t) \subset \mathcal{F}_t$. However, considering special cases of representation (4.29) allows to characterize different classes of models which will be discussed in more detail in the remainder of the book:

(i) For $g_1 = g_2$, we obtain an intensity model where g_1 does not only multiplicatively interact with λ_0 but enters λ_0 also as an argument. Moreover, if we assume that $\lambda_0(t; g_1(\mathcal{F}_t))$ can be expressed as $\lambda_0(t; g_1(\mathcal{F}_t)) = \lambda_0(t/g_1(\mathcal{F}_t))$, we obtain a class we refer to as *accelerated intensity models* as g_1 scales (and thus accelerates or decelerates) the time argument in λ_0. This model is the intensity counterpart to an *accelerated failure time model* as discussed below.

(ii) If we assume that $g_1(\mathcal{F}_t) = g_1(\mathcal{F}_{\breve{N}(t)})$ as well as $g_2(\mathcal{F}_t) = g_2(\mathcal{F}_{\breve{N}(t)})$, and $\lambda_0(t; g_2(\mathcal{F}_{\breve{N}(t)})) = \lambda_0(x(t); g_2(\mathcal{F}_{\breve{N}(t)}))$, then $\lambda(t; \mathcal{F}_t)$ only depends on information prevailing at the beginning of the spell, i.e., $\lambda(t; \mathcal{F}_t) = \lambda_0(x(t); g_2(\mathcal{F}_{\breve{N}(t)}))g_1(\mathcal{F}_{\breve{N}(t)})$, and the resulting intensity corresponds to a *conditional hazard* function

$$\lambda(t; \mathcal{F}_t) = h(x(t), g_1(\mathcal{F}_{\breve{N}(t)}), g_2(\mathcal{F}_{\breve{N}(t)}))$$
$$= \frac{f_x(x(t); g_2(\mathcal{F}_{\breve{N}(t)}), g_1(\mathcal{F}_{\breve{N}(t)}))}{1 - F_x(x(t); g_2(\mathcal{F}_{\breve{N}(t)}), g_1(\mathcal{F}_{\breve{N}(t)}))}. \tag{4.30}$$

As the information set only changes at each event but not during a spell, the hazard function can be represented in terms of the inter-event durations $x_{\breve{N}+1} = x_i$, i.e.,

$$h(x_i; g_2(\mathcal{F}_{t_{i-1}}), g_1(\mathcal{F}_{t_{i-1}})) = \frac{f(x_i; g_2(\mathcal{F}_{t_{i-1}}), g_1(\mathcal{F}_{t_{i-1}}))}{1 - F(x_i; g_2(\mathcal{F}_{t_{i-1}}), g_1(\mathcal{F}_{t_{i-1}}))}. \tag{4.31}$$

According to the definitions stated in Sect. 4.1.2, we will refer to *intensity processes* as long as possibly time-varying covariates have to be taken into account and the process has to be updated *during* a spell. If, however, point processes are only driven by information updated at each event time but not during a spell, we will refer to *hazard processes.*

(iii) If we assume that $g_2 = 0$ and λ_0 is only a function of the time elapsed since the last event, i.e., $\lambda_0(t; \mathcal{F}_t) = \lambda_0(x(t)) = h_0(x(t))$, we obtain the class of *proportional intensity models.* If, in addition, $g_1(\mathcal{F}_t) = g_1(\mathcal{F}_{\breve{N}(t)})$, then $\lambda(t; \mathcal{F}_t) = h_0(x(t))g_1(\mathcal{F}_{\breve{N}(t)})$ corresponds to a *proportional hazard* (PH) model.

(iv) Assuming $g_1(\mathcal{F}_t) = g_2(\mathcal{F}_t) = g_1(\mathcal{F}_{\breve{N}(t)})$ and in addition $\lambda_0(t; g_1(\mathcal{F}_{\breve{N}(t)})) = \lambda_0(t/g_1(\mathcal{F}_{\breve{N}(t)}))$, then $\lambda(t; \mathcal{F}_t) = \lambda_0(t/g_1(\mathcal{F}_{\breve{N}(t)}))g_1(\mathcal{F}_{\breve{N}(t)})$ belongs to the class of so-called *accelerated failure time* (AFT) models where the time scale is accelerated (or decelerated) by observable characteristics. As shown below, AFT models are naturally linked to *duration models.*

(v) The functional form of λ_0 can be chosen parametrically or can be left unspecified (and can be estimated nonparametrically). For instance, assuming $\lambda_0(t; g_2(\mathcal{F}_t)) = \lambda_0(x(t)) = \omega x(t)^{a-1}$ for $\omega > 0$, $a > 0$, allows the intensity to monotonically increase (for $a > 1$) or decrease (for $a < 1$) during the spell. Under the additional assumption that $g_1(\mathcal{F}_t) = g_1(\mathcal{F}_{\breve{N}(t)})$, the resulting intensity process $\lambda(t; \mathcal{F}) = \omega x(t)^{a-1} g_1(\mathcal{F}_{\breve{N}(t)})$ generates durations x_i which are Weibull distributed with shape parameter a and scale parameter g_1. As illustrated below, this special case of a Weibull (proportional) hazard model can be alternatively represented as a specific AFT model, which is obviously driven by the specific parametric form of λ_0. Finally, the case of a completely unspecified baseline function λ_0 is mostly relevant in the class of proportional hazard models yielding the class of *semiparametric proportional hazard models*.

Models of the intensity function are discussed in a dynamic setting in Chap. 11. The special cases of (non-dynamic) PH and AFT models will be discussed in more detail in the next subsection which will build the basis for dynamic extensions discussed in the following chapters.

4.2.2 *Hazard Models*

4.2.2.1 Proportional Hazard (PH) Models

The proportional hazard model has its origin in cross-sectional survival analysis in biostatistics[7] and labor economics.[8]

As stated in the previous section, the PH model is specified as the product of a baseline hazard function $h_0(x_i) > 0$ and a strictly positive function of a vector of covariates $\mathbf{z}_{i-1}$ with coefficients γ. It is given by

$$h(x_i; \mathbf{z}_{i-1}) = h_0(x_i) \exp(-\mathbf{z}'_{i-1}\gamma). \tag{4.32}$$

The baseline hazard $h_0(x)$ corresponds to $h(x; \mathbf{z} = 0)$. That is, if the regressors are centered, h_0 has an interpretation as the intensity function for the mean values of $\mathbf{z}$. The key property of the PH model is that $\partial \ln h(x; \mathbf{z})/\partial \mathbf{z} = \gamma$, thus, γ is interpreted as the constant proportional effect of $\mathbf{z}$ on the conditional arrival rate.

The most common examples of a (parametric) PH model are the exponential and Weibull regression model. In these models, the baseline hazard h_0 is specified according to the exponential or Weibull hazard function. While the Weibull model allows only for monotonically increasing or decreasing hazard shapes, more flexible parameterizations are obtained, for instance, by the gamma, generalized gamma, Burr or generalized F distribution (see appendix). Alternatively, non-standard PH models are obtained by deviating from classical gamma or Burr shapes and parameterizing the baseline hazard function, for instance, as a linear spline function. Assume that the observed durations are partitioned into K categories, where $\bar{x}_k$, $k = 1, \ldots, K-1$ denote the chosen category bounds. Then, a spline baseline

[7]See, e.g., Kalbfleisch and Prentice (1980), Cox and Oakes (1984) or the recent survey by Oakes (2001).

[8]A well known example is the analysis of the length of unemployment spells which is studied by a wide range of theoretical and empirical papers, see e.g. Lancaster (1979), Nickell (1979), Heckmann and Singer (1984), Moffitt (1985), Honoré (1990), Meyer (1990), Han and Hausman (1990), Gritz (1993), McCall (1996) or van den Berg and van der Klaauw (2001) among many others.

hazard is obtained by

$$h_0(x) = \exp\left[\nu_0 + \sum_{k=1}^{K-1} \nu_{0,m} \mathbb{1}_{\{x_i \leq \bar{x}_k\}}(\bar{x}_k - x_i)\right], \tag{4.33}$$

where ν_0 is a constant and $\nu_{0,k}$ are coefficients associated with the nodes of the spline function. Parametric PH models are easily estimated by maximum likelihood (ML) adapting the log likelihood function (4.24). Note that in a full information ML approach, a consistent estimation of γ requires the correct specification of h_0. However, often parametric distributions are not sufficient to capture the actually observed hazard shape. In such a case, a full information ML estimation of the PH model leads to inconsistent estimates of γ.

A more flexible PH model is proposed by Cox (1972). In this framework, the baseline hazard h_0 remains completely unspecified and can be estimated non-parametrically. Cox (1975) illustrates that the estimation of γ does not require a specification of the baseline hazard, and thus, the estimation of γ and h_0 can be separated. In order to estimate γ and h_0, he suggests a two-step approach where γ is consistently estimated using a *partial likelihood* approach while the estimation of the baseline hazard follows from a modification of the estimator by Kaplan and Meier (1958) as proposed by Breslow (1972, 1974).[9]

Since the partial likelihood approach proposed by Cox (1975) is based on the order statistics of the durations, this method is of limited use in the case of ties, i.e., when several observations have the same outcome. Then, only approximative procedures can be used (see, e.g., Breslow 1974). In this case, a valuable alternative way to estimate the PH model, is to apply a categorization approach as proposed by Han and Hausman (1990) and Meyer (1990). The main idea behind this procedure is to exploit the relationship between ordered response specifications and models for grouped durations (see Sueyoshi 1995). Since a categorization approach will play an important role in the context of (censored) autoregressive proportional hazard models (see Chap. 10), it will be discussed here in more detail.

The main idea is to write the PH model in terms of the (log) integrated baseline hazard. According to the implications of Theorem 4.2, the integrated hazard function

$$H(x_i) := \int_0^{x_i} \lambda(s)ds \tag{4.34}$$

is i.i.d. standard exponentially distributed. Thus, the PH model is rewritten as

$$H_0(x_i) = \exp(\mathbf{z}_{i-1}'\gamma)H(x_i), \tag{4.35}$$

where

[9]For more details, see Kalbfleisch and Prentice (1980) or in the survey of Kiefer (1988).

$$H_0(x_i) := \int_0^{x_i} \lambda_0(s)ds$$

denotes the *integrated baseline hazard*. By assuming the validity of the "exponential formula", (4.26), the integrated baseline hazard is written as $H_0(x_i) = -\ln S_0(x_i)$, where $S_0(\cdot)$ denotes the baseline survivor function associated with the baseline hazard h_0. Hence, (4.35) can be formulated in terms of the baseline survivor function as

$$\ln S_0(x_i) = -\exp(\mathbf{z}'_{i-1}\gamma)H(x_i). \tag{4.36}$$

Since $H(x_i)$ is standard exponentially distributed, it is straightforward to verify that

$$\mathbb{E}[H_0(x_i)] = \exp(\mathbf{z}'_{i-1}\gamma)$$
$$\mathbb{V}[H_0(x_i)] = \exp(2\mathbf{z}'_{i-1}\gamma).$$

Equivalently, the PH model can be expressed in terms of the log integrated hazard leading to

$$\ln H_0(x_i) = \ln(-\ln S_0(x_i)) = \mathbf{z}'_{i-1}\gamma + \varepsilon_i^*, \quad i = 1, \ldots, n, \tag{4.37}$$

where $\varepsilon_i^* := \ln H(x_i)$ follows an i.i.d. standard extreme value type I distribution (standard Gumbel (minimum) distribution) with mean $\mathbb{E}[\varepsilon_i^*] = -0.5772$ (corresponding to the Euler–Mascheroni constant), variance $\mathbb{V}[\varepsilon_i^*] = \pi^2/6$ and density function

$$f_{\varepsilon^*}(s) = \exp(s - \exp(s)). \tag{4.38}$$

Note that this approach does not require to specify the baseline hazard h_0, however it requires a complete parameterization of the *log integrated* baseline hazard. Thus, the PH model boils down to a regression model for $\ln H_0$. The function H_0 can be interpreted as a transformation of the underlying duration x_i, where the transformation is known when h_0 is completely specified, otherwise it is unknown. For example, by assuming a standard Weibull specification, i.e., $h_0(x) = ax^{a-1}$, the PH model yields

$$\ln x_i = \mathbf{z}'_{i-1}\frac{\gamma}{a} + \frac{1}{a}\varepsilon_i^*, \quad i = 1, \ldots, n. \tag{4.39}$$

In this case, the PH model is a log-linear model with standard extreme value distributed errors. The coefficient vector γ/a can be consistently (however, not efficiently) estimated by OLS.[10] Alternatively, the (fully parametric) PH model is consistently and efficiently estimated by ML.

---

[10]In this case, an intercept term has to be included in (4.39) since ε_i^* has a nonzero mean.

When h_0 is non-specified, H_0 is unknown and can be interpreted as a latent variable. Han and Hausman (1990) propose treating this model as a special type of ordered response model leading to a semiparametric estimation of h_0. The central idea is to introduce a categorization of the duration x_i and to consider (4.37) as a latent process which is observable only at discrete points associated with the chosen category bounds. Then, by adapting the categorization introduced above, we define

$$\mu_k^* := \ln H_0(\bar{x}_k), \qquad k = 1, \dots, K-1, \tag{4.40}$$

as the value of the latent variable $\ln H_0(\cdot)$ at the (observable) category bound $\bar{x}_k$. In this formulation, the PH model is interpreted as an ordered response model based on a standard extreme value distribution. Hence, the thresholds of the latent model correspond to the log integrated baseline hazard calculated at the points of the corresponding categorized durations. The direct relationship between the latent thresholds and the log integrated baseline hazard function is one of the main advantages of this approach. Thus, the unknown baseline survivor function S_0 can be estimated at the $K-1$ discrete points by a nonlinear function of the estimated thresholds μ_k^*, thus

$$S_0(\bar{x}_k) = \exp(-\exp(\mu_k^*)), \qquad k = 1, \dots, K-1. \tag{4.41}$$

Based on the discrete points of the baseline survivor function, we can estimate a *discrete* baseline hazard $\bar{h}_0$, corresponding to the conditional failure probability given the elapsed time since the last event,

$$\begin{aligned} \bar{h}_0(\bar{x}_k) &:= \mathbb{P}r\left[\bar{x}_k \le x_i < \bar{x}_{k+1} \,|x_i \ge \bar{x}_k\right] \\ &= \frac{S_0(\bar{x}_k) - S_0(\bar{x}_{k+1})}{S_0(\bar{x}_k)}, \qquad k = 0, \dots, K-2, \end{aligned} \tag{4.42}$$

where $\bar{x}_0 := 0$. This formulation serves as an approximation of the baseline hazard if divided by the length of the discretization interval. If the length of the intervals goes to zero, the approximation converges to the original definition, (4.1), thus

$$h_0(\bar{x}_k) \approx \frac{\bar{h}_0(\bar{x}_k)}{\bar{x}_{k+1} - \bar{x}_k}. \tag{4.43}$$

Note that the result of Cox (1975) also holds in this discretization framework. Hence, γ can be estimated consistently without an explicit specification of h_0, which means that the consistency of $\hat{\gamma}$ does not depend on the chosen categorization and the number of the categories.[11] This result will be exploited in a dynamic framework in Chap. 10.

[11] Nevertheless, the efficiency of the estimator is affected by the chosen categorization.

The log likelihood function associated with the PH model represented by (4.37), (4.38) and (4.40) has the well known form of an ordered response specification based on the standard extreme value distribution

$$\ln \mathcal{L}(\mathbf{Y};\boldsymbol{\theta}) = \sum_{i=1}^{n}\sum_{k=1}^{K-1} \mathbb{1}_{\{x_i \in (\bar{x}_{k-1},\bar{x}_k]\}} \ln \int_{\mu_{k-1}^{*}-\mathbf{z}_{i-1}'\gamma}^{\mu_k^{*}-\mathbf{z}_{i-1}'\gamma} f_{\varepsilon^*}(s)ds. \tag{4.44}$$

4.2.2.2 Accelerated Failure Time Models

As illustrated above, in the PH model, explanatory variables act multiplicatively with the baseline intensity. In contrast, in the AFT model, it is assumed that $\mathbf{z}_i$ accelerates (or decelerates) the time to failure. Thus, the covariates alter the rate at which one proceeds along the time axis, leading to a hazard function that is given by

$$h(x_i;\mathbf{z}_{i-1}) = h_0\left(x_i \exp(-\mathbf{z}_{i-1}'\gamma)\right)\exp(-\mathbf{z}_{i-1}'\gamma), \tag{4.45}$$

which can be alternatively written as[12]

$$\ln x_i = \mathbf{z}_{i-1}'\gamma + \varepsilon_i, \quad i = 1,\ldots,n, \tag{4.46}$$

where ε_i is an error term that follows some continuous distribution. Hence, while the PH model implies a linear model with an unknown left-hand variable and a standard extreme value distributed error term, the AFT model leads to a log-linear representation with x_i as the left-hand variable and an unknown distribution of the error term. While in the PH model, the covariates act multiplicatively with the hazard function, in the AFT model, the covariates act multiplicatively with the log duration. For the special case when x_i is Weibull distributed with parameter a (see Sect. 4.2.2.1), we obtain a model that belongs to both model families, implying a multiplicative relationship between the covariates and the intensity function, as well as between the covariates and the log duration. In this case, (4.46) corresponds to (4.39), hence, $a\varepsilon_i = \varepsilon_i^*$ is standard extreme value distributed.

A more general class of models that nests both the PH model and the AFT model is introduced by Ridder (1990) and is called the *Generalized Accelerated Failure Time (GAFT) model*. It is given by

$$\ln g(x_i) = \mathbf{z}_{i-1}'\gamma + \varepsilon_i, \tag{4.47}$$

where $g(\cdot)$ is an arbitrary non-decreasing function defined on $[0,\infty)$ and ε_i follows an unknown continuous distribution with p.d.f $f(\cdot)$. The GAFT model nests the PH

[12] See Kalbfleisch and Prentice (1980).

model if $f(\cdot) = f_{\varepsilon^*}$ and nests the AFT model if $g(x) = x$. Ridder (1990) illustrates that the GAFT model can be identified non-parametrically.

4.2.3 Duration Models

Instead of specifying the intensity function, a point process can be alternatively directly described by the process of durations between subsequent points. In order to ensure non-negativity, the most simple approach is to specify the durations in terms of a log-linear regression model,

$$\ln x_i = \mathbf{z}'_{i-1}\gamma + \varepsilon_i, \tag{4.48}$$

where ε_i is some i.i.d. error term. Such a model is easily estimated by OLS. As discussed in Sect. 4.2.2.2, this type of model belongs to the class of AFT models or – for the special case when ε_i is standard extreme value distributed – also to the class of PH models. However, generally, regression models for log durations are associated with AFT representations. A more detailed discussion of (dynamic) duration models will be given in Chap. 5.

4.2.4 Count Data Models

An alternative representation of a point process model is obtained by specifying the joint distribution of the number of points in equally spaced intervals of length Δ. Denote N_j^Δ as the number of events in the interval $[j\Delta, (j+1)\Delta)$, i.e.,

$$N_j^\Delta := N((j+1)\Delta) - N(j\Delta), \qquad j = 1, 2, \ldots \tag{4.49}$$

The specification of a count data models require to aggregate the data which naturally induces a certain loss of information. Denote $\mathbf{z}_j^\Delta$ as a covariate vector associated with N_j^Δ, then, the most simple count data model is given by the Poisson model by assuming that

$$N_j^\Delta | \mathbf{z}_j^\Delta \sim Po(\lambda). \tag{4.50}$$

More general models are obtained by using the NegBin distribution (see, e.g., Cameron and Trivedi 1998) or the double Poisson distribution introduced by Efron (1986). A more detailed discussion of (dynamic) count data models and corresponding distributions will be given in Chap. 12.

4.3 Censoring and Time-Varying Covariates

In this section, we discuss two phenomena which typically occur in traditional duration data (like unemployment data) and also play a role in financial point processes: The effects of censoring and of time-varying covariates. As will be illustrated, both phenomena are easily incorporated in an intensity framework.

4.3.1 Censoring

A typical property of economic duration data is the occurrence of censoring leading to incomplete spells. Therefore, a wide strand of econometric duration literature focuses on the consideration of censoring mechanisms.[13] In the context of financial point processes, censoring occurs if there exist intervals in which the point process cannot be observed directly. This might be, for example, due to non-trading periods, like nights, weekends or holidays. Assume in the following that it is possible to identify whether a point t_i lies within such a censoring interval. Consider, for example, a point process where the points are associated with the occurrence of a cumulative price change of given size.[14] Then, prices move during non-trading periods as well (due to trading on other markets), but can be observed at the earliest at the beginning of the next trading day. In this case, we only know *that* a price event occurred but we do not know *when* it exactly occurred.[15] Hence, we can identify only the minimum length of the corresponding duration (i.e., the time from the previous point t_{i-1} to the beginning of the censoring interval) and the maximum length of the spell (i.e., the time from the end of the censoring interval to the next point t_{i+1}).

In the following, t_i^l and t_i^u with $t_i^l \leq t_i \leq t_i^u$ are defined as the boundaries of a potential censoring interval around t_i, and c_i is defined as the indicator variable that indicates whether t_i occurs within a censoring interval, i.e.,

$$c_i := \begin{cases} 1 & \text{if } t_i \in (t_i^l; t_i^u), \\ 0 & \text{if } t_i = t_i^l = t_i^u. \end{cases}$$

In the case of left-censoring, right-censoring or left-right-censoring, the non-observed duration x_i can be isolated by the boundaries $x_i \in [x_i^l, x_i^u]$ where the lower and upper boundary x_i^l and x_i^u are computed corresponding to

---

[13]See, e.g., Horowitz and Neumann (1987, 1989), the survey by Neumann (1997), Gorgens and Horowitz (1999) or Orbe et al. (2002).

[14]Such processes will be discussed in more details in Chap. 10.

[15]However, note that we cannot identify whether more than one price movement occurred during the non-trading period.

$$x_i \in \begin{cases} [t_i - t^u_{i-1}; t_i - t^l_{i-1}] & \text{if } c_{i-1} = 1, c_i = 0 \quad \text{(left-censoring)} \\ [t^l_i - t_{i-1}; t^u_i - t_{i-1}] & \text{if } c_{i-1} = 0, c_i = 1 \quad \text{(right-censoring)} \\ [t^l_i - t^u_{i-1}; t^u_i - t^l_{i-1}] & \text{if } c_{i-1} = 1, c_i = 1 \quad \text{(left-right-censoring).} \end{cases} \tag{4.51}$$

A common assumption that is easily fulfilled in the context of financial data is the assumption of *independent censoring*. This assumption means that the censoring mechanism is determined exogenously and is not driven by the duration process itself. For a detailed exposition and a discussion of different types of censoring mechanisms, see e.g. Neumann (1997).

Under the assumption of independent censoring, the likelihood can be decomposed into

$$\mathcal{L}(\mathbf{Y};\boldsymbol{\theta}|c_1,\ldots,c_n) = \mathcal{L}(\mathbf{Y};\boldsymbol{\theta},c_1,\ldots,c_n)\cdot\mathcal{L}(\mathbf{Y};c_1,\ldots,c_n). \tag{4.52}$$

Hence, the second factor does not depend on the parameters of the model, and thus the parameter vector $\boldsymbol{\theta}$ is estimated by maximizing the first factor of (4.52). Therefore, the corresponding log likelihood function is given by

$$\begin{aligned} \ln \mathcal{L}(\mathbf{Y};c_1,\ldots,c_n;\boldsymbol{\theta}) = &\sum_{i=1}^{n}(1-c_i)(1-c_{i-1})\left(-\Lambda(t_{i-1},t_i) + \ln\lambda(t_i;\mathcal{F}_{t_i})\right) \\ &+\sum_{i=1}^{n} c_{i-1}(1-c_i)\ln\left(S(t_i - t^u_{i-1}) - S(t_i - t^l_{i-1})\right) \\ &+\sum_{i=1}^{n}(1-c_{i-1})c_i\ln\left(S(t^l_i - t_{i-1}) - S(t^u_i - t_{i-1})\right) \\ &+\sum_{i=1}^{n} c_{i-1}c_i\ln\left(S(t^l_i - t^u_{i-1}) - S(t^u_i - t^l_{i-1})\right). \end{aligned} \tag{4.53}$$

The first term in (4.53) is the log likelihood contribution of a non-censored duration while the following terms are the contributions of left-censored, right-censored and left-right-censored observations.

4.3.2 Time-Varying Covariates

In the following, we discuss the treatment of time-varying covariates $\mathbf{z}(t)$. Here, the impact of the covariates on the intensity is not constant during the time from t_{i-1} to t_i, but is time-varying. However, the formulation of a continuous-time model for the covariate path is rather difficult, since it requires the identification of the precise timing of events on the continuous path. In order to circumvent these difficulties, one typically builds a model based on a discrete-time framework. Hence, it is common to proceed by assuming that the occurrence of events can occur at discrete points of time only. This leads to a discretization of the intensity concept.

By following the notation in Lancaster (1997), denote $\mathcal{Z}_i(x)$ as the path of $\mathbf{z}(t)$ from $\mathbf{z}(t_{i-1})$ to $\mathbf{z}(t_{i-1}+x)$ and $\mathcal{Z}_i(x_1, x_2)$ as the path of $\mathbf{z}(t)$ from $\mathbf{z}(t_{i-1}+x_1)$ to $\mathbf{z}(t_{i-1}+x_2)$. Furthermore, according to the framework outlined in Sect. 4.2.2.1, the observed durations are divided into M intervals and it is assumed that the durations x_i can only take the discrete values $\bar{x}_m$, $m = 1, \ldots, M-1$. Moreover, it is assumed that the discretization is sensitive enough to capture each particular point, i.e., $\min\{x_i\} \geq \bar{x}_1 \;\forall\, i$.

Then, following Lancaster (1997), a discrete formulation of the intensity function at the points $t_{i-1} + \bar{x}_m$ is obtained by

$$\bar{\lambda}(t_{i-1} + \bar{x}_m; \mathcal{Z}_i(\bar{x}_m)) = \mathbb{P}r\left[x_i = \bar{x}_m \,|x_i \geq \bar{x}_m; \mathcal{Z}_i(\bar{x}_m)\right] \tag{4.54}$$

and can be computed based on

$$\begin{aligned}\mathbb{P}r\left[x_i \geq \bar{x}_m, \mathcal{Z}_i(\bar{x}_m)\right] &= \prod_{j=1}^{m-1}\left[1 - \bar{\lambda}(t_{i-1} + \bar{x}_j; \mathcal{Z}_i(\bar{x}_j))\right] \\ &\quad\times \prod_{j=1}^{m-1} \mathbb{P}r\left[\mathcal{Z}_i(\bar{x}_{j-1}, \bar{x}_j)\,\middle|x_i \geq \bar{x}_j, \mathcal{Z}_i(\bar{x}_{j-1})\right]\end{aligned} \tag{4.55}$$

and

$$\begin{aligned}\mathbb{P}r[x_i = \bar{x}_m, x_i \geq \bar{x}_m, \mathcal{Z}_i(\bar{x}_m)] &= \bar{\lambda}(t_{i-1} + \bar{x}_m; \mathcal{Z}_i(\bar{x}_m)) \\ &\quad\times \prod_{j=1}^{m-1}\left[1 - \bar{\lambda}(t_{i-1} + \bar{x}_j; \mathcal{Z}_i(\bar{x}_j))\right] \\ &\quad\times \prod_{j=1}^{m} \mathbb{P}r\left[\mathcal{Z}_i(\bar{x}_{j-1}, \bar{x}_j)\,\middle|x_i \geq \bar{x}_j, \mathcal{Z}_i(\bar{x}_{j-1})\right].\end{aligned} \tag{4.56}$$

The expressions (4.55) and (4.56) simplify in the case of a so-called *exogenous covariate process*, i.e., following the definition in Lancaster (1997), if and only if

$$\begin{aligned}&\mathbb{P}r\left[\mathcal{Z}_i(x, x+\Delta)\,|\,x_i \geq x + \Delta, \mathcal{Z}_i(x)\right] \\ &\quad = \mathbb{P}r\left[\mathcal{Z}_i(x, x+\Delta)\,|\mathcal{Z}_i(x)\right], \quad \forall\, x \geq 0, \Delta > 0.\end{aligned} \tag{4.57}$$

Hence, exogeneity of the covariate process means that the information that no further event has been observed until $x + \Delta$ has no predictability for the further path of the covariates from x to $x + \Delta$. In this case, (4.55) and (4.56) simplify to

$$\mathbb{P}r\left[x_i \geq \bar{x}_m, \mathcal{Z}_i(\bar{x}_m)\right] = \prod_{j=1}^{m-1}\left[1 - \bar{\lambda}(t_{i-1} + \bar{x}_j; \mathcal{Z}_i(\bar{x}_j))\right] \tag{4.58}$$

and

$$\mathbb{P}r[x_i = \bar{x}_m, x_i \geq \bar{x}_m, \mathcal{Z}_i(\bar{x}_m)] = \bar{\lambda}(t_{i-1} + \bar{x}_m; \mathcal{Z}_i(\bar{x}_m)) \times \prod_{j=1}^{m-1}\left[1 - \bar{\lambda}(t_{i-1} + \bar{x}_j; \mathcal{Z}_i(\bar{x}_j))\right]. \tag{4.59}$$

Only when the covariates are exogenous, (4.55) and (4.56) can be interpreted as conditional probabilities, given the covariate path. However, even when they are valid probabilities they can *never* be interpreted as the values of a conditional survivor function or probability density function of x_i given $\mathcal{Z}_i(\bar{x}_j)$ at the point $\bar{x}_m$. This is due to the fact that $\mathcal{Z}_i(\bar{x}_j)$ is itself a function of t and the conditioning event of the conditional survivor function or conditional p.d.f. changes when the argument itself changes.

Nevertheless, even though it cannot be interpreted as the value of a (discrete) p.d.f., it can be used to draw statistical inference. Hence, the intensity function, given the covariate path, is always defined even when there exists no counterpart that is interpretable as a conditional density or conditional survivor function. These relationships illustrate the importance of the intensity concept in modelling duration data.

4.4 An Outlook on Dynamic Extensions

The implementation of autoregressive structures in point processes can be performed in alternative ways. According to the three possible specifications of a point process, dynamics can be introduced either in the intensity process, the duration process or in the counting process. A priori, it is quite unclear which way should be preferred and whether one specification is superior to another. Ultimately, the particular concepts have to be judged by their ability to result in well specified empirical models that provide a satisfying fit to the data and whose coefficients may be readily economically interpretable. Nonetheless, in the context of a dynamic framework, it is necessary to have a closer look at the fundamental differences between the particular approaches and at their strengths and weaknesses with regard to the specific properties of financial point processes.

The most simple and probably most intuitive way to model autoregressive point processes is to specify an autoregressive process in terms of the durations. While a renewal process is specified via independently identically distributed intervals, more general types of point processes arise by allowing for dynamics in successive event intervals $\{x_i\}$, $i = 1, \ldots, n$. For instance, the class of so-called *Wold processes* (see, e.g., Daley and Vere-Jones 2005) is obtained when the process $\{x_i\}$ forms a Markov chain, i.e., the distribution of x_{i+1} given $x_i, x_{i-1}, \ldots$, depends only on x_i. In fact, a (stationary) AR(1) process for inter-event durations,

$$x_i = c + \phi x_{i-1} + \varepsilon_i,$$

with ε_i being an i.i.d. non-negative random variable and $c > 0$ and $0 < \phi < 1$ satisfying non-negativity restrictions would serve as a natural candidate of a Wold process. A so called *Semi-Markov process* according to Cox and Isham (1980) is obtained by specifying a stochastic sequence of distribution functions for the duration sequence. By supposing M distribution functions $F^{(1)}(\cdot), \ldots, F^{(M)}(\cdot)$ associated with M particular states, the process is said to be in state j in t if the current distribution function of $x_{\breve{N}(t)+1}$ is $F^{(j)}(\cdot)$. The transitions are determined by a transition matrix P^* where the ijth element is the transition probability from state i to state j. For more details, see Cox and Isham (1980).

By relaxing the Markovian structure, more general dynamic point processes are obtained. Allowing for higher-order (ARMA-type) dynamics in duration processes yields the class of *autoregressive duration processes*. As discussed in more detail in Chap. 5, these processes are characterized by an intensity function containing an autoregressive component in terms of past durations.

A particular class of *autoregressive duration models* is proposed by Engle and Russell (1997, 1998) and Engle (1996, 2000) and is considered in more detail in Chaps. 5 and 6 of this book. The major advantage of such an approach is its practicability since standard time series packages for ARMA or GARCH models can be more or less directly applied to the duration data. Presumably for this reason, an autoregressive duration model is the most common type of financial point process specification and is commonly used in the recent financial econometrics literature.

Autoregressive duration models, however, also reveal major drawbacks. First, they are not easily extended to a multivariate framework as the individual processes occur asynchronously in time. Consequently, there exist no joint points that can be used to couple the processes making it difficult to estimate contemporaneous correlations between the individual processes. Second, the treatment of censoring effects is rather difficult. In the case of censoring, the exact timing of particular points of the process cannot be observed and can only be approximated by a corresponding interval. Such effects induce problems in an autoregressive framework because the information about the exact length of a spell is needed for the sequel of the time series. One possible solution is, for example, to build a modified model where non-observed durations are replaced by a function of their corresponding conditional expectations. However, such modifications are not straightforward in a time series context. Third, in a discrete-time duration framework, it is difficult to account for time-varying covariates. As discussed in Sect. 4.3.2, an intensity representation is more useful and convenient in such a context.

The key advantage of *dynamic intensity models* compared to autoregressive duration approaches is that they allow to model point processes in a continuous-time framework. Thus, while the duration between two points is by definition observed at the individual points t_i themselves, the intensity function is defined at *any* point in time. This property plays an important role in the modelling of time-varying covariates and in the context of autoregressive multivariate point process models. In an intensity framework, it is possible to couple the particular processes at *each* point of the pooled process, and thus addressing the problem of asynchronicity. *Multivariate autoregressive intensity* models are discussed in more detail in Chap. 12.

A particular type of dynamic point processes is the class of *self-exciting processes*. Hawkes (1971) introduces a special class of linear self-exciting processes where the intensity is a weighted function of the backward recurrence time to all previous points. The advantage of this approach is to estimate the nature of the dependence of the intensity without imposing a strong a priori time series structure. For more details and applications of Hawkes processes to financial data, see Chaps. 11 and 12.

An alternative to self-exciting processes is the class of *autoregressive point processes*. In this context, renewal processes are augmented by time series structures that may be specified in terms of the intensity function, the integrated hazard function, the duration between subsequent points, or in terms of the counting function. Detailed illustrations of such types of point processes will be given in the sequel of this book.

However, as in a duration framework, the treatment of censoring effects is rather difficult. For a censored observation, the occurrence time, and thus the intensity function at the particular time, is not exactly measurable. Thus, as long as these (unobserved) realizations are needed in the time series recursion of a dynamic model, approximations have to be used. Therefore, for the modelling of point processes that are subject to strong censoring mechanisms, alternative approaches are necessary. One alternative is to build a dynamic model based on a function of the survivor function. The survivor function plays an important role in the context of censoring (see Sect. 4.3.1), since it allows to isolate the non-observed occurrence times in terms of survivor probabilities with respect to corresponding censoring bounds. Hence, by exploiting the relationship between the survivor function and the integrated hazard function (see (4.26)), an alternative dynamic point process model can be built based on the integrated hazard function. Moreover, as illustrated in Chap. 10, *autoregressive integrated hazard models* are valuable approaches for dynamic extensions of semiparametric proportional hazard models. In this context, the discretization approach presented in Sect. 4.2.2.1 will be applied to obtain a semiparametric estimation of a non-specified baseline hazard in a dynamic framework.

A further possibility to model autoregressive point processes is to specify an *autoregressive count data* model. The strength of this approach is that these models are based (per assumption) on equi-distant time intervals. For this reason, they are easily extended to a multivariate framework (see Davis et al. 2001 or Heinen and Rengifo 2007). However, the treatment of censoring mechanisms and the inclusion of time-varying covariates is rather difficult in this context. These types of models are discussed in more detail in Chap. 13.

References

Bauwens L, Hautsch N (2009) Modelling Financial High Frequency Data Using Point Processes. In: Andersen TG, Davis RA, Kreiss J-P, Mikosch T (eds) Handbook of financial time series. Springer, Berlin, Heidelberg

Bowsher CG (2007) Modelling security markets in continuous time: intensity based, multivariate point process models. J Econom 141:876–912

Brémaud P (1981) Point processes and queues, Martingale dynamics. Springer, New York

Breslow N (1972) Contribution to the Discussion of the Paper by D. R. Cox. J R Stat Soc Series B 34:216–17

Breslow NE (1974) Covariance analysis of censored survival data. Biometrics 30:89–100

Brock W, Dechert WD, Scheinkman J, LeBaron B (1996) A test for independence based on the correlation dimension. Econom Rev 15:197–235

Brown TC, Nair MG (1988) A simple proof of the multivariate random time change theorem for point processes. J Appl Probab 25:210–214

Cameron AC, Trivedi PK (1998) Regression analysis of count data. Cambridge University Press, Cambridge

Cox DR (1972) Regression Models and Life Tables. J R Stat Soc Series B 34:187–220

Cox DR (1975) Partial likelihood. Biometrika 62:269

Cox DR, Isham V (1980) Point processes. Chapman and Hall, London

Cox DR, Oakes D (1984) Analysis of survival data. Chapman and Hall, London

Cox DR, Snell EJ (1968) A general definition of residuals. J R Stat Soc Series B 30:248–265

Daley D, Vere-Jones D (2005) An introduction to the theory of point processes. Volume I: Elementary theory and methods. Springer, New York

Davis RA, Rydberg TH, Shephard N, Streett SB (2001) The CBin model for counts: testing for common features in the speed of trading, quote changes, limit and market order arrivals. Discussion paper, Nuffield College, Oxford

Efron B (1986) Double exponential families and their use in generalized linear regression. J Am Stat Assoc 81:709–721

Engle RF (2000) The econometrics of ultra-high-frequency data. Econometrica 68(1):1–22

Engle RF, Russell JR (1998) Autoregressive conditional duration: a new model for irregularly spaced transaction data. Econometrica 66:1127–1162

Gorgens T, Horowitz JL (1999) Semiparametric estimation of a censored regression model with an unknown transformation of the dependent variable. J Econom 90:155–191

Gritz RM (1993) The impact of training on the frequency and duration of employment. J Econom 57:21–51

Han A, Hausman JA (1990) Flexible parametric estimation of duration and competing risk models. J Appl Econom 5:1–28

Hawkes AG (1971) Spectra of some self-exciting and mutually exciting point processes. Biometrika 58:83–90

Heckmann JJ, Singer B (1984) Econometrics duration analysis. J Econom 24:63–132

Heinen A, Rengifo E (2007) Multivariate autoregressive modelling of time series count data using copulas. Empir Financ 14:564–583

Honoré BE (1990) Simple estimation of duration models with unobserved heterogeneity. Econometrica 58:453–473

Horowitz JL, Neumann GR (1987) Semiparametric estimation of employment duration models. Econometric Review 6(1):5–40

Kalbfleisch JD, Prentice RL (1980) The statistical analysis of failure time data. Wiley, New York

Kaplan EL, Meier P (1958) Nonparametric estimation from incomplete observations. J Am Stat Assoc 53:457–481

Karr AF (1991) Point processes and their statistical inference. Dekker, New York

Kiefer NM (1988) Economic duration data and hazard functions. J Econ Lit 26:646–679

Lancaster T (1979) Econometric methods for the duration of unemployment. Econometrica 47(4):939–956

Lancaster T (1997) The econometric analysis of transition data. Cambridge University Press

McCall BP (1996) Unemployment insurance rules, joblessness and part-time work. Econometrica 64(3):647–682

Meyer PA (1971) Démonstration simplifiée d'un théorème Knight. In Lecture Notes in Mathematics vol. 191. Springer, pp. 191–195

Meyer BD (1990) Unemployment insurance and unemployment spells. Econometrica 58(4):757–782

Moffitt R (1985) Unemployment insurance and the distribution of unemployment spells. J Econom 28:85–101

Neumann G (1997) Search models and duration data. In: Pesaran MH (ed) Handbook of Applied Econometrics, chap. 7. Basil Blackwell, Oxford, pp 300–351

Nickell S (1979) Estimating the probability of leaving unemployment. Econometrica 47(5):s 1249–1266

Oakes D (2001) Biometrika centenary: survival analysis. Biometrika 88:99–142

Orbe J, Ferreira E, Nunez-Anton V (2002) Length of time spent in chapter 11 bankruptcy: a censored partial regression model. Appl Econ 34:1949–1957

Ridder G (1990) The non-parametric identification of generalized accelerated failure-time models. Rev Econ Stud 57:167–182

Snyder DL, Miller MI (1991) Random point processes and time and space, 2nd edn. Springer, New York,

Sueyoshi GT (1995) A class of binary response models for grouped duration data. J Appl Econom 10:411–431

van den Berg GJ, B. van der Klaauw (2001) Combining micro and macro unemployment duration data. J Econom 102:271–309

Yashin A, Arjas E (1988) A note on random intensities and conditional survival functions. J Appl Probab 25:630–635

Chapter 5
Univariate Multiplicative Error Models

The terminology *multiplicative error model* (MEM) has been introduced by Engle (2002b) for a general class of time series models for positive-valued random variables which are decomposed into the product of their conditional mean and a positive-valued error term. Such models might be alternatively classified as *autoregressive conditional mean* models where the conditional mean of a distribution is assumed to follow a stochastic process. The idea of a MEM is well known in financial econometrics and originates from the structure of the *autoregressive conditional heteroscedasticity* (ARCH) model introduced by Engle (1982) or the *stochastic volatility* (SV) model proposed by Taylor (1982) where the conditional *variance* is dynamically parameterized and multiplicatively interacts with an innovation term. In high-frequency econometrics, a MEM has been firstly introduced by Engle and Russell (1997, 1998) to model the dynamic behavior of the time between trades and was referred to as *autoregressive conditional duration* (ACD) model. Thus, the ACD model is a special type of MEM applied to financial durations.

In this chapter, we discuss univariate MEMs. In order not to confuse the reader with different terminologies for the same model, we discuss the model in the context of financial durations and thus use the terminology *ACD* models in the following sections. Needless to say that the model can be directly applied to any other positive-valued (continuous) process, such as trading volumes, market depth, bid-ask spreads or the number of trades (if the latter are sufficiently continuous). For instance, in case of the modelling of trading volumes, Manganelli (2005) calls the corresponding model an autoregressive conditional volume (ACV) process. To avoid an inflation of terminology we, however, refrain from using these different labels.

Note that MEMs can be not only applied to different types of variables but also to different sampling schemes. In case of financial durations, the process is observed in event time and thus observations are irregularly spaced in time. Conversely, the process is defined in calendar time if a MEM is used to model aggregated trading activity based on (equi-distant) time intervals.

Since we discuss MEMs in the context of financial duration modelling, we also stress their role as a model for financial point processes and their link to concepts in point process theory as discussed in Chap. 4. In this context, we

N. Hautsch, *Econometrics of Financial High-Frequency Data*,
DOI 10.1007/978-3-642-21925-2_5, © Springer-Verlag Berlin Heidelberg 2012

illustrate the representation of an ACD model in terms of an intensity process. These interpretations, however, are not meaningful if other variables than durations are modelled.

In Sect. 5.1, we discuss autoregressive models for log durations as a natural starting point. In Sect. 5.2, we present the basic form of the autoregressive conditional duration (ACD) model proposed by Engle and Russell (1997, 1998). Because it is the most common type of autoregressive duration model and is extensively considered in recent econometrics literature, we discuss theoretical properties and estimation issues in more detail. Section 5.3 discusses (quasi) maximum likelihood estimation of ACD models. In Sect. 5.4, the inclusion of covariates and consideration of intraday periodicities is illustrated. A logarithmic specification of the ACD model – the so-called Log ACD model – is discussed in Sect. 5.5. Section 5.6 is devoted to specification tests for the ACD model. Here, we focus on Portmanteau tests, independence tests, distribution tests as well as Lagrange Multiplier and (integrated) conditional moment tests.

5.1 ARMA Models for Log Variables

A natural starting point for an autoregressive model for positive-valued variables $x_i > 0$ is to specify an (autoregression) model for log variables. Since log variables are not subject to non-negativity restrictions, traditional time series models are easily applicable. Accordingly, a simple ARMA model for log variables is given by

$$\ln x_i = \omega + \sum_{j=1}^{P} \alpha_j \ln x_{i-j} + \sum_{j=1}^{Q} \beta_j \tilde{\varepsilon}_{i-j} + \tilde{\varepsilon}_i, \quad i = 1, \ldots, n, \tag{5.1}$$

where $\tilde{\varepsilon}_i$ is a white noise variable. If x_i is a financial duration, the model belongs to the class of AFT models (see Chap. 4). Then, covariates, including, in this context, past durations, accelerate or decelerate the time to failure. A quasi maximum likelihood (QML) estimator for $\boldsymbol{\theta} = (\omega, \alpha, \beta)'$ is obtained by estimating the model under the normality assumption for $\tilde{\varepsilon}_i$, implying a conditional log normal distribution for x_i. Based on QML estimates, the empirical distribution of the residuals $\hat{\tilde{\varepsilon}}_i$ yields a nonparametric estimate of the underlying distribution, and thus, the baseline hazard.

More sophisticated specifications are obtained by ARMA-GARCH type models. For instance, the conditional mean function of $\ln x_i$ can be specified according to (5.1), while its conditional variance, h_i, follows a standard GARCH process. Thus,

$$\begin{aligned} \tilde{\varepsilon}_i &= \sqrt{h_i} u_i, \qquad u_i \sim \mathcal{N}(0,1), \\ h_i &= \omega_h + \sum_{j=1}^{P_h} \alpha_{h,j} \tilde{\varepsilon}_{i-j}^2 + \sum_{j=1}^{Q_h} \beta_{h,j} h_{i-j}. \end{aligned} \tag{5.2}$$

In case of trade durations or volume durations, h_i is interpreted as a duration volatility which is economically associated with liquidity risk (see, for example, Ghysels et al. 1998). By exploiting the asymptotic properties of the QML estimator of the Gaussian GARCH model (see Bollerslev and Wooldridge 1992), the autoregressive parameters of (5.1) and (5.2) are estimated consistently.[1]

The separability of the conditional mean and the conditional variance of $\ln x_i$ is obviously implied by the normality assumption. However, such a separation of the two first moments is not straightforward for plain variables. In general, distributions defined on positive support typically imply a strict dependence between the first moment and higher order moments and do not allow to disentangle the conditional mean and variance function. Then, a parameterization of the conditional mean implies per (distributional) assumption also a parameterization of higher order conditional moments. Ghysels et al. (1998) argue that such distributional assumptions are too restrictive and are not flexible enough to model duration dynamics. For a deeper discussion of this issue, see Chap. 6.

However, researchers are often not interested in models (and forecasts) of $\ln x_i$ but of x_i. Dufour and Engle (2000) illustrate that the forecast performance of autoregressions in log-durations perform rather poorly compared to more sophisticated ACD specifications as presented in the following subsection. Thus, an alternative specification is given by a simple ARMA model for x_i where the innovations follow a distribution defined on positive support:

$$x_i = \omega + \sum_{j=1}^{P} \alpha_j x_{i-j} + \sum_{j=1}^{Q} \beta_j \tilde{\varepsilon}_{i-j} + \tilde{\varepsilon}_i, \tag{5.3}$$

where $\omega > 0, \alpha_j \geq 0, \beta_j \geq 0$. A QML estimator for $\boldsymbol{\theta} = (\omega, \alpha, \beta)'$ is obtained by assuming a standard exponential distribution for $\tilde{\varepsilon}_i$ leading to the quasi maximum likelihood function based on data $\mathbf{X}$ and parameters $\boldsymbol{\theta}$

$$\ln \mathcal{L}_{QML}(\mathbf{X}; \boldsymbol{\theta}) = -\sum_{i=1}^{n} \tilde{\varepsilon}_i = -\sum_{i=1}^{n} \left[x_i - \omega - \sum_{j=1}^{P} \alpha_j x_{i-j} - \sum_{j=1}^{Q} \beta_j \tilde{\varepsilon}_{i-j} \right], \tag{5.4}$$

which is the true log likelihood if the p.d.f. of $\tilde{\varepsilon}_i$ is the exponential density. According to QML theory (see, e.g., Gouriéroux et al. 1984), correct specifications of the conditional mean function ensure consistent estimation of $\boldsymbol{\theta}$.

A drawback of this approach is that in this case, the marginal distribution of the resulting duration process is not exponential. Thus, in difference to Gaussian ARMA models, the relationship between the conditional distribution and the marginal

[1]For more details, see Sect. 5.3.1, where the asymptotic properties of the GARCH QML estimator are carried over to ACD models.

distribution of x_i is not obvious. Lawrence and Lewis (1980) propose an exponential ARMA (EARMA) model which is based on i.i.d. exponential innovations and leads to an exponential marginal distribution. This result is achieved by specifying a linear autoregressive model for a stationary variable that is based on a probabilistic choice between different linear combinations of independent exponentially distributed random variables.

5.2 A MEM for Durations: The ACD Model

The most popular autoregressive duration approach and most common type of MEM is proposed by Engle (1996, 2000) and Engle and Russell (1997, 1998). The basic idea of the autoregressive conditional duration (ACD) model is a dynamic parameterization of the conditional mean function

$$\Psi_i := \Psi_i(\boldsymbol{\theta}) = \mathbb{E}[x_i | \mathcal{F}_{t_{i-1}}; \boldsymbol{\theta}], \tag{5.5}$$

where $\boldsymbol{\theta}$ denotes a $p \times 1$ parameter vector. Here, $\mathcal{F}_{t_{i-1}}$ denotes the information set up to the most recent (irregularly spaced) observation t_{i-1}. As long as the exact timing of the observation at t_i is not important, it is sufficient to index all observations by i and to denote the information set by $\mathcal{F}_i$.

It is assumed that the standardized durations

$$\varepsilon_i := \frac{x_i}{\Psi_i} \tag{5.6}$$

follow an i.i.d. process defined on positive support with $\mathbb{E}[\varepsilon_i] = 1$. Obviously, the ACD model can be regarded as a GARCH model for duration data. Different types of ACD models can be divided either by the choice of the functional form used for the conditional mean function Ψ_i or by the choice of the distribution for ε_i.

The basic ACD specification is based on a linear parameterization of the conditional mean function

$$\Psi_i = \omega + \sum_{j=1}^{P} \alpha_j x_{i-j} + \sum_{j=1}^{Q} \beta_j \Psi_{i-j}. \tag{5.7}$$

A sufficient, however not necessary, condition for non-negativity of the process is $\omega > 0$ and $\alpha \geq 0$, $\beta \geq 0$.

The ACD process can be rewritten in terms of an intensity representation

$$\lambda(t; \mathcal{F}_t) = \tilde{\lambda}_{\varepsilon} \left(\frac{x(t)}{\Psi_{\tilde{N}(t)+1}} \right) \frac{1}{\Psi_{\tilde{N}(t)+1}}, \tag{5.8}$$

where $\tilde{\lambda}_\varepsilon(s)$ denotes the hazard function of the ACD innovation ε_i. This representation shows that the ACD model belongs to the class of AFT models since past dynamics influence the rate of failure time. Changes of the intensity function during a spell are only induced by the hazard shape of ε_i while new information enters the model exclusively at the particular points t_i. As suggested by Hamilton and Jorda (2002), (5.8) can be used as the starting point for generalized specifications by directly parameterizing the intensity function and allowing for news arrival within a spell. As soon as time-varying covariates are taken into account, such specifications require to switch from a duration setting to an intensity setting. This is discussed in more detail in Chap. 11.

The basic idea of the ACD model is to (dynamically) parameterize the conditional duration mean rather than the intensity function itself. Thus, the complete dynamic structure as well as the influence of covariates is captured by the function Ψ_i which per construction can only be updated at the points t_i.

The conditional mean of the ACD model is given by definition as $\mathbb{E}[x_i|\mathcal{F}_{i-1}] = \Psi_i$, whereas the unconditional mean and the conditional variance are

$$\mathbb{E}[x_i] = \mathbb{E}[\Psi_i] \cdot \mathbb{E}[\varepsilon_i] = \frac{\omega}{1 - \sum_{j=1}^{P} \alpha_j - \sum_{j=1}^{Q} \beta_j}, \tag{5.9}$$

$$\mathbb{V}[x_i|\mathcal{F}_{i-1}] = \Psi_i^2 \cdot \mathbb{V}[\varepsilon_i]. \tag{5.10}$$

For the case $P = Q = 1$, the unconditional variance of x_i is given by[2]

$$\mathbb{V}[x_i] = \mathbb{E}[x_i]^2 \mathbb{V}[\varepsilon_i] \left[\frac{1 - \beta^2 - 2\alpha\beta}{1 - (\alpha + \beta)^2 - \alpha^2 \mathbb{V}[\varepsilon_i]} \right]. \tag{5.11}$$

Correspondingly, the autocorrelation function is derived as

$$\rho_1 := \mathrm{Cov}(x_i, x_{i-1}) = \frac{\alpha_1(1 - \beta_1^2 - \alpha_1\beta_1)}{1 - \beta_1^2 - 2\alpha_1\beta_1}, \tag{5.12}$$

$$\rho_n = (\alpha + \beta)\rho_{n-1} \quad \text{for } n \geq 2. \tag{5.13}$$

Accordingly, covariance stationarity conditions of the ACD model are similar to the covariance stationarity conditions of the GARCH model (see Bollerslev 1986) and are satisfied by

$$(\alpha + \beta)^2 - \alpha^2 \mathbb{V}[\varepsilon_i] < 1.$$

The corresponding results for higher-order ACD models are similar but more cumbersome to compute. It is easy to see that $\mathbb{V}[x_i] > \mathbb{E}[x_i]^2$, thus, the ACD model implies excess dispersion, i.e., the unconditional standard deviation exceeds

[2] In the case $P = Q = 1$, we set $\alpha := \alpha_1$ and $\beta := \beta_1$.

the unconditional mean. This property might be regarded as the counterpart to the "overkurtosis property" of the Gaussian GARCH model. General formulations of lower and upper bounds for the p.d.f. of the duration process implied by an ACD (P, Q) model are given by Fernandes (2004).

By introducing the martingale difference $\eta_i := x_i - \Psi_i$, the ACD(P, Q) model can be written in terms of an ARMA(max(P,Q),Q) model for plain durations

$$x_i = \omega + \sum_{j=1}^{max(P,Q)} (\alpha_j + \beta_j)x_{i-j} - \sum_{j=1}^{Q} \beta_j \eta_{i-j} + \eta_i. \tag{5.14}$$

5.3 Estimation of the ACD Model

5.3.1 QML Estimation

A natural choice for the distribution of ε_i is the exponential distribution. As discussed in Chap. 4, the exponential distribution is the central distribution for stochastic processes defined on positive support and can be seen as the counterpart to the normal distribution for random variables defined on $\mathbb{R}$. Consequently, the specification of an Exponential-ACD (EACD) model is a natural starting point. Though the assumption of an exponential distribution is quite restrictive for many applications,[3] it has the major advantage of leading to a QML estimator for the ACD parameters. The exponential quasi log likelihood function is given by

$$\ln \mathcal{L}(\mathbf{X}; \boldsymbol{\theta}) = \sum_{i=1}^{n} l_i(\boldsymbol{\theta}) = -\sum_{i=1}^{n} \left[\ln \Psi_i + \frac{x_i}{\Psi_i} \right], \tag{5.15}$$

where $l_i(\boldsymbol{\theta})$ denotes the log likelihood contribution of the ith observation. The score and the Hessian are given by

$$\mathbf{s}(\boldsymbol{\theta}) := \frac{\partial \ln \mathcal{L}(\mathbf{X}; \boldsymbol{\theta})}{\partial \boldsymbol{\theta}} = \sum_{i=1}^{n} \mathbf{s}_i(\boldsymbol{\theta}) = \sum_{i=1}^{n} \frac{\partial \Psi_i}{\partial \boldsymbol{\theta}} \cdot \frac{1}{\Psi_i} \left[\frac{x_i}{\Psi_i} - 1 \right], \tag{5.16}$$

$$\begin{aligned} \mathcal{H}(\boldsymbol{\theta}) :&= \frac{\partial^2 \ln \mathcal{L}(\mathbf{X}; \boldsymbol{\theta})}{\partial \boldsymbol{\theta}\, \partial \boldsymbol{\theta}'} = \sum_{i=1}^{n} \mathbf{h}_i(\boldsymbol{\theta}) \\ &= \sum_{i=1}^{n} \left\{ \frac{\partial}{\partial \boldsymbol{\theta}'} \left[\frac{1}{\Psi_i} \frac{\partial \Psi_i}{\partial \boldsymbol{\theta}} \right] \left(\frac{x_i}{\Psi_i} - 1 \right) - \frac{1}{\Psi_i} \frac{\partial \Psi_i}{\partial \boldsymbol{\theta}} \frac{\partial \Psi_i}{\partial \boldsymbol{\theta}'} \frac{x_i}{\Psi_i^2} \right\}, \end{aligned} \tag{5.17}$$

[3] See also the descriptive statistics in Chap. 3.

where $\mathbf{s}_i(\boldsymbol{\theta})$ is a $p \times 1$ vector denoting the ith contribution to the score vector and $\mathbf{h}_i(\boldsymbol{\theta})$ is a $p \times p$ matrix denoting the ith contribution to the Hessian matrix. Under correct specification of the model, i.e., $\Psi_i = \Psi_{i,0}$, where $\Psi_{i,0} := \Psi_i(\boldsymbol{\theta}_0) = \mathbb{E}[x_i \,|\, \mathcal{F}_{i-1}; \boldsymbol{\theta}_0]$ denotes the "true" conditional mean function, it follows that $\varepsilon_i = x_i / \Psi_i$ is stochastically independent of Ψ_i and has an expectation of one. Hence, the score $\mathbf{s}_i(\boldsymbol{\theta})$ is a martingale difference with respect to the information set $\mathcal{F}_{i-1}$ and

$$\mathbb{E}[\mathcal{H}(\boldsymbol{\theta}_0)|\mathcal{F}_{i-1}] = \sum_{i=1}^{n} \tilde{\mathbf{h}}_i(\boldsymbol{\theta}_0) = -\sum_{i=1}^{n} \frac{1}{\Psi_{i,0}^2} \frac{\partial \Psi_{i,0}}{\partial \boldsymbol{\theta}} \frac{\partial \Psi_{i,0}}{\partial \boldsymbol{\theta}'}, \tag{5.18}$$

where $\tilde{\mathbf{h}}_i(\boldsymbol{\theta}) := \mathbb{E}[\mathbf{h}_i(\boldsymbol{\theta}) | \mathcal{F}_{i-1}]$.

The correct specification of the conditional mean function is an essential prerequisite to establish the QML property of the EACD estimator. Engle (2000) illustrates that the results of Bollerslev and Wooldridge (1992) can be directly applied to the EACD model. These results are summarized in the following theorem:

Theorem 5.1. *Assume the following regularity conditions:*

(i) $\boldsymbol{\Theta}$ *is a compact parameter space and has nonempty interior;* $\boldsymbol{\Theta}$ *is a subset of* $\mathbb{R}^p$.

(ii) For some $\boldsymbol{\theta}_0 \in int(\boldsymbol{\Theta})$, $\mathbb{E}[x_i \,|\, \mathcal{F}_{i-1}; \boldsymbol{\theta}_0] = \Psi_i(\boldsymbol{\theta}_0) := \Psi_{i,0}$.

(iii) (a) $\Psi_i(\boldsymbol{\theta}) := \Psi_i$ *is measurable for all* $\boldsymbol{\theta} \in \boldsymbol{\Theta}$ *and is twice continuously differentiable on int* $\boldsymbol{\Theta}$ *for all* x_i;
(b) Ψ_i *is positive with probability one for all* $\boldsymbol{\theta} \in \boldsymbol{\Theta}$.

(iv) (a) $\boldsymbol{\theta}_0$ *is the identifiable unique maximizer of* $n^{-1}\sum_{i=1}^{n} \mathbb{E}[l_i(\boldsymbol{\theta}) - l_i(\boldsymbol{\theta}_0)]$;
(b) $\{l_i(\boldsymbol{\theta}) - l_i(\boldsymbol{\theta}_0)\}$ *satisfies the UWLLN* $\forall\, i = 1, 2, \ldots, n$.[4]

(v) (a) $\{\mathbf{h}_i(\boldsymbol{\theta}_0)\}$ *and* $\{\tilde{\mathbf{h}}_i(\boldsymbol{\theta}_0)\}$ *satisfy the UWLLN;*
(b) $\{\mathbf{h}_i(\boldsymbol{\theta}) - \mathbf{h}_i(\boldsymbol{\theta}_0)\}$ *satisfies the UWLLN;*
(c) $\mathbf{A}^\circ := n^{-1}\sum_{i=1}^{n} \mathbb{E}[\tilde{\mathbf{h}}_i(\boldsymbol{\theta}_0)]$ *is uniformly positive definite.*

(vi) (a) $\{\mathbf{s}_i(\boldsymbol{\theta}_0)\mathbf{s}_i(\boldsymbol{\theta}_0)'\}$ *satisfies the UWLLN;*
(b) $\mathbf{B}^\circ := n^{-1}\sum_{i=1}^{n} \mathbb{E}[\mathbf{s}_i(\boldsymbol{\theta}_0)\mathbf{s}_i(\boldsymbol{\theta}_0)']$ *is uniformly positive definite;*
(c) $\mathbf{B}^{\circ -1/2} n^{-1/2} \sum_{i=1}^{n} \mathbf{s}_i(\boldsymbol{\theta}_0)\mathbf{s}_i(\boldsymbol{\theta}_0)' \xrightarrow{d} \mathcal{N}(\mathbf{0}, \mathbf{I}_p)$ *with* I_p *denoting the p-dimensional identity matrix.*

(vii) (a) $\{\tilde{\mathbf{h}}_i(\boldsymbol{\theta}) - \tilde{\mathbf{h}}_i(\boldsymbol{\theta}_0)\}$ *satisfies the UWLLN;*
(b) $\{\mathbf{s}_i(\boldsymbol{\theta})\mathbf{s}_i(\boldsymbol{\theta})' - \mathbf{s}_i(\boldsymbol{\theta}_0)\mathbf{s}_i(\boldsymbol{\theta}_0)'\}$ *satisfies the UWLLN.*

Then,

$$\left[\mathbf{A}^{\circ -1}\mathbf{B}^\circ\mathbf{A}^{\circ -1}\right]^{-1/2} \sqrt{n}(\hat{\boldsymbol{\theta}} - \boldsymbol{\theta}_0) \xrightarrow{d} \mathcal{N}(\mathbf{0}, \mathbf{I}_p).$$

---

[4]UWLLN: Uniform Weak Law of Large Numbers.

Furthermore,

$$\hat{\mathbf{A}}^\circ - \mathbf{A}^\circ \xrightarrow{p} \mathbf{0} \qquad \textit{and} \qquad \hat{\mathbf{B}}^\circ - \mathbf{B}^\circ \xrightarrow{p} \mathbf{0},$$

where

$$\hat{\mathbf{A}}^\circ = \frac{1}{n}\sum_{i=1}^{n} \tilde{\mathbf{h}}_i(\hat{\boldsymbol{\theta}}) \qquad \textit{and} \qquad \hat{\mathbf{B}}^\circ = \frac{1}{n}\sum_{i=1}^{n} \mathbf{s}_i(\hat{\boldsymbol{\theta}})\mathbf{s}_i(\hat{\boldsymbol{\theta}})'.$$

Proof: See Bollerslev and Wooldridge (1992), p. 167. □

This theorem illustrates that the maximization of the quasi log likelihood function (5.15) leads to a consistent estimate of $\boldsymbol{\theta}$ without specifying the density function of the disturbances. As pointed out by Bollerslev and Wooldridge (1992), the matrix $\mathbf{A}^{\circ-1}\mathbf{B}^\circ\mathbf{A}^{\circ-1}$ is a consistent estimator of the White (1982) robust asymptotic variance covariance of $\sqrt{n}(\hat{\boldsymbol{\theta}} - \boldsymbol{\theta}_0)$. A variance covariance estimator that is robust not only against distributional misspecification but also against dynamic misspecification in the ACD errors is obtained by following Newey and West (1987) and estimating $\mathbf{B}^\circ$ by

$$\hat{\mathbf{B}}^\circ = \hat{\gamma}_0 + \sum_{j=1}^{J}\left(1 - \frac{j}{J+1}\right)(\hat{\gamma}_j + \hat{\gamma}_j'),$$

where

$$\hat{\gamma}_j := n^{-1}\sum_{i=j+1}^{n} \mathbf{s}_i(\hat{\boldsymbol{\theta}})\mathbf{s}_i(\hat{\boldsymbol{\theta}})'$$

and J denotes the exogenously given truncation lag order. The assumptions of Bollerslev and Wooldridge (1992) are quite strong since they require asymptotic normality of the score vector and uniform weak convergence of the likelihood and its second derivative. Moreover, as shown in Chap. 3, typical high-frequency duration processes (like, for example, trade duration processes) are highly persistent and nearly integrated. In the integrated case, the unconditional mean of Ψ_i is not finite leading to non-normal limiting distributions of the estimator. In order to cope with these problems, Lee and Hansen (1994) establish the asymptotic properties for the IGARCH QML estimator under weaker assumptions than in Bollerslev and Wooldridge (1992). Engle and Russell (1998) illustrate that these results are easily carried over to the EACD(1,1) case. The main results of Lee and Hansen (1994) are summarized as follows:

Theorem 5.2. *Assume the following regularity conditions:*

(i) $\boldsymbol{\theta}_0 = (\omega_0, \alpha_0, \beta_0)' \in int(\boldsymbol{\Theta})$.

(ii) $\mathbb{E}[x_i | \mathcal{F}_{i-1}; \boldsymbol{\theta}_0] := \Psi_i(\boldsymbol{\theta}_0) = \Psi_{i,0} = \omega_0 + \alpha_0 x_{i-1} + \beta_0 \Psi_{i-1}$.

(iii) $\varepsilon_i = x_i/\Psi_{i,0}$ *is strictly stationary and ergodic.*
(iv) ε_i *is non-degenerate.*
(v) $\mathbb{E}\left[\varepsilon_i^2 \middle| \mathcal{F}_{i-1}\right] < \infty$ *a.s.*
(vi) $\sup_i \mathbb{E}\left[\ln \beta_0 + \alpha_0 \varepsilon_i \middle| \mathcal{F}_{i-1}\right] < 0$ *a.s.*
(vii) $\ln \mathcal{L}(\mathbf{X};\boldsymbol{\theta}) = \sum_{i=1}^n l_i(\boldsymbol{\theta}) = -\sum_{i=1}^n \left[\ln \Psi_i + \frac{x_i}{\Psi_i}\right]$, *where* $\Psi_i = \omega + \alpha x_{i-1} + \beta \Psi_{i-1}$.
(viii) $\boldsymbol{\theta}_0$ *is the identifiable unique maximizer of* $n^{-1}\sum_{i=1}^n \mathbb{E}[l_i(\boldsymbol{\theta}) - l_i(\boldsymbol{\theta}_0)]$.

Then,

$$\left[\mathbf{A}^{\circ -1}\mathbf{B}^{\circ}\mathbf{A}^{\circ -1}\right]^{-1/2}\sqrt{n}(\hat{\boldsymbol{\theta}} - \boldsymbol{\theta}_0) \xrightarrow{d} \mathcal{N}(0, \mathbf{I}_p)$$

and

$$\hat{\mathbf{A}}^{\circ} - \mathbf{A}^{\circ} \xrightarrow{p} 0 \quad \textit{and} \quad \hat{\mathbf{B}}^{\circ} - \mathbf{B}^{\circ} \xrightarrow{p} 0.$$

Proof: See Lee and Hansen (1994), p. 47ff. □

These results are also valid in case of $\alpha + \beta = 1$. Moreover, the standardized durations are not necessarily assumed to follow an i.i.d. process. It is rather required that they are strictly stationary and ergodic. This property generalizes the results of Bollerslev and Wooldridge (1992) to a broader class of models, including, for example, the class of so-called semiparametric ACD models introduced by Drost and Werker (2004) (see Chap. 6). Note that these results are based on the linear EACD(1,1) model and cannot necessarily carried over to more general cases, like (nonlinear) EACD(P,Q) models. In any case, a crucial requirement for the QML estimation of the ACD model is the validity of the conditional mean restriction, i.e., the correct specification of the conditional mean function Ψ_i. This assumption will be explored in more detail in Sect. 5.6.

The consistency of the Exponential QML estimator is inherently related to the fact that its score function corresponds to an unconditional moment function with the property

$$\mathbb{E}[s_i(\boldsymbol{\theta}_0)] = \mathbf{0}, \quad \boldsymbol{\theta}_0 \in \boldsymbol{\Theta}, \quad i = 1, \ldots, n. \tag{5.19}$$

This allows to construct corresponding moment estimators. As discussed by Drost and Werker (2004) and Engle and Gallo (2006), this relationship can be alternatively achieved by choosing a Gamma distribution. From (5.16) it is easily seen that the Exponential QML score contribution of a single observation i can be written as

$$\mathbf{s}_i(\boldsymbol{\theta}) = (\varepsilon_i - 1)\frac{\partial}{\partial \boldsymbol{\theta}} \ln \Psi_i. \tag{5.20}$$

Then, assuming the ACD innovations ε_i to follow a normalized Gamma distributed, i.e., $\varepsilon_i \sim \mathcal{G}(m, m)$ with $\mathbb{E}[\varepsilon_i] = 1$ (see appendix A) and p.d.f.

$$f_m(\varepsilon_i) = \frac{\varepsilon_i^{m-1}\exp(-\varepsilon_i/m)}{m^m \Gamma(m)}, \qquad m > 0,$$

it is easily shown that the resulting score function can be written as

$$1 + \varepsilon_i \frac{\partial f_m(\varepsilon_i)}{\partial m} \frac{1}{f_m(\varepsilon_i)} = m(1 - \varepsilon_i), \tag{5.21}$$

which is proportional to (5.20) yielding the same underlying moment function (5.19). Consequently, the estimator based on a Gamma distribution is identical to the one obtained from an exponential quasi maximum likelihood procedure. Its main advantage is to allow for more flexibility due to an additional parameter m. However, note that the Gamma distribution for $m \neq 1$ is only defined for positive random variables whereas the exponential distribution (i.e., the case $m = 1$) is also defined for zero realizations. Hence, in cases, where zero or near-to-zero observations or present in the data, the maximization of a gamma log likelihood might cause numerical difficulties. As a further alternative, Brownlees et al. (2011) recommend estimating the model by the method of moments using optimal instruments. In particular, they show that the efficient moment estimator of $\boldsymbol{\theta}$ solves the moment conditions

$$\frac{1}{n}\sum_{i=1}^{n} \frac{\partial \Psi_i}{\partial \boldsymbol{\theta}} \Psi_i^{-1} \left(\frac{x_i}{\Psi_i} - 1 \right) = \mathbf{0} \tag{5.22}$$

with the asymptotic covariance given by

$$\mathrm{AV}(\hat{\boldsymbol{\theta}}) = \frac{\mathbb{V}[\varepsilon_i]}{n} \left\{ \lim_{n\to\infty} \frac{1}{n} \sum_{i=1}^{n} \mathrm{E}\left[\frac{\partial \Psi_i}{\partial \boldsymbol{\theta}} \frac{\partial \Psi_i}{\partial \boldsymbol{\theta}'} \Psi_i^{-2} \right] \right\}^{-1}, \tag{5.23}$$

which is consistently estimated by

$$\widehat{\mathrm{AV}}(\hat{\boldsymbol{\theta}}) = \hat{\mathbb{V}}[\varepsilon_i] \left[\sum_{i=1}^{n} \frac{\partial \hat{\Psi}_i}{\partial \boldsymbol{\theta}} \frac{\partial \hat{\Psi}_i}{\partial \boldsymbol{\theta}'} \hat{\Psi}_i^{-2} \right]^{-1},$$

$$\hat{\mathbb{V}}[\varepsilon_i] = \frac{1}{n} \sum_{i=1}^{n} (x_i/\hat{\Psi}_i - 1)^2.$$

As illustrated by Engle and Russell (1998), an important implication of the strong analogy between the Gaussian GARCH model and the Exponential ACD model is that the ACD model can be estimated by GARCH software. In particular, the ACD parameter vector $\boldsymbol{\theta}$ can be estimated by taking $\sqrt{x_i}$ as the dependent variable in a GARCH regression where the conditional mean is set to zero. Therefore, given the QML properties of the EACD model, the parameter estimates $\hat{\boldsymbol{\theta}}$ are consistent, however not necessarily efficient.

Indeed, summary statistics of financial durations (see Chap. 3) show that the assumption of an exponential distribution for the standardized durations typically does not hold. Thus, QML parameter estimates can be biased in finite samples. Grammig and Maurer (2000) analyze the performance of different ACD specifications based on Monte Carlo studies and show that the QML estimation of these models may perform poorly in finite samples, even in quite large samples such as 15,000 observations. For this reason, they propose to specify the ACD model based on more general distributions. Bauwens et al. (2004) investigate the predictive performance of ACD models by using density forecasts. They illustrate that the predictions can be significantly improved by allowing for more flexible distributions. In these cases, the ACD model is not estimated by QML but by standard ML.

5.3.2 ML Estimation

In duration literature, a standard way to obtain more flexible distributions is to use mixture models. In this framework, a specific parametric family of distributions is mixed with respect to a heterogeneity variable leading to a mixture distribution.[5] The most common mixture model is obtained by multiplying the integrated hazard rate by a random heterogeneity term. In most applications, a gamma distributed random variable is used leading to a mixture model which is analytically tractable and allows for the derivation of simple results.

Below, we give a small classification over the most common types of mixture models leading to a flexible family of distributions. Consider a Weibull distribution which can be written in the form

$$\frac{x}{\lambda} = \frac{u}{a}, \quad a > 0,$$

where λ and a are the scale and shape parameter of the distribution, respectively, and u is a random variable which follows an unit exponential distribution. Note that in this case, $(x/\lambda)^a$ is the integrated hazard function. A gamma mixture of Weibull distributions is obtained by multiplying the integrated hazard by a random heterogeneity term following a gamma distribution $v \sim \mathcal{G}(\eta, \eta)$ with mean $\mathbb{E}[v] = 1$ and variance $\mathbb{V}[v] = \eta^{-1}$. Then, the resulting mixture model follows a Burr distribution with parameters λ, a and η.

More flexible models are obtained by a generalization of the underlying Weibull model and by assuming that $u \sim \mathcal{G}(m, m)$ leading to a duration model based on the

[5] For an overview of mixture distributions, see, e.g., Lancaster (1997).

generalized gamma distribution. The generalized gamma family of density functions nests the Weibull family when $m = 1$ and the (two-parameter) gamma distribution when $a = 1$.

Both types of distributions have been already successfully applied in the ACD framework leading to the Burr ACD model (Grammig and Maurer 2000) and the generalized gamma ACD model (Lunde 2000). However, both models belong to different distribution families and are not nested. An extension of the generalized gamma ACD model which also nests one member of the Burr family is based on a gamma mixture of the generalized gamma distribution leading to the generalized F distribution (see Kalbfleisch and Prentice 1980 or Lancaster 1997). The generalized F distribution is obtained by assuming the generalized gamma model as basic duration model and multiplying the integrated hazard by a gamma variate $v \sim \mathcal{G}(\eta, \eta)$. Then, the marginal density function of x is given by

$$f(x) = \frac{a x^{am-1}[\eta + (x/\lambda)^a]^{(-\eta-m)}\eta^\eta}{\lambda^{am}\mathcal{B}(m,\eta)}, \tag{5.24}$$

where $\mathcal{B}(\cdot)$ describes the complete Beta function with $\mathcal{B}(m,\eta) = \frac{\Gamma(m)\Gamma(\eta)}{\Gamma(m+\eta)}$. The moments of the generalized F distribution are given by

$$\mathbb{E}[x^s] = \lambda\eta^{1/a}\frac{\Gamma(m+s/a)\Gamma(\eta-s/a)}{\Gamma(m)\Gamma(\eta)}, \quad s < a\eta. \tag{5.25}$$

Hence, the generalized F ACD model as introduced by Hautsch (2003) is based on three parameters a, m and η, and thus, nests the generalized gamma ACD model for $\eta \to \infty$, the Weibull ACD model for $m = 1, \eta \to \infty$ and the log-logistic ACD model for $m = \eta = 1$. For a discussion of even more special cases of the Generalized F distribution, see Karanasos (2008). Figures 5.1–5.3 show the hazard functions implied by the generalized F distribution based on different parameter combinations. If an ACD process is specified based on a distribution with mean unequal to one (i.e., other than the exponential distribution), it can be without loss of generality written as

$$x_i = \Psi_i\zeta^{-1}\tilde{\varepsilon}_i =: \Phi_i\tilde{\varepsilon}_i, \tag{5.26}$$

where $\tilde{\varepsilon}_i$ denotes the error term with $\zeta := \mathbb{E}[\tilde{\varepsilon}_i] \neq 1$ and

$$\begin{aligned}\Phi_i := \Psi_i/\zeta &= \frac{\omega}{\zeta} + \frac{\alpha}{\zeta}x_{i-1} + \beta\Phi_{i-1}\\ &= \tilde{\omega} + \tilde{\alpha}x_{i-1} + \beta\Phi_{i-1}\end{aligned} \tag{5.27}$$

with $\tilde{\omega} := \omega/\zeta$ and $\tilde{\alpha} := \alpha/\zeta$.

Hence, to specify an ACD model, for instance, based on the generalized F distribution, λ and $\mathbb{E}[\Psi_i] = \omega/(1-\alpha-\beta)$ are not simultaneously identifiable.

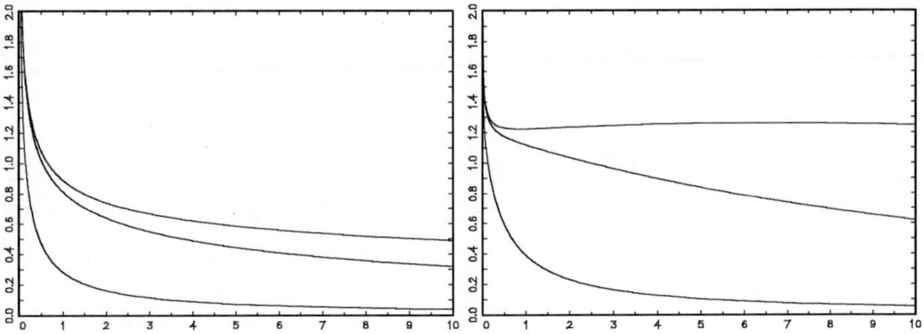

Fig. 5.1 Hazard functions implied by the generalized F distribution. *Left*: $m = 0.8$, $a = 0.8$, $\lambda = 1.0$, *upper*: $\eta = 100$, *middle*: $\eta = 10$, *lower*: $\eta = 0.5$. *Right*: $m = 0.8$, $a = 1.1$, $\lambda = 1.0$, *upper*: $\eta = 100$, *middle*: $\eta = 10$, *lower*: $\eta = 0.5$

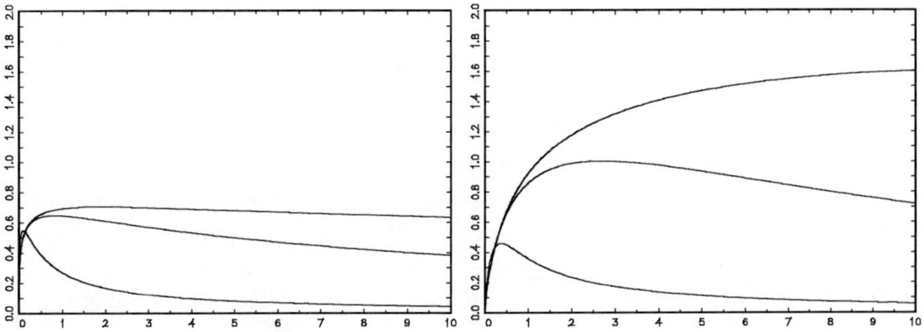

Fig. 5.2 Hazard functions implied by the generalized F distribution. *Left*: $m = 1.4$, $a = 0.9$, $\lambda = 1.0$, *upper*: $\eta = 100$, *middle*: $\eta = 10$, *lower*: $\eta = 0.5$. *Right*: $m = 1.4$, $a = 1.2$, $\lambda = 1.0$, *upper*: $\eta = 100$, *middle*: $\eta = 10$, *lower*: $\eta = 0.5$

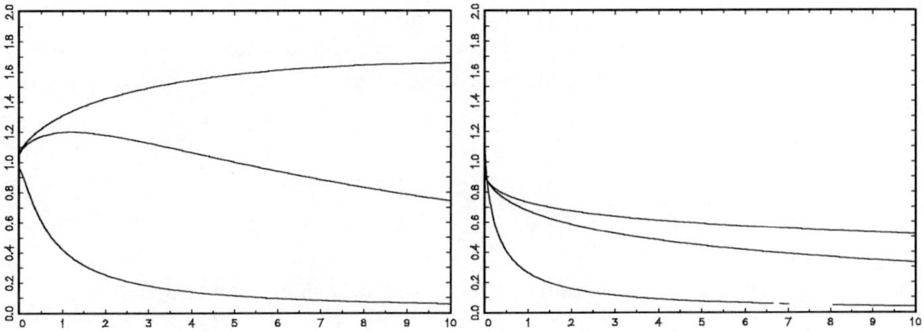

Fig. 5.3 Hazard functions implied by the generalized F distribution. *Left*: $m = 1.2^{-1}$, $a = 1.2$, $\lambda = 1.0$, *upper*: $\eta = 100$, *middle*: $\eta = 10$, *lower*: $\eta = 0.5$. *Right*: $m = 1.2$, $a = 1.2^{-1}$, $\lambda = 1.0$, *upper*: $\eta = 100$, *middle*: $\eta = 10$, *lower*: $\eta = 0.5$

Therefore, either λ or ω have to be fixed. It is more natural to fix $\lambda = 1$ and to choose ζ as

$$\zeta = \frac{\eta^{1/a}\Gamma(m+1/a)\Gamma(\eta-1/a)}{\Gamma(m)\Gamma(\eta)}.$$

The generalized F ACD specification is given by (5.26) with $\tilde{\varepsilon}_i$ being generalized F distributed with $\lambda = 1$. Then, applying the change of variables theorem the density of x_i and thus the corresponding log likelihood function is given by

$$\ln \mathcal{L}(\mathbf{X};\boldsymbol{\theta}) = \sum_{i=1}^{n} \ln \frac{\Gamma(m+\eta)}{\Gamma(m)\Gamma(\eta)} + \log a - am \log \Phi_i + (am-1)\ln x_i \\ - (\eta+m)\ln\left[\eta + (\Phi_i^{-1}x_i)^a\right] + \eta \ln(\eta). \tag{5.28}$$

This specification is equivalent to setting $\lambda = \Phi$ and dynamically parameterizing it according to (5.27) by setting $\Psi_i = \Phi_i\zeta$. Alternative distributions employed for ACD models are the Pareto distribution (see Luca and Zuccolotto 2006) or generalizations of the generalized F distribution (see Karanasos 2004). Note that any generalizations of the exponential distribution require the strict positiveness of x_i. Hence, in case of zero observations, the model has to be estimated either by QML based on the exponential distribution, by (G)MM or by ML employing sufficiently flexible distributions which explicitly account for zero observations. In this context, Hautsch et al. (2010) propose a zero-augmented generalized F distribution and implement it in a MEM framework.

Alternatively, even more flexible distributions for ACD models are obtained by mixture distributions with time-varying weights. De Luca and Gallo (2009) propose a mixture of exponentials with time-varying weights and show that the resulting distribution successfully captures also the occurrence of duration observations in the tails.

Finally, besides the family of gamma mixture distributions, another natural distribution for positive-valued random variables is the log-normal distribution with p.d.f.

$$f(x) = \frac{1}{x\sigma\sqrt{2\pi}} \exp\left(-\frac{1}{2}\left(\frac{\ln x - \mu}{\sigma}\right)^2\right), \tag{5.29}$$

where μ and σ denote the corresponding location and scale parameters, respectively, implying $\mathbb{E}[x] = \exp\left(\mu + \sigma^2/2\right)$ and $\mathbb{V}[x] = \exp\left(2\mu + \sigma^2\right)\left(\exp(\sigma^2) - 1\right)$. As shown in Sect. 5.5, specifying an ACD model based on a log-normal distribution is particularly useful in cases where the conditional mean function follows a logarithmic specification as well and allows to carry over Gaussian QML properties.

5.4 Seasonalities and Explanatory Variables

As illustrated in Chap. 3, financial duration processes are typically subject to strong intraday periodicities. One common solution in the ACD framework is to generate seasonally adjusted series by partialling out the time-of-day effects. In this context, the durations are decomposed into a deterministic and stochastic component. Engle and Russell (1998) assume that deterministic seasonality effects act multiplicatively, thus

$$x_i = \breve{x}_i s_i, \tag{5.30}$$

where $\breve{x}_i$ denotes the seasonal adjusted duration and s_i the seasonality component at i. Then, the conditional mean is given by

$$\Psi_i = \mathbb{E}[\breve{x}_i \,|\, \mathcal{F}_{i-1}]\, s_i =: \breve{\Psi}_i s_i. \tag{5.31}$$

The deterministic seasonality function s_i can be specified in different ways. Obviously, the most simple way to account for seasonalities is to include appropriate dummy variables, i.e.,

$$s_i = \sum_{j=1}^{Q} \delta_j \mathbb{1}_{\tau_{j-1} \leq t_i < \tau_j}, \tag{5.32}$$

where δ_j are parameters to be estimated and τ_j, $j = 0, 1, \ldots, \tau_Q$, denote exogenously chosen (calendar) time points splitting the trading day into $Q + 1$ intervals (with τ_Q chosen *before* the end of the trading day to ensure identification of the constant ω in Ψ_i).

While seasonality dummies capture the deterministic pattern in terms of piecewise constant functions, a *linear spline function* allows for a piecewise linear function defined over the knots τ_j. It is given by

$$s_i = 1 + \sum_{j=1}^{Q} \delta_j (t_i - \tau_j) \mathbb{1}_{\{t_i > \tau_j\}} \tag{5.33}$$

and is normalized to one at the beginning of the trading day to ensure the identification of ω. The linear spline function can be extended to a quadratic or cubic spline function where the (piecewise) linear functions are replaced by second or third order polynomials, respectively. See, e.g., de Boor (1978).

An alternative and common specification is to use the flexible Fourier series approximation proposed by Andersen and Bollerslev (1998b) based on the work of Gallant (1981). Assuming a polynomial of degree Q, the non-stochastic seasonal trend term is of the form

$$s_i = s(\delta, \bar{t}_i, Q) = \delta \cdot \bar{t}_i + \sum_{j=1}^{Q} \left(\delta_{c,j} \cos(\bar{t}_i \cdot 2\pi j) + \delta_{s,j} \sin(\bar{t}_i \cdot 2\pi j)\right), \quad (5.34)$$

where δ, $\delta_{c,j}$, and $\delta_{s,j}$ are the seasonality coefficients to be estimated and $\bar{t}_i \in [0, 1]$ is the normalized calendar time associated with t_i computed as the number of seconds from opening of the exchange until the intraday calendar time of t_i divided by the length of the trading day.

In the high-frequency literature, it is common to apply a two-step estimation approach. In the first step, durations are seasonally filtered, whereas in the second step, the model is estimated based on seasonally standardized durations. The major advantage of this procedure is a significant reduction of the number of parameters and thus of the computational burden in the second step (Q)ML or GMM estimation. This is particularly important if the number of observations is high.

Nevertheless, two-step estimation is inefficient and in some cases even inconsistent. In the latter case, it is inevitable to jointly estimate all parameters. Veredas et al. (2008) propose a semiparametric estimator, where the seasonality components are jointly estimated non-parametrically with the parameters of the ACD model, and show that joint estimation leads to efficiency gains and improved forecasting properties. Brownlees and Gallo (2011) develop a shrinkage estimator to jointly estimate the ACD component and a flexibly specified deterministic component. The basic principle of the shrinkage estimator is to penalize the log likelihood by a penalty function which is quadratic in the number of parameters of the flexible component. This implies that these parameters are shrinked toward zero with the amount of penalization (and thus shrinkage) controlled by a smoothing parameter. Brownlees and Gallo (2011) show that the shrinkage estimates yield improved forecasts compared to the case of basic ML estimation.

Explanatory variables can be included in two different ways. The first possibility is to include them in form of a function $g(\cdot)$ which enters the conditional mean function *additively*, i.e.,

$$\Psi_i = \omega + \sum_{j=1}^{P} \alpha_j x_{i-j} + \sum_{j=1}^{Q} \beta_j \Psi_{i-j} + g(\mathbf{z}_{i-1}'\gamma), \qquad \text{or} \quad (5.35)$$

$$\Psi_i - g(\mathbf{z}_{i-1}'\gamma) = \omega + \sum_{j=1}^{P} \alpha_j x_{i-j} + \sum_{j=1}^{Q} \beta_j (\Psi_{i-j} - g(\mathbf{z}_{i-1-j}'\gamma)). \quad (5.36)$$

In the most simple form, $g(\cdot)$ is just chosen as $g(y) = y$ implying an additive but not necessarily non-negative inclusion of covariates. Alternatively, $g(\cdot)$ might be chosen as a non-negative function, e.g., $g(y) = \exp(y)$ which ensures that covariates affect conditional durations in a non-negative, however, obviously non-linear fashion.

Specification (5.35) implies a dynamic inclusion of explanatory variables, i.e., in this specification, the covariate effects enter the ACD specification in terms of an infinite lag structure (see, e.g. Hendry 1995). In contrast, (5.36) implies a static inclusion of regressor effects. A priori it is unclear which specification should be preferred in practice. However, in some cases, a dynamic inclusion according to (5.35) makes the interpretation of estimates of γ difficult. This is the case whenever the regressors $\mathbf{z}$ are connected to certain time periods (for instance, to capture the effect of a news event occurring at a fixed time). In such a case, (5.36), is preferred.

Alternatively, explanatory variables might be included *multiplicatively* as an additional scaling function. Then, define $\breve{x}_i$ as the duration standardized by both seasonality and covariate effects. Thus

$$\breve{x}_i := \frac{x_i}{s_i g(\mathbf{z}'_{i-1}\gamma)} \tag{5.37}$$

and Ψ_i is given by

$$\Psi_i = \breve{\Psi}_i s_i g(\mathbf{z}'_{i-1}\gamma). \tag{5.38}$$

5.5 The Log-ACD Model

Bauwens and Giot (2000) and Lunde (2000) propose a logarithmic ACD model[6] that ensures the non-negativity of durations without any parameter restrictions and is obtained by $x_i = \Psi_i \varepsilon_i$ with

$$\begin{aligned} \ln \Psi_i &= \omega + \alpha \ln \varepsilon_{i-1} + \beta \ln \Psi_{i-1}, \\ &= \omega + \alpha \ln x_{i-1} + (\beta - \alpha) \ln \Psi_{i-1}, \end{aligned} \tag{5.39}$$

where ε_i is i.i.d. with mean one. In order to distinguish the model from an alternative logarithmic specification introduced below, we call the model in line with Bauwens and Giot (2000) Logarithmic ACD model of type I (LACD_1).

In contrast to the linear ACD model, the LACD model implies a concave relationship between ε_{i-1} and x_i (the so-called *news impact curve*). I.e., the difference in the impact of innovations with $\varepsilon_i < 1$ ("negative" surprises) on x_i is larger than in the case of innovations with $\varepsilon_i > 1$ ("positive" surprises).

Similarly to the linear ACD model presented in the previous sections, in case of innovation distributions with a mean non-equal to one, the process can be represented as

$$\begin{aligned} x_i &= \Psi_i \tilde{\varepsilon}_i / \zeta =: \Phi_i \tilde{\varepsilon}_i \\ \ln \Phi_i &= \tilde{\omega} + \alpha \ln \varepsilon_{i-1} + \beta \ln \Phi_{i-1}, \end{aligned}$$

[6] In some studies, this model is also called "Nelson type" ACD model since it resembles the EGARCH specification proposed by Nelson (1991).

where $\tilde{\omega} := \omega + (\beta - 1)\ln\zeta$ and $\varepsilon_i := \tilde{\varepsilon}_i/\zeta$ with $\mathbb{E}[\tilde{\varepsilon}_i] := \zeta \neq 1$.

Similarly to the basic ACD model, the Log-ACD model can be also represented by an ARMA specification. Allen et al. (2008) show that a Log-ACD model as given by (5.39) can be represented as an ARMA(R,R) specification for $\ln x_i$,

$$\ln x_i = \breve{\omega} + \sum_{j=1}^{R} \delta_j \ln x_{i-j} + \sum_{j=1}^{R} \theta_j \xi_{i-j} + \xi_i, \tag{5.40}$$

where $\xi_i := (\ln\varepsilon_i - \mathbb{E}[\ln\varepsilon_i]) \sim iid(0, \sigma_\xi^2)$, $\breve{\omega} := \omega + \sum_{j=1}^{R} \theta_j \mathbb{E}[\ln\varepsilon_i] + \mathbb{E}[\ln\varepsilon_i]$, and $R := \max(P, Q)$.

Bauwens and Giot (2000) also propose an alternative Log-ACD specification given by

$$\begin{aligned} \ln\Psi_i &= \omega + \alpha\varepsilon_{i-1} + \beta\ln\Psi_{i-1} \\ &= \omega + \alpha(x_{i-1}/\Psi_{i-1}) + \beta\ln\Psi_{i-1}, \end{aligned} \tag{5.41}$$

which implies a convex news impact curve and is referred to as Log ACD type II (LACD_2) model. Both Log-ACD specifications can be compactly written as

$$\ln\Psi_i = \omega + \sum_{j=1}^{P} \alpha_j g(\varepsilon_{i-j}) + \sum_{j=1}^{Q} \beta_j \ln\Psi_{i-j}, \tag{5.42}$$

where $g(\varepsilon_i) = \ln\varepsilon_i$ (type I) or $g(\varepsilon_i) = \varepsilon_i$ (type II). Bauwens et al. (2008) and Karanasos (2008) derive the unconditional moments for both Log ACD specifications. In the case $P = Q = 1$, the mth moment of x_i is given by

$$\mathbb{E}[x_i^m] = \mu_m \exp\left(\frac{m\omega}{1-\beta}\right) \prod_{j=1}^{\infty} \mathbb{E}\left[\exp\left\{m\alpha\beta^{j-1} g(\varepsilon_i)\right\}\right], \tag{5.43}$$

where m is an arbitrary positive integer m and the conditions $\mu_m = \mathbb{E}|x_i^m| < \infty$, $|\beta| < 1$ and $\mathbb{E}[\exp(m\alpha\beta^{j-1} g(\varepsilon_i))] < \infty$ have to be satisfied.

Correspondingly, under the conditions $|\beta| < 1$, $\mathbb{E}[\exp(2\alpha g(\varepsilon_i))] < \infty$, the autocorrelation function is given by

$$\begin{aligned} \rho_n = & \frac{\mu \mathbb{E}\left[\varepsilon_i e^{\alpha\beta^{n-1} g(\varepsilon_i)}\right] \prod_{j=1}^{n-1} \mathbb{E}\left[e^{\alpha\beta^{j-1} g(\varepsilon_i)}\right] \prod_{j=1}^{\infty} \mathbb{E}\left[e^{\alpha(1+\beta^n)\beta^{j-1} g(\varepsilon_i)}\right]}{\mu_2 \prod_{j=1}^{\infty} \mathbb{E}\left[e^{2\alpha\beta^{j-1} g(\varepsilon_i)}\right] - \mu^2 \left(\prod_{j=1}^{\infty} \mathbb{E}\left[e^{\alpha\beta^{j-1} g(\varepsilon_i)}\right]\right)^2} \\ & - \frac{\mu^2 \left(\prod_{j=1}^{\infty} \mathbb{E}\left[e^{\alpha\beta^{j-1} g(\varepsilon_i)}\right]\right)^2}{\mu_2 \prod_{j=1}^{\infty} \mathbb{E}\left[e^{2\alpha\beta^{j-1} g(\varepsilon_i)}\right] - \mu^2 \left(\prod_{j=1}^{\infty} \mathbb{E}\left[e^{\alpha\beta^{j-1} g(\varepsilon_i)}\right]\right)^2}. \end{aligned} \tag{5.44}$$

From these results it follows that the LACD model is covariance stationary if

$$\beta < 1, \quad \mathbb{E}[\varepsilon_i \exp\{\alpha g(\varepsilon_i)\}] < \infty, \quad \mathbb{E}[\exp\{2\alpha g(\varepsilon_i)\}] < \infty.$$

Generalizations of these results are given in Bauwens et al. (2008) and Karanasos (2008). In practice, it is suggested to truncate the infinite sums after a sufficiently large number.

As pointed out by Allen et al. (2008), the Exponential QML properties of the linear ACD model cannot be straightforwardly carried over to the Log-ACD case. Loosely speaking, the underlying reason is that in the Log-ACD case, the innovation ε_{i-1} affects Ψ_i and x_i in a nonlinear way which perishes the validity of the QML score function in case of distributional misspecification. This is similar to the result that the QML property of the Gaussian GARCH model can*not* be carried over to Nelson's (1991) Exponential GARCH.

As an alternative, Allen et al. (2008) propose estimating the LACD model using the log-normal distribution. Assuming $\ln x_i$ to be normally distributed with mean Ψ_i and variance σ^2, the corresponding log likelihood function is given by

$$\ln \mathcal{L}(\mathbf{X};\boldsymbol{\theta}) = -\frac{1}{2}\ln 2\pi - \frac{1}{2}\ln\sigma^2 - \ln x_i - \frac{1}{2}\frac{(\ln x_i - \ln\Psi_i)^2}{\sigma^2}. \tag{5.45}$$

Then, Allen et al. (2008) show the consistency and asymptotic normality of $\hat{\boldsymbol{\theta}}$,

$$\sqrt{n}(\hat{\boldsymbol{\theta}} - \boldsymbol{\theta}_0) \xrightarrow{d} \mathcal{N}(\mathbf{0}, \mathbf{A}(\boldsymbol{\theta}_0)^{-1}\mathbf{B}(\boldsymbol{\theta}_0)\mathbf{A}(\boldsymbol{\theta}_0)^{-1}), \tag{5.46}$$

where

$$\mathbf{A}(\boldsymbol{\theta}_0) = \lim_{n\to\infty} n^{-1}\mathbb{E}[\mathcal{H}(\boldsymbol{\theta})]\Big|_{\boldsymbol{\theta}=\boldsymbol{\theta}_0} = \lim_{n\to\infty} n^{-1}\mathbb{E}\left[\frac{\partial^2 \ln\mathcal{L}(\mathbf{X};\boldsymbol{\theta})}{\partial\boldsymbol{\theta}\,\partial\boldsymbol{\theta}'}\right]\Bigg|_{\boldsymbol{\theta}=\boldsymbol{\theta}_0}, \tag{5.47}$$

$$\begin{aligned}\mathbf{B}(\boldsymbol{\theta}_0) &= \lim_{n\to\infty} n^{-1}\,\mathbb{E}\left[\mathbf{s}(\boldsymbol{\theta})\mathbf{s}(\boldsymbol{\theta})'\right]\big|_{\boldsymbol{\theta}=\boldsymbol{\theta}_0} \\ &= \lim_{n\to\infty} n^{-1}\mathbb{E}\left[\frac{\partial \ln\mathcal{L}(\mathbf{X};\boldsymbol{\theta})}{\partial\boldsymbol{\theta}}\frac{\partial \ln\mathcal{L}(\mathbf{X};\boldsymbol{\theta})}{\partial\boldsymbol{\theta}'}\right]\Bigg|_{\boldsymbol{\theta}=\boldsymbol{\theta}_0}.\end{aligned} \tag{5.48}$$

An important advantage of the LACD model is that Ψ_i is straightforwardly extended by covariates without violating the non-negativity restrictions. In this context, the discussion of Sect. 5.4 applies.

5.6 Testing the ACD Model

In this section, we show several procedures to test the ACD model. Section 5.6.1 illustrates Portmanteau tests for ACD residuals. Here, we discuss the classical Box–Pierce and Ljung–Box tests as well as refinements thereof accounting for the peculiar features of ACD residuals. Section 5.6.2 discusses independence tests, such as the BDS tests proposed by Brock et al. (1996) as well as Hong and Lee's (2003) omnibus test based on generalized spectral densities. In Sect. 5.6.3, we focus on explicit tests on the distribution of ACD residuals. Besides density evaluations using Rosenblatt's (1952) probability integral transformations, we focus on non-parametric tests against distributional misspecification as proposed by Fernandes and Grammig (2005) based on Aït-Sahalia (1996). Section 5.6.4 concentrates on Lagrange Multiplier (LM) tests on correct functional form of the conditional mean specification. These tests have optimal power against local alternatives but only low power against more general alternatives. Therefore, Sect. 5.6.5 discusses conditional moment (CM) tests as originally introduced by Newey (1985). These tests are consistent against a finite number of possible alternatives since they rely on a finite number of conditional moment restrictions. In Sect. 5.6.5.2 we illustrate the use of integrated conditional moment (ICM) tests proposed by Bierens (1982, 1990). By employing an infinite number of conditional moments, this test possesses the property of consistency against all possible alternatives, and thus, is a generalization of the CM test. Finally, in Sect. 5.6.6, we present a Monte Carlo study where we analyze the size and the power of different forms of LM, CM and ICM tests on the basis of various types of augmented ACD models as data generating processes.

5.6.1 Portmanteau Tests

One obvious way to evaluate the goodness-of-fit of the ACD model is to investigate the dynamic and distributional properties of the ACD residuals

$$e_i := \hat{\varepsilon}_i = x_i / \hat{\Psi}_i, \quad i = 1, \ldots, n. \tag{5.49}$$

Under correct model specification, the series must be i.i.d. Hence, Portmanteau statistics as proposed by Box and Pierce (1970) and Ljung and Box (1978) based on the ACD residuals can be used to analyze whether the specification is able to account for the inter-temporal dependence in the duration process. In particular, if the residuals e_i are i.i.d., the Box–Pierce statistic is given by

$$Q_{BP}(k) = n \sum_{j=1}^{k} \hat{\rho}_j^2 \overset{a}{\sim} \chi^2_{k-s}, \tag{5.50}$$

where s denotes the number of autoregressive parameters to be estimated underlying e_i, i.e. $s := \max\{P, Q\}$ in case of an ACD model, and $\hat{\rho}_j$ denotes the jth lag sample autocorrelation

$$\hat{\rho}_j = \frac{\sum_{i=1}^{n-j}(e_i - \bar{e})(e_{i+j} - \bar{e})}{\sum_{i=1}^{n}(e_i - \bar{e})^2}, \tag{5.51}$$

where $\bar{e}$ is the mean of e_i which equals one under correct specification. To improve the poor finite sample properties of the Box–Pierce test when the sample size ranges from small to moderate, Ljung and Box (1978) suggest using the statistic

$$Q_{LB}(k) = n(n+2)\sum_{j=1}^{k} \frac{\hat{\rho}_j^2}{n-j} \overset{a}{\sim} \chi^2_{k-s}. \tag{5.52}$$

Both the Box–Pierce and Ljung–Box test statistics rely on the result that if the underlying series is normally distributed with a (true) mean zero, the asymptotic distribution of ρ_j is normal with mean zero and variance $(n-j)/(n(n+2))$ where the latter expression is approximated by n^{-1} if n is large.

Various studies analyze the effects of an unknown, possibly non-zero mean and deviations from normality. Dufour and Roy (1985) show that in case of an unknown mean, the finite sample performance of Q_{LB} can be improved if ρ_j is not standardized by $(n-j)/(n(n+2))$ but by its exact first and second moments. Kwan et al. (2005) illustrate that the standard Box–Pierce and Ljung–Box tests can be clearly undersized if the underlying data generating process is skewed. In such a situation, a non-parametric Portmanteau test as proposed by Dufour and Roy (1986) relying on rank autocorrelations provides a better performance in finite samples. Further modifications of Portmanteau statistics based on variance-stabilizing transformations have been proposed by Kwan and Sim, (1996a,b).

As stressed by Pacurar (2008), the asymptotic χ^2 distributions which are derived for Box–Pierce or Ljung–Box tests based on residuals arising from an ARMA model cannot be directly carried over to ACD residuals. In fact, for the case of an Exponential ACD model, Li and Yu (2003) show that the asymptotic distribution of $\sqrt{n}\hat{\boldsymbol{\rho}}$ with $\hat{\boldsymbol{\rho}} := (\hat{\rho}_1, \ldots, \hat{\rho}_k)'$ and

$$\hat{\rho}_j = \frac{\sum_{i=1}^{n-j}(e_i - 1)(e_{i+j} - 1)}{\sum_{i=1}^{n}(e_i - 1)^2},$$

or, alternatively (as $\sum_{i=1}^{n}(e_i-1)^2 \overset{p}{\to} 1$)

$$\hat{\rho}_j = \sum_{i=1}^{n-j}(e_i - 1)(e_{i+j} - 1), \tag{5.53}$$

is multivariate normal with mean zero and covariance matrix $\mathbf{I}_k - \mathbf{Z}\mathbf{H}^{-1}\mathbf{Z}'$ with $\mathbf{I}_k$ denoting the k-dimensional identity matrix, $\mathbf{H} := -\mathbb{E}\left[n^{-1}\mathcal{H}(\boldsymbol{\theta})\right]$, $\ln \mathcal{L}(\mathbf{X};\boldsymbol{\theta}) = -\sum_{i=1}^{n}(\ln \Psi_i + \frac{x_i}{\Psi_i})$ and

$$\mathbf{Z} := \begin{bmatrix} \frac{1}{n}\sum_{i=2}^{n}\frac{x_i}{\Psi_i^2}(e_{i-1}-1) & \frac{1}{n}\sum_{i=2}^{n}\frac{x_i x_{i-1}}{\Psi_i^2}(e_{i-1}-1) \\ \vdots & \vdots \\ \frac{1}{n}\sum_{i=k+1}^{n}\frac{x_i}{\Psi_i^2}(e_{i-k}-1) & \frac{1}{n}\sum_{i=k+1}^{n}\frac{x_i x_{i-1}}{\Psi_i^2}(e_{i-k}-1) \end{bmatrix}_{k\times 2}. \tag{5.54}$$

Then, the statistic is

$$Q_{LY}(k) = n\hat{\boldsymbol{\rho}}'(\mathbf{I}_k - \mathbf{Z}\mathbf{H}^{-1}\mathbf{Z}')^{-1}\hat{\boldsymbol{\rho}} \overset{a}{\sim} \chi^2_k. \tag{5.55}$$

This result explicitly holds for the case of an underlying exponential distribution but can be easily generalized, e.g., to the case of a Weibull distribution. A similar finding has been shown by Li and Mak (1994) for Portmanteau tests for squared standardized residuals in a GARCH model.

Hence, Box–Pierce and Ljung–Box tests can be still used as approximative tests indicating the model's dynamic fit to perform (relative) model comparisons. However, whenever one is interested in possibly exact inference on the dynamic properties of ACD residuals, modified Portmanteau tests as discussed above should be used.

Note that the independence of e_i is tested only based on the first k autocorrelations implying testing against the specific alternative hypothesis $H_1 : \rho_j \neq 0$ for $1 \leq j \leq k$. However, Box–Pierce and Ljung–Box tests are straightforwardly applicable to test not only for correlations but also for dependencies in higher order moments, e.g., in *squared* ACD residuals. An application of the Ljung–Box test based on squared residuals was proposed by McLeod and Li (1983) and is commonly referred to as McLeod–Li test.

5.6.2 Independence Tests

While the Portmanteau tests discussed above can only detect autocorrelations in the employed series but do not automatically detect dependencies in higher order moments, they are not applicable to explicitly test for the independence of ACD residuals. Brock et al. (1996) propose a nonparametric test which has power against many alternatives to i.i.d. processes, including chaotic effects. The major idea is to evaluate the "nearness" of m-period histories of residuals. In particular, define

$$\mathbb{1}(e_i, e_j, d) := \begin{cases} 1 & \text{if} |e_i - e_j| < d, \\ 0 & \text{otherwise,} \end{cases} \tag{5.56}$$

and

$$\mathbb{1}_m(e_i, e_j, d) := \prod_{k=0}^{m-1} \mathbb{1}(e_{i+k}, e_{j+k}, d), \tag{5.57}$$

where d is some distance. Hence, $\mathbb{1}_m(\cdot)$ is one whenever the two m-period histories $\{e_i, e_{i+1}, \ldots, e_{i+m-1}\}$ and $\{e_j, e_{j+1}, \ldots, e_{j+m-1}\}$ are near each other in the sense that for each term, $|e_{i+k} - e_{j+k}| < d$. The proportion of m-period histories that are near each other is estimated by the correlation integral

$$C_m(n,d) := \frac{2}{(n-m)(n-m+1)} \sum_{i=1}^{n-m} \sum_{j=i+1}^{n-m+1} \mathbb{1}_m(x_i, x_j, d) \tag{5.58}$$

with limit

$$C_m(d) := \operatorname*{plim}_{n\to\infty} C_m(n,d).$$

If the observations are i.i.d., then $C_m(d) = C_1(d)^m$. Conversely, if the observations are from a chaotic process, then $C_m(d) > C_1(d)^m$. Exploiting the asymptotic distribution of the correlation integral, Brock et al. (1996) construct a test statistic as

$$BDS(m,d) := \sqrt{n}\frac{C_m(d) - C_1(d)^m}{\hat{V}_m^{1/2}} \xrightarrow{d} \mathcal{N}(0,1), \tag{5.59}$$

where $\hat{V}_m := \hat{\mathbb{V}}[\sqrt{n}\{C_m(n,d) - C_1(n,d)^m\}]$ denotes the asymptotic variance which can be estimated by

$$\hat{V}_m = 4\left(K^m + (m-1)^2 C^{2m} - m^2 K C^{2m-2} + 2\sum_{j=1}^{m-1} K^{m-j} C^{2j}\right), \tag{5.60}$$

where $C := C_1(n,d)$ and

$$K := \frac{6}{(n-m-1)(n-m)(n-m+1)} \times \sum_{i=1}^{n-m} \left(\left[\sum_{j=1}^{i-1} \mathbb{1}_m(e_j, e_i)\right] \left[\sum_{k=i+1}^{n-m+1} \mathbb{1}_m(e_i, e_k)\right]\right).$$

The BDS test has the property to be nuisance-parameter-free in the sense that any $\sqrt{n}$-consistent parameter estimator has no impact on its null limit distribution under a class of conditional mean models. However, as stressed by Hong and Lee (2003), the BDS test requires choosing the parameters m and d. In finite samples, this choice might affect the power and size properties of the test. Moreover, it is shown that the BDS test is not an omnibus test in the sense that it has power against any possible alternative.

As an alternative, Hong and Lee (2003) propose a misspecification test for ACD models based on the generalized spectrum test introduced by Hong (1999). It is based on the covariance between empirical characteristic functions of e_i and e_{i-j} given by

$$\sigma_j(u,v) := \mathrm{Cov}[e^{\iota u e_i}, e^{\iota v e_{i-j}}], \tag{5.61}$$

where $\iota := \sqrt{-1}$ and $j = 0, \pm 1, \ldots$ Then, $\sigma_j(u,v) = \varphi_j(u,v) - \varphi(u)\varphi(v)$, where $\varphi_j(u,v) := \mathbb{E}[e^{\iota(ue_i + ve_{i-j})}]$ and $\varphi(u) := \mathbb{E}[e^{\iota u e_i}]$ are the joint and marginal characteristic functions of (e_i, e_{i-j}). Consequently, $\sigma_j(u,v) = 0$ for all $(u,v) \in \mathbb{R}^2$ if e_i and e_j are independent. The Fourier transform of $\sigma_j(u,v)$ is given by

$$f(w,u,v) = \frac{1}{2\pi} \sum_{j=-\infty}^{\infty} \sigma_j(u,v) e^{-\iota j w}, \quad w \in [-\pi, \pi]. \tag{5.62}$$

As the negative partial derivative of $f(w,u,v)$ with respect to (u,v) at $(0,0)$ yields the "conventional" spectral density, Hong (1999) refers it to as a "generalized spectral density" of $\{e_i\}$. It can capture any type of pairwise dependence across various lags in $\{e_i\}$. If $\{e_i\}$ is i.i.d., $f(w,u,v)$ becomes a flat spectrum

$$f_0(w,u,v) = \frac{1}{2\pi}\sigma_0(u,v). \tag{5.63}$$

Hong and Lee (2003) suggest estimating $f(w,u,v)$ using a kernel estimator,

$$\begin{aligned}
\hat{f}_n(w,u,v) &= \frac{1}{2\pi} \sum_{j=1-n}^{n-1} (1 - |j|/n)^{1/2} K(j/p_n) \hat{\sigma}_j(u,v) e^{-\iota j w}, \\
\hat{\sigma}_j(u,v) &= \hat{\varphi}_j(u,v) - \hat{\varphi}_j(u,0)\hat{\varphi}_j(0,v), \quad j = 0, \pm 1, \ldots, \pm(n-1), \\
\hat{\varphi}_j(u,v) &= \begin{cases} (n-j)^{-1} \sum_{i=1+j}^{n} e^{\iota(ue_i + ve_{i-j})} & \text{if } j \geq 0, \\ (n+j)^{-1} \sum_{i=1-j}^{n} e^{\iota(ue_{i+j} + ve_i)} & \text{if } j < 0, \end{cases}
\end{aligned} \tag{5.64}$$

where $K : \mathbb{R} \to [-1, 1]$ is a symmetric kernel with bandwidth b_n such that $b_n \to \infty$, $b_n/n \to 0$ as $n \to \infty$, and $(1 - |j|/n)^{1/2}$ is a finite-sample adjustment factor. Correspondingly, $f_0(w,u,v)$ is estimated by

$$\hat{f}_0(w,u,v) = \frac{1}{2\pi}\hat{\sigma}_0(u,v). \tag{5.65}$$

Hong and Lee (2003) propose a test statistic by comparing $\hat{f}_n(w,u,v)$ and $\hat{f}_0(w,u,v)$ via an L_2-norm, see Hong and Lee (2003) for details. For the choice of the kernel, Hong and Lee suggest using the Daniell kernel as optimal to maximize the asymptotic power of the test. However, as stressed by Meitz and Teräsvirta (2006), its computation is demanding since it has an unbounded support. This is particularly true if the underlying sample size is huge. Therefore, as an alternative, Meitz and Teräsvirta (2006) propose using a Parzen kernel.

Note that both the BDS test as well as the Hong and Lee (2003) test are not applicable to exclusively test the conditional mean function of ACD models. As correct specifications of the mean function do not necessarily rule out higher order dependence, both tests would indicate a rejection of the ACD mean specification though it may be correct. For such situations, Hong and Lee (2011) construct a test on the dynamics in the conditional mean function based on a partial derivative of the generalized spectrum. They show that the test can detect a wide class of neglected linear and nonlinear dynamic structures in conditionally expected durations. Duchesne and Pacurar (2008) propose a test based on a kernel spectral density estimator of ACD residuals yielding a generalized version of Hong's (1996) test.

5.6.3 Distribution Tests

The residual series $\{e_i\}$ should have a mean of one and a distribution which should correspond to the specified distribution of the ACD errors. Hence, graphical checks of the residual series can be performed based on quantile-quantile plots (QQ plots). Alternatively, moment conditions implied by the specific distributions might be investigated to evaluate the goodness-of-fit. In the case of an exponential distribution, a simple moment condition implies the equality of the mean and the standard deviation. As discussed in Chap. 4, Engle and Russell (1998) propose the statistic $\sqrt{n}((\hat{\sigma}_e^2 - 1)/\sigma_\varepsilon)$ to test for excess dispersion, where $\hat{\sigma}_e^2$ is the sample variance of e_i and σ_ε is the standard deviation of $(\varepsilon_i - 1)^2$ which equals $\sqrt{8}$ under the exponential null hypothesis. Under the null, this test statistic is asymptotically standard normally distributed.

Another way to evaluate the goodness-of-fit is to evaluate the in-sample density forecasts implied by the model. Diebold et al. (1998) propose an evaluation method based on Rosenblatt's (1952) probability integral transform

$$q_i := \int_{-\infty}^{x_i} f(s)ds. \tag{5.66}$$

They show that under the null hypothesis, i.e., correct model specification, the distribution of the q_i series is i.i.d. uniform. Hence, testing the q_i series against the uniform distribution allows to evaluate the performance of density forecasts.[7] In this context, Pearson's goodness-of-fit test might be performed by categorizing the probability integral transforms q_i and computing a χ^2-statistic based on the frequencies of the individual categories,

$$\sum_{m=1}^{M} \frac{(n_m - n\hat{p}_m^*)^2}{n\hat{p}_m^*} \overset{a}{\sim} \chi_M^2,$$

[7]For more details, see e.g., Bauwens et al. (2004) or Dufour and Engle (2000), who apply this concept to the comparison of alternative financial duration models.

where M denotes the number of categories, n_m the number of observations in category m and $\hat{p}_m^*$ the estimated probability to observe a realization of q_i in category m.

More sophisticated tests against distributional misspecification are proposed by Fernandes and Grammig (2005) based on the work of Aït-Sahalia (1996). They suggest a nonparametric testing procedure that is based on the distance between the estimated parametric density function and its non-parametric estimate. This test is very general, since it tests for correctness of the complete (conditional) density.

In the following we illustrate the concept to test the distribution of ACD *residuals* which are consistent estimates of ε_i as long as the conditional mean function Ψ_i is correctly specified. Fernandes and Grammig (2005) show that there are no asymptotic costs in substituting the true errors ε_i by their $\sqrt{n}$-consistent estimates e_i. Define the c.d.f. and p.d.f. of the ACD residuals e_i as $F(e, \boldsymbol{\theta})$ and $f(e, \boldsymbol{\theta})$, respectively, with $\boldsymbol{\theta}$ denoting the underlying parameter vector. The principle of the test proposed by Fernandes and Grammig (2005) is to test whether there is any value $\boldsymbol{\theta}_0$ of the parameter vector such that the true and parametric density functions of e_i coincide almost everywhere. Consider the null hypothesis

$$H_0 : \exists\, \boldsymbol{\theta}_0 \in \Theta \text{ such that } f(e, \boldsymbol{\theta}_0) = f(e),$$

where Θ denotes the underlying parameter space. The so-called D-test is based on the distance

$$\Xi_f := \int_e \mathbb{1}_{\{e \in \mathcal{S}\}} \{f(e, \boldsymbol{\theta}) - f(e)\}^2 dF(e), \tag{5.67}$$

where the integral is over the support of f and $\mathcal{S}$ defines the subset of regions in which non-parametric (kernel) density estimation is stable. Then, the sample counterpart of (5.67) is given by

$$\Xi_{\hat{f}} = \frac{1}{n} \sum_{i=1}^{n} \mathbb{1}_{\{e_i \in \mathcal{S}\}} \{f(e_i, \hat{\boldsymbol{\theta}}) - \hat{f}(e_i)\}^2, \tag{5.68}$$

where $\hat{\boldsymbol{\theta}}$ and $\hat{f}$ denote pointwise consistent estimates of $\boldsymbol{\theta}_0$ and f, respectively. Hence, $\Xi_{\hat{f}}$ evaluates the difference between the parametric and nonparametric estimates of the density function f. As the parametric estimate is consistent only under correct specification, $\Xi_{\hat{f}}$ converges to zero under the null.

To construct a formal test, consider a kernel density estimate $\hat{f}(e)$ of $f(e)$ given by

$$\hat{f}(e) = \frac{1}{n b_n} \sum_{i=1}^{n} K_{e,b_n}(e_i), \tag{5.69}$$

where $K(e_i) := K_{e,b_n}(e_i)$ is a continuously differentiable kernel function $K(\cdot) \geq 0$ with $\int K(u)du = 1$ and bandwidth b_n. In case of using a fixed kernel, Fernandes and Grammig (2005) propose a test statistic for $\Xi_{\hat{f}}$ of the form

$$\hat{\tau}_n^D = \frac{nb_n^{1/2}\Xi_{\hat{f}} - b_n^{-1/2}\hat{\delta}_D}{\hat{\sigma}_D}, \tag{5.70}$$

where $b_n = o(n^{-1/(2s+1)})$ denotes the bandwidth of the fixed kernel and s is the order of the kernel. Moreover, $\hat{\delta}_D$ and $\hat{\sigma}_D^2$ are consistent estimates of $\delta_D := I_K \mathbb{E}[\mathbb{1}_{\{e \in \mathcal{S}\}} f(e)]$ and $\sigma_D^2 := J_K \mathbb{E}[\mathbb{1}_{\{e \in \mathcal{S}\}} f^3(e)]$, where I_K and J_K depend on the form of the (fixed) kernel and are defined by

$$I_K := \int_u K^2(u)du,$$

$$J_K := \int_u \left\{ \int_u K(u)K(u+v)du \right\}^2 dv.$$

The parameters δ_D and σ_D^2 can be consistently estimated using the empirical distribution to compute the expectation and then plugging in the corresponding (fixed) kernel density estimate:

$$\hat{\delta}_D = I_K \frac{1}{n} \sum_{i=1}^{n} \mathbb{1}_{\{e_i \in \mathcal{S}\}} \hat{f}(e_i),$$

$$\hat{\sigma}_D^2 = J_K \frac{1}{n} \sum_{i=1}^{n} \mathbb{1}_{\{e_i \in \mathcal{S}\}} \hat{f}(e_i)^3.$$

In case of the optimal uniform kernel according to Gosh and Huang (1991) given by

$$K^u(u) = \begin{cases} (2\sqrt{3})^{-1}, & \text{for } |u| \leq \sqrt{3}, \\ 0, & \text{for } |u| > \sqrt{3}, \end{cases} \tag{5.71}$$

f is estimated by substituting $K_{e,b_n}(e_i) = K^u\left(\frac{e-e_i}{b_n}\right)$ in (5.69), and I_K and J_K are given by $I_K = (2\sqrt{3})^{-1}$ and $J_K = (3\sqrt{3})^{-1}$.

Under a set of regularity conditions (see Fernandes and Grammig 2005), the statistic $\hat{\tau}_n^D$ is asymptotically normally distributed,

$$\hat{\tau}_n^D \xrightarrow{d} \mathcal{N}(0, 1).$$

However, as the ACD residuals have a support which is bounded from below, the test statistic $\hat{\tau}_n^D$ may perform poorly due to the well-known boundary bias induced by fixed kernels. This problem is caused by the fact that a fixed kernel assigns weight outside the support of the density when smoothing occurs near the boundary. An alternative is to consider asymmetric kernels such as the flexible gamma kernel proposed by Chen (2000). This kernel is based on the density of the gamma distribution with shape parameter $x/b_n + 1$ and the bandwidth b_n serving as scale parameter. It is given by

$$K^{\gamma}_{x/b_n+1,b_n}(u) := \frac{u^{x/b_n}\exp(-u/b_n)}{b_n^{x/b_n+1}\Gamma(x/b_n+1)}\mathbb{1}_{\{u\geq 0\}}, \tag{5.72}$$

where $b_n = o(n^{-4/9})$. Then, the corresponding gamma kernel density estimate is given by

$$\tilde{f}(e) = \frac{1}{nb_n}\sum_{i=1}^{n} K^{\gamma}_{e/b_n+1,b_n}(e_i). \tag{5.73}$$

Since the support of these kernels is the positive real line, they never assign weight outside the support of the underlying density. In addition to being free of boundary bias, the kernel shape changes according to the position of the observations, which modifies the amount of smoothing.

Using the gamma kernel as underlying kernel estimator, Fernandes and Grammig (2005) propose a modified test statistic of the form

$$\tilde{\tau}_n^D := \frac{nb_n^{1/4}\Xi_{\hat{f}} - b_n^{-1/4}\tilde{\delta}_G}{\tilde{\sigma}_G} \xrightarrow{d} \mathcal{N}(0,1), \tag{5.74}$$

where $\tilde{\delta}_G$ and $\tilde{\sigma}_G^2$ are consistent estimates of $\delta_G := \frac{1}{2\sqrt{\pi}}\mathbb{E}[\mathbb{1}_{\{e\in\mathcal{S}\}}e^{-1/2}f(e)]$ and $\sigma_G^2 := \frac{1}{\sqrt{2\pi}}\mathbb{E}[\mathbb{1}_{\{e\in\mathcal{S}\}}e^{-1/2}f^3(e)]$. Similarly to above, δ_G and σ_G^2 can be consistently estimated by

$$\tilde{\delta}_G = \frac{1}{2\sqrt{\pi}n}\sum_{i=1}^{n}\mathbb{1}_{\{e_i\in\mathcal{S}\}}e_i^{-1/2}\tilde{f}(e_i),$$

$$\tilde{\sigma}_G^2 = \frac{1}{\sqrt{2\pi}n}\sum_{i=1}^{n}\mathbb{1}_{\{e_i\in\mathcal{S}\}}e_i^{-1/2}\tilde{f}(e_i)^3.$$

To choose the bandwidth b_n, Fernandes and Grammig (2005) adapt Silverman's (1986) rule of thumb and choose

$$b_n = \frac{1}{\ln n}(\hat{\lambda}/4)^{-1/5}(2-\hat{\lambda})^{-4/5}n^{-4/9},$$

where $\hat{\lambda}$ is a consistent estimate of the parameter λ of an exponential distribution which is straightforwardly estimated by the sample mean of the ACD residuals or just set to one.

In a similar fashion, Fernandes and Grammig (2005) propose also a so-called H-test based on the distance between parametric and nonparametric estimates of the hazard function. See Fernandes and Grammig (2005) for more details.

The appealing features of the gamma kernel, such as the reduced variance in the interior part of the support, come at a cost. Compared to symmetric kernels, the gamma kernel estimator features a somewhat higher bias when moving away from the boundary (see, e.g., Hagmann and Scaillet 2007). That property makes it necessary to implement an effective technique for bias correction. A simple technique for multiplicative bias correction for fixed kernels is suggested by Hjort and Glad (1995) and extended to asymmetric kernels by Hagmann and Scaillet (2007) as a special case of local bias correction methods. This approach is semiparametric in the sense that the density is being estimated nonparametrically while using a parametric start. Hautsch et al. (2010) employ these techniques to construct a generalization of the Fernandes and Grammig (2005) test which allows to test against discrete-continuous mixture distributions explicitly accounting also for zero observations.

5.6.4 Lagrange Multiplier Tests

Note that the tests discussed in the previous sections are not appropriate for explicitly testing the conditional mean restriction of ACD models. Clearly, these tests have power against misspecifications of the conditional mean function, but they do not allow to identify whether a possible rejection is due to a violation of distributional assumptions caused by misspecifications of higher order conditional moments or due to a violation of the conditional mean restriction. Especially in the context of QML estimation of the ACD model, one is mainly interested in the validity of the conditional mean restriction but not necessarily in the correct specification of the complete density. Moreover, the distribution tests illustrated in Sect. 5.6.3 built on consistent estimates of the ACD errors, which require a correct specification of the conditional mean function. Procedures that explicitly test this particular conditional moment restriction are discussed in the following subsections.

In the econometric literature, the Lagrange Multiplier test has proven to be a useful diagnostic tool to detect model misspecifications. See for example, Breusch (1978), Breusch and Pagan (1979, 1980), Godfrey (1978a,b), or Engle (1984). In case of QML estimation of the ACD model (see Sect. 5.3.1), the LM test statistic is computed as

$$LM = \mathbf{f}_0'\mathbf{z}_0(\mathbf{z}_0'\mathbf{z}_0)^{-1}\mathbf{z}_0'\mathbf{f}_0, \tag{5.75}$$

where

$$\mathbf{f}_0 := \left(\frac{x_1}{\Psi_1} - 1, \ldots, \frac{x_n}{\Psi_n} - 1\right)', \qquad \mathbf{z}_0 := \left(\frac{1}{\Psi_1}\frac{\partial \Psi_1}{\partial \boldsymbol{\theta}_0}, \ldots, \frac{1}{\Psi_n}\frac{\partial \Psi_n}{\partial \boldsymbol{\theta}_0}\right)',$$

both evaluated under the null. It is easy to show that this test statistic corresponds to the uncentered R^2 from a regression of $\mathbf{f}_0$ on $\mathbf{z}_0$ and is commonly computed as nR^2 where R^2 is the uncentered R^2 from a regression of a vector of ones on the scores of the model.

To perform the LM test, it is necessary to specify a general model which encompasses the model under the null. Consider a more general form of LM test which allows to test for misspecifications of the conditional mean function of unknown form. Assume that the ACD specification under the null is a special case of a more general (additive) model of the form

$$\Psi_i = \Psi_i^0 + \boldsymbol{\theta}_a' \mathbf{z}_{ai}, \tag{5.76}$$

where Ψ_i^0 denotes the conditional mean function under the null depending on the parameter vector $\boldsymbol{\theta}_0$, while $\boldsymbol{\theta}_a$ and $\mathbf{z}_{ai}$ denote the vectors of additional parameters and missing variables, respectively. Then, we can test for the correct specification of the null model by testing the parameter restriction $\boldsymbol{\theta}_a = 0$. Following the idea of Engle and Ng (1993), $\mathbf{z}_{ai}$ might be specified in terms of so-called sign bias variables $\mathbb{1}_{\{\varepsilon_{i-1}<1\}}$, $\mathbb{1}_{\{\varepsilon_{i-1}<1\}}\varepsilon_{i-1}$ and $\mathbb{1}_{\{\varepsilon_{i-1}\geq 1\}}\varepsilon_{i-1}$, and extensions thereof. Such specifications allow to investigate whether the specification is appropriate to capture possible nonlinearities in the news impact function. The resulting LM test is formulated based on the auxiliary regression

$$e_i = \mathbf{z}_{0i}' \tilde{\boldsymbol{\beta}}_0 + \mathbf{z}_{ai}' \tilde{\boldsymbol{\beta}}_a + u_i, \tag{5.77}$$

where u_i is a zero mean i.i.d. error term, $\tilde{\boldsymbol{\beta}}_0$ and $\tilde{\boldsymbol{\beta}}_a$ are vectors of regression coefficients, $\mathbf{z}_{0i} = 1/\Psi_i^0 \cdot \partial \Psi_i^0 / \partial \boldsymbol{\theta}_0$ and $\mathbf{z}_{ai} = 1/\Psi_i^0 \cdot \partial \Psi_i^0 / \partial \boldsymbol{\theta}_a$ evaluated at $\boldsymbol{\theta}_a = 0$ and at the QML estimator under the null. Then, the statistic is given by n times the R^2 from the regression (5.77) and follows asymptotically a $\chi^2(m)$ distribution where m denotes the number of restrictions.

However, as discussed in Meitz and Teräsvirta (2006), this test is not robust if the ACD errors ε_i are not exponentially distributed. They suggest to follow the approach by Wooldridge (1990) and to apply the following procedure:

1. Compute the residuals, r_i, from a regression of $\mathbf{z}_{ai}$ on $\mathbf{z}_{0i}$.
2. Regress a vector of ones on $r_i(x_i/\Psi_i - 1)$ and compute the sum of squared residuals, SSR.
3. Compute the asymptotically $\chi^2(m)$ distributed test statistic as n times SSR.

As illustrated by Wooldridge (1990), this procedure leads to a consistent test which is (asymptotically) not affected by violations of the underlying distributional assumptions.

More specific LM tests against particular parametric alternatives are derived by Meitz and Teräsvirta (2006). In particular, they consider two types of misspecification: the conditional duration is either additively or multiplicatively misspecified. In these cases, it is given by

$$x_i = (\Psi_i + \varphi_i)\varepsilon_i \tag{5.78}$$

or

$$x_i = \Psi_i \varphi_i \varepsilon_i, \tag{5.79}$$

where $\Psi_i = \Psi(\boldsymbol{\theta}_1)$ denotes the conditional mean depending on parameters $\boldsymbol{\theta}_1$ and $\varphi_i = \varphi_i(\boldsymbol{\theta}_1, \boldsymbol{\theta}_2)$ is the misspecification depending not only on $\boldsymbol{\theta}_1$ but also on additional parameters $\boldsymbol{\theta}_2$. Define

$$\mathbf{a}_i(\boldsymbol{\theta}_1) := \frac{1}{\Psi_i(\boldsymbol{\theta}_1)} \frac{\partial \Psi_i(\boldsymbol{\theta}_1)}{\partial \boldsymbol{\theta}_1},$$
$$\mathbf{b}_i(\boldsymbol{\theta}_1, \boldsymbol{\theta}_2) := \frac{1}{\Psi_i(\boldsymbol{\theta}_1)} \frac{\partial \varphi_i(\boldsymbol{\theta}_1, \boldsymbol{\theta}_2)}{\partial \boldsymbol{\theta}_2},$$
$$\mathbf{c}_i(\boldsymbol{\theta}_1) := \frac{x_i}{\Psi_i(\boldsymbol{\theta}_1)} - 1.$$

Assume that under the null hypothesis $H_0 : \boldsymbol{\theta}_2 = \boldsymbol{\theta}_2^0$, the function φ_i satisfies $\varphi_i(\boldsymbol{\theta}_1, \boldsymbol{\theta}_2^0) = 0$, where the superscript '0' denotes the true parameter. Then, as shown by Meitz and Teräsvirta (2006), under the null $H_0 : \boldsymbol{\theta}_2 = \boldsymbol{\theta}_2^0$, the LM test statistic for a test against a general additive alternative is given by

$$LM = \left\{\sum_{i=1}^n \hat{\mathbf{c}}_i \hat{\mathbf{b}}_i'\right\} \left\{\sum_{i=1}^n \hat{\mathbf{b}}_i \hat{\mathbf{b}}_i' - \left(\sum_{i=1}^n \hat{\mathbf{b}}_i \hat{\mathbf{a}}_i'\right) \left(\sum_{i=1}^n \hat{\mathbf{a}}_i \hat{\mathbf{a}}_i'\right)^{-1} \left(\sum_{i=1}^n \hat{\mathbf{a}}_i \hat{\mathbf{b}}_i'\right)\right\}^{-1}$$
$$\times \left\{\sum_{i=1}^n \hat{\mathbf{c}}_i \hat{\mathbf{b}}_i\right\} \stackrel{a}{\sim} \chi^2_{dim(\boldsymbol{\theta}_2)}. \tag{5.80}$$

Likewise, assume that under the null hypothesis $H_0 : \boldsymbol{\theta}_2 = \boldsymbol{\theta}_2^0$, the function φ_i satisfies $\varphi_i(\boldsymbol{\theta}_1, \boldsymbol{\theta}_2^0) = 1$. Then, under the null $H_0 : \boldsymbol{\theta}_2 = \boldsymbol{\theta}_2^0$, a test against a general multiplicative alternative is obtained by

$$LM = \left\{\sum_{i=1}^n \hat{\varphi}_i \hat{\mathbf{c}}_i \hat{\mathbf{b}}_i'\right\} \left\{\sum_{i=1}^n \hat{\varphi}_i^2 \hat{\mathbf{b}}_i \hat{\mathbf{b}}_i' - \left(\sum_{i=1}^n \hat{\varphi}_i \hat{\mathbf{b}}_i \hat{\mathbf{a}}_i'\right) \left(\sum_{i=1}^n \hat{\mathbf{a}}_i \hat{\mathbf{a}}_i' s\right)^{-1} \left(\sum_{i=1}^n \hat{\varphi}_i \hat{\mathbf{a}}_i \hat{\mathbf{b}}_i'\right)\right\}^{-1}$$
$$\times \left\{\sum_{i=1}^n \hat{\varphi}_i \hat{\mathbf{c}}_i \hat{\mathbf{b}}_i\right\} \stackrel{a}{\sim} \chi^2_{dim(\boldsymbol{\theta}_2)}. \tag{5.81}$$

As stressed by Meitz and Teräsvirta (2006), misspecifications of the conditional distribution of the durations may affect the properties of the LM test statistics as they implicitly build on distributional assumptions. For instance, it is necessary that the conditional variance of the durations is correctly specified under the null hypothesis. To avoid distortions of the test statistics due to distributional misspecifications, Meitz and Teräsvirta (2006) suggest a procedure building on Wooldridge (1991) resulting in "robust" versions of the test statistics with their asymptotic behavior being unaffected by possible distributional misspecifications. In case of additive misspecification, it is suggested to compute the test statistic as follows:

1. Using the QML estimate of $\boldsymbol{\theta}_1$ under the null hypothesis and compute
$$\hat{\mathbf{a}}_i' = \frac{1}{\Psi_i(\hat{\boldsymbol{\theta}}_1)}\frac{\partial \Psi_i(\hat{\boldsymbol{\theta}}_1)}{\partial \boldsymbol{\theta}_1'}, \quad \hat{\mathbf{b}}_i' = \frac{1}{\Psi_i(\hat{\boldsymbol{\theta}}_1)}\frac{\partial \varphi_i(\hat{\boldsymbol{\theta}}_1, \boldsymbol{\theta}_2^0)}{\partial \boldsymbol{\theta}_2'}, \quad \hat{\mathbf{c}}_i = \frac{x_i}{\Psi_i(\hat{\boldsymbol{\theta}}_1)} - 1,$$
for $i = 1\ldots, n$.
2. Regress $\hat{\mathbf{b}}_i'$ on $\hat{\mathbf{a}}_i'$, $i = 1, \ldots, n$, and compute the corresponding $dim(\boldsymbol{\theta}_2) \times 1$ residual vectors $\hat{r}_i$.
3. Regress 1 on $\hat{\mathbf{c}}_i \hat{r}_i$, $i = 1, \ldots, n$, and compute the sum of squared residuals (SSR).
4. Compute the test statistic as $nR^2 = n - SSR$ which is asymptotically χ^2 distributed with $dim(\boldsymbol{\theta}_2)$ degrees of freedom under the null hypothesis.

In case of the LM test against multiplicative misspecification, the same procedure applies with $\hat{\mathbf{b}}_i$ replaced by $\hat{\varphi}_i \hat{\mathbf{b}}_i$. Applications of this framework to test against specific types of misspecifications are illustrated in Meitz and Teräsvirta (2006).

5.6.5 Conditional Moment Tests

5.6.5.1 Adapting Newey's Conditional Moment Test

The LM test discussed in the previous subsection has optimal power against local (parametric) alternatives and is a special case of a conditional moment test. The main idea behind the conditional moment (CM) test is to test the validity of conditional moment restrictions implied by the data which hold when the model is correctly specified. In the ACD framework, a natural moment condition is obtained by the conditional mean restriction.

Define $\boldsymbol{\rho}_i := \boldsymbol{\rho}_i(\boldsymbol{\theta})$, $i = 1\ldots, n$, as the $s \times 1$ vector of conditional moment functions with the property $\mathbb{E}[\boldsymbol{\rho}_i|\mathbf{w}_i] = \mathbf{0}$, where $\mathbf{w}_i$ is a $s \times q$ matrix of instruments. Correspondingly, we obtain the $q \times 1$ vector of unconditional moment functions as $\boldsymbol{\tau}_i := \boldsymbol{\tau}_i(\boldsymbol{\theta}) := \mathbf{w}_i'\boldsymbol{\rho}_i$. Moreover, we define the $q \times 1$ vector of sample moments $\boldsymbol{\varphi}_n := n^{-1}\sum_{i=1}^n \boldsymbol{\tau}_i$. In the ACD framework, natural choices for $\boldsymbol{\rho}_i$ are $(x_i - \Psi_i)$ or $(x_i/\Psi_i - 1)$ allowing to test the null hypotheses for $s = 1$,
$$H_0: \quad \mathbb{E}[x_i - \Psi_i|\mathbf{w}_i] = 0 \quad \text{or} \quad H_0^*: \quad \mathbb{E}[x_i/\Psi_i - 1|\mathbf{w}_i] = 0.$$

We assume that $\boldsymbol{\theta}$ is estimated by exponential QML. Correspondingly, we denote the $p \times 1$ vector $\mathbf{s}_i := \mathbf{s}_i(\boldsymbol{\theta})$ as the score associated with the ith log likelihood contribution. Accordingly, we define the $n \times p$ matrix $\mathbf{s} := \mathbf{s}(\boldsymbol{\theta}) := (\mathbf{s}_1', \ldots, \mathbf{s}_n')$ and $\mathcal{H}(\boldsymbol{\theta}) := \frac{\partial \ln \mathcal{L}(\boldsymbol{\theta})}{\partial \boldsymbol{\theta} \partial \boldsymbol{\theta}'}$ as the Hessian of the pseudo log likelihood. Furthermore, we make the following assumptions:

(A1) $\boldsymbol{\tau}_i(\boldsymbol{\theta}_0)$ follows a stationary and ergodic process with $\boldsymbol{\theta}_0$ defining the true parameter.

(A2) $\boldsymbol{\tau}_i$ is continuously differentiable in $\boldsymbol{\theta}$ with $\mathbb{E}[\boldsymbol{\tau}_i(\boldsymbol{\theta})] < \infty$.

(A3) $\boldsymbol{\varphi}_n \xrightarrow{p} \mathbb{E}[\boldsymbol{\tau}_i]$ and $n^{-1} \sum_{i=1}^{n} \partial \boldsymbol{\tau}_i / \partial \boldsymbol{\theta}' \xrightarrow{p} \mathbb{E}\left[\partial \boldsymbol{\tau}_i(\boldsymbol{\theta}_0)/\partial \boldsymbol{\theta}'\right]$.

(A4) $n^{1/2} \begin{bmatrix} n^{-1} \sum_{i=1}^{n} \boldsymbol{\tau}_i \\ n^{-1} \sum_{i=1}^{n} \mathbf{s}_i \end{bmatrix} \xrightarrow{d} \mathcal{N}(\mathbf{0}, \boldsymbol{\Sigma})$ with $\boldsymbol{\Sigma}$ denoting a positive semi-definite covariance matrix of dimension $p + q$.

(A5) For some neighborhood $\mathfrak{N}$ of $\boldsymbol{\theta}_0$: $\mathbb{E}[\sup_{\boldsymbol{\theta} \in \mathfrak{N}} ||\mathcal{H}(\boldsymbol{\theta})||] < \infty$.

In the following, a modified form of Newey's (1985) conditional moment test is illustrated which allows for non-i.i.d. data and is robust to any misspecification other than violations of the conditional mean restriction, as, e.g., distributional misspecification or conditional heteroscedasticity in the scores. The asymptotic distribution of $n^{1/2}\hat{\boldsymbol{\varphi}}_n$ is derived by expanding $\hat{\boldsymbol{\varphi}}_n$ around $\boldsymbol{\theta}_0$ using the mean value theorem,

$$n^{1/2}\hat{\boldsymbol{\varphi}}_n = n^{1/2} \left[n^{-1} \sum_{i=1}^{n} \boldsymbol{\tau}_i(\boldsymbol{\theta}_0) + \left(\operatorname*{plim}_{n\to\infty} n^{-1} \sum_{i=1}^{n} \partial \boldsymbol{\tau}_i(\boldsymbol{\theta}^*)/\partial \boldsymbol{\theta} \right) (\hat{\boldsymbol{\theta}} - \boldsymbol{\theta}_0) \right], \tag{5.82}$$

where $\boldsymbol{\theta}^* := \boldsymbol{\theta}_0 + \lambda(\hat{\boldsymbol{\theta}} - \boldsymbol{\theta}_0)$, $0 \leq \lambda \leq 1$. With $\hat{\boldsymbol{\theta}}$ being a QML estimator, we have

$$n^{1/2}(\hat{\boldsymbol{\theta}} - \boldsymbol{\theta}_0) = -\left[n^{-1}\mathcal{H}(\boldsymbol{\theta}^*)\right]^{-1} n^{-1/2} \sum_{i=1}^{n} \mathbf{s}_i(\boldsymbol{\theta}_0).$$

Substituting back into (5.82) yields

$$n^{1/2}\hat{\boldsymbol{\varphi}}_n = n^{-1/2} \sum_{i=1}^{n} \boldsymbol{\tau}_i(\boldsymbol{\theta}_0) - \left(\operatorname*{plim}_{n\to\infty} n^{-1} \sum_{i=1}^{n} \partial \boldsymbol{\tau}_i(\boldsymbol{\theta}^*)/\partial \boldsymbol{\theta} \right) \times n^{1/2}\mathcal{H}(\boldsymbol{\theta}^*)^{-1} \sum_{i=1}^{n} \mathbf{s}_i(\boldsymbol{\theta}_0).$$

This expression can be re-written as

$$n^{1/2}\hat{\boldsymbol{\varphi}}_n = \mathbf{B} \begin{bmatrix} n^{-1/2} \sum_{i=1}^{n} \boldsymbol{\tau}_i(\boldsymbol{\theta}_0) \\ n^{-1/2} \sum_{i=1}^{n} \mathbf{s}_i(\boldsymbol{\theta}_0) \end{bmatrix}, \tag{5.83}$$

where the $q \times (p+q)$ matrix $\mathbf{B}$ is given by

$$\mathbf{B} = \left[\mathbf{I}_q \;\; \vdots \;\; \left(\underset{n\to\infty}{\text{plim}}\, n^{-1} \sum_{i=1}^{n} \partial \boldsymbol{\tau}_i(\boldsymbol{\theta}^*)/\partial \boldsymbol{\theta} \right) \left(n^{-1}\mathcal{H}(\boldsymbol{\theta}^*)\right)^{-1} \right], \tag{5.84}$$

and $\mathbf{I}_q$ denotes a $(q \times q)$ identity matrix. Then, we yield $n^{1/2}\hat{\varphi}_n \xrightarrow{d} \mathcal{N}(0, \mathbf{B}\boldsymbol{\Sigma}\mathbf{B}')$ and thus

$$n[\hat{\varphi}_n'(\mathbf{B}\boldsymbol{\Sigma}\mathbf{B}')^{-1}\hat{\varphi}_n] \overset{a}{\sim} \chi^2_q. \tag{5.85}$$

Under the given assumptions, we have $\boldsymbol{\Sigma} := \sum_{j=-n}^{n} \gamma_j = \gamma_0 + \sum_{j=1}^{n}(\gamma_j + \gamma_j')$, where $\gamma_j := \mathbb{E}[\boldsymbol{\phi}_i(\boldsymbol{\theta}_0)\boldsymbol{\phi}_{i-j}(\boldsymbol{\theta}_0)']$ and $\boldsymbol{\phi}_i := \boldsymbol{\phi}(x_i, \boldsymbol{\theta}_0) = (\boldsymbol{\tau}_i(\boldsymbol{\theta}_0), \mathbf{s}_i(\boldsymbol{\theta}_0))'$ is the $(q+p)\times 1$ vector of moment restrictions and scores in i. Then, $\boldsymbol{\Sigma}$ can be consistently estimated by a kernel-based estimator

$$\hat{\boldsymbol{\Sigma}} = \sum_{j=-n+1}^{n-1} K(j/q_n)\hat{\gamma}_j,$$

where $K(\cdot)$ is a kernel function and q_n is a bandwidth depending on n. Natural choices are Bartlett kernels, quadratic spectral kernels or Parzen kernels as, e.g., suggested by Newey and West (1987) and Andrews (1991).

Estimating the matrix $\mathbf{B}$ requires consistently estimating $\mathcal{H}(\boldsymbol{\theta})$ by the empirical Hessian which is ensured by the dominance condition (A5). Moreover, $\underset{n\to\infty}{\text{plim}}\, n^{-1}\sum_i \partial\boldsymbol{\tau}_i(\boldsymbol{\theta}^*)/\partial\boldsymbol{\theta}$ can be consistently estimated by

$$n^{-1}\sum_{i=1}^{n} \partial\hat{\boldsymbol{\tau}}_i/\partial\boldsymbol{\theta} = n^{-1}\sum_{i=1}^{n}(\mathbf{w}_i\partial\hat{\boldsymbol{\tau}}_i/\partial\boldsymbol{\theta} + \hat{\boldsymbol{\tau}}_i\partial\mathbf{w}_i/\partial\boldsymbol{\theta}),$$

where

$$\partial\hat{\boldsymbol{\tau}}_i/\partial\boldsymbol{\theta} = \begin{cases} -x_i\hat{\mathbf{s}}_i/(x_i - \hat{\Psi}_i) & \text{in case of } H_0, \\ -\hat{\mathbf{s}}_i\hat{\Psi}_i^2/(x_i - \hat{\Psi}_i) & \text{in case of } H_0^*. \end{cases}$$

Note that in case of i.i.d. observations, $\boldsymbol{\Sigma}$ is consistently estimated by $n^{-1}\hat{\boldsymbol{\phi}}_i, \hat{\boldsymbol{\phi}}_i'$, whereas $\underset{n\to\infty}{\text{plim}}\, \sum_i \partial\boldsymbol{\tau}(\cdot)/\partial\boldsymbol{\theta}$ can be consistently estimated by the outer product between score and moment vector (see Tauchen 1985, or Newey 1985). Then, we get the well-known expression (see, e.g., Pagan and Vella 1989)

$$n[\hat{\boldsymbol{\phi}}_n'(\mathbf{B}\boldsymbol{\Sigma}\mathbf{B}')^{-1}\hat{\boldsymbol{\phi}}_n] = \boldsymbol{\iota}'\mathbf{R}(\mathbf{R}'\mathbf{R} - \mathbf{R}'\mathbf{s}(\mathbf{s}'\mathbf{s})^{-1}\mathbf{s}'\mathbf{R})^{-1}\mathbf{R}'\boldsymbol{\iota}, \tag{5.86}$$

where $\boldsymbol{\iota}$ is a $(n \times 1)$ vector of ones and $\mathbf{R}$ is the $n \times q$ matrix with $\hat{\boldsymbol{\tau}}_i'$ as ith element.

Valuable choices for the weighting functions $w_j(\cdot)$ are lagged sign bias variables (as discussed in Sect. 5.6.4) and/or functionals (e.g. moments) of past durations.

A well known result is that the power of the CM test depends heavily on the choice of the weighting functions. Newey (1985) illustrates how to obtain an optimal conditional moment test with maximal local power. It is shown that the LM test corresponds to an optimal CM test in the case of a particular local alternative. However, since the CM test is based on a finite number of conditional moment restrictions, it cannot be consistent against *all* possible alternatives.

Generalized moment tests are proposed by Chen and Hsieh (2010). They construct moment functions which allow not only to test the validity of the conditional mean function but also to test for independence and distributional misspecification. Conditional mean and independence tests are constructed based on moment restrictions implied by the exponential QML method. Correspondingly, the distribution test relies on the ML method and the assumption of independent error terms.

5.6.5.2 Integrated Conditional Moment Tests

Bierens (1990) illustrates that any CM test of functional form can be converted into a chi-square test that possesses the property of consistency against all possible alternatives. The main idea behind the consistent conditional moment test is based on the following lemma:

Lemma 5.1 (Bierens 1990). *Let ϱ be a random variable satisfying the condition $\mathbb{E}|\varrho| < \infty$ and let z be a bounded random variable in $\mathbb{R}$ with $\mathbb{P}r[\mathbb{E}(\varrho|z) = 0] < 1$. Then the set $S = \{t \in \mathbb{R} : \mathbb{E}[\varrho \exp(tz)] = 0\}$ is countable and thus has Lebesgue measure zero.* □

Bierens shows that $\mathbb{E}[\varrho \exp(tz)] \neq 0$ in a neighborhood of $t = t_0$ where t_0 is such that $\mathbb{E}[\varrho \exp(t_0 z)] = 0$ and $\mathbb{P}r\,[\mathbb{E}[\varrho \exp(tz)|z] = 0] < 1$. de Jong (1996) extends Bierens' test towards the case of serially dependent data. In the following, we assume that the duration process is stationary and obeys the concept of ν-stability. Moreover it is supposed that $\mathbb{E}|\varepsilon_i - 1| < \infty$. By assuming that the model is misspecified, i.e. $\mathbb{P}r[\mathbb{E}[\rho_i(\boldsymbol{\theta}_0)|\mathcal{F}_{i-1}] = 0] < 1$ and replacing the conditioning information by $\xi(x_{i-1}), \xi(x_{i-2}), \ldots$, where $\xi(\cdot)$ is a bounded one-to-one mapping from $\mathbb{R}$ into $\mathbb{R}$, the set

$$S = \left\{ t \in \mathbb{R}^d : \mathbb{E}\left[\rho_i(\boldsymbol{\theta}_0) \exp\left(\sum_{j=1}^{d} t_j \xi(x_{i-j}) \right) \right] = 0 \right\}$$

with $d = \min(i-1, c)$ has Lebesgue measure zero. Therefore, de Jong (1996) suggests a consistent CM test based on the unconditional moment restriction

$$\hat{M}_n(t) = n^{-1/2} \sum_{i=1}^{n} \rho_i(\boldsymbol{\theta}_0) \exp\left(\sum_{j=1}^{d} t_j \xi(x_{i-j}) \right), \tag{5.87}$$

where $\boldsymbol{\theta}_0$ is estimated consistently by QML. de Jong points out that a conditional moment restriction test based on $d = c < n$ does not allow us to consistently test the hypotheses H_0 and H_1 for an infinite number of lags and thus $d = i - 1$ should be preferred. Equation (5.87) has the property that under the alternative hypothesis H_1, $\operatorname{plim}_{n\to\infty} \hat{M}_n(t) \neq 0$ for all t except in a set with Lebesgue measure zero. Therefore, the principle of the consistent conditional moment test is to employ a class of weighting functions which are indexed by a continuous nuisance parameter (vector) $\mathbf{t}$. Since this nuisance parameter is integrated out, this test is called integrated conditional moment (ICM) test. The lemma above implies that by choosing a vector $\mathbf{t}' \notin S$, a consistent CM test is obtained. However, S depends on the distribution of the data, and thus it is impossible to choose a fixed vector $\mathbf{t}$ for which the test is consistent. As suggested by de Jong (1996), a solution to this problem is to achieve test consistency by maximizing a functional of $\hat{M}_n(\mathbf{t})$ over a compact subset Ξ of $\mathbb{R}^c$. The main idea is that a vector $\mathbf{t}'$ which maximizes a test statistic based on $\hat{M}_n(\mathbf{t})$ cannot belong to the set S. By defining a space for infinite sequences $\{t_1, t_2, \ldots\}$ as

$$\Xi = \{\mathbf{t} : \ a_j \leq t_j \leq b_j \ \forall \, j \, ; \ t_j \in \mathbb{R}\}, \tag{5.88}$$

where $a_j < b_j$ and $|a_j|$, $|b_j| \leq Bj^{-2}$ for some constant B, de Jong suggests to consider the use of a functional of $\sup_{\mathbf{t}\in\Xi} |\hat{M}_n(\mathbf{t})|$ as test statistic. Consequently, a difficulty arises by the fact that the limiting distribution of the test statistic $\sup_{\mathbf{t}\in\Xi} |\hat{M}_n(\mathbf{t})|$ is case-dependent which prevents the use of generally applicable critical values. For this reason de Jong introduces a simulation procedure based on a conditional Monte Carlo approach. In particular, he shows that under the null, the moment restriction (5.87) has the same asymptotic finite-dimensional distribution as

$$\hat{\hat{M}}_n(\mathbf{t}) = n^{-1/2} \sum_{i=1}^{n} \sigma_i \rho_i(\boldsymbol{\theta}_0) \exp\left(\sum_{j=1}^{d} \mathbf{t}\xi(x_{i-j}) \right) \tag{5.89}$$

pointwise in $\mathbf{t}$, where σ_i are bounded i.i.d. random variables independent of x_i and $\rho_i(\boldsymbol{\theta}_0)$ with $\mathbb{E}[\sigma_i^2] = 1$. Thus the distribution of $\hat{M}_n(\mathbf{t})$ can be approximated based on the simulation of n-tuples of σ_i. de Jong proves that the critical regions obtained by the simulation of $\hat{\hat{M}}_n(\mathbf{t})$ are asymptotically valid (see de Jong 1996, Theorem 5). Nevertheless, a further difficulty is that a consistent ICM test rests on the statistic $\sup_{\mathbf{t}\in\Xi} |\hat{M}_n(\mathbf{t})|$. The calculation of this test statistic is quite cumbersome since it requires the maximization over a parameter space of dimension $n-1$. For this reason de Jong suggests to find another continuous functional of $\hat{M}_n(\mathbf{t})$ that possesses the same consistency property but is more easily calculated. Then, de Jong proposes to use the functional

$$\Lambda = n^{-1} \int_{\Xi} \hat{M}_n(\mathbf{t})^2 \varphi_1(t_1) d\varphi_1 \ldots \varphi_j(t_j) dt_j \ldots, \tag{5.90}$$

where the integrations run over an infinite number of t_j. According to (5.88), each t_j is integrated over the subset of $\mathbb{R}$, such that $a_j \leq |t_j| \leq b_j$. $\varphi_j(t)$ denote a sequence of density functions that integrate to one over the particular subsets. de Jong shows that the use of this functional leads to a consistent test. Since $\hat{M}_n(\mathbf{t})$ can be written as a double summation and the integrals can be calculated one at a time, we obtain a functional which is easier to calculate than $\sup_{\mathbf{t}\in\Xi} |\hat{M}_n(\mathbf{t})|$. By choosing a uniform distribution, i.e. $\varphi_j(t) = \mathbf{t}^{-1}$, the ICM test statistic Λ results in

$$\Lambda = n^{-1} \sum_{i=1}^{n} \sum_{j=1}^{n} \rho_i(\boldsymbol{\theta}_0)\rho_j(\boldsymbol{\theta}_0) \prod_{s=1}^{d} \left\{ \frac{1}{b_j - a_j} \left[\xi(x_{i-s}) + \xi(x_{j-s})\right]^{-1} \right.$$
$$\left. \times \left[\exp(b_j(\xi(x_{i-s}) + \xi(x_{j-s}))) - \exp(a_j(\xi(x_{i-s}) + \xi(x_{j-s})))\right] \right\}. \quad (5.91)$$

Summarizing, the implementation of ICM tests to ACD models requires the following steps:

1. *(Q)ML estimation of the ACD model:* Estimate the particular ACD model by (Q)ML and calculate the conditional moment restriction $\rho_i(\boldsymbol{\theta}_0)$.
2. *Choice of a_j and b_j:* Choose values for a_j and b_j, defining the parameter space Ξ. de Jong (1996) suggests to use $a_j = Aj^{-2}$ and $b_j = Bj^{-2}$ where the values A and B $(0 < A < B)$ can be chosen arbitrarily. Asymptotically the choice of A and B should have no influence on the power of the test, however in finite samples it probably has. Monte Carlo simulations of de Jong (1996) suggest to choose a small range, for example $A = 0$ and $B = 0.5$.
3. *Choice of $\xi(\cdot)$:* According to the lemma above, the function $\xi(\cdot)$ must be a bounded one-to-one mapping from $\mathbb{R}$ into $\mathbb{R}$. Asymptotically, the choice of the function $\xi(\cdot)$ is irrelevant, however, Bierens (1990) proposes to use $\xi(x) = \arctan(x)$. We suggest $\xi(x) = \arctan(0.01 \cdot x) \cdot 100$ which is also a bounded function but has the advantage that it is nearly linear in the relevant region which improves the small sample properties of the test.
4. *Choice of d:* Note that in the case of dependent data, the test consistency is only ensured by accounting for all feasible lags, $d = i - 1$, i.e., the dimension of the parameter space under consideration grows with the sample size. An alternative which does not require as much computer time, would be to choose a fixed value $d < n$. However, in this case the test does not allow us to consistently test the moment condition for an infinite number of conditioning variables.
5. *Simulation of n-tuples of σ_i:* Simulate R n-tuples of (bounded) i.i.d. random variables $\sigma_{i,r}$, $i = 1, \ldots, n$, with $\mathbb{E}[\sigma_{i,r}^2] = 1$ for $r = 1, \ldots, R$. Following de Jong (1996), we generate the σ_i variables such that $\mathbb{E}[\sigma_i = 1] = \mathbb{E}[\sigma_i = -1] = 0.5$.
6. *Computation of the test statistic and simulating of the critical values:*
   - Compute the test statistic Λ according to (5.91).
   - For each n-tuple of σ_i, compute the simulated test statistic

$$\tilde{\Lambda}_m = n^{-1} \sum_{i=1}^{n} \sum_{j=1}^{n} (\sigma_{i,r} \rho_i(\boldsymbol{\theta}_0))(\sigma_{j,r} \rho_j(\boldsymbol{\theta}_0))$$

$$\times \prod_{s=1}^{d} \Big\{ (b_j - a_j)^{-1} \left[\xi(x_{i-s}) + \xi(x_{j-s}) \right]^{-1}$$

$$\times \left[\exp(b_j(\xi(x_{i-s}) + \xi(x_{j-s}))) - \exp(a_j(\xi(x_{i-s}) + \xi(x_{j-s}))) \right] \Big\},$$

for $r = 1, \ldots, R$.

7. *Computation of simulated p-values:* Since the critical region of the test has the form $(C, \infty]$, we compute the simulated p-value of the ICM test as

$$pv_{ICM} = \frac{1}{R} \sum_{r=1}^{R} \mathbb{1}_{\{\tilde{\Lambda}_r \leq \Lambda\}}. \tag{5.92}$$

5.6.6 *Monte Carlo Evidence*

The following Monte Carlo study provides insights into the size and power properties of various conditional moment tests. Samples of size 3,000 are drawn which is still relatively small for high-frequency financial data and allows us to study the finite-sample properties. Each Monte Carlo experiment is repeated 500 times. The following five data generating processes (DGPs) ensuring $\mathbb{E}[\Psi_i] = 1$ are considered:

$$\Psi_i = 0.1 + 0.1 x_{i-1} + 0.8 \Psi_{i-1} \tag{5.93}$$

$$\Psi_i = \exp(0.137 + 0.3 \varepsilon_{i-1} + 0.8 \ln \Psi_{i-1}) \tag{5.94}$$

$$\Psi_i = (0.05 \Psi_{i-1} + 0.5) \varepsilon_{i-1} + 0.8 \Psi_{i-1} \tag{5.95}$$

$$\Psi_i = \exp(-0.18 + 0.5 \varepsilon_{i-1} - 0.48 |\varepsilon_{i-1} - 1| + 0.8 \ln \mu_{i-1}) \tag{5.96}$$

$$\Psi_i = \begin{cases} 0.05 + 0.20 x_{i-1} + 0.85 \Psi_{i-1} & \text{if } x_{i-1} \leq 0.25, \\ 0.10 + 0.05 x_{i-1} + 0.90 \Psi_{i-1} & \text{if } x_{i-1} \in (0.25, 1.5], \\ 0.20 + 0.03 x_{i-1} + 0.80 \Psi_{i-1} & \text{if } x_{i-1} > 1.5, \end{cases} \tag{5.97}$$

where $x_i = \Psi_i \varepsilon_i$, $\varepsilon_i \sim Exp(1)$. Equation (5.94) is a logarithmic ACD (LACD) specification as discussed in Sect. 5.5.

Specifications (5.95) to (5.97) are nonlinear and extended ACD models which are discussed in more detail in Chap. 6. In particular, specification (5.95) includes innovations both multiplicatively *and* additively. Specification (5.96) implies a news impact function which is kinked at $\varepsilon_{i-1} = 1$. Such a model has been proposed by

Table 5.1 Choice of weighting functions w_i in the CM tests

	Conditioning information
	$\mathbf{z}_{i,1} = (\mathbf{1}_{\{\varepsilon_{i-1}<1\}}, \ \mathbf{1}_{\{\varepsilon_{i-1}<1\}}\varepsilon_{i-1}, \ \mathbf{1}_{\{\varepsilon_{i-1}\geq 1\}}\varepsilon_{i-1})'$
	$\mathbf{z}_{i,2} = (\mathbf{z}'_{i,1}, \ \mathbf{1}_{\{\varepsilon_{i-2}<1\}}, \ \mathbf{1}_{\{\varepsilon_{i-2}<1\}}\varepsilon_{i-2}, \ \mathbf{1}_{\{\varepsilon_{i-2}\geq 1\}}\varepsilon_{i-2})'$
	$\mathbf{z}_{i,3} = (\mathbf{1}_{\{x_{i-1}<1\}}, \ \mathbf{1}_{\{x_{i-1}<1\}}x_{i-1}, \ \mathbf{1}_{\{x_{i-1}\geq 1\}}x_{i-1})'$
	$\mathbf{z}_{i,4} = (\mathbf{z}'_{i,3}, \ \mathbf{1}_{\{x_{i-2}<1\}}, \ \mathbf{1}_{\{x_{i-2}<1\}}x_{i-2}, \ \mathbf{1}_{\{x_{i-2}\geq 1\}}x_{i-2})'$
	CM tests
CM_1	$\mathbf{w}_{i,1} = (x_{i-1}, \ x^2_{i-1}, \ x^3_{i-1}, \ \varepsilon_{i-1}, \ \varepsilon^2_{i-1}, \ \varepsilon^3_{i-1})'$
CM_2	$\mathbf{w}_{i,2} = (w'_{i,1}, \ x_{i-2}, \ x^2_{i-2}, \ x^3_{i-2}, \ \varepsilon_{i-2}, \ \varepsilon^2_{i-2}, \ \varepsilon^3_{i-2})'$
CM_3	$\mathbf{w}_{i,3} = (x_{i-1}, \ \mathbf{z}'_{i,1})'$
CM_4	$\mathbf{w}_{i,4} = (x_{i-1}, \ x_{i-2}, \ \mathbf{z}'_{i,2})'$
CM_5	$\mathbf{w}_{i,5} = (\varepsilon_{i-1}, \ \mathbf{z}'_{i,3})'$
CM_6	$\mathbf{w}_{i,6} = (\varepsilon_{i-1}, \ \varepsilon_{i-2}, \ \mathbf{z}'_{i,4})'$
CM_7	$\mathbf{w}_{i,7} = (\mathbf{z}'_{i,1}, \ \mathbf{z}'_{i,3})'$
CM_8	$\mathbf{w}_{i,8} = (\mathbf{z}'_{i,2}, \ \mathbf{z}'_{i,4})'$
CM_9	$\mathbf{w}_{i,9} = (x_{i-1}, \ x_{i-2}, \ \ldots \ , x_{i-10})'$
CM_{10}	$\mathbf{w}_{i,10} = (\varepsilon_{i-1}, \ \varepsilon_{i-2}, \ \ldots \ , \varepsilon_{i-10})'$
CM_{11}	bins for ε_{i-1} and ε_{i-2}: $[0, 0.1), [0.1, 0.2), [0.2, 0.5), [0.5, 0.8), [0.8, 1), [1.2, 1.5), [1.5, 2), [2, 3), [3, \infty)$
CM_{12}	bins for x_{i-1} and x_{i-2}: $[0, 0.1), [0.1, 0.2), [0.2, 0.5), [0.5, 0.8), [0.8, 1), [1.2, 1.5), [1.5, 2), [2, 3), [3, \infty)$

			ICM tests			
ICM_1	$\rho_i = x_i - \Psi_i$,	$\mathbf{w}(\cdot) = 1$,	$A = 0$,	$B = 0.5$,	$K = 100$	$d = 1$
ICM_2	$\rho_i = x_i - \Psi_i$,	$\mathbf{w}(\cdot) = 1$,	$A = 0$,	$B = 0.5$,	$K = 100$	$d = 2$
ICM_3	$\rho_i = x_i - \Psi_i$,	$\mathbf{w}(\cdot) = 1$,	$A = 0$,	$B = 0.5$,	$K = 100$	$d = 5$
ICM_4	$\rho_i = x_i - \Psi_i$,	$\mathbf{w}(\cdot) = 1$,	$A = 0$,	$B = 0.5$,	$K = 100$	$d = 10$

Dufour and Engle (2000) and allows large innovations ε_i ("positive" surprises in durations) to have a different impact on future durations than small innovations ("negative" surprises). Since it is in line with Nelson's EGARCH model, it is referred to as an EXACD specification. Finally, (5.97) corresponds to a threshold ACD (TACD) model as proposed by Zhang et al. (2001). For more details on these specifications, see Chap. 6.

For each data generating process (DGP), we estimate a (linear) ACD(1,1) specification $\Psi_i = \omega + \alpha x_{i-1} + \beta \Psi_{i-1}$ or a LACD(1,1) specification $\ln \Psi_i = \omega + \alpha \varepsilon_{i-1} + \beta \ln \Psi_{i-1}$, respectively. We use the conditional moment function $\rho_i = x_i/\Psi_i - 1$ and 12 weighting functions $w_{i,j}$, $j = 1, \ldots, 12$, based on functions of past realizations, innovations, and indicator variables indicating possible nonlinear news impact effects. As benchmarks we compute different specifications of the ICM test based on the conditional moment function $\rho_i = x_i - \Psi_i$ (see Table 5.1).

Table 5.2 gives the rejection rates of the individual tests when a linear ACD(1,1) specification is estimated. Correspondingly, Table 5.3 displays the results based on estimations of the LACD(1,1) model. The first column shows the size since the estimated model and the DGP coincide. The CM tests tend to be slightly oversized

Table 5.2 Rejection frequencies of the individual (I)CM tests (see Table 5.1). Size of simulated samples: 3,000. Number of replications: 500. Estimated model: ACD(1,1)

	DGP (5.93)		DGP (5.94)		DGP (5.95)		DGP (5.96)		DGP (5.97)	
	5%	10%	5%	10%	5%	10%	5%	10%	5%	10%
CM_1	0.066	0.126	1.000	1.000	0.498	0.605	1.000	1.000	0.140	0.212
CM_2	0.076	0.142	0.994	1.000	0.526	0.670	1.000	1.000	0.132	0.210
CM_3	0.074	0.146	1.000	1.000	0.454	0.591	1.000	1.000	0.156	0.250
CM_4	0.070	0.148	1.000	1.000	0.443	0.584	1.000	1.000	0.162	0.246
CM_5	0.068	0.138	1.000	1.000	0.464	0.581	1.000	1.000	0.168	0.254
CM_6	0.064	0.136	1.000	1.000	0.436	0.584	1.000	1.000	0.162	0.266
CM_7	0.074	0.116	1.000	1.000	0.485	0.591	1.000	1.000	0.182	0.282
CM_8	0.076	0.130	1.000	1.000	0.447	0.567	1.000	1.000	0.186	0.284
CM_9	0.072	0.120	0.998	1.000	0.488	0.601	1.000	1.000	0.168	0.274
CM_{10}	0.068	0.122	0.996	1.000	0.440	0.564	1.000	1.000	0.188	0.286
CM_{11}	0.064	0.104	1.000	1.000	0.519	0.615	1.000	1.000	0.210	0.314
CM_{12}	0.066	0.126	1.000	1.000	0.450	0.574	1.000	1.000	0.222	0.338
ICM_1	0.010	0.022	0.840	0.872	0.175	0.251	0.930	0.952	0.014	0.034
ICM_2	0.008	0.020	0.822	0.860	0.203	0.275	0.918	0.940	0.014	0.034
ICM_3	0.010	0.022	0.824	0.860	0.199	0.306	0.908	0.930	0.012	0.030
ICM_4	0.006	0.022	0.818	0.852	0.199	0.316	0.912	0.934	0.016	0.030

Table 5.3 Rejection frequencies of the individual (I)CM tests (see Table 5.1). Size of simulated samples: 3,000. Number of replications: 500. Estimated model: LACD(1,1)

	DGP (5.94)		DGP (5.93)		DGP (5.95)		DGP (5.96)		DGP (5.97)	
	5%	10%	5%	10%	5%	10%	5%	10%	5%	10%
CM_1	0.096	0.140	0.224	0.346	0.743	0.843	0.860	0.918	0.062	0.113
CM_2	0.082	0.148	0.250	0.374	0.701	0.808	0.838	0.896	0.063	0.113
CM_3	0.100	0.156	0.306	0.458	0.910	0.948	0.942	0.964	0.058	0.122
CM_4	0.090	0.150	0.324	0.470	0.887	0.941	0.872	0.916	0.058	0.132
CM_5	0.060	0.130	0.342	0.496	0.893	0.946	0.960	0.976	0.072	0.135
CM_6	0.062	0.128	0.326	0.492	0.843	0.908	0.936	0.964	0.070	0.147
CM_7	0.078	0.132	0.438	0.590	0.960	0.979	0.994	0.994	0.077	0.130
CM_8	0.074	0.134	0.422	0.558	0.935	0.964	0.968	0.982	0.080	0.142
CM_9	0.072	0.126	0.320	0.464	0.935	0.962	0.874	0.918	0.067	0.127
CM_{10}	0.068	0.124	0.312	0.446	0.881	0.946	0.866	0.914	0.067	0.132
CM_{11}	0.080	0.146	0.410	0.564	0.969	0.981	0.912	0.948	0.070	0.137
CM_{12}	0.084	0.134	0.412	0.528	0.950	0.977	0.856	0.910	0.077	0.140
ICM_1	0.026	0.074	0.066	0.134	0.065	0.128	0.042	0.098	0.008	0.028
ICM_2	0.030	0.064	0.074	0.166	0.044	0.088	0.046	0.096	0.003	0.023
ICM_3	0.034	0.066	0.074	0.166	0.048	0.094	0.042	0.102	0.005	0.015
ICM_4	0.038	0.084	0.078	0.172	0.046	0.109	0.044	0.110	0.003	0.022

for the given sample size whereas the ICM test is strongly undersized. The power of the CM tests is generally quite high and increases with the strength of the deviation from linearity in Ψ_i. Consequently, the tests have very high power to evaluate the linear ACD model against the DGPs (5.94) or (5.96). Lower rejection rates are shown for tests against additive stochastic components (DGP (5.95)) and regime switching behavior (DGP (5.97)). Both forms of misspecification are hard to detect since the deviation from a linear ACD is not too severe. A similar picture is revealed by Table 5.3. Nevertheless, the test's power against an underlying linear ACD specification when a LACD is estimated is clearly lower than in the reversed case. Not surprisingly, the tests have lower power to distinguish between an LACD and an alternative specification implying also a concave news impact function. The highest power is shown for conditional moment tests based on weighting functions which are particularly sensitive against nonlinearities in the news response function (e.g., CM_{11} and CM_{12}). These specifications have power against a wide range of possible misspecifications. This is still true even when we take into account that the tests tend to be oversized. In contrast, the power properties of the ICM tests are very poor regardless the choice of the underlying nuisance parameters. This is a general finding for omnibus tests and is also confirmed by Meitz and Teräsvirta (2006) using Hong and Lee's (2003) spectral density test illustrated above.

In conclusion, the results indicate that an appropriate choice of the weighting functions induces consistency against a wide range of misspecifications while preserving reasonable (size-adjusted) power properties in finite samples. Consequently, in real applications, CM tests are valuable complements to LM type tests. Both kind of tests serve as constructive tests in the sense of Godfrey (1996) allowing to detect possible sources of model misspecification. Moreover, the proposed framework is straightforwardly applied to test also multivariate MEM processes or restrictions on higher order moments.

References

Aït-Sahalia Y (1996) Testing continuous-time models of the spot interest rate. Rev Financ Stud 9:385–426

Allen D, Chan F, McAleer M, Peiris S (2008) Finite sample properties of the QMLE for the Log-ACD model: application to Australian stocks. J Econom 147:163–185

Andersen TG, Bollerslev T (1998b) Deutsche mark-dollar volatility: intraday activity patterns, macroeconomic announcements, and longer run dependencies. J Finance 53:219–265

Andrews D (1991) Heteroscedasticity and autocorrelation consistent covariance matrix estimation. Econometrica 59:817–858

Bauwens L, Giot P, Grammig J, Veredas D (2004) A comparison of financial duration models via density forecasts. Int J Forecast 20:589–609

Bauwens L, Galli F, Giot P (2008) The moments of Log-ACD models. Quant Qual Anal Soc Sci 2:1–28

Bauwens L, Giot P (2000) The logarithmic ACD model: an application to the bid/ask quote process of two NYSE stocks. Annales d'Economie et de Statistique 60:117–149

Bierens HJ (1982) Consistent model specification tests. J Econom 20:105–134

Bierens HJ (1990) A consistent conditional moment test of functional form. Econometrica 58:1443–1458

Bollerslev T (1986) Generalized autoregressive conditional heteroskedasticity. J Econom 31:307–327

Bollerslev T, Wooldridge J (1992) Quasi-maximum likelihood estimation and inference in dynamic models with time varying covariances. Econom Rev 11:143–172
Box GEP, Pierce DA (1970) Distribution of residual autocorrelations in the autoregressive-integrated moving average time series models. J Am Stat Assoc 65:1509–1526
Breusch TS (1978) Testing for autocorrelation in dynamic linear models. Aust Econ Pap 17:334–355
Breusch TS, Pagan RA (1979) A simple test for heteroskedasticity and random coefficient variation. Econometrica 47:203–207
Brock W, Dechert WD, Scheinkman J, LeBaron B (1996) A test for independence based on the correlation dimension. Econom Rev 15:197–235
Brownlees C, Cipollini F, Gallo GM (2011) Journal of Financial Econometrics 9:489–518
Brownlees C, Gallo GM (2011) International Journal of Forecasting 27:365–378
Chen X (2000) A beta kernel estimation for the density functions. Comput Stat Data Anal 31:131–145
Chen Y-T, Hsieh C-S (2010) Generalized moment tests for autoregressive conditional duration models. J Financ Econom 8:345–391
de Boor C (1978) A practical guide to splines. Springer Verlag, Berlin, Heidelberg
de Jong RM (1996) The Bierens test under data dependence. J Econom 72:1–32
De Luca G, Gallo G (2009) Time-varying mixing weights in mixture autoregressive conditional duration models. Econom Rev 28:101–120
Diebold FX, Gunther TA, Tay AS (1998) Evaluating density forecasts, with applications to financial risk management. Int Econ Rev 39:863–883
Drost FC, Werker BJM (2004) Semiparametric duration models. J Bus Econ Stat 22:40–50
Duchesne P, Pacurar M (2008) Evaluating financial time series models for irregularly spaced data: a spectral density approach. Comput Oper Res 35:130–155
Dufour A, Engle RF (2000) The ACD model: predictability of the time between consecutive trades. Working Paper, ISMA Centre, University of Reading
Dufour JM, Roy R (1985) Some robust exact results on sample autocorrelations and tests of randomnes. J Econom 29:257–273
Dufour JM, Roy R (1986) Generalized Portmanteau statistics and tests of randomness. Commun Stat Theory Methods 15:2953–2972
Engle RF (2000) The econometrics of ultra-high-frequency data. Econometrica 68(1):1–22
Engle RF (2002b) New frontiers for ARCH models. J Appl Econom 17:425–446
Engle RF (1982) Autoregressive conditional heteroscedasticity with estimates of the variance of United Kingdom inflation. Econometrica 50:987–1006
Engle RF (1984) Wald, likelihood ratio and lagrange multiplier tests in econometrics. In: Griliches Z, Intriligator MD (eds) Handbook of econometrics, vol. II, chap. 13. Elsevier Science, pp. 775–826
Engle RF (1996) The Econometrics of ultra-high frequency data. Discussion Paper 96-15, University of California San Diego
Engle RF, Gallo GM (2006) A multiple indicators model for volatility using intra-daily data. J Econom 131:3–27
Engle RF, Ng VK (1993) Measuring and testing the impact of news on volatility. J Finance 48:1749–1778
Engle RF, Russell JR (1997) Forecasting the frequency of changes in quoted foreign exchange prices with the autoregressive conditional duration model. J Empir Financ 4:187–212
Engle RF, Russell JR (1998) Autoregressive conditional duration: a new model for irregularly spaced transaction data. Econometrica 66:1127–1162
Fernandes M (2004) Bounds for the probability distribution function of the linear ACD process. Stat Probab Lett 68:169–176
Fernandes M, Grammig J (2005) Non-parametric specification tests for conditional duration models. J Econom 127:35–68
Gallant RA (1981) On the bias in flexible functional forms and an essential unbiased form: The Fourier flexible form. J Econom 15:211–245

Gosh BK, Huang W-M (1991) The power and optimal kernel of the Bickel–Rosenblatt test for goodness-of-fit. Ann Stat 19:999–1009

Ghysels E, Gouriéroux C, Jasiak J (1998) Stochastic volatility duration models. Discussion paper, CIRANO

Godfrey LG (1978) Testing against general autoregressive and moving average error models when the regressors include lagged dependent variables. Econometrica 46:1293–1302

Godfrey LG (1996) Misspecification tests and their use in econometrics. J Stat Plan Inference 49:241–260

Gouriéroux C, Monfort A, Trognon A (1984) Pseudo maximum likehood methods: theory. Econometrica 52:681–700

Grammig J, Maurer K-O (2000) Non-monotonic hazard functions and the autoregressive conditional duration model. Econom J 3:16–38

Hagmann M, Scaillet O (2007) Local multiplicative bias correction for asymmetric kernel density estimators. J Econom 141:213–249

Hamilton JD, Jorda O (2002) A model of the federal funds rate target. J Polit Econ 110:1135

Hautsch N (2003) Assessing the risk of liquidity suppliers on the basis of excess demand intensities. J Financ Econom 1:189–215

Hautsch N, Malec P, Schienle M (2010) Capturing the zero: a new class of zero-augmented distributions and multiplicative error processes. Discussion Paper 2010-055, Humboldt-Universität zu Berlin

Hendry DF (1995) Dynamic econometrics. Oxford University Press, Oxford

Hjort NL, Glad IK (1995) Nonparametric density estimation with a parametric start. Ann Stat 23:882–904

Hong Y (1996) Consistent testing for serial correlation of unknown form. Econometrica 64:837–864

Hong Y (1999) Hypothesis testing in time series via the empirical characteristic function: a generalized spectral density approach. J Am Stat Assoc 84:1201–1220

Hong Y, Lee T-H (2003) Diagnostic checking for the adequacy of nonlinear time series models. Econom Theory 19:1065–1121

Hong Y, Lee Y-J (2011) Journal of Time Series Analysis 32:1–32

Kalbfleisch JD, Prentice RL (1980) The statistical analysis of failure time data. Wiley, New York

Karanasos M (2004) The statistical properties of long-memory ACD models. WSEAS Trans Bus Econ 2:169–175

Karanasos M (2008) The statistical properties of exponential ACD models. Quant Qual Anal Soc Sci 2:29–49

Kwan ACC, Sim A-B (1996a) On the finite-sample distribution of modified Portmanteau tests for randomness of a Gaussian time series. Biometrika 83:938–943

Kwan ACC, Sim A-B (1996b) Portmanteau tests of randomness and Jenkins' variance-stabilizing transformation. Econ Lett 50:41–49

Kwan ACC, Sim A-B, Wu Y (2005) A comparative study of the finite-sample performance of some Portmanteau tests for randomness of a time series. Comput Stat Data Anal 48:391–413

Lancaster T (1997) The econometric analysis of transition data. Cambridge University Press

Lawrence AL, Lewis PA (1980) The exponential autoregressive-moving average EARMA(P,Q) model. J R Stat Soc Series B 42:150–161

Lee S, Hansen B (1994) Asymptotic theory for the GARCH(1 1) quasi-maximum likelihood estimator. Econom Theory 10:29–52

Li WK, Mak TK (1994) On the squared residual autorrelations in non-linear time series with conditional heteroscedasticity. J Time Series Anal 15:627–636

Li WK, Yu LH (2003) On the residual autocorrelation of the autoregressive conditional duration model. Econ Lett 79:169–175

Ljung GM, Box GEP (1978) On a measure of lack of fit in time series models. Biometrika 65:297–303

Luca GD, Zuccolotto P (2006) Regime-switching Pareto distributions for ACD models. Comput Stat Data Anal 51:2179–2191

Lunde A (2000) A generalized gamma autoregressive conditional duration model. Discussion paper, University of Aarhus

Manganelli S (2005) Duration, volume and volatility impact of trades. J Finan Markets 8:377–399

McLeod AI, Li WK (1983) Diagnostic checking arma time series models using squared residual autocorrelations. J Time Series Anal 4:269–273

Meitz M, Teräsvirta T (2006) Evaluating models of autoregressive conditional duration. J Bus Econ Stat 24:104–124

Nelson D (1991) Conditional heteroskedasticity in asset returns: a new approach. J Econom 43:227–251

Newey WK (1985) Maximum likelihood specification testing and conditional moment tests. Econometrica 5:1047–1070

Newey WK, West KD (1987) A simple, positive semidefinite, heteroskedasticity and autocorrelation consistent covariance matrix. Econometrica 55:703–708

Pacurar M (2008) Autoregressive conditional duration models in finance: a survey of the theoretical and empirical literature. J Econ Surveys 22:711–751

Pagan A, Vella F (1989) Diagnostic tests for models based on individual data: a survey. J Appl Econom 4:29–59

Rosenblatt M (1952) Remarks on a multivariate transformation. Ann Math Stat 23:470–472

Silverman BW (1986) Density estimation for statistics and data analysis. Chapman and Hall, London

Tauchen G (1985) Diagnostic testing and evaluation of maximum likelihood models. J Econom 30:415–443

Taylor SJ (1982) Financial returns modelled by the product of two stochastic processes – a study of daily sugar prices. In: Anderson OD (ed) Time series analysis: theory and practice, North-Holland, Amsterdam

Veredas D, Rodriguez-Poo J, Espasa A (2008) Semiparametric estimation for financial durations. In: Bauwens WPL, Veredas D (eds) High frequency financial econometrics. Physica-Verlag, Heidelberg, pp 225–251

White H (1982) Maximum likelihood estimation of misspecified models. Econometrica 50(1):1–25

Wooldridge JM (1990) A unified approach to robust, regression-based specification tests. Econom Theory 6:17–43

Wooldridge JM (1991) Specification testing and quasi-maximum-likelihood estimation. J Econom 48:29–55

Zhang MY, Russell JR, Tsay RS (2001) A nonlinear autoregressive conditional duration model with applications to financial transaction data. J Econom 104:179–207

Chapter 6
Generalized Multiplicative Error Models

In this chapter, we present generalizations of the basic multiplicative error model as introduced in Chap. 5. Section 6.1 discusses a class of ACD models which can be presented in terms of a generalized polynomial random coefficient model according to Carrasco and Chen (2002). We illustrate various special cases, discuss the theoretical properties and show empirical illustrations. In Sect. 6.2, we consider regime-switching ACD models allowing for parameters which might change in dependence of observable or unobservable characteristics. We concentrate on threshold ACD models, smooth transition ACD models as well as Markov Switching ACD specifications. Section 6.3 focuses on ACD models accommodating long range dependence in the data. In this context, we discuss different possibilities to capture long memory. In Sect. 6.4, we focus on mixture and component models. We discuss two types of mixture models, where the conditional mean function is driven by a dynamic latent component. The stochastic conditional duration model proposed by Bauwens and Veredas (2004), as discussed in Sect. 6.4.1, assumes the conditional mean function to follow a latent AR(1) process. The stochastic multiplicative error model introduced by Hautsch (2008) shown in Sect. 6.4.2 generalizes this idea and allows for both latent as well as observation-driven dynamics. Likewise, a component MEM specification as proposed by Brownlees et al. (2011) allows to combine intradaily dynamics with daily dynamics. Further generalizations, such as semiparametric ACD models and stochastic volatility duration models are discussed in Sect. 6.5.

6.1 A Class of Augmented ACD Models

In this section, we discuss a general class of ACD models which can be presented in terms of a generalized polynomial random coefficient ACD model as analyzed by Carrasco and Chen (2002). Building on the notation introduced in Chap. 5, this class of models is given by

$$\vartheta(\Psi_i) = \mathcal{A}(\varepsilon_i)\vartheta(\Psi_{i-1}) + \mathcal{C}(\varepsilon_i), \tag{6.1}$$

N. Hautsch, *Econometrics of Financial High-Frequency Data*,
DOI 10.1007/978-3-642-21925-2_6,

where $\vartheta(\cdot)$ is a continuous function with domain $[0,\infty)$, $\mathcal{A}$ and $\mathcal{B}$ are polynomial functions, and $\varepsilon_i = x_i/\Psi_i$ denotes an i.i.d. innovation term with $\mathbb{E}[\varepsilon_i] = 1$.

6.1.1 *Special Cases*

The class of generalized polynomial random coefficient ACD models contains various extensions of the basic linear ACD model allowing for additive as well as multiplicative stochastic components, i.e., specifications, where lagged innovations enter the conditional mean function additively and/or multiplicatively. Moreover, it contains parameterizations capturing not only linear but also more flexible news impact curves. For simplicity of exposition, the following discussions are restricted to models with a lag order of $P = Q = 1$.

6.1.1.1 Additive and Multiplicative ACD (AMACD) Model

A simple extension of the basic ACD specification incorporating both an *additive* and *multiplicative* innovation component is given by

$$\Psi_i = \omega + (\alpha\Psi_{i-1} + \nu)\varepsilon_{i-1} + \beta\Psi_{i-1}, \tag{6.2}$$

where ν is a parameter. This specification implies a news impact curve with a slope given by $\alpha\Psi_{i-1} + \nu$. Thus, the lagged innovation enters the conditional mean function additively, as well as multiplicatively. In this sense, the (so called) Additive and Multiplicative ACD (AMACD) model is more flexible and nests the linear ACD model for $\nu = 0$.

6.1.1.2 Box–Cox ACD (BACD) Model

Hautsch (2003) suggests an additive ACD model based on power transformations of Ψ_i and ε_i:

$$\Psi_i^{\delta_1} = \omega + \alpha\varepsilon_{i-1}^{\delta_2} + \beta\Psi_{i-1}^{\delta_1}, \tag{6.3}$$

where $\delta_1, \delta_2 > 0$. It is easy to see that this model can be written in terms of Box–Cox transformations. Thus,

$$\frac{\Psi_i^{\delta_1} - 1}{\delta_1} = \tilde{\omega} + \tilde{\alpha}(\varepsilon_{i-1}^{\delta_2} - 1)/\delta_2 + \beta\frac{\Psi_{i-1}^{\delta_1} - 1}{\delta_1}, \tag{6.4}$$

where

$$\tilde{\omega} := \frac{\omega + \alpha + \beta - 1}{\delta_1} \quad \text{and} \quad \tilde{\alpha} := \frac{\alpha\delta_2}{\delta_1}.$$

This specification allows for concave, convex, as well as linear news impact functions. It nests the AMACD model for $\delta_1 = \delta_2 = 1$, the LACD_1 model for $\delta_1 \to 0$, $\delta_2 \to 0$ and the LACD_2 model for $\delta_1 \to 0$, $\delta_2 = 1$ (see Chap. 5). For $\delta_1 \to 0$, it coincides with a Box–Cox ACD specification proposed by Dufour and Engle (2000).

6.1.1.3 EXponential ACD (EXACD) Model

Alternative parameterizations are obtained by the assumption of piece-wise linear news impact functions. Dufour and Engle (2000) introduce the so-called EXponential[1] ACD model capturing features of the EGARCH specification proposed by Nelson (1991). This model allows for a linear news impact function that is kinked at $\varepsilon_{i-1} = 1$:

$$\ln \Psi_i = \omega + \alpha\varepsilon_{i-1} + c|\varepsilon_{i-1} - 1| + \beta \ln \Psi_{i-1}. \tag{6.5}$$

For durations shorter than the conditional mean ($\varepsilon_{i-1} < 1$), the news impact curve has a slope $\alpha - c$ and an intercept $\omega + c$, while for durations longer than the conditional mean ($\varepsilon_{i-1} > 1$), slope and intercept are $\alpha + c$ and $\omega - c$, respectively.

6.1.1.4 Augmented Box–Cox ACD (ABACD) Model

While the EXACD model allows for news impact curves that are kinked at $\varepsilon_{i-1} = 1$ only, a valuable generalization is to parameterize also the position of the kink. Using the parameterization for modelling asymmetric GARCH processes introduced by Hentschel (1995), we obtain a specification that we call augmented Box–Cox ACD model:

$$\Psi_i^{\delta_1} = \omega + \alpha(|\varepsilon_{i-1} - b| + c(\varepsilon_{i-1} - b))^{\delta_2} + \beta\Psi_{i-1}^{\delta_1}. \tag{6.6}$$

In this specification, the parameter b gives the position of the kink while δ_2 determines the shape of the piecewise functions around the kink. For $\delta_2 > 1$, the shape is convex and for $\delta_2 < 1$, it is concave. It nests the BACD model for $b = 0$ and $c = 0$. This model does not encompass the basic ACD model since the ABACD specification is based on an *additive* stochastic component.

Even though this specification of the news impact function allows for more flexibility, it has one major drawback since the parameter restriction $|c| <= 1$ has to be imposed in order to circumvent complex values whenever $\delta_2 \neq 1$. This restriction is binding in the case where the model has to be fitted to data that imply

[1]They use this notation to prevent confusion with the Exponential ACD model (EACD) based on an exponential distribution.

an upward kinked concave news impact function. Such a pattern is quite typical for financial durations (see also the empirical results in Sect. 6.1.3) and is only possible for $c < -1$ (and $\alpha < 0$). Hence, in such a case, a restricted version of the BACD model must be estimated. In particular, for cases when c converges to the boundary, either δ_2 or, alternatively, $|c|$ has to be fixed to 1. Setting $\delta_2 = 1$ implies a piecewise linear news impact function that is kinked at b. In that case, the model is reformulated as

$$\frac{\Psi_i^{\delta_1} - 1}{\delta_1} = \tilde{\omega} + \tilde{\alpha}\varepsilon_{i-1} + \tilde{c}|\varepsilon_{i-1} - b| + \beta\frac{\Psi_{i-1}^{\delta_1} - 1}{\delta_1}, \tag{6.7}$$

where

$$\tilde{\omega} := \frac{\omega - \alpha c b + \beta - 1}{\delta_1}, \qquad \tilde{\alpha} := \frac{\alpha c}{\delta_1}, \qquad \text{and} \qquad \tilde{c} := \frac{\alpha}{\delta_1}.$$

Such a specification nests the EXACD model for $\delta_1 \to 0$ and $b = 1$. Alternatively, by setting $c = 1$, the specification becomes

$$\Psi_i^{\delta_1} = \omega + \tilde{\alpha}(\varepsilon_{i-1} - b)^{\delta_2}\mathbb{1}_{\{\varepsilon_{i-1} \geq b\}} + \beta\Psi_{i-1}^{\delta_1},$$

where $\tilde{\alpha} = 2^{\delta_2}\alpha$. Thus, the news impact function is zero for $\varepsilon_{i-1} \leq b$ and follows a concave (convex) function for $\delta_2 < 1$ ($\delta_2 > 1$). Correspondingly, setting $c = -1$ leads to

$$\Psi_i^{\delta_1} = \omega + \tilde{\alpha}(b - \varepsilon_{i-1})^{\delta_2}\mathbb{1}_{\{\varepsilon_{i-1} \leq b\}} + \beta\Psi_{i-1}^{\delta_1}.$$

Note that the latter two restricted ABACD specifications do not nest the EXACD model but the BACD model for $b = 0$ ($b \to \infty$) whenever c is set to 1 (-1).

6.1.1.5 Hentschel ACD (HACD) Model

An alternative nonlinear ACD model is proposed by Fernandes and Grammig (2006). They introduce a specification that is the direct counterpart to the augmented GARCH process proposed by Hentschel (1995). Though Fernandes and Grammig call it augmented ACD model, here, we call it H(entschel)-ACD model in order to avoid confusion with other augmented ACD processes considered in this section. The HACD mode is given by

$$\Psi_i^{\delta_1} = \omega + \alpha\Psi_{i-1}^{\delta_1}(|\varepsilon_{i-1} - b| + c(\varepsilon_{i-1} - b))^{\delta_2} + \beta\Psi_{i-1}^{\delta_1}. \tag{6.8}$$

This specification is quite similar to the ABACD model. The main difference is that the HACD model is based on a *multiplicative* stochastic component (since $\Psi_{i-1}^{\delta_1}$

acts multiplicatively with a function of ε_{i-1}) while the ABACD model is based on an *additive* stochastic component. Therefore, the HACD model imposes the same parameter restriction for c as in the ABACD case. Since the HACD model includes a multiplicative stochastic component, it nests the basic linear ACD model for $\delta_1 = \delta_2 = 1$, $b = c = 0$. It coincides with a special case of the ABACD model for $\delta_1 \to 0$. Therefore, it nests also the LACD$_1$ model for $\delta_1 \to 0$, $\delta_2 \to 0$, $b = c = 0$ and the LACD$_2$ model for $\delta_1 \to 0$, $\delta_2 = 1$, $b = c = 0$. Moreover, it corresponds to the Box–Cox specification introduced by Dufour and Engle (2000) for $\delta_1 \to 0$, $b = c = 0$ and to the EXACD model for $\delta_1 \to 0$, $\delta_2 = b = 1$. However, in general, it does not encompass the AMACD, BACD and ABACD model since it is based on a multiplicative stochastic component.

6.1.1.6 Augmented Hentschel ACD (AHACD) Model

An encompassing model that nests all specifications outlined above is given by

$$\begin{aligned}\Psi_i^{\delta_1} = \omega + \alpha \Psi_{i-1}^{\delta_1}(|\varepsilon_{i-1} - b| \\ + c(\varepsilon_{i-1} - b))^{\delta_2} + \nu(|\varepsilon_{i-1} - b| + c(\varepsilon_{i-1} - b))^{\delta_2} + \beta \Psi_{i-1}^{\delta_1}.\end{aligned} \tag{6.9}$$

We call this specification augmented Hentschel ACD model since it combines the HACD model with the ABACD specification. This ACD model allows for both additive and multiplicative stochastic coefficients, and therefore implies an additive as well as a multiplicative impact of past shocks on the conditional mean function. It encompasses all specifications nested by the HACD model, as well as all other models based on additive stochastic components. In particular, it nests the AMACD model for $\delta_1 = \delta_2 = 1$, $b = c = 0$, the BACD model for $\alpha = b = c = 0$ and the ABACD model for $\alpha = 0$. The AGACD, the HACD and the ABACD models coincide for $\delta_1 \to 0$. Therefore, the parameters α and ν are only separately identifiable whenever $\delta_1 > 0$. A further generalization would be to specify different news impact parameters b, c and δ_2 for the additive and the stochastic component. However, in this case the estimation of the model becomes tricky due to numerical complexity.

6.1.1.7 Spline News Impact ACD (SNIACD) Model

A further type of ACD specification is obtained by modelling the news response in terms of a piecewise linear function. In the spirit of Engle and Ng (1993), the news impact curve might be parameterized as a linear spline function with nodes at given break points of ε_{i-1}. In particular, the range of ε_{i-1} is divided into M intervals where M^- (M^+) denotes the number of intervals in the range $\varepsilon_{i-1} < 1$ ($\varepsilon_{i-1} > 1$) with $M = M^- + M^+$. By denoting the breakpoints by $\{\bar{\varepsilon}_{M^-}, \ldots, \bar{\varepsilon}_{-1}, \bar{\varepsilon}_0, \bar{\varepsilon}_1, \ldots, \bar{\varepsilon}_{M^+}\}$, the SNIACD model is given by

$$\Psi_i = \omega + \sum_{m=0}^{M^+} \alpha_m^+ \mathbb{1}_{\{\varepsilon_{i-1} \geq \bar{\varepsilon}_m\}} (\varepsilon_{i-1} - \bar{\varepsilon}_m)$$

$$+ \sum_{m=0}^{M^-} \alpha_m^- \mathbb{1}_{\{\varepsilon_{i-1} < \bar{\varepsilon}_m\}} (\varepsilon_{i-1} - \bar{\varepsilon}_m) + \beta \Psi_{i-1}, \tag{6.10}$$

where α_m^+ and α_m^- denote the coefficients associated with the piecewise linear spline. Alternatively, the model can also be specified in terms of a logarithmic transformation of Ψ_i. In this form, the model is more easy to estimate since it does not require any non-negativity restrictions. The intervals must not be equally sized, nor do we need the same number of intervals on each side of $\bar{\varepsilon}_0$. As pointed out by Engle and Ng (1993), a slow increase in M as a function of the sample size should asymptotically give a consistent estimate of the news impact curve. This specification allows for extremely flexible (nonlinear) news responses, but does not necessarily nest the other (parametric) ACD specifications.

6.1.2 Theoretical Properties

By recalling the formulation of ACD models in terms of a generalized polynomial coefficient model, (6.1), we can classify all ACD specifications discussed above by using a corresponding parameterization for $\vartheta(\cdot)$, $\mathcal{A}(\cdot)$ and $\mathcal{C}(\cdot)$. This is shown in Table 6.1. The theoretical properties for this class of models are derived by Carrasco and Chen (2002) building on results by Mokkadem (1990). They provide sufficient conditions ensuring β-mixing, strict stationarity and the existence of higher order moments. We reproduce the main findings in the following proposition:

Proposition 6.1. *Assume that the durations x_i, $i = 1, \ldots, n$, follow the process $x_i = \Psi_i \varepsilon_i$, where Ψ_i is given by one of the processes in Table 6.1. Assume that ε_i is an i.i.d. random variable that is independent of Ψ_i and presume that the marginal probability distribution of ε_i is absolutely continuous with respect to the Lebesgue measure on $(0, \infty)$. Moreover, ε_i is independent of $\sigma(\vartheta(\Psi_{i-1}, \ldots, \vartheta(\Psi_1))$. Let $\varrho(\mathcal{A}(\cdot))$ be the largest eigenvalue in absolute value of the polynomial function $\mathcal{A}(\cdot)$. Moreover, assume that*

(i) $\mathcal{A}(\cdot)$ and $\mathcal{C}(\cdot)$ are polynomial functions that are measurable with respect to the sigma field generated by ε_i.
(ii) $\varrho(\mathcal{A}(0)) < 1$.
(iii) $|\mathcal{A}(0)| < 1$, $\mathbb{E}[\mathcal{A}(\varepsilon_i)]^s < 1$ and $\mathbb{E}[\mathcal{C}(\varepsilon_i)]^s < \infty$ for some integers $s \geq 1$.

Then, the process $\{\Psi_i\}$ is Markov geometrically ergodic and $\mathbb{E}[\Psi_i]^s < \infty$. If the processes $\{\Psi_i\}_{i=1}^n$ and $\{x_i\}_{i=1}^n$ are initialized from their ergodic distribution, then they are strictly stationary and β-mixing with exponential decay.

Table 6.1 Classification of ACD models

Typ	$\vartheta(\Psi_i)$	$\mathcal{A}(\varepsilon_i)$	$\mathcal{C}(\varepsilon_i)$
Linear and logarithmic ACD models			
ACD	Ψ_i	$\alpha\varepsilon_{i-1}+\beta$	ω
LACD$_1$	$\ln\Psi_i$	β	$\omega+\alpha\ln\varepsilon_{i-1}$
LACD$_2$	$\ln\Psi_i$	β	$\omega+\alpha\varepsilon_{i-1}$
Nonlinear ACD models			
BACD	$\Psi_i^{\delta_1}$	β	$\omega+\alpha\varepsilon_{i-1}^{\delta_2}$
EXACD	$\ln\Psi_i$	β	$\omega+\alpha\varepsilon_{i-1}+c\lvert\varepsilon_{i-1}-1\rvert$
ABACD	$\Psi_i^{\delta_1}$	β	$\omega+\alpha(\lvert\varepsilon_{i-1}-b\rvert+c(\varepsilon_{i-1}-b))^{\delta_2}$
HACD	$\Psi_i^{\delta_1}$	$\alpha(\lvert\varepsilon_{i-1}-b\rvert+\beta$ $+c(\varepsilon_{i-1}-b))^{\delta_2}$	ω
Augmented ACD models			
AMACD	Ψ_i	$\alpha\varepsilon_{i-1}+\beta$	$\omega+\nu\varepsilon_{i-1}$
AHACD	$\Psi_i^{\delta_1}$	$\alpha(\lvert\varepsilon_{i-1}-b\rvert+\beta$ $+c(\varepsilon_{i-1}-b))^{\delta_2}$	$\omega+\nu(\lvert\varepsilon_{i-1}-b\rvert+c(\varepsilon_{i-1}-b))^{\delta_2}$
Spline news impact ACD model			
SNIACD	Ψ_i or $\ln\Psi_i$	β	$\omega+\sum_{m=0}^{M^+}\alpha_m^+\mathbf{1}_{\{\varepsilon_{i-1}\geq\bar{\varepsilon}_m\}}(\varepsilon_{i-1}-\bar{\varepsilon}_m)$ $+\sum_{m=0}^{M^-}\alpha_m^-\mathbf{1}_{\{\varepsilon_{i-1}<\bar{\varepsilon}_m\}}(\varepsilon_{i-1}-\bar{\varepsilon}_m)$

Proof: See Carrasco and Chen (2002), Proposition 2 and 5. □

Hence, establishing stationarity and ergodicity conditions for the individual ACD specifications requires imposing restrictions on the functions $\mathcal{A}(\cdot)$ and $\mathcal{C}(\cdot)$. Corresponding examples in the context of individual GARCH specifications are given by Carrasco and Chen (2002). Generalizations of these results are given by Meitz and Saikkonen (2008) in the framework of Markov models.[2]

6.1.3 Empirical Illustrations

Table 6.4 shows estimates of different ACD specifications based on (de-seasonalized) trade durations and \$0.05 midquote change durations for the Coca-Cola and GE stock traded at the NYSE covering the period from 01/02/01 to 05/31/01. For each type of financial duration, seven different ACD specifications are estimated, the basic ACD model, the LACD$_2$ model, the BACD model, the EXACD model, the ABACD model and a logarithmic version of the SNIACD model. The individual time series are re-initialized every trading day and are estimated by

[2]See also Carrasco and Chen (2005) and Meitz and Saikkonen (2008) for a correction of some of the results provided by Carrasco and Chen (2002). These corrections, however, do not affect the proposition presented above.

QML. The lag orders are chosen according to the Bayes Information Criterion (BIC) suggesting ACD(2,1) or ACD(2,2) models as the preferred specification. Note that for the individual types of nonlinear ACD specifications, the inclusion of a second lag is not straightforward because it also requires a parameterization of the second lag news impact function. Hence, it doubles the corresponding news response parameters which is not practicable for the highly parameterized ABACD and SNIACD models. For this reason, a parameterization is chosen that allows, on the one hand, to account for higher order dynamics, but on the other hand, ensures model parsimony. In particular, the news impact of the second lag is modelled in the same way as for the first lag, i.e., based on the same parameters δ_1, δ_2, c, with only the autoregressive parameters α, β, c and ν being doubled. Then, the ABACD(2,2) model is given by

$$\Psi_i^{\delta_1} = \omega + \sum_{j=1}^{2} \alpha_j (|\varepsilon_{i-j} - b| + c_j(\varepsilon_{i-j} - b))^{\delta_2} + \sum_{j=1}^{2} \beta_j \Psi_{i-j}^{\delta_1}.$$

Correspondingly, the SNIACD(2,2) model is given by

$$\begin{aligned}\Psi_i &= \omega + \sum_{m=0}^{M^+} \alpha_m^+ \mathbb{1}_{\{\varepsilon_{i-1} > \bar{\varepsilon}_m\}}(\varepsilon_{i-1} - \bar{\varepsilon}_m) + \sum_{m=0}^{M^-} \alpha_m^- \mathbb{1}_{\{\varepsilon_{i-1} < \bar{\varepsilon}_m\}}(\varepsilon_{i-1} - \bar{\varepsilon}_m) \\ &+ \sum_{m=0}^{M^+} (\alpha_m^+ + a^+) \mathbb{1}_{\{\varepsilon_{i-2} \geq \bar{\varepsilon}_m\}}(\varepsilon_{i-2} - \bar{\varepsilon}_m) \\ &+ \sum_{m=0}^{M^-} (\alpha_m^- + a^-) \mathbb{1}_{\{\varepsilon_{i-2} < \bar{\varepsilon}_m\}}(\varepsilon_{i-2} - \bar{\varepsilon}_m) + \sum_{j=1}^{2} \beta_j \Psi_{i-j}. \end{aligned} \tag{6.11}$$

Tables 6.2–6.5 give the estimation results and diagnostics. Two difficulties have to be considered in this context. First, for nonlinear models involving absolute value functions in the news impact function, the estimation of the Hessian matrix is often cumbersome due to numerical problems. For this reason, the asymptotic standard errors are estimated by the OPG estimator of the variance covariance matrix. Second, in several specifications, the coefficients δ_1 and c converge to their boundaries. Then, the coefficients are set to their boundaries and correspondingly restricted models are re-estimated. In this context, see the discussion in Sect. 6.1.1.4.

Analyzing the estimation results, the following findings can be summarized. First, the innovation parameters are quite low, while the persistence parameters are close to one. Note that no explicit non-negativity restrictions on the autoregressive parameters are imposed. For this reason, even negative values for α_1 are obtained in several regressions. However, in these cases, they are overcompensated by positive

Table 6.2 QML estimates of various types of augmented ACD models for Coca-Cola trade durations. Sample period 03/19/01 to 03/30/01. 15,174 observations. Standard errors based on OPG estimates (in parantheses). SNIACD model estimated based on the category bounds $(0.1, 0.2, 0.5, 1.0, 1.5, 2.0, 3.0)$ with $\bar{\varepsilon}_0 = 1.0$. Diagnostics: Log Likelihood (LL), Bayes Information Criterion (BIC), mean ($\bar{\hat{\varepsilon}}_i$), standard deviation (S.D.) and Ljung–Box statistic with respect to 20 lags (LB) of ACD residuals

	ACD	LACD	BACD	EXACD	ABACD		SNIACD
ω	0.122	−0.038	0.100	−0.041	0.010	ω	−0.036
	(0.016)	(0.003)	(0.043)	(0.004)	(0.010)		(0.011)
α_1	−0.024	−0.025	−0.026	−0.075	0.066		
	(0.004)	(0.004)	(0.038)	(0.009)	(0.011)		
α_2	0.065	0.063	0.058	0.104	−0.062		
	(0.005)	(0.005)	(0.087)	(0.009)	(0.011)		
β_1	0.837	0.869	0.872	0.868	0.871	β_1	0.899
	(0.019)	(0.017)	(0.018)	(0.018)	(0.018)		(0.014)
δ_1			0.483		—[a]	α_1^+	0.062
			(0.736)		—[a]		(0.021)
δ_2			0.672		1.000[b]	α_2^+	−0.071
			(0.089)		—[b]		(0.039)
b					1.420	α_3^+	0.031
					(0.203)		(0.031)
c_1				0.080	−0.773	α_4^+	−0.033
				(0.012)	(0.119)		(0.013)
c_2				−0.068	−1.393	α_1^-	−0.124
				(0.013)	(0.187)		(0.020)
ν_1						α_2^-	−0.153
							(0.038)
ν_2						α_3^-	0.245
							(0.112)
						α_4^-	−0.863
							(0.335)
						a^+	0.230
							(0.026)
						a^-	0.009
							(0.004)
LL	−15,059	−15,060	−15,056	−15,047	−15,045		−15,038
BIC	−15,079	−15,079	−15,085	−15,075	−15,079		−15,095
$\bar{\hat{\varepsilon}}_i$	1.000	1.000	1.000	1.000	1.000		1.000
S.D.	1.245	1.246	1.249	1.245	1.246		1.242
LB	17.146	15.823	16.083	15.254	15.761		17.527

[a]Estimation based on a logarithmic specification
[b]Parameter set to boundary

values of α_2.[3] Nonetheless, based on the estimates we do not find any violations of the non-negativity restriction of Ψ_i.

[3]Note that in a nonlinear ACD specification, an upward kinked concave news impact function actually implies a negative value for α; see also Sect. 6.1.1.4.

Table 6.3 QML estimates of various types of augmented ACD models for Coca-Cola \$0.05 price durations. Sample period 01/02/01 to 05/31/01. 12,971 observations. Standard errors based on OPG estimates (in parantheses). SNIACD model estimated based on the category bounds $(0.1, 0.2, 0.5, 1.0, 1.5, 2.0, 3.0)$ with $\bar{\varepsilon}_0 = 1.0$. Diagnostics: Log Likelihood (LL), Bayes Information Criterion (BIC), mean ($\bar{\hat{\varepsilon}}_i$), standard deviation (S.D.) and Ljung–Box statistic with respect to 20 lags (LB) of ACD residuals

	ACD	LACD	BACD	EXACD	ABACD		SNIACD
ω	0.012	−0.050	−0.004	−0.036	0.035	ω	0.020
	(0.002)	(0.003)	(0.027)	(0.003)	(0.007)		(0.009)
α_1	0.079	0.075	0.037	0.183	−0.170		
	(0.005)	(0.005)	(0.061)	(0.011)	(0.032)		
α_2	−0.020	−0.025	−0.019	−0.113	0.132		
	(0.003)	(0.004)	(0.031)	(0.010)	(0.025)		
β_1	0.930	0.988	0.987	0.990	0.991	β_1	0.989
	(0.004)	(0.002)	(0.002)	(0.002)	(0.002)		(0.002)
δ_1			0.081		—[a]	α_1^+	0.021
			(0.135)		—[a]		(0.018)
δ_2			0.275		1.096	α_2^+	0.010
			(0.044)		(0.137)		(0.034)
b					1.016	α_3^+	−0.034
					(0.103)		(0.029)
c_1				−0.179	−1.000[b]	α_4^+	0.006
				(0.013)	—[b]		(0.014)
c_2				0.136	−0.801	α_1^-	0.226
				(0.013)	(0.035)		(0.020)
ν_1						α_2^-	0.246
							(0.038)
ν_2						α_3^-	0.324
							(0.101)
						α_4^-	−0.157
							(0.199)
						a^+	−0.394
							(0.028)
						a^-	0.005
							(0.003)
LL	−12,072	−12,081	−12,016	−11,996	−11,996		−11,991
BIC	−12,091	−12,100	−12,044	−12,024	−12,029		−12,048
$\bar{\hat{\varepsilon}}_i$	1.008	1.008	1.033	1.005	1.005		1.004
S.D.	1.208	1.210	1.234	1.196	1.196		1.194
LB	30.816	31.776	26.809	23.016	24.431		23.217

[a]Estimation based on a logarithmic specification
[b]Parameter set to boundary

Second, by comparing the goodness-of-fit of the specifications based on the BIC values, we find the best performance for EXACD and BACD models. Especially for price durations, the more simple (linear and logarithmic) models are rejected in favor of the (A)BACD and EXACD model. For trade durations, no clear picture is revealed. For the Coca-Cola stock, the EXACD model is the best specification, while

Table 6.4 QML estimates of various types of augmented ACD models for GE trade durations. Sample period 03/19/01 to 03/30/01. 25,101 observations. Standard errors based on OPG estimates (in parantheses). SNIACD model estimated based on the category bounds $(0.1, 0.2, 0.5, 1.0, 1.5, 2.0, 3.0)$ with $\bar{\varepsilon}_0 = 1.0$. Diagnostics: Log Likelihood (LL), Bayes Information Criterion (BIC), mean ($\bar{\hat{\varepsilon}}_i$), standard deviation (S.D.) and Ljung–Box statistic with respect to 20 lags (LB) of ACD residuals

	ACD	LACD	BACD	EXACD	ABACD		SNIACD
ω	0.036	−0.038	−0.003	−0.037	0.017	ω	−0.007
	(0.005)	(0.003)	(0.030)	(0.003)	(0.024)		(0.007)
α_1	−0.001	−0.001	−0.000	−0.014	0.016		
	(0.006)	(0.006)	(0.006)	(0.010)	(0.018)		
α_2	0.043	0.038	0.042	0.056	0.023		
	(0.006)	(0.006)	(0.033)	(0.010)	(0.024)		
β_1	0.922	0.965	0.964	0.966	0.960	β_1	0.962
	(0.007)	(0.005)	(0.005)	(0.005)	(0.005)		(0.005)
δ_1			0.705		0.497	α_1^+	0.039
			(0.558)		(0.528)		(0.014)
δ_2			0.758		0.595	α_2^+	−0.021
			(0.107)		(0.078)		(0.027)
b					0.444	α_3^+	−0.014
					(0.030)		(0.023)
c_1				0.023	−0.956	α_4^+	−0.008
				(0.014)	(0.143)		(0.012)
c_2				−0.031	1.000[a]	α_1^-	−0.030
				(0.014)	–[a]		(0.019)
ν_1						α_2^-	−0.048
							(0.030)
ν_2						α_3^-	−0.227
							(0.096)
						α_4^-	0.636
							(1.290)
						a^+	0.097
							(0.031)
						a^-	0.006
							(0.005)
LL	−24,813	−24,815	−24,811	−24,813	−24,803		−24,801
BIC	−24,834	−24,836	−24,841	−24,843	−24,844		−24,862
$\bar{\hat{\varepsilon}}_i$	1.001	1.000	0.999	1.000	1.000		1.001
S.D.	1.007	1.007	1.006	1.007	1.006		1.005
LB	38.695	37.281	36.779	37.039	35.826		36.380

[a] Parameter set to boundary

for the GE stock, the basic ACD model leads to the highest BIC. Nonetheless, for all specifications, we observe the strongest increase of the log-likelihood function when the (L)ACD model is extended to a BACD or EXACD model. This result illustrates that for both types of financial durations, it is crucial to account for nonlinear news impact effects. Not surprisingly, the most flexible SNIACD model leads to the overall highest log-likelihood values for all series indicating the best fit to the data.

Table 6.5 QML estimates of various types of augmented ACD models for GE \$0.05 price durations. Sample period 01/02/01 to 05/31/01. 16,008 observations. Standard errors based on OPG estimates (in parantheses). SNIACD model estimated based on the category bounds $(0.1, 0.2, 0.5, 1.0, 1.5, 2.0, 3.0)$ with $\bar{\varepsilon}_0 = 1.0$. Diagnostics: Bayes Information Criterion (BIC), mean ($\bar{\hat{\varepsilon}}_i$), standard deviation (S.D.) and Ljung–Box statistic with respect to 20 lags (LB) of ACD residuals

	ACD	LACD	BACD	EXACD	ABACD	SNIACD	
ω	0.009	−0.030	−0.093	−0.034	−0.016	ω	0.012
	(0.002)	(0.004)	(0.017)	(0.005)	(0.003)		(0.006)
α_1	0.143	0.122	0.372	0.174	−0.389		
	(0.009)	(0.007)	(0.049)	(0.011)	(0.062)		
α_2	−0.094	−0.092	−0.269	−0.128	0.325		
	(0.007)	(0.006)	(0.036)	(0.010)	(0.054)		
β_1	1.227	1.532	1.434	1.433	1.408	β_1	1.391
	(0.073)	(0.057)	(0.063)	(0.067)	(0.066)		(0.068)
β_2	−0.284	−0.538	−0.442	−0.442	−0.417	β_2	−0.400
	(0.066)	(0.056)	(0.062)	(0.066)	(0.065)		(0.067)
δ_1			—[a]		—[a]	α_1^+	0.020
			—[a]		—[a]		(0.012)
δ_2			0.407		1.000[b]	α_2^+	0.051
			(0.051)		—[b]		(0.024)
b					0.355	α_3^+	0.002
					(0.032)		(0.020)
c_1				−0.100	−1.235	α_4^+	0.018
				(0.014)	(0.043)		(0.009)
c_2				0.082	−1.179	α_1^-	0.211
				(0.014)	(0.041)		(0.017)
ν_1						α_2^-	0.223
							(0.029)
ν_2						α_3^-	0.120
							(0.072)
α_1^+						α_4^-	0.512
							(0.181)
α_2^+						a^+	−0.377
							(0.028)
α_3^+						a^-	−0.039
							(0.006)
BIC	−14,729	−14,730	−14,699	−14,708	−14,699		−14,725
$\bar{\hat{\varepsilon}}_i$	1.002	1.001	1.000	1.001	1.000		1.000
S.D.	1.060	1.057	1.059	1.059	1.056		1.057
LB	33.643	23.238	35.240	29.765	32.469		33.801

[a]Estimation based on a logarithmic specification
[b]Parameter set to boundary

Third, the estimated Box–Cox parameters $\hat{\delta}_1$ and $\hat{\delta}_2$ are almost always lower than one for price durations, while for trade durations, values greater and less than one are obtained. These results are in conflict to the linear ACD and, in most cases, also to the LACD model. Hence, we notice that price duration processes imply concave news impact curves, i.e., the adjustments of the conditional expected mean are stronger in periods of smaller than expected price durations (volatility shocks) than in periods with unexpectedly low price intensities. Corresponding results are found based on the EXACD model as the mostly highly significant negative parameters for c imply upward kinked concave shaped news response curves. For trade durations, the picture is less clear since we obtain evidence for concave, as well as convex news impact curves.

Fourth, in most cases only restricted versions of the augmented ACD models are estimated because either δ_1 tends to zero and/or $|c|$ tends to one. Since most duration series seem to imply a concave news impact function, it is not surprising that the second restriction is binding for nearly all series. In the first case, the model is estimated under the restriction $\delta \rightarrow 0$, which is performed by estimating the model based on a logarithmic transformation. Note that this has consequences for the estimates of ω, α and ν as a logarithmic model belongs to the class of Box–Cox models and not to the class of power ACD models, which imply parameter transformations from ω, α and c to $\tilde{\omega}$, $\tilde{\alpha}$ and $\tilde{c}$ as illustrated in Sect. 6.1.1.4. In the second case, two restricted versions of the model are re-estimated: one specification under the restriction $|c| = 1$ and one model under $\delta_2 = 1$. Then, the specification is chosen that leads to the higher log-likelihood value. It turns out that generally neither of the restricted models outperforms the other.

Fifth, the SNIACD model is estimated using the categorization $\{0.1, 0.2, 0.5, 1.0, 1.5, 2.0, 3.0\}$ with $\bar{\varepsilon}_0 = 1.0$ which allows for more flexibility in case of very small and very large innovations. For most of the thresholds, we obtain significant estimates. However, since these coefficients belong to a spline function, it is not useful to interpret them separately. The mostly significant parameters a^+ and a^- indicate that it is also useful to account for flexible news impact effects for the second lag.

Figure 6.1 depicts the corresponding news impact functions implied by the estimates of the SNIACD model. The shape of the estimated news impact curves confirm the estimation results of the parametric ACD models. The news response curves for trade durations reveal high nonlinearities, especially for very small innovations. For Coca-Cola, we even observe a news response function that implies a downward shape for low values of ε_{i-1}. Here, for extremely small innovations, the first order autocorrelation is negative rather than positive. This finding illustrates that small durations induce significantly different adjustments of the expected mean as long durations which has to be taken into account in the econometric modelling. This result is in line with the findings of Zhang et al. (2001) who provide evidence for similar effects based on estimations of the TACD model.

The news response function for price durations reveals a significantly different shape. The larger (positive) slope of the curve indicates a higher (positive) autocorrelation for price durations. Nonetheless, we notice a nonlinear news impact curve with a strongly increasing pattern for $\varepsilon_{i-1} < 1$ and a nearly flat function for $\varepsilon_{i-1} \geq 1$. Hence, we observe different adjustment processes for unexpectedly small price durations, i.e., in periods of unexpectedly high volatility.

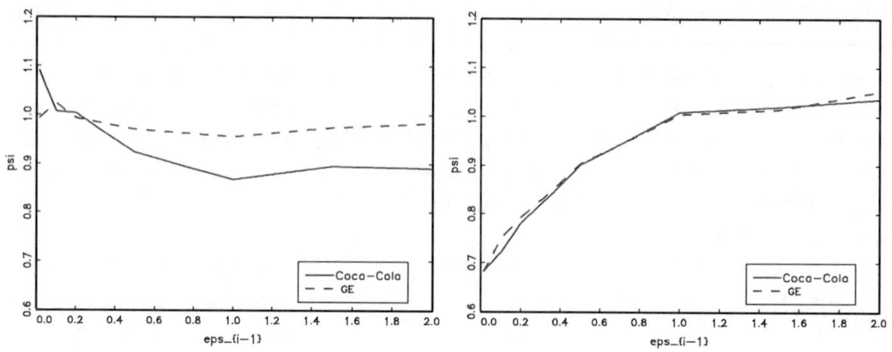

Fig. 6.1 Estimated news impact curves for trade durations (*left*) and price durations (*right*) of the Coca-Cola and GE stock

Table 6.6 gives the (Q)ML estimates of ACD, EXACD and SNIACD models for de-seasonalized trade durations, 10 bp price durations, and 30-s as well as 2-min aggregated volumes for JP Morgan traded at NYSE, June 2009. We observe the following effects: First, the parameter estimates indicate that trade durations and price durations reveal a lower persistence than time aggregated trading volumes. Second, as indicated by the EXACD estimates, we observe clear evidence for asymmetric news impacts. The estimates of c are highly significant and reveal concave news impact functions. These effects are mostly evident based on aggregated data. Third, in terms of the BIC, the EXACD model yields the best goodness-of-fit. Hence, allowing for a kinked news impact function seems to be a parsimonious way to capture nonlinearities in MEM/ACD dynamics. These effects are confirmed by Fig. 6.2 depicting the estimated news impact curves based on the SNIACD model.

6.2 Regime-Switching ACD Models

In the following subsection, we summarize several further functional extensions of the ACD model. All models have in common that they allow for regime-dependence of the conditional mean function. The regimes are determined either based on observable variables, like in threshold ACD models and smooth transition ACD models or based on unobservable factors, as in Markov switching ACD models.

6.2.1 Threshold ACD Models

Zhang et al. (2001) introduce a threshold ACD (TACD) model which allows the expected duration to depend nonlinearly on past information variables. The TACD model can be seen as a generalization of the threshold GARCH models introduced by Rabemananjara and Zakoian (1993) and Zakoian (1994). Assume a

Table 6.6 QML estimates of the ACD model for de-seasonalized trade durations, 10bp price durations, and 30-s as well as 2-min aggregated volumes for JP Morgan traded at NYSE, June 2009. Category bounds for SNIACD model: $(0.1, 0.2, 0.5, 1.0, 1.5, 2.0, 3.0)$ with $\bar{\varepsilon}_0 = 1.0$

Par.	Est.	p-val.	Est.	p-val.	Est.	p-val.	Est.	p-val.
	Trade dur.		10bp dur.		30-s vol.		2-min vol.	
ACD								
ω	0.019	0.000	0.009	0.000	0.031	0.000	0.007	0.000
α	0.055	0.000	0.078	0.000	0.245	0.000	0.245	0.000
β	0.925	0.000	0.913	0.000	0.738	0.000	0.760	0.000
BIC	−38,238		−9,939		−16,204		−3,325	
LB	2.40		48.6		22.5		31.1	
EXACD								
ω	−0.0494	0.000	−0.0624	0.000	−0.163	0.000	−0.196	0.000
α	0.0631	0.000	0.1060	0.000	0.238	0.000	0.257	0.000
c	0.9789	0.000	0.9801	0.000	0.961	0.000	0.985	0.000
β	−0.0204	0.000	−0.0596	0.000	−0.112	0.000	−0.133	0.000
BIC	−38,226		−9,915		−16,233		−3,324	
LB	2.53		43.1		27.3		36.6	
SNIACD								
ω	0.014	0.021	0.035	0.007	0.027	0.074	0.051	0.180
β	0.979	0.000	0.981	0.000	0.961	0.000	0.986	0.000
α_1^-	0.460	0.034	0.238	0.134	-0.183	0.446	−1.317	0.692
α_2^-	−0.020	0.530	0.196	0.173	−0.162	0.364	0.250	0.701
α_3^-	−0.042	0.268	−0.237	0.138	0.200	0.284	−0.094	0.849
α_4^-	0.113	0.000	0.215	0.003	0.221	0.009	0.375	0.082
α_1^+	0.061	0.008	0.039	0.413	0.268	0.000	0.190	0.227
α_2^+	−0.011	0.807	0.041	0.653	−0.116	0.255	−0.183	0.585
α_3^+	0.005	0.888	0.004	0.953	0.073	0.413	0.207	0.506
α_4^+	−0.041	0.021	−0.068	0.043	−0.182	0.000	−0.154	0.407
BIC	−38,229		−9,934		−16,210		−3,325	
LB	2.49		42.3		25.3		38.3	

categorization of durations into M categories with $\bar{x}_m$, $m = 1, \ldots, M$, denoting the corresponding category bounds. Then, an M-regime TACD(P,Q) model is given by

$$\begin{cases} x_i = \Psi_i \varepsilon_i^{(m)} \\ \Psi_i = \omega^{(m)} + \sum_{j=1}^{P} \alpha_j^{(m)} x_{i-j} + \sum_{j=1}^{Q} \beta_j^{(m)} \Psi_{i-j}, \end{cases} \quad \text{if } x_{i-1} \in (\bar{x}_{m-1}, \bar{x}_m] \tag{6.12}$$

where $\omega^{(m)} > 0$, $\alpha_j^{(m)} \geq 0$ and $\beta_j^{(m)} \geq 0$ are regime-switching ACD parameters. The i.i.d. error term $\varepsilon_i^{(m)}$ is assumed to follow some distribution $f_\varepsilon(s; \theta_\varepsilon^{(m)})$ with $\mathbb{E}[\varepsilon_i^{(m)}] = 1$ depending on regime-switching distribution parameters $\boldsymbol{\theta}_\varepsilon^{(m)}$.

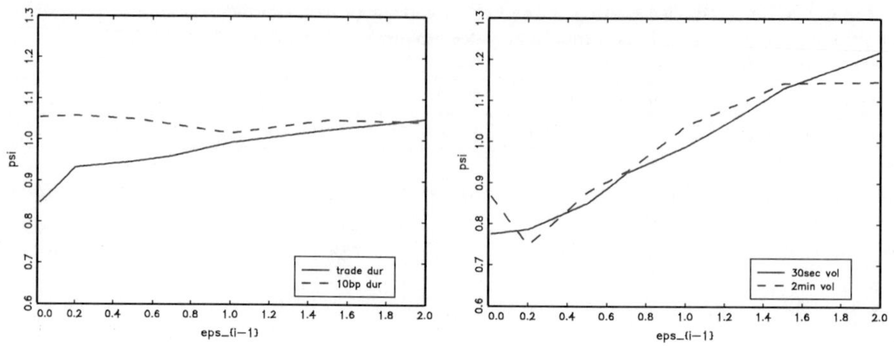

Fig. 6.2 Estimated news impact curves for trade durations, 10bp price durations, and 30-s as well as 2-min aggregated volumes for JP Morgan (NYSE). June 2009

Correspondingly to the basic ACD model, the model can be rewritten as

$$x_i = \omega^{(m)} + \sum_{j=1}^{\max(P,Q)} (\alpha_j^{(m)} + \beta_j^{(m)}) x_{i-j} - \sum_{j=1}^{Q} \beta_j^{(m)} \eta_{i-j} + \eta_i, \tag{6.13}$$

with $\eta_i := x_i - \Psi_i$. Hence, the TACD model corresponds to a threshold ARMA(M; max(P, Q), Q) process.[4] Obviously, the TACD model is strongly related to the SNIACD model. The main difference is that in the SNIACD model only the impact of lagged innovations is regime-dependent while in the TACD model all autoregressive and distributional parameters are allowed to be regime-switching. Hence, in this sense, the TACD model is more general and allows for various forms of nonlinear dynamics. However, the derivation of theoretical properties is not straightforward in this framework. Zhang et al. (2001) derive conditions for geometric ergodicity and the existence of moments for the TACD(1,1) case. Moreover, in the SNIACD model, the threshold values associated with the particular regimes are given exogenously while in the TACD model they are endogenous and are estimated together with the other parameters. The estimation has to be performed by a grid search algorithm over the thresholds $\bar{x}_m$ and maximizing the conditional likelihood for each combination of the grid thresholds. Clearly, for higher numbers of M, this procedure becomes computationally quite cumbersome.

6.2.2 Smooth Transition ACD Models

Note that in the TACD model, as well as in the SNIACD model, the transition from one state to another state is not smooth but follows a jump process. An alternative is

[4] For more details concerning threshold autoregressive (TAR) models, see, for example, Tong (1990).

to allow for a smooth transition which is driven by some transition function. Smooth transition autoregressive (STAR) models are considered, for example, by Granger and Teräsvirta (1993), Teräsvirta (1994,1998) for the conditional mean of financial return processes and by Hagerud (1997) and Gonzalez-Rivera (1998) for conditional variance processes. Meitz and Teräsvirta (2006) propose a smooth transition ACD (STACD) model of the form

$$\begin{aligned} \Psi_i &= \omega + \sum_{j=1}^{P} \alpha_j x_{i-j} + \sum_{j=1}^{P} (\omega_j^* + \alpha_j^* x_{i-j}) G(\cdot) + \sum_{j=1}^{Q} \beta_j \Psi_{i-j}, \\ &= \omega + \sum_{j=1}^{P} \omega_j^* G(\cdot) + \sum_{j=1}^{P} (\alpha_j + \alpha_j^* G(\cdot)) x_{i-j} + \sum_{j=1}^{Q} \beta_j \Psi_{i-j}, \end{aligned} \tag{6.14}$$

where $G(\cdot) = G(\ln x_i; \gamma, c)$ denotes a transition function given by

$$G(\ln x_{i-j}; \gamma, c) = \left(1 + \exp\left\{-\gamma \prod_{m=1}^{M} (\ln x_{i-j} - c_m)\right\}\right)^{-1}, \tag{6.15}$$

for $c_1 \leq \ldots \leq c_M, \gamma > 0$. The choice of $G(\cdot)$ as a logistic function ensures that it is bounded and non-negative. As it is defined on the entire real axis, $\ln x_{i-j}$ is used as the transition variable. As discussed by Meitz and Teräsvirta (2006), an alternative would be to choose a cumulative distribution function of a random variable with positive support as transition function. However, due to its non-decreasing nature, such a transition function would impose strong a priori structures preventing possible non-linearities. The parameter M determines the shape of the transition function. For $M = 1$, the transition function increases in x_{i-j}. Alternatively, choosing $M = 2$ allows very short and long durations having a different news impact than more moderate durations. Figure 6.3 shows the transition function for the case $M = 2$, $c_1 = -1$, $c_2 = 1$ and for different values for γ. It is shown that the smoothness of the transition declines with γ. Hence, in the limit, $\gamma \rightarrow \infty$, the transition is not smooth anymore and the model corresponds to a special case of the TACD model as proposed by Zhang et al. (2001). In this case, ω and α would jump between a regime for very short and long durations and a regime for moderate durations (as approximately given by $0.5 \geq x_i \leq 2.5$ in Fig. 6.3). In particular, a two-regime TACD(1,1) model with $\beta^{(1)} = \beta^{(2)}$ and $\varepsilon_i^{(m)} \sim Exp(1)$, $m = 1, 2$, is the limiting case of a STACD(1,1) process for $M = 1$ and $\gamma \rightarrow \infty$. Correspondingly, a three-regime TACD model with $\beta^{(1)} = \beta^{(2)} = \beta^{(3)}$, $\omega^{(1)} = \omega^{(3)}$, $\alpha^{(1)} = \alpha^{(3)}$ and $\varepsilon_i^{(m)} \sim Exp(1)$, $m = 1, 2, 3$, is the limiting case of a STACD(1,1) process with $M = 2$ and $\gamma \rightarrow \infty$.

Correspondingly, Meitz and Teräsvirta (2006) also suggest a smooth transition Log ACD (STLACD) model given by

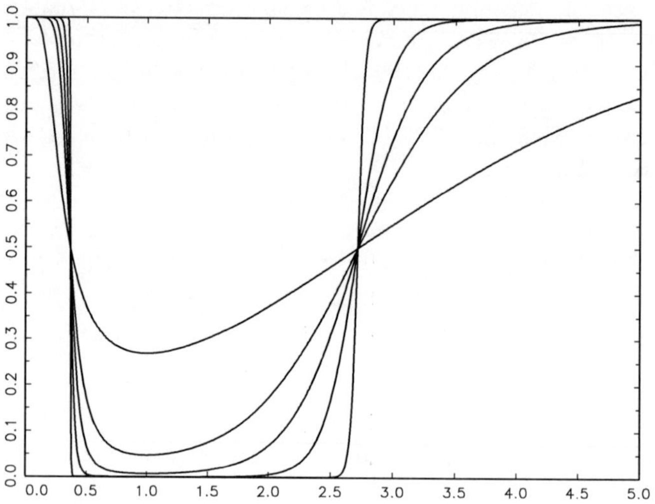

Fig. 6.3 The transition function $G(\ln x_i; \gamma, c)$ as a function of x_i for $M = 2$, $c_1 = -1$, $c_2 = 1$ and for $\gamma \in \{1, 3, 5, 10, 50\}$. The smoothest (*steepest*) curves are associated with $\gamma = 1$ ($\gamma = 50$)

$$\begin{aligned}\ln \Psi_i &= \omega + \sum_{j=1}^{P} \alpha_j \ln x_{i-j} + \sum_{j=1}^{P} (\omega_j^* + \alpha_j^* \ln x_{i-j}) G(\cdot) + \sum_{j=1}^{Q} \beta_j \ln \Psi_{i-j}, \quad (6.16)\\ &= \omega + \sum_{j=1}^{P} \omega_j^* G(\cdot) + \sum_{j=1}^{P} (\alpha_j + \alpha_j^* G(\cdot)) \ln x_{i-j} + \sum_{j=1}^{Q} \beta_j \ln \Psi_{i-j},\end{aligned}$$

with $G(\cdot) = G(\ln x_{i-j}; \gamma, c)$. Clearly, there are numerous alternative possibilities to choose the transition variable. Instead of specifying $G(\cdot)$ in terms of x_i, it could be also specified in terms of the ACD innovation $\varepsilon_i = x_i/\Psi_i$. Then, regime switches would be driven by durations which are scaled by their conditional expectations, and thus "surprises" in durations. Alternatively, regime switching of ACD parameters could be also driven by deterministic time patterns. For instance, Meitz and Teräsvirta (2006) propose specifying $G(\cdot)$ in terms of a standardized calendar time variable $\bar{t}_i \in [0, 1]$ reflecting, e.g., the time of the day. Then, choosing the transition function as $G^*(\bar{t}_{i-1}; \gamma, c) = G(\bar{t}_{i-1}; \gamma, c) - 0.5$ yields a so-called time-varying ACD (TVACD) specification,

$$\begin{aligned}\Psi_i &= \omega + \sum_{j=1}^{P} \alpha_j x_{i-j} + \sum_{j=1}^{Q} \beta_j \Psi_{i-j} + \sum_{j=1}^{P} \left(\omega_j^* + \alpha_j^* x_{i-j} + \sum_{j=1}^{Q} \beta_j^* \Psi_{i-j} \right) G^*(\cdot)\\ &= (\omega + \omega^* G^*(\cdot)) + \sum_{j=1}^{P} (\alpha_j + \alpha_j^* G^*(\cdot)) x_{i-j} + \sum_{j=1}^{P} (\beta_j + \beta_j^* G^*(\cdot)) \Psi_{i-j},\end{aligned} \quad (6.17)$$

with $G^*(\cdot) = G^*(\bar{t}_{i-1}; \gamma, c))$. Here, not only ω and α but also the persistence parameter β vary.

6.2.3 Markov Switching ACD Models

Hujer et al. (2002) propose a Markov switching ACD model where the conditional mean function depends on an unobservable stochastic process, which follows a Markov chain. Define in the following M_i^* as a discrete valued stochastic process which indicates the state of the unobservable Markov chain and takes the values $1, 2, \ldots, M$. Then, the MSACD model is given by

$$x_i = \Psi_i \varepsilon_i, \qquad \varepsilon_i \sim f_\varepsilon(s) \tag{6.18}$$

$$\Psi_i = \sum_{m=1}^{M} \mathbb{P}r\left[M_i^* = m \,|\mathcal{F}_{t_{i-1}}\right] \cdot \Psi_i^{(m)}, \tag{6.19}$$

$$f_\varepsilon(s) = \sum_{m=1}^{M} \mathbb{P}r\left[M_i^* = m \,|\mathcal{F}_{t_{i-1}}\right] \cdot f_\varepsilon(s; \theta_\varepsilon^{(m)}), \tag{6.20}$$

where $\Psi_i^{(m)}$ denotes the regime-specific conditional mean function

$$\Psi_i^{(m)} = \omega^{(m)} + \sum_{j=1}^{P} \alpha_j^{(m)} x_{i-j} + \sum_{j=1}^{Q} \beta_j^{(m)} \Psi_{i-j}^{(m)} \tag{6.21}$$

and $f_\varepsilon(s; \theta_\varepsilon^{(m)})$ is the regime-dependent conditional density for ε_i. The Markov chain is characterized by a transition matrix P^* with elements $p_{kl}^* := \mathbb{P}r[M_i^* = k| M_{i-1}^* = l]$. Hence, the marginal density of x_i is given by

$$f(x_i|\mathcal{F}_{i-1}) = \sum_{m=1}^{M} \mathbb{P}r\left[M_i^* = m \,|\mathcal{F}_{i-1}\right] \cdot f(x_i|M_i^* = m; \mathcal{F}_{i-1}). \tag{6.22}$$

Hujer et al. (2002) estimate the MSACD model based on the Expectation-Maximization (EM) algorithm proposed by Dempster et al. (1977). They show that the MSACD outperforms the basic (linear) ACD model and leads to a better description of the underlying duration process. In this sense, they confirm the results of Zhang et al. (2001) that nonlinear ACD specifications are more appropriate to model financial duration processes. Hujer et al. (2002) illustrate that the unobservable regime variable can be interpreted in light of market microstructure theory and allows for more sophisticated tests of corresponding theoretical hypotheses.

6.3 Long Memory ACD Models

As illustrated in Chap. 3, trading variables typically exhibit a strong persistence implying ACD parameters α and β to sum nearly to one. A typical indication for the existence of long range dependence is an autocorrelation function which displays no exponential decay but a slow, hyperbolic rate of decay. Formally,

$$\lim_{k\to\infty} \rho_k/[ck^{-\alpha}] = 1,$$

with $\alpha \in (0,1)$ and $c > 0$ (see, e.g. Beran, 1994). A consequence of this result is that the variance of sample averages computed over sub-samples of the length n, $\bar{x}$, decline with a rate $n^{-\alpha}$, i.e.

$$\mathbb{V}[\bar{x_n}] \approx c_v n^{-\alpha}, \quad c_v > 0.$$

Figure 6.4 shows the plots of $\ln n$ vs. $\ln \mathbb{V}[\bar{x_n}]$ for trade durations, price durations as well as the number of trades and cumulated volumes in 30-s intervals for JPM, Microsoft and Deutsche Telekom traded at the NYSE, NASDAQ and XETRA, respectively. The estimated slopes underlying the corresponding bivariate regressions yield clear evidence for long range dependence.

While the existence of long memory patterns in return and volatility series have been already explored in much detail,[5] only a few approaches pay attention to such effects in high-frequency series. Engle (2000) applies a two-component model as proposed by Ding and Granger (1996), given by

$$\Psi_i = w\Psi_{1,i} + (1-w)\Psi_{2,i}, \tag{6.23}$$

$$\Psi_{1,i} = \alpha_1 x_{i-1} + (1-\alpha_1)\Psi_{1,i-1}, \tag{6.24}$$

$$\Psi_{2,i} = \omega + \alpha_2 x_{i-1} + \beta_2 \Psi_{2,i-1}. \tag{6.25}$$

Hence, Ψ_i consists of the weighted sum of two components, $\Psi_{1,i}$ and $\Psi_{2,i}$ with weights w and $1-w$, respectively. Equation (6.24) is an integrated ACD(1,1) specification capturing long-term movements, while (6.25) is a standard ACD(1,1) specification that captures short-term fluctuations in financial durations. As illustrated by Ding and Granger (1996), the resulting two-component process is covariance stationary, but allows for a slower decay of the ACF compared to the corresponding standard model. Engle (2000) shows that the two component ACD model improves the goodness-of-fit and captures the duration dynamics in a better way than the basic ACD model. Ding and Granger (1996) analyze the limiting case of a multi-component model by increasing the number of components towards infinity. They show that such a model implies autocorrelation functions that reflect

[5] For an overview, see Beran (1994) or Baillie (1996).

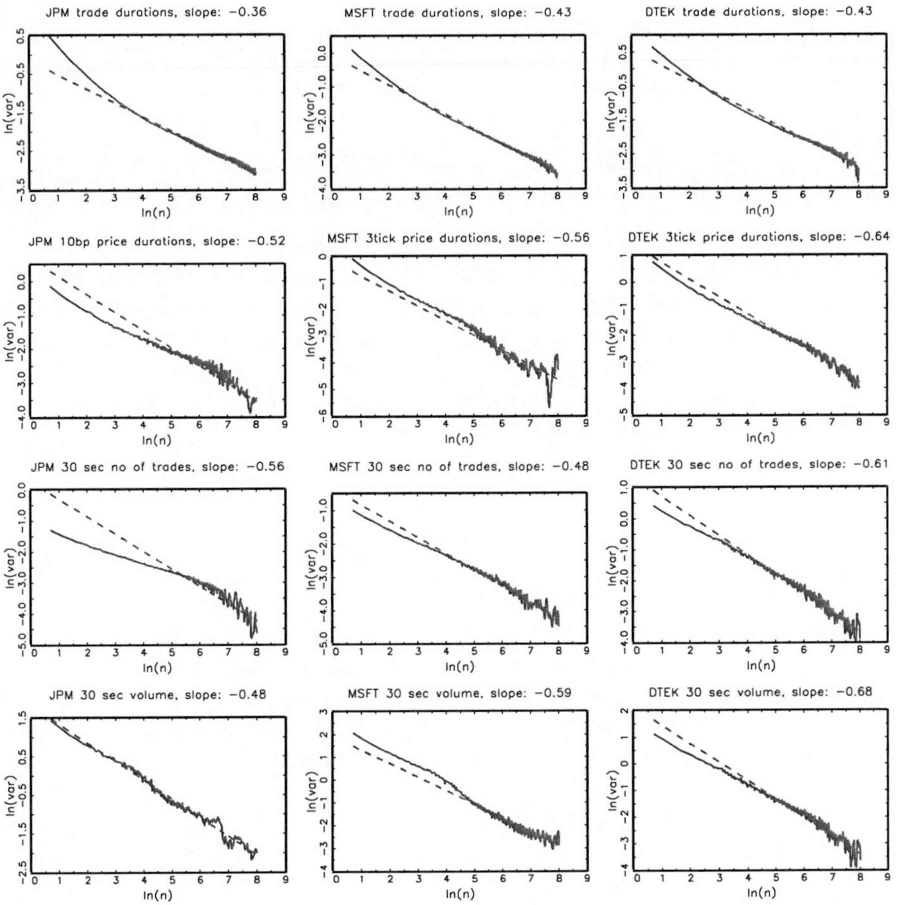

Fig. 6.4 Plots of $\ln n$ vs. $\ln \mathbb{V}[\bar{x}_n]$ for trade durations, midquote change durations, the number of trades as well as cumulated trading volumes during 30-s intervals for JP Morgan (NYSE), Microsoft (NASDAQ) and Deutsche Telekom (XETRA). Aggregation levels for price durations in basis points of the average price level for JP Morgan and in minimum tick size for Microsoft and Deutsche Telekom. Sample period: June 2009 for JP Morgan and Microsoft and September 2010 for Deutsche Telekom

the typical characteristics of long memory processes. Moreover, it is shown that the resulting model is closely related to the fractionally integrated GARCH (FIGARCH) model proposed by Baillie et al. (1996). Following this string of the literature, Jasiak (1998) proposes a fractionally integrated ACD (FIACD) model which is the counterpart to the FIGARCH model. The FIACD model is based on a fractionally integrated process for Ψ_i and is given by

$$[1-\beta(L)]\Psi_i = \omega + [1-\beta(L)-\phi(L)(1-L)^d]x_i, \qquad (6.26)$$

where $\phi(L) := 1 - \alpha(L) - \beta(L)$, $\alpha(L) := \sum_{j=1}^{P} \alpha_j L^j$ and $\beta(L) := \sum_{j=1}^{Q} \beta_j L^j$ denote polynomials in terms of the lag operator L. The fractional integration operator $(1-L)^d$ (with $0 < d < 1$) is given by

$$(1-L)^d = \sum_{j=0}^{\infty} \binom{d}{j} (-1)^j L^j = \sum_{j=0}^{\infty} \frac{\Gamma(j-d)}{\Gamma(-d)\Gamma(j+1)} L^j, \tag{6.27}$$

where $\Gamma(\cdot)$ denotes the gamma function. The FIACD model is not covariance stationary and implies infinite first and second unconditional moments of x_i. Following the results of Bougerol and Picard (1992), it can be shown that it is strictly stationary and ergodic for $0 \leq d \leq 1$. For $d = 0$, the model corresponds to the basic ACD model, whereas for $d = 1$, it corresponds to an integrated ACD process as the counterpart to an integrated GARCH model.

An alternative long memory specification is provided by Karanasos (2004) and is given by

$$x_i = \omega + \frac{1-\beta(L)}{\phi(L)(1-L)^d} \eta_i = \omega + \sum_{j=0}^{\infty} \omega_j \eta_{i-j}, \tag{6.28}$$

where $\eta_i := x_i - \Psi_i$. As in the FIACD model, we have the case of long-range dependence and thus non-summable autocovariance functions for $d > 0$ with the model being weakly stationary for $d < 0.5$. As shown by Karanasos (2004), the model can be written as an infinite sum of lagged values of η_i,

$$x_i = \omega + \sum_{j=0}^{\infty} \omega_j \eta_{i-j}, \tag{6.29}$$

where

$$\omega_j := \sum_{r=1}^{n} \lambda_r^+ \sum_{l=0}^{j} \binom{-d}{j-l} \pi_{rl} (-1)^{j-1},$$

$$\pi_{rl} := \sum_{j=0}^{\min\{l,m\}} \lambda_r^{l-j} (-\beta_j) \quad (\beta_0 := -1)$$

$$\lambda_r^+ := \frac{\lambda_r^{n-1}}{\prod_{l=1, l \neq r}^{n} (\lambda_r - \lambda_l)}.$$

Moreover, Ψ_i can be expressed as

$$\Psi_i := \left(1 - \frac{(1-L)^d \phi(L)}{1-\beta(L)}\right) x_i = \sum_{j=1}^{\infty} \psi_j x_{i-j}, \tag{6.30}$$

with $\psi_1 > 0$, $\psi_j \geq 0$ for $j \geq 2$.

In contrast to Jasiak's FIACD model, (6.28) implies finite first and second moments of x_i with $\mathbb{E}[x_i] = \omega < \infty$ and $\mathbb{E}[x_i^2] < \infty$ if

$$\left(1 - \frac{1}{\mathbb{E}[\varepsilon_i^2]}\right) \sum_{j=0}^{\infty} \omega_j^2 < 1.$$

An analytical expression for the autocorrelation function is given by Karanasos (2004).

A similar type of long memory ACD model is proposed by Koulikov (2003) and is given by

$$\Psi_i = \omega + \alpha \sum_{j=1}^{\infty} \theta_{j-1} \eta_{i-j}, \tag{6.31}$$

where $\{\theta_j : j \geq 0\} \subseteq \mathbb{R}_{0+}$ is an infinite sequence of coefficients with $\theta_0 = 1$ and ω and α are model parameters. The process (6.31) is covariance stationary and ergodic as long as ε_i is a (zero mean) martingale difference, and the coefficients θ_j are square-summable (see Koulikov 2003). Following Granger and Joyeux (1980), or Hosking (1981), a possible parameterization of the coefficients θ_j is given by a power series expansion of $(1-\beta z)^{-1}(1-z)^{-d}$ as given by

$$\theta_j := \sum_{k=0}^{j} \beta^k \theta_{j-k}^*, \tag{6.32}$$

where

$$\theta_j^* := \frac{\Gamma(d+j)}{\Gamma(d)\Gamma(1+j)} \quad \forall j \geq 0, \tag{6.33}$$

are the coefficients of the expansion of $(1-z)^{-d}$, β is a model parameter with $|\beta| < 1$ and $\Gamma(\cdot)$ denotes the gamma function. For $d \in (0,1)$, (6.33) produces a non-summable autocovariance function, and thus long memory. For $0<d<\frac{1}{2}$, the power series expansion implies a sequence of square-summable hyperbolic decaying coefficients and thus ensures covariance stationarity. By replacing z by the lag operator L, we can re-write Ψ_i as

$$\Psi_i = \omega + \alpha(1-\beta L)^{-1}(1-L)^{-d}\eta_{i-1} \tag{6.34}$$

with $(1-L)^{-d}\eta_{i-1} = \sum_{j=1}^{\infty} \theta_{j-1}\eta_{i-j}$.

Table 6.7 gives QML estimates of the FIACD and the LMACD specification proposed by Karanasos (2004) for de-seasonalized trade durations, 10bp price durations, and 30-s number of trades for JP Morgan traded at NYSE. As revealed by the estimates of the long memory parameter d, there is strong evidence for the

Table 6.7 QML estimates of the FIACD and the LMACD specification proposed by Karanasos (2004) for de-seasonalized trade durations, 10 bp price durations, and 30-s number of trades for JP Morgan traded at NYSE. Sampling period: 01/06/09-05/06/09 for trade durations and June 2009 for price durations and trade counts. Diagnostics: Bayes Information Criterion (BIC) and Ljung-Box (LB) statistics of ACD residuals associated with 20 lags

par.	est.	p-val.	est.	p-val.	est.	p-val.
	trade dur.		10bp price dur.		30-s no. of trades	
FIACD						
ω	0.006	0.000	0.003	0.000	0.004	0.274
α	0.021	0.000	0.029	0.000	0.021	0.181
β	0.967	0.000	0.962	0.000	0.955	0.000
d	0.092	0.000	0.124	0.000	0.247	0.000
BIC	−36,657		−9,927		−16,041	
LB	4.71		15.39		34.94	
LMACD						
ω	1.053	0.000	1.086	0.000	0.994	0.000
α	0.023	0.000	0.031	0.000	0.021	0.174
β	0.965	0.000	0.960	0.000	0.955	0.000
d	0.082	0.000	0.120	0.000	0.245	0.000
BIC	−36,645		−9,929		−16,040	
LB	4.41		14.57		34.87	

existence of long range dependence. In all cases, d is between $(0, 0.5)$ indicating the processes being stationary. The significance of α and β reveals also the presence of short memory dynamics. In terms of goodness-of-fit, as revealed by BIC and Ljung–Box statistics, all specifications perform relatively similar.

A counterpart to a long memory stochastic volatility model is proposed by Deo et al. (2010). This specification is discussed in more detail in Sect. 6.4.1. Conditions for the propagation of the long memory parameter from durations to counts are analyzed by Deo et al. (2009). In particular, they establish conditions for weakly stationary long range dependent duration processes with memory parameter d ensuring that the resulting counting process $N(t)$ satisfies $\mathbb{V}[N(t)] \sim Ct^{2d+1}$, $C > 0$ if $t \to \infty$.

6.4 Mixture and Component Multiplicative Error Models

6.4.1 The Stochastic Conditional Duration Model

In the Markov Switching ACD model the latent variable follows a Markov process, i.e., the state of the process depends only on the state of the previous observation. An alternative specification has been proposed by Bauwens and Veredas (2004).[6]

[6]See also Meddahi et al. (1998) who introduce a similar specification.

They assume that the conditional mean function Ψ_i given the history $\mathcal{F}_{i-1}$ is not deterministic but is driven by a latent AR(1) process. Hence, the Stochastic Conditional Duration (SCD) model is given by

$$\ln \Psi_i = \omega + \beta \ln \Psi_{i-1} + v_i, \tag{6.35}$$

where $x_i = \Psi_i \varepsilon_i$, ε_i denotes the usual i.i.d. ACD innovation term and v_i is conditionally normally distributed, i.e., $v_i | \mathcal{F}_{i-1} \sim$ i.i.d. $\mathcal{N}(0, \sigma^2)$ independently of ε_i. Hence, the marginal distribution of x_i is determined by a mixture of a log-normal distribution of Ψ_i and the distribution of ε_i. Economically, the latent factor can be interpreted as information flow (or general state of the market) that cannot be observed directly but drives the duration process. In this sense, the SCD model is the counterpart to the stochastic volatility model introduced by Taylor (1982). Bauwens and Veredas (2004) analyze the theoretical properties of the SCD model and illustrate that the SCD model, even with restrictive distributional assumptions for ε_i, is quite flexible and allows for a wide range of different (marginal) hazard functions of the durations. In a comparison study with the Log-ACD model they find a better fit of the SCD model. The model is estimated by simulated maximum likelihood (see Sect. 7.2) or, as proposed by Bauwens and Veredas (2004), based on QML by applying the Kalman filter.[7]

A long memory version of the SCD model is proposed by Deo et al. (2010). In their so-called long memory stochastic duration (LMSD) model, Ψ_i is given by

$$\ln \Psi_i = \omega + (1 - L)^{-d} v_i \tag{6.36}$$

with $d \in [0, 0.5)$. Deo et al. (2010) propose estimating the model by QML using the Whittle (1962) approximation. A further extension of the SCD model based on non-Gaussian state-space models and allowing for asymmetric behavior of the expected duration is suggested by Feng et al. (2004).

6.4.2 Stochastic Multiplicative Error Models

The major difference between the ACD and SCD model is that in the ACD framework, dynamics are *observation-driven* whereas in the SCD setting, dynamics are *parameter-driven*. This implies that $\mathbb{E}[x_i | \mathcal{F}_{i-1}]$ is deterministic in the ACD model, while it is stochastic in the SCD model. Hautsch (2008) proposes combining both types of dynamics resulting in an ACD model (or MEM, respectively) which is augmented by a dynamic latent factor. It is given by

$$x_i = \Psi_i e^{\delta \lambda_i} \varepsilon_i, \tag{6.37}$$

[7]Alternatively, Strickland et al. (2006) propose estimating the SCD model based on Monte Carlo Markov Chain (MCMC) techniques. Bauwens and Galli (2009) employ efficient importance sampling techniques as discussed in Sect. 7.2.

with $\delta > 0$ being a parameter, λ_i denoting a latent factor, ε_i being i.i.d. and following a distribution on positive support with $\mathbb{E}[\varepsilon_i] = 1$. The conditional mean Ψ_i can be specified according to various ACD specifications as discussed in the previous sections. Then, $\Psi_i e^{\delta\lambda_i}$ corresponds to the conditional mean given $\mathcal{F}_{i-1}$ and λ_i. Accordingly, $\Psi_i = \mathbb{E}[x_i|\mathcal{F}_{i-1}] = x_i/(e^{\delta\lambda_i}\varepsilon_i)$ is stochastic. The latent factor λ_i is assumed to follow a zero mean AR(1) process, given by

$$\lambda_i = a\lambda_{i-1} + \nu_i, \quad \nu_i \sim \text{i.i.d.}\,\mathcal{N}(0,1), \tag{6.38}$$

where ν_i is assumed to be independent of ε_i. Note that because of the symmetry of the distribution of ν_i, the sign of δ is not identified.

Hautsch (2008) calls the model a *stochastic MEM* (SMEM) or – in the ACD case – a stochastic ACD (SACD) model since it nests both the ACD model for $\delta = 0$ and the SCD model for Ψ being constant (i.e., $\alpha = \beta = 0$ in case of a linear ACD model). Correspondingly, for $\alpha \neq 0$ and $\beta = 0$ an SCD model is obtained which is mixed with a further random variable. Hence, it allows to explicitly test for both types of dynamics. The SACD model can also be seen as a special case of a semiparametric ACD model with $x_i = \Psi_i\tilde{\varepsilon}_i$ and $\tilde{\varepsilon}_i = e^{\lambda_i}\varepsilon_i$ following a dynamic process. Moreover, due to its two-factor structure, the SMEM allows capturing dynamics not only in first conditional moments but also in higher order conditional moments. In this sense, the SMEM serves as an alternative to the SVD model (see Sect. 6.5.3). Due to its latent dynamic component, the model cannot be estimated by (Q)ML and requires simulation-based techniques. More information on estimation details as well as statistical properties are given in Sect. 7.2, where the SMEM is discussed in a multivariate setting.

As an illustration, Table 6.8 reproduces some of the estimation results of Hautsch (2008) for SMEMs using a Log-ACD specification of Ψ_i, i.e., $\ln\Psi_i = \omega + \sum_{j=1}^{P}\alpha_j\varepsilon_{i-j} + \sum_{j=1}^{Q}\beta_j\ln\Psi_{i-j}$ for 5 min averaged trade sizes and the number of trades for Boeing, traded at NYSE, 2001. The lag order, P and Q are restricted to $P \leq 2$ and $Q \leq 2$. As revealed by the estimates of a, there is significant evidence for the existence of a persistent latent component. Moreover, it is shown that both the parameter driven dynamics as well as the observation driven dynamics interact. The persistence parameter is driven toward one when the latent factor is taken into account. As discussed in Hautsch (2008), this finding indicates that news tend to enter the model primarily through the latent component, which is in line with the idea that the latter serves as a proxy for unobserved information driving the trading process. It is shown that the inclusion of the unobservable factor increases the goodness-of-fit of the model. Nevertheless, neither the parameter driven component nor the observation driven component can be rejected illustrating that trading dynamics are not sufficiently captured by a one-factor model but rather by a two-factor model.

Table 6.8 Maximum likelihood efficient importance sampling estimates of different parameterizations of SMEMS for 5 min average trade sizes and number of trades for Boeing traded at the NYSE, sample period 02/01/01 to 31/05/01. Standard errors are computed based on the inverse of the estimated Hessian. The ML-EIS estimates are computed using $R = 50$ Monte Carlo replications based on 5 EIS iterations (for details on the EIS approach, see Sect. 7.2). Diagnostics: Log likelihood function (LL), Bayes Information Criterion (BIC), mean, standard deviation and Ljung–Box statistics (LB) based on 20 lags of the filtered residuals. Significance at the 1%, 5% and 10% levels are denoted by ***, ** and *, respectively

	(1)	(2)	(3)	(4)	(5)
5 min average trade sizes					
ω	−0.403***	−0.416***	−5.841***	−0.070***	−0.057***
α_1	0.001***	0.002***		0.011***	−0.006**
α_2		−0.000			0.014***
β_1	0.928***	0.761***		0.981***	1.074***
β_2		0.165			−0.091
p	0.517***	0.519***	0.450***	1.202***	1.188***
m	8.032***	7.991***	12.753***	4.276***	4.321***
Latent component					
a			0.954***	0.260***	0.391***
δ			0.113***	0.506***	0.494***
LL	−6,211	−6,200	−6,019	−5,954	−5,943
BIC	−6,234	−6,232	−6,041	−5,985	−5,983
Mean	1.011	1.010	1.017	1.024	1.027
S.D.	0.871	0.870	0.890	0.901	0.909
LB	39.166***	23.256	38.896***	19.486	15.183
5 min number of trades					
ω	−0.202***	−0.025***	−0.426***	−0.035***	−0.042***
α_1	0.077***	0.105***		0.026***	0.046***
α_2		−0.092***			−0.018
β_1	0.953***	1.704***		0.991***	0.562***
β_2		−0.708***			0.427*
p	1.632***	1.752***	2.170***	2.563***	2.366***
m	3.674***	3.237***	2.639***	2.100***	2.317***
Latent component					
a			0.957***	0.766***	0.821***
δ			0.067***	0.110***	0.092***
LL	−1,982	−1,957	−1,975	−1,925	−1,921
BIC	−2,005	−1,988	−1,998	−1,956	−1,961
Mean	0.999	0.999	1.002	1.001	1.002
S.D.	0.323	0.322	0.324	0.322	0.322
LB	55.018***	33.285**	54.158***	36.095**	24.503

6.4.3 Component Multiplicative Error Models

A further type of mixture MEM is proposed by Brownlees et al. (2011) to model the behavior of intraday volumes. To capture not only intra-day dynamics but also daily dynamics, they propose a so-called *component MEM* given by

$$x_{d,i} = \Psi_{d,i}\mu_d\varepsilon_{d,i}, \quad \varepsilon_{d,i} \overset{iid}{\sim} (1, \sigma^2), \tag{6.39}$$

where d indexes the corresponding trading day and i the intraday observation. Furthermore, μ_d denotes a *daily* component varying from day to day but being constant during a day. By restricting the lag order for simplicity to one, it is given by

$$\mu_d = \omega^d + \alpha^d x_{d-1} + \beta^d \mu_{d-1}, \tag{6.40}$$

where ω^d, α^d and β^d are parameters, and x_d is the standardized daily volume,

$$x_d = \frac{1}{n_d}\sum_{i=1}^{n_d} \frac{x_{d,i}}{\Psi_{d,i}}, \tag{6.41}$$

with n_d denoting the number of observations on day d. Accordingly, $\Psi_{d,i}$ denotes the *intraday* component given by a standard MEM dynamic based on standardized intradaily volume,

$$\Psi_{d,i} = \omega + \alpha\frac{x_{d,i-1}}{\mu_d} + \beta\Psi_{d,i-1}. \tag{6.42}$$

In order to make the model identifiable, the restriction $\omega = 1 - \alpha - \beta$ is imposed ensuring that $\mathbb{E}[\Psi_{d,i}] = 1$. Hence, the component μ_d adjusts the mean level of the series on a daily basis, whereas $\Psi_{d,i}$ captures intraday variations around this level. Given that both types of dynamics are observation-driven, the model is straightforwardly estimated by QML or GMM.

Alternatively, the daily component might be specified in terms of a latent process with

$$\ln\mu_d = \omega + \alpha^d \ln\mu_{d-1} + u_d, \quad u_d \sim \mathcal{N}(0, \sigma^2). \tag{6.43}$$

Such a specification is proposed by Brownlees and Vannucci (2010) and is similar to Hautsch's (2008) SMEM. It also multiplicatively combines an observation driven dynamic as well as a parameter driven dynamic where, however, the latter is updated on a daily basis. Hence, as the SMEM, it cannot be estimated using (Q)ML and requires simulation-based techniques. Brownlees and Vannucci (2010) propose estimating the model using Bayesian inference and develop a corresponding MCMC algorithm.

6.5 Further Generalizations of Multiplicative Error Models

6.5.1 Competing Risks ACD Models

Bauwens and Giot (2003) propose a two-state competing risks ACD specification to model the duration process in dependence of the direction of the contemporaneous midquote change. They specify two competing ACD models (for upwards midquote movements vs. downwards midquote movements) with regime-switching parameters that are determined by the direction of lagged price movements. Then, at each observation the end of only one of the two spells can be observed (depending on whether one observes an increase or decrease of the midquote) while the other spell is censored. Engle and Lunde (2003) propose a similar competing risks model to study the interdependencies between the trade and quote arrival process. However, they impose an asymmetric treatment of both processes since they assume that the trade process is the "driving process" which is completely observed while the quote process is subject to censoring mechanisms.

6.5.2 Semiparametric ACD Models

The assumption of independently, identically distributed innovations ε_i is a standard assumption in the class of MEMs. However, Drost and Werker (2004) argue that the i.i.d. assumption for standardized durations could be too restrictive to describe financial duration processes accurately. In the semiparametric ACD model, the conditional p.d.f. of ACD innovations $\varepsilon_i = x_i/\Psi_i$ is characterized by

$$f(\varepsilon_i|\mathcal{F}_{i-1}) = f(\varepsilon_i|\mathcal{H}_{i-1}), \tag{6.44}$$

where $\mathcal{H}_i \subset \mathcal{F}_i$. The terminology *semi*parametric ACD model stems from the fact that the conditional density $f(\varepsilon_i|\mathcal{H}_{i-1})$ is not specified. The set $\mathcal{H}_i$ defines the relevant past variables driving the conditional distribution of ε_i and formalizes the dependence among the innovations ε_i. Clearly, if $\mathcal{H}_i$ equals the trivial sigma field, the innovations are assumed to be independent and the model coincides with the basic ACD model. Conversely, if $\mathcal{H}_i = \mathcal{F}_i$, the dependence structure of the innovations ε_i is completely unrestricted. In case of $\mathcal{H}_i = \sigma(\varepsilon_i)$ the innovations follow a Markov structure.

Drost and Werker (2004) show how to construct semiparametrically efficient estimators in case of (6.44) and $f(\varepsilon_i|\mathcal{H}_{i-1})$ being non-specified. The key concept in this context is the semiparametric score function, i.e. the semiparametric counterpart to the score function $\mathbb{E}[\mathbf{s}_i(\boldsymbol{\theta}_0)] = \mathbf{0}, \boldsymbol{\theta}_0 \in \boldsymbol{\Theta}, i = 1,\ldots,n$. The semiparametric score function is computed using techniques based on tangent spaces (see Bickel et al. 1993) and is given by

$$\begin{aligned}\mathbf{s}_i^*(\boldsymbol{\theta}) = {} & \frac{\varepsilon_i - 1}{\mathbb{V}[\varepsilon_i|\mathcal{H}_{i-1}]}\mathbb{E}\left[\frac{\partial}{\partial\boldsymbol{\theta}}\ln\Psi_i\,\middle|\,\mathcal{H}_i\right] \\ & - \left(1+\varepsilon_i\frac{\partial f(\varepsilon_i|\mathcal{H}_{i-1})/\partial\boldsymbol{\theta}}{f(\varepsilon_i|\mathcal{H}_{i-1})}\right)\left[\frac{\partial}{\partial\boldsymbol{\theta}}\ln\Psi_i - \mathbb{E}\left\{\frac{\partial}{\partial\boldsymbol{\theta}}\ln\Psi_i\,\middle|\,\mathcal{H}_{i-1}\right\}\right],\end{aligned} \tag{6.45}$$

where $f(\varepsilon_i|\mathcal{H}_{i-1})$ (as well its derivative) has to be estimated by kernel techniques. Then, the semiparametric estimator is constructed as the one-step Newton-Raphson improvement from an arbitrary consistent parametric estimator. In particular, let $\tilde{\boldsymbol{\theta}}_n$ denote this arbitrary consistent estimator which is naturally chosen as the exponential QML estimator. Then, the semiparametrically efficient estimator is obtained by

$$\hat{\boldsymbol{\theta}}_n = \tilde{\boldsymbol{\theta}}_n + \left(\frac{1}{n}\sum_{i=1}^{n}\mathbf{s}_i^*(\tilde{\boldsymbol{\theta}}_n)\mathbf{s}_i^*(\tilde{\boldsymbol{\theta}}_n)'\right)^{-1}\frac{1}{n}\sum_{i=1}^{n}\mathbf{s}_i^*(\tilde{\boldsymbol{\theta}}_n). \tag{6.46}$$

Applying this concept to ACD models based on different types of innovation processes, Drost and Werker (2004) illustrate that even small dependencies in the innovations can induce sizeable efficiency gains of the efficient semiparametric procedure over the QML procedure.

6.5.3 Stochastic Volatility Duration Models

One drawback of autoregressive duration models (like EARMA or ACD models) is that they do not allow for separate dynamic parameterizations of higher order moments. This is due to the fact that typical duration distributions, like the exponential distribution, imply a direct relationship between the first moment and higher order moments. To address this problem, Ghysels et al. (1998) propose a two factor model which allows one to estimate separate dynamics for the conditional variance of durations (the so-called duration volatility).

Consider an i.i.d. exponential model with gamma heterogeneity, $x_i = u_i/\lambda_i$, where u_i is an i.i.d. standard exponential variate and the hazard λ_i is assumed to depend on some heterogeneity component v_i. Thus,

$$\lambda_i = a v_i, \qquad v_i \sim \mathcal{G}(\kappa,\kappa) \tag{6.47}$$

with v_i assumed to be independent of u_i. Ghysels et al. (1998) consider the equation $x_i = u_i/(av_i)$ as a two factor formulation and rewrite this expression in terms of Gaussian factors. Hence, x_i is expressed as

$$x_i = \frac{\mathcal{G}^{-1}(1,\Phi(m_1))}{a\mathcal{G}^{-1}(\kappa,\Phi(m_2))} := \frac{G(1,m_1)}{aG(\kappa,m_2)}, \tag{6.48}$$

where m_1 and m_2 are i.i.d. standard normal random variables, $\Phi(\cdot)$ is the c.d.f. of the standard normal distribution and $\mathcal{G}^{-1}(m,\alpha)$ is the α-quantile function of the $\mathcal{G}(m,m)$ distribution. The function $G(1,m_1)$ can be solved as $G(1,m_1) = -\log(1-\Phi(m_1))$ while $G(\kappa,m_2)$ does not allow for a simple analytical solution and has to be approximated numerically (for more details, see Ghysels et al. 1998).

This (static) representation of an exponential duration model with gamma heterogeneity as the function of two independent Gaussian random variables is used as starting point for a dynamic specification of the stochastic volatility duration (SVD) model. Ghysels et al. (1998) propose specifying the process $\mathbf{m}_i := (m_{1,i}\, m_{2,i})'$ in terms of a VAR representation, where the marginal distribution is constrained to be a $\mathcal{N}(\mathbf{0}, \mathbf{I})$ distribution, where $\mathbf{I}$ denotes the identity matrix. This restriction ensures that the marginal distribution of the durations belongs to the class of exponential distributions with gamma heterogeneity. Thus,

$$\mathbf{m}_i = \sum_{j=1}^{P} \boldsymbol{\Lambda}_j \mathbf{m}_{i-j} + \boldsymbol{\varepsilon}_i, \tag{6.49}$$

where $\boldsymbol{\Lambda}_j$ denotes the matrix of autoregressive VAR parameters and $\boldsymbol{\varepsilon}_i$ is a vector of Gaussian white noise random variables with variance-covariance matrix $\boldsymbol{\Sigma}(\boldsymbol{\Lambda})$ which ensures that $\mathbb{V}(\mathbf{m}_i) = \mathbf{I}$.

The SVD model belongs to the class of nonlinear dynamic factor models for which the likelihood typically is difficult to evaluate. Therefore, Ghysels et al. (1998) propose a two step procedure. In the first step, one exploits the property that the marginal distribution of x_i is a Pareto distribution depending on the parameters a and κ (see appendix). Thus, a and κ can be estimated by (Q)ML. In the second step, the autoregressive parameters $\boldsymbol{\Lambda}_j$ are estimated by using the methods of simulated moments (for more details, see Ghysels et al. 1998).

Ghysels et al. (1998) apply the SVD model to analyze trade durations of the Alcatel stock traded at the Paris Stock Exchange and find empirical evidence for strong dynamics in both factors. However, the higher dynamic flexibility of the SVD model induces strong distributional assumptions. Bauwens et al. (2004) show that the assumption of a Pareto distribution for the marginal distribution of x_i is inappropriate for specific types of financial durations (particularly volume durations) which makes the estimation of the Pareto parameters a and κ extremely cumbersome. Moreover, they find a relatively poor prediction performance of the SVD model compared to alternative ACD specifications.

References

Baillie RT (1996) Long memory processes and fractional integration in econometrics. J Econom 73:5–59

Baillie RT, Bollerslev T, Mikkelsen H-O (1996) Fractionally integrated generalized autoregressive conditional heteroskedasticity. J Econom 52:91–113

Bauwens L, Galli F (2009) Efficient importance sampling for ML estimation of SCD models. Comput Stat Data Anal 53:1974–1992

Bauwens L, Giot P (2003) Asymmetric ACD models: introducing price information in ACD models with a two state transition model. Empir Econ 28:1–23

Bauwens L, Veredas D (2004) The stochastic conditional duration model: a latent factor model for the analysis of financial durations. J Econom 119:381–412

Bauwens L, Giot P, Grammig J, Veredas D (2004) A comparison of financial duration models via density forecasts. Int J Forecast 20:589–609

Beran J (1994) Statistics for long-memory processes. Chapman and Hall, New York

Bickel PJ, Klaasen CAJ, Ritov Y, Wellner AJ (1993) Efficient and addaptive statistical inference for semiparametric models. John Hopkins University Press, Baltimore

Brownlees C, Cipollini F, Gallo GM (2011) Intra-daily volume modeling and prediction for algorithmic trading. J Financ Econom 9:489–518

Brownlees C, Vannucci M (2010) A Bayesian approach for capturing daily heterogeneity in intra-daily durations time series: the mixed autoregressive conditional duration model. Working Paper, Stern School of Business, New York University

Bougerol P, Picard N (1992) Stationarity of GARCH processes. J Econom 52:115–127

Carrasco M, Chen X (2002) Mixing and moment properties of various GARCH and stochastic volatility models. Econom Theory 18(1):17–39

Carrasco M, Chen X (2005) Erratum: mixing and moment properties of various GARCH and stochastic volatility models. Unpublished Manuscript

Dempster AP, Laird NM, Rubin DB (1977) Maximum likelihood from incomplete data via the EM algorithm. J R Stat Soc Series B 39:1–38

Deo R, Hsieh M, Hurvich CM (2010) Long memory in intertrade durations, counts and realized volatility of NYSE stocks. J Stat Plan Inference 140:3715–3733

Deo R, Hurvich CM, Soulier P, Wang Y (2009) Conditions for the propagation of memory parameter from durations to counts and realized volatility. Econom Theory 25:764–792

Ding Z, Granger CWJ (1996) Modeling volatility persistence of speculative returns: a new approach. J Econom 73:185–215

Drost FC, Werker BJM (2004) Semiparametric duration models. J Bus Econ Stat 22:40–50

Dufour A, Engle RF (2000) The ACD model: predictability of the time between consecutive trades. Working Paper, ISMA Centre, University of Reading

Engle RF (2000) The econometrics of ultra-high-frequency data. Econometrica 68(1):1–22

Engle RF, Lunde A (2003) Trades and quotes: a bivariate point process. J Financ Econom 11:159–188

Engle RF, Ng VK (1993) Measuring and testing the impact of news on volatility. J Finance 48:1749–1778

Feng D, Jiang GJ, Song PX-K (2004) Stochastic conditional duration models with 'leverage effect' for financial transaction data. J Financ Econom 2:390–421

Fernandes M, Grammig J (2006) A family of autoregressive conditional duration models. J Econom 130:1–23

Ghysels E, Gouriéroux C, Jasiak J (1998) Stochastic volatility duration models. Discussion paper, CIRANO

Gonzalez-Rivera G (1998) Smooth transition GARCH models. Stud Nonlinear Dynamics Econometrics 3:61–78

Granger CWJ, Joyeux R (1980) An introduction to long-memory time series models and fractional differencing. J Time Series Anal 1:15–39

Granger CWJ, Teräsvirta T (1993) Modelling nonlinear economic relationships. Oxford University Press, Oxford, UK

Hagerud GE (1997) A new non-linear GARCH model. Discussion paper, EFI, Stockholm School of Economics

Hautsch N (2003) Assessing the risk of liquidity suppliers on the basis of excess demand intensities. J Financ Econom 1:189–215

Hautsch N (2008) Capturing common components in high-frequency financial time series: a multivariate stochastic multiplicative error model. J Econ Dyn Control 32:3978–4009

Hentschel L (1995) All in the family: nesting symmetric and asymmetric GARCH models. J Finan Econ 39:71–104

Hosking JRM (1981) Fractional differencing. Biometrika 68:165–176

Hujer R, Vuletic S, Kokot S (2002) The Markov switching ACD model. Finance and Accounting Working Paper 90, Johann Wofgang Goethe-University, Frankfurt. Available at SSRN: http://ssrn.com/abstract=332381

Jasiak J (1998) Persistence in intratrade durations. Finance 19:166–195

Karanasos M (2004) The statistical properties of long-memory ACD models. WSEAS Trans Bus Econ 2:169–175

Koulikov D (2003) Modeling sequences of long memory non-negative covariance stationary random variables. Discussion Paper 156, CAF

Meddahi N, Renault E, Werker BJ (1998) Modelling high-frequency data in continuous time. Discussion paper, University de Montréal

Meitz M, Saikkonen P (2008) Ergodicity, mixing, and existence of moments of a class of Markov models with applications to GARCH and ACD models. Econom Theory 24:1291–1320

Meitz M, Teräsvirta T (2006) Evaluatinsg models of autoregressive conditional duration. J Bus Econ Stat 24:104–124

Mokkadem A (1990) Propriétés de mélange des modèles autoregressifs polynomiaux. Annales de l'Institut Henri Poincaré 26:219–260

Nelson D (1991) Conditional heteroskedasticity in asset returns: a new approach. J Econom 43:227–251

Rabemananjara J, Zakoian JM (1993) Threshold ARCH models and asymmetries in volatility. J Appl Econom 8:31–49

Strickland CM, Forbes CS, Martin MG (2006) Bayesian analysis of the stochastic conditional duration model. Comput Stat Data Anal 50:2247–2267

Taylor SJ (1982) Financial returns modelled by the product of two stochastic processes – a study of daily sugar prices. In: Anderson OD (ed) Time series analysis: theory and practice, North-Holland, Amsterdam

Teräsvirta T (1994) Specification, estimation, and evaluation of smooth transition autoregressive models. J Am Stat Assoc 89:208–218

Terärsvirta T (1998) Modelling economic relationships with smooth transition regressions. In Ullah A, Giles DEA, (eds) Handbook of applied economic statistics Marcel Dekker, New York, pp 507–552

Tong H (1990) Non-linear time series: a dynamical system approach. Oxford University Press, Oxford

Whittle P (1962) Gaussian estimation in stationary time series. Bulletin de l'Institute International de Statistique 39:105–129

Zakoian JM (1994) Threshold heteroskedastic models. J Econ Dyn Control 18:931–955

Zhang MY, Russell JR, Tsay RS (2001) A nonlinear autoregressive conditional duration model with applications to financial transaction data. J Econom 104:179–207

Chapter 7
Vector Multiplicative Error Models

This chapter focusses on multivariate extensions of multiplicative error models. The basic multivariate (or vector) multiplicative error model is introduced in Sect. 7.1.1. We discuss specification, statistical inference and provide empirical illustrations. Section 7.2 is devoted to stochastic vector MEMs corresponding to multivariate versions of univariate stochastic MEMs as presented in Chap. 5. Here, the idea is to augment a VMEM process by a common latent component which jointly affects all individual processes. We illustrate how to estimate this class of models using simulated maximum likelihood and illustrate applications to the modelling of trading processes.

7.1 VMEM Processes

7.1.1 The Basic VMEM Specification

Consider a K-dimensional positive-valued time series, denoted by $\{\mathbf{x}_i\}$, $i = 1\ldots,n$, with $\mathbf{x}_i :\ = (x_i^1,\ldots,x_i^K)$. The so-called vector MEM (VMEM) for $\mathbf{x}_i$ is defined by

$$\mathbf{x}_i = \boldsymbol{\Psi}_i \odot \boldsymbol{\varepsilon}_i,$$

where $\boldsymbol{\Psi}_i := \mathbb{E}[\mathbf{x}_i|\mathcal{F}_{i-1}]$ is a $K \times 1$ vector, $\odot$ denotes the Hadamard product (element-wise multiplication) and $\boldsymbol{\varepsilon}_i$ is a K-dimensional vector of mutually and serially i.i.d. innovation processes, where the jth element is given by

$$\varepsilon_i^j|\mathcal{F}_{i-1} \sim \text{i.i.d. } \mathcal{D}(1,\sigma_j^2), \quad j = 1,\ldots,K.$$

The VMEM is a straightforward extension of the univariate linear MEM/ACD model and is specified by Manganelli (2005) as

DOI 10.1007/978-3-642-21925-2_7,

$$\boldsymbol{\Psi}_i = \boldsymbol{\omega} + \mathbf{A}_0\mathbf{x}_i + \sum_{j=1}^{P} \mathbf{A}_j\mathbf{x}_{i-j} + \sum_{j=1}^{Q} \mathbf{B}_j\boldsymbol{\Psi}_{i-j}, \tag{7.1}$$

where $\boldsymbol{\omega}$ is a $K \times 1$ vector, and $\mathbf{A}_0$, $\mathbf{A}_j$ as well as $\mathbf{B}_j$ are $K \times K$ parameter matrices. The matrix $\mathbf{A}_0$ captures contemporaneous relationships between the elements of $\mathbf{x}_i$ and is specified as a matrix where only the upper triangular elements are non-zero. This triangular structure excludes simultaneity between the individual variables and implies that x_i^j is causal for x_i^m, $m > j$, but x_i^m is not causal for x_i^j. Consequently, x_i^m is conditionally i.i.d. given $\{x_i^j, \mathcal{F}_{i-1}\}$ for $j < m$.

A VMEM approach is obviously only meaningful if the multivariate time series are synchronized in time. In case of financial trading variables, the model is applicable, e.g., to simultaneously model trading characteristics (trade durations, trade sizes, bid-ask spreads, trade-to-trade returns etc.). This approach is pursued by Manganelli (2005). In case of processes which do not occur synchronously in time (such as trading activities across different assets), a time synchronization of the data is necessary. Then, most naturally, aggregated variables over equi-distant time intervals as discussed in Chap. 3 are used.

Analogously to the univariate case, the linear VMEM can be alternatively presented in terms of a vector ARMA (VARMA) process for $\mathbf{x}_i$. By introducing the vector of martingale differences $\boldsymbol{\eta}_i := \mathbf{x}_i - \boldsymbol{\Psi}_i$, and for simplicity of illustration restricting our attention to the case $P = Q = 1$, the VMEM(1,1) model can be written as

$$\mathbf{x}_i = (\mathbf{I} - \mathbf{A}_0)^{-1}(\boldsymbol{\omega} + (\mathbf{A}_1 + \mathbf{B}_1)\mathbf{x}_{i-1} - \mathbf{B}_1\boldsymbol{\eta}_{i-1} + \boldsymbol{\eta}_i. \tag{7.2}$$

Invertibility of $(\mathbf{I} - \mathbf{A}_0)$ is ensured by the triangular structure of $\mathbf{A}_0$ ruling out simultaneous relationships between the variables. Then, the unconditional mean is straightforwardly given by

$$\mathbb{E}[\mathbf{x}_i] = (\mathbf{I} - \mathbf{A}_0 - \mathbf{A}_1 - \mathbf{B}_1)^{-1}\boldsymbol{\omega}, \tag{7.3}$$

with the conditions for weak stationarity given by all eigenvalues of $|(\mathbf{I} - \mathbf{A}_0)^{-1}(\mathbf{A}_1 + \mathbf{B}_1)|$ having modulus smaller than one.

The advantage of this specification is that contemporaneous relationships between the variables are taken into account without requiring multivariate distributions for $\boldsymbol{\varepsilon}_i$. Furthermore, the theoretical properties of univariate MEMs as discussed in the previous chapter can be straightforwardly extended to the multivariate case. However, an obvious drawback is the requirement to impose an explicit ordering of the variables in $\mathbf{x}_i$ induced by the triangular structure. The order is typically chosen in accordance with a specific research objective or following economic reasoning.

An alternative way to capture contemporaneous relationships is to allow for mutual correlations between the innovation terms ε_i^j. Then, the innovation term vector follows a density function which is defined over non-negative K-dimensional support $[0, +\infty)^K$ with unit mean $\boldsymbol{\iota}$ and covariance matrix $\boldsymbol{\Sigma}$, i.e.,

$$\boldsymbol{\varepsilon}_i | \mathcal{F}_{i-1} \sim \text{i.i.d. } \mathcal{D}(\boldsymbol{\iota}, \boldsymbol{\Sigma})$$

implying

$$\mathbb{E}[\mathbf{x}_i | \mathcal{F}_{i-1}] = \boldsymbol{\Psi}_i,$$

$$\mathbb{V}[\mathbf{x}_i | \mathcal{F}_{i-1}] = \boldsymbol{\Psi}_i \boldsymbol{\Psi}_i' \odot \boldsymbol{\Sigma}.$$

Fining an appropriate multivariate distribution defined on positive support is a difficult task. As discussed by Cipollini et al. (2007), a possible candidate is a multivariate gamma distribution which however imposes severe restrictions on the contemporaneous correlations between the errors ε_i^j.

Alternatively, the dependence structure can be captured by copula approaches. Bodnar and Hautsch (2011) propose modelling the dependence in $\boldsymbol{\varepsilon}_i$ by a Gaussian copula with dynamic correlation matrix. Define

$$\boldsymbol{\varepsilon}_i^* = (\Phi^{-1}(F_1(\varepsilon_{1,i})), ..., \Phi^{-1}(F_K(\varepsilon_{K,i})))',$$

where $\Phi(.)$ denotes the c.d.f. of the univariate standard normal distribution and $F_j(.)$ denotes the marginal distribution function of ε_i^j. The assumption of the Gaussian copula implies that the transformed residuals $\boldsymbol{\varepsilon}_i^*$ are conditionally normally distributed with conditional correlation matrix $\mathbf{R}_i$. The transformation from $\boldsymbol{\varepsilon}_i$ to $\boldsymbol{\varepsilon}_i^*$ is monotone though non-linear. Therefore, the series $\{\boldsymbol{\varepsilon}_i^*\}$ as well as $\{\boldsymbol{\varepsilon}_i^{*2}\}$ might be autocorrelated while the $\{\boldsymbol{\varepsilon}_i\}$ themselves are uncorrelated. To capture these effects, Bodnar and Hautsch propose an VARMA-(M)GARCH parameterization given by

$$\boldsymbol{\varepsilon}_i^* = \sum_{j=1}^{Q^*} \mathbf{C}_j \boldsymbol{\varepsilon}_{i-j}^* + \boldsymbol{\nu}_i, \tag{7.4}$$

$$\boldsymbol{\nu}_i = \sqrt{\mathbf{h}_i} \odot \boldsymbol{\xi}_i, \tag{7.5}$$

$$\mathbf{h}_i = \omega_h + \sum_{j=1}^{P^h} \mathbf{A}_j (\boldsymbol{\nu}_{i-j} \odot \boldsymbol{\nu}_{i-j}) + \sum_{j=1}^{Q^h} \mathbf{B}_j \mathbf{h}_{i-j}, \tag{7.6}$$

where $\mathbf{C}_j$, $\mathbf{A}_j$ and $\mathbf{B}_j$ are $K \times K$ parameter matrices with $\boldsymbol{\varepsilon}_i^* \sim \mathcal{N}(\mathbf{0}, \mathbf{R}_i)$. Correspondingly, $\mathbf{h}_i$ is a $K \times 1$ vector of conditional variances of $\boldsymbol{\nu}_i$ with $\boldsymbol{\xi}_i$ denoting a vector of i.i.d. $\mathcal{N}(0, 1)$ innovations. Then, the conditional correlation matrix $\mathbf{R}_i$ is modelled according to Engle's (2002) Dynamic Conditional Correlation (DCC) model and is given by

$$\mathbf{R}_i = \mathbf{Q}_i^{*\,-1} \mathbf{Q}_i \mathbf{Q}_i^{*\,-1}, \tag{7.7}$$

$$\mathbf{Q}_i = \left(1 - \sum_{j=1}^{P^R} \gamma_j - \sum_{j=1}^{Q^R} \delta_j\right) \bar{\mathbf{Q}} + \sum_{j=1}^{P^R} \gamma_j \boldsymbol{\xi}_{i-j} \boldsymbol{\xi}'_{i-j} + \sum_{j=1}^{Q^R} \delta_j \mathbf{Q}_{i-j}, \quad (7.8)$$

where $\bar{\mathbf{Q}}$ is the unconditional covariance matrix of $\boldsymbol{\xi}_i$. Hence, the Gaussian copula implies a transformation of $\boldsymbol{\varepsilon}_i$ into a multivariate normal distribution with dynamic conditional mean and conditional covariance matrix. Bodnar and Hautsch (2011) suggest a two-stage maximum likelihood approach, where the MEM parameters are estimated in a first step, while the VARMA-GARCH-DCC parameters are estimated in a second step. Modelling the trading process of various NYSE stocks, Bodnar and Hautsch (2011) show that the assumption of normality of the components $\boldsymbol{\varepsilon}_i^*$ is well supported by the data. Moreover, significant evidence for serial dependencies in conditional variances and correlations is shown.

Obviously, the conditional mean function $\boldsymbol{\Psi}_i$ can be specified in various alternative ways. For instance, a logarithmic VMEM specification is obtained by

$$\ln \boldsymbol{\Psi}_i = \boldsymbol{\omega} + \mathbf{A}_0 \ln \mathbf{x}_i + \sum_{j=1}^{P} \mathbf{A}_j g(\boldsymbol{\varepsilon}_{i-j}) + \sum_{j=1}^{Q} \mathbf{B}_j \ln \boldsymbol{\Psi}_{i-j}, \quad (7.9)$$

where $g(\boldsymbol{\varepsilon})_{i-j}) = \boldsymbol{\varepsilon}_{i-j}$ or $g(\boldsymbol{\varepsilon}_{i-j}) = \ln \boldsymbol{\varepsilon}_{i-j}$, respectively (see also Sect. 5.5). Generalized VMEMs can be specified in accordance to the approaches discussed in the previous chapter.

7.1.2 *Statistical Inference*

Define $f(x_i^1, x_i^2, \ldots, x_i^K | \mathcal{F}_{i-1}; \boldsymbol{\theta})$ as the joint conditional density given $\mathcal{F}_{i-1}$ and the parameter vector $\boldsymbol{\theta} = (\boldsymbol{\theta}^1, \ldots, \boldsymbol{\theta}^K)'$. Due to the triangular structure of $\mathbf{A}_0$, the joint density can be decomposed into

$$\begin{aligned} f(x_i^1, x_i^2, \ldots, x_i^K | \mathcal{F}_{i-1}; \boldsymbol{\theta}) &= f(x_i^1 | x_i^2, \ldots, x_i^K; \mathcal{F}_{i-1}, \boldsymbol{\theta}) \\ &\quad \times f(x_i^2 | x_i^3, \ldots, x_i^K; \mathcal{F}_{i-1}, \boldsymbol{\theta}) \\ &\quad \vdots \\ &\quad \times f(x_i^K | \mathcal{F}_{i-1}; \boldsymbol{\theta}). \end{aligned}$$

Then, the log likelihood function is given as

$$\ln \mathcal{L}(\mathbf{Y}; \boldsymbol{\theta}) = \sum_{i=1}^{n} \sum_{j=1}^{K} \ln f(x_i^j | x_i^{j+1}, \ldots, x_i^K; \mathcal{F}_{i-1}). \quad (7.10)$$

Constructing the likelihood based on an exponential distribution leads to the quasi likelihood function with components

$$\ln f(x_i^j|x_i^{j+1},\ldots,x_i^K;\mathcal{F}_{i-1}) = -\sum_{i=1}^{n}\left(\ln\Psi_i^j + x_i^j/\Psi_i^j\right),$$

where the elements of the score and Hessian are given by

$$\frac{\partial \ln f(x_i^j|x_i^{j+1},\ldots,x_i^K;\mathcal{F}_{i-1})}{\partial \boldsymbol{\theta}^j} = -\sum_{i=1}^{n}\frac{\partial\Psi_i^j}{\partial\boldsymbol{\theta}^j}\frac{1}{\Psi_i^j}\left(\frac{x_i^j}{\Psi_i^j}-1\right),$$

$$\frac{\partial^2 \ln f(x_i^j|x_i^{j+1},\ldots,x_i^K;\mathcal{F}_{i-1})}{\partial \boldsymbol{\theta}^j\partial\boldsymbol{\theta}^{j'}} = \sum_{i=1}^{n}\left\{\frac{\partial}{\partial\boldsymbol{\theta}^{j'}}\left(\frac{1}{\Psi_i^j}\frac{\partial\Psi_i^j}{\partial\boldsymbol{\theta}^j}\right)\left(\frac{x_i^j}{\Psi_i^j}-1\right) - \frac{1}{\Psi_i^j}\frac{\partial\Psi_i^j}{\partial\boldsymbol{\theta}^j}\frac{\partial\Psi_i^j}{\partial\boldsymbol{\theta}^{j'}}\frac{x_i^j}{(\Psi_i^j)^2}\right\}.$$

The model can be estimated equation by equation as long as the likelihood can be decomposed into

$$f(x_i^1,x_i^2,\ldots,x_i^K|\mathcal{F}_{i-1};\boldsymbol{\theta}) = \prod_{j=1}^{K} f(x^j|x^{j+1},\ldots,x^K;\mathcal{F}_{i-1},\boldsymbol{\theta}^j). \tag{7.11}$$

This requires the parameters of the system to be variation free according to Engle et al. (1983). In case of the linear specification (7.1), this is naturally ensured. In case of the logarithmic specification (7.9), it is ensured only if $g(\boldsymbol{\varepsilon}_{i-j}) = \ln\boldsymbol{\varepsilon}_{i-j}$ as this specification can be re-written in terms of a logarithmic version of (7.1).[1] In more general cases, the decomposition of the likelihood according to (7.11) is not necessarily possible which requires estimating all parameters simultaneously.

7.1.3 Applications

Hautsch and Jeleskovic (2008) apply the VMEM to jointly model 1-min squared returns, average trade sizes, number of trades as well as average trading costs based on data of the electronic trading of the Australian Stock Exchange (ASX). The data stem from completely reconstructed order books for the stocks BHP Billiton Limited (BHP) and National Australian Bank (NAB) during the trading period July and August 2002 covering 45 trading days. The log returns are pre-adjusted to account for the bid-ask bounce and correspond to the residuals of an MA(1)

[1] Recall the discussion of Log-ACD models in Chap. 5.

filter. The trading costs are computed as the hypothetical trading costs of an order of the size of 10,000 shares in excess to the trading costs which would prevail if investors could trade at the mid-quote. As reported by Hautsch and Jeleskovic (2008), resulting average excess trading costs are 60 ASD for BHP and 188 ASD for NAB during the analyzed trading period.

Hautsch and Jeleskovic (2008) propose modelling this process by a four-dimensional augmented Log-VMEM process which accounts for the occurrence of zeros and is given by

$$\begin{aligned}\ln \boldsymbol{\Psi}_i &= \boldsymbol{\omega} + \mathbf{A}_0[(\ln \mathbf{x}_i) \odot \mathbf{I}_{\{\mathbf{x}_i>0\}}] + \mathbf{A}_0^0 \odot \mathbf{I}_{\{\mathbf{x}_i=0\}} \\ &+ \sum_{j=1}^{P} \mathbf{A}_j [g(\boldsymbol{\varepsilon}_{i-j}) \odot \mathbf{I}_{\{\mathbf{x}_{i-1}>0\}}] + \sum_{j=1}^{P} \mathbf{A}_j^0 \odot \mathbf{I}_{\{\mathbf{x}_{i-1}=0\}} \\ &+ \sum_{j=1}^{Q} \mathbf{B}_j \ln \boldsymbol{\Psi}_{i-j}, \qquad (7.12)\end{aligned}$$

where $\mathbf{I}_{\{\mathbf{x}_i>0\}}$ and $\mathbf{I}_{\{\mathbf{x}_i=0\}}$ denote 4×1 vectors of indicator variables indicating non-zero and zero realizations, respectively, and $\boldsymbol{\mathcal{A}}_j^0$, $j = 0, \ldots, p$, are corresponding 4×4 parameter matrices.[2]

Table 7.1 re-produces the estimation results of Hautsch and Jeleskovic (2008) for a Log-VMEM(1,1) specification for BHP and NAB based on a specification with fully parameterized matrix $\mathbf{A}_1$ and diagonal matrix $\mathbf{B}_1$ for seasonally adjusted (pre-filtered) squared log returns, trade sizes, number of trades and transaction costs standardized by their corresponding seasonality components.

The innovation terms are chosen as $g(\boldsymbol{\varepsilon}_i) = \boldsymbol{\varepsilon}_i$. For the process of squared returns, $x_i^1 = r_i^2$, it is assumed that $x_i^1|(x_i^2, \ldots, x_i^4, \mathcal{F}_{i-1}) \sim \mathcal{N}(0, \Psi_i^1)$. Accordingly, for x_i^j, $j \in \{2, 3, 4\}$, it is assumed $x_i^j|x_i^{j+1}, \ldots, x_i^4, \mathcal{F}_{i-1} \sim Exp(\Psi_i^j)$. As zeros only (simultaneously) occur in trade sizes and the number of trades, only the $(2, 3)$-element in $\mathbf{A}_0^0$ and one of the two middle columns in $\mathbf{A}_1^0$ can be identified. Consequently, all other parameters in $\mathbf{A}_0^0$ and $\mathbf{A}_1^0$ are set to zero.

The following results can be summarized: First, there exist significant mutual correlations between nearly all variables. Volatility is positively correlated with liquidity demand and liquidity supply. Hence, active trading as driven by high volumes and high trading intensities is accompanied by high volatility. The significantly negative estimates of A_{24}^0 and A_{34}^0 indicate that these are trading periods which are characterized by low transaction costs.

Second, the diagonal elements in $\mathbf{A}_1$ and the elements in $\mathbf{B}_1$ reveal that all trading components are strongly positively autocorrelated but are not very persistent. The persistence is highest for trade sizes and trading intensities.

[2]For a more sophisticated approach to model positive-valued (continuous) random variables which reveal a non-trivial part of zero outcomes, see Hautsch et al. (2010).

Table 7.1 Maximum likelihood estimation results of a Log VMEM for seasonally adjusted (i) squared (MA(1) filtered) log returns, (ii) average trade sizes, (iii) number of trades, and (iv) average trading costs per 1-min interval. Standard errors are computed based on the OPG covariance matrix. ASX trading, July–August 2002. Diagnostics: Log likelihood function (LL) and Bayes Information Criterion (BIC). Reproduced from Hautsch and Jeleskovic (2008)

	BHP		NAB	
Par.	Coeff.	Std. err.	Coeff.	Std. err.
ω_1	−0.0673	0.0663	0.0023	0.0302
ω_2	0.1921	0.0449	0.1371	0.0254
ω_3	−0.4722	0.1009	−0.1226	0.0432
ω_4	−0.4914	0.1066	−0.5773	0.0485
$A_{0,12}$	0.0549	0.0092	0.1249	0.0056
$A_{0,13}$	0.3142	0.0173	0.6070	0.0122
$A_{0,14}$	0.4685	0.0489	0.7876	0.0094
$A_{0,23}$	0.0673	0.0074	0.0531	0.0070
$A_{0,24}$	−0.1002	0.0289	0.0176	0.0093
$A_{0,34}$	−0.2181	0.0618	−0.0235	0.0123
$A^0_{0,12}$	−3.8196	0.0402	−1.5086	0.0176
$A_{1,11}$	0.1446	0.0080	0.0804	0.0038
$A_{1,12}$	0.0043	0.0090	0.0804	0.0041
$A_{1,13}$	−0.0939	0.0173	0.2036	0.0125
$A_{1,14}$	0.1487	0.0602	−0.0833	0.0214
$A_{1,21}$	0.0004	0.0034	−0.0002	0.0015
$A_{1,22}$	0.0488	0.0049	0.0259	0.0025
$A_{1,23}$	−0.0377	0.0115	−0.0116	0.0093
$A_{1,24}$	−0.1911	0.0398	−0.1329	0.0226
$A_{1,31}$	0.0100	0.0053	−0.0022	0.0020
$A_{1,32}$	0.0095	0.0071	0.0045	0.0031
$A_{1,33}$	0.1088	0.0152	0.0894	0.0109
$A_{1,34}$	0.3420	0.0932	0.0341	0.0377
$A_{1,41}$	0.0064	0.0113	0.0044	0.0067
$A_{1,42}$	0.0091	0.0163	0.0081	0.0081
$A_{1,43}$	0.0524	0.0321	0.0537	0.0249
$A_{1,44}$	0.4256	0.0898	0.5105	0.0431
$A^0_{1,21}$	1.1467	0.0911	−0.5181	0.0204
$A^0_{1,22}$	0.1497	0.0212	0.0341	0.0134
$A^0_{1,23}$	0.0946	0.0318	0.0985	0.0132
$A^0_{1,24}$	−0.0006	0.0755	0.0115	0.0579
$B_{1,11}$	0.4027	0.0252	0.2616	0.0078
$B_{1,22}$	0.7736	0.0179	0.9109	0.0081
$B_{1,33}$	0.9731	0.0074	0.9673	0.0070
$B_{1,44}$	0.5369	0.1024	0.7832	0.0374
LL	−60,211		−58,622	
BIC	−60,378		−58,790	

Third, liquidity variables Granger cause future volatility. In particular, high trade sizes predict high future return volatilities. Conversely, the impact of trading intensities and trading costs on future volatility is less clear revealing contradictive results for both stocks. Obviously, there is no prediction power of return volatility for future liquidity demand and supply.

Fourth, trading intensities and trading costs negatively influence future trade sizes. Thus, a high speed of trading tends to reduce trade sizes over time. Likewise, increasing trading costs seem to lower the incentive for high order sizes but on the other hand increase the speed of trading. These results might be induced by the fact that investors tend to break up large orders into sequences of small orders if liquidity supply is low.

7.2 Stochastic Vector Multiplicative Error Models

7.2.1 *Stochastic VMEM Processes*

A further generalization of VMEM processes and multivariate extension of the stochastic MEM has been introduced by Hautsch (2008). The major idea is to capture mutual (time-varying) dependencies by a subordinated common (latent) factor jointly driving the individual processes. Economically, this process might be associated with the underlying (latent) information process jointly influencing the multivariate trading process. The so-called *stochastic* VMEM (SVMEM) can be compactly represented as

$$\mathbf{x}_i = \boldsymbol{\Psi}_i \odot \boldsymbol{\lambda}_i \odot \boldsymbol{\varepsilon}_i, \tag{7.13}$$

where $\boldsymbol{\lambda}_i$ is a $(K \times 1)$ vector with elements $\{\lambda_i^{\delta_i}\}$, $i = 1, \ldots, K$,

$$\ln \lambda_i = a \ln \lambda_{i-1} + \nu_i, \quad \nu_i \sim \text{i.i.d. } \mathcal{N}(0,1), \tag{7.14}$$

and ν_i is assumed to be independent of $\boldsymbol{\varepsilon}_i$. In this multivariate setting, the component λ_i is interpreted as a common dynamic factor with process-specific impacts δ_i (requiring the identification condition $\mathbb{V}[\nu_i] = 1$). The elements of $\boldsymbol{\Psi}_i$ represent "genuine" (e.g., trade-driven) effects given the latent factor.

Hautsch (2008) applies the SVMEM to the three-dimensional process of intraday returns y_i, trade sizes v_i and trading intensities ρ_i (thus $K = 3$) as given by

$$y_i = \mathbb{E}[y_i | \mathcal{F}_{i-1}] + \xi_i, \tag{7.15}$$

$$\xi_i = \sqrt{h_i e^{\delta_1 \lambda_i} s_{h,i}} w_i, \qquad w_i \sim \text{i.i.d.} \mathcal{N}(0,1), \tag{7.16}$$

$$v_i = \Phi_i e^{\delta_2 \lambda_i} s_{v,i} u_i, \qquad u_i \sim \text{i.i.d.} \mathcal{GG}(p_2, m_2), \tag{7.17}$$

$$\rho_i = \Psi_i e^{\delta_3 \lambda_i} s_{\rho,i} \varepsilon_i, \qquad \varepsilon_i \sim \text{i.i.d.} \mathcal{GG}(p_3, m_3), \tag{7.18}$$

where h_i, Φ_i and Ψ_i denote the so-called observation-driven dynamic components, w_i, u_i and ε_i are process-specific innovation terms, which are assumed to be independent, and $s_{h,i}, s_{v,i}, s_{\rho,i} > 0$ capture deterministic time-of-day effects in volatilities, trade sizes, and trading intensities, respectively. The volatility innovations w_i are assumed to follow a standard normal distribution whereas the volume and trading intensity innovations u_i and ε_i follow standard generalized gamma distributions with parameters p_2, m_2 and p_3, m_3, respectively.

The process-specific impact of the latent factor is given by $\lambda_{ij} := \delta_j \lambda_i$ with $\lambda_{ij} = a\lambda_{i-1,j} + \delta_j \nu_i$, and thus $\frac{d\lambda_{ij}}{d\nu_i} > (<)\, 0$ for $\delta_j > (<)\, 0$ with $j = 1, 2, 3$. Since the distribution of ν_i is symmetric, the sign of the parameters δ_j are not individually identified and require to restrict the sign of one of the parameters δ_j. Then, the signs of δ_k with $k \neq j$ are identified.

The process-specific components h_i, Φ_i and Ψ_i are parameterized in terms of a three-dimensional version of (7.9),

$$\mathbf{\Psi}_i = \omega + \mathbf{A}_0 \mathbf{z}_{0,i} + \sum_{j=1}^{P} \mathbf{A}_j \mathbf{z}_{i-j} + \sum_{j=1}^{Q} \mathbf{B}_j \mathbf{\Psi}_{i-j}, \tag{7.19}$$

where

$$\mathbf{\Psi}_i := (\ln h_i,\ \ln \Phi_i,\ \ln \Psi_i)', \tag{7.20}$$

$$\mathbf{z}_{0,i} := (0,\ \ln v_i,\ \ln \rho_i)', \tag{7.21}$$

$$\mathbf{z}_i := \left(\frac{|\xi_i|}{\sqrt{h_i s_{h,i}}},\ \frac{v_i}{\Phi_i s_{v,i}},\ \frac{\rho_i}{\Psi_i s_{\rho,i}} \right)' = \left(|w_i| e^{\delta_1 \lambda_i / 2},\ u_i e^{\delta_2 \lambda_i},\ \varepsilon_i e^{\delta_3 \lambda_i} \right)'. \tag{7.22}$$

As stressed by Hautsch (2008), the fact that the innovations $\mathbf{z}_i$ do not depend on the latent component ensures that h_i, Φ_i and Ψ_i are completely *observation-driven* and eases the estimation of the model. On the other hand, since the latent variable is not integrated out of the innovations, a shock in λ_i influences $\{h_i, \Phi_i, \Psi_i\}$ not only directly (in period i), but also indirectly (through $\mathbf{z}_i$) in the subsequent periods. Therefore, λ_i can generate cross-dependencies between the observation-driven processes h_i, Φ_i and Ψ_i even when $\mathbf{A}_0 = \mathbf{A}_j = \mathbf{0}$. As illustrated by Hautsch (2008), due to this feature the model allows to parsimoniously capture cross-dependencies.

If we set for simplicity $\mathbf{A}_0 = \mathbf{0}$, $P = Q = 1$, $s_{h,i} = s_{v,i} = s_{\rho,i} = 1$, and diagonal parameterizations of $\mathbf{A}_1$ and $\mathbf{B}_1$, the model is rewritten as

$$\xi_i = \sqrt{\tilde{h}_i} w_i, \qquad \tilde{h}_i = h_i e^{\delta_1 \lambda_i},$$

$$v_i = \tilde{\Phi}_i u_i, \qquad \tilde{\Phi}_i = \Phi_i e^{\delta_2 \lambda_i},$$

$$\rho_i = \tilde{\Psi}_i \varepsilon_i, \qquad \tilde{\Psi}_i = \Psi_i e^{\delta_3 \lambda_i},$$

where

$$\ln \tilde{h}_i - \delta_1 \lambda_i = \omega_1 + \alpha_1^{11} \frac{|\xi_{i-1}|}{\sqrt{h_{i-1}}} + \beta_1^{11} (\ln \tilde{h}_{i-1} - \delta_1 \lambda_{i-1}),$$

$$\ln \tilde{\Phi}_i - \delta_2 \lambda_i = \omega_2 + \alpha_1^{22} \frac{v_{i-1}}{\Phi_{i-1}} + \beta_1^{22} (\ln \tilde{\Phi}_{i-1} - \delta_2 \lambda_{i-1}),$$

$$\ln \tilde{\Psi}_i - \delta_3 \lambda_i = \omega_3 + \alpha_1^{33} \frac{\rho_{i-1}}{\Psi_{i-1}} + \beta_1^{33} (\ln \tilde{\Psi}_{i-1} - \delta_3 \lambda_{i-1}).$$

Hence, λ_i serves as an additional (static) regressor which is driven by its own dynamics according to (7.14). More details on the statistical properties of the multivariate SMEM are given in Hautsch (2008).

7.2.2 *Simulation-Based Inference*

Let $\mathbf{Y}$ denote the entire data matrix with $\mathbf{Y}_i := \{y_j\}_{j=1}^{i}$ and define $\boldsymbol{\theta}$ to be the vector of SVMEM parameters. The conditional likelihood, given the realizations of the latent variable $\boldsymbol{\Lambda}_i$, is given by

$$\mathcal{L}(\mathbf{Y}; \boldsymbol{\theta} | \boldsymbol{\Lambda}_n) = \prod_{i=1}^{n} \frac{1}{\sqrt{2\tilde{h}_i \pi}} \exp\left[-\frac{\xi_i^2}{2\tilde{h}_i}\right] \frac{p_2 v_i^{p_2 m_2 - 1}}{\Gamma(m_2) \tilde{\Phi}_i^{p_2 m_2}} \exp\left[-\left(\frac{v_i}{\tilde{\Phi}_i}\right)^{p_2}\right]$$
$$\times \frac{p_3 \rho_i^{p_3 m_3 - 1}}{\Gamma(m_3) \tilde{\Psi}_i^{p_3 m_3}} \exp\left[-\left(\frac{\rho_i}{\tilde{\Psi}_i}\right)^{p_3}\right], \qquad (7.23)$$

where

$$\tilde{h}_i := h_i e^{\delta_1 \lambda_i} s_{h,i},$$

$$\tilde{\Phi}_i := \Phi_i e^{\delta_2 \lambda_i} s_{v,i},$$

$$\tilde{\Psi}_i := \Psi e^{\delta_3 \lambda} s_{\rho,i}.$$

Accordingly, the integrated likelihood function is given by

$$\mathcal{L}(\mathbf{Y}; \boldsymbol{\theta}) = \int \prod_{i=1}^{n} \frac{1}{\sqrt{2\tilde{h}_i \pi}} \exp\left[-\frac{\xi_i^2}{2\tilde{h}_i}\right] \frac{p_2 v_i^{p_2 m_2 - 1}}{\Gamma(m_2) \tilde{\Phi}_i^{p_2 m_2}} \exp\left[-\left(\frac{v_i}{\tilde{\Phi}_i}\right)^{p_2}\right]$$
$$\times \frac{p_3 \rho_i^{p_3 m_3 - 1}}{\Gamma(m_3) \tilde{\Psi}_i^{p_3 m_3}} \exp\left[-\left(\frac{\rho_i}{\tilde{\Psi}_i}\right)^{p_3}\right] \frac{1}{\sqrt{2\pi}} \exp\left[-\frac{1}{2} (\lambda_i - \mu_{0,i})^2\right] d\boldsymbol{\Lambda}$$

$$= \int \prod_{i=1}^{n} g(y_i|\lambda_i, \mathbf{Y}_{i-1}; \boldsymbol{\theta}) p(\lambda_i|\boldsymbol{\Lambda}_{i-1}; \boldsymbol{\theta}) d\boldsymbol{\Lambda}$$

$$= \int \prod_{i=1}^{n} f(y_i, \lambda_i|\mathbf{Y}_{i-1}, \boldsymbol{\Lambda}_{i-1}; \boldsymbol{\theta}) d\boldsymbol{\Lambda},$$

where $\mu_{0,i} := \mathbb{E}[\lambda_i|\boldsymbol{\Lambda}_{i-1}]$, $g(\cdot)$ denotes the conditional density of y_i given $(\lambda_i, \mathbf{Y}_{i-1})$ and $p(\cdot)$ denotes the conditional density of λ_i given $\boldsymbol{\Lambda}_{i-1}$. The computation of the n-dimensional integral in (7.24) is done numerically using the efficient importance sampling (EIS) method proposed by Richard and Zhang (2007) and requires rewriting the integral (7.24) as

$$\mathcal{L}(\mathbf{Y}; \boldsymbol{\theta}) = \int \prod_{i=1}^{n} \frac{f(y_i, \lambda_i|\mathbf{Y}_{i-1}, \boldsymbol{\Lambda}_{i-1}; \boldsymbol{\theta})}{m(\lambda_i|\boldsymbol{\Lambda}_{i-1}, \boldsymbol{\phi}_i)} \prod_{i=1}^{n} m(\lambda_i|\boldsymbol{\Lambda}_{i-1}, \boldsymbol{\phi}_i) d\boldsymbol{\Lambda}, \tag{7.24}$$

where $\{m(\lambda_i|\boldsymbol{\Lambda}_{i-1}, \boldsymbol{\phi}_i)\}_{i=1}^n$ denotes a sequence of auxiliary importance samplers indexed by auxiliary parameters $\boldsymbol{\phi}_i$. The importance sampling estimate of the likelihood is obtained by

$$\mathcal{L}(\mathbf{Y}; \boldsymbol{\theta}) \approx \hat{\mathcal{L}}_R(\mathbf{Y}; \boldsymbol{\theta}) = \frac{1}{R} \sum_{r=1}^{R} \prod_{i=1}^{n} \frac{f(y_i, \lambda_i^{(r)}|\mathbf{Y}_{i-1}, \boldsymbol{\Lambda}_{i-1}^{(r)}; \boldsymbol{\theta})}{m(\boldsymbol{\lambda}_i^{(r)}|\boldsymbol{\Lambda}_{i-1}^{(r)}, \boldsymbol{\phi}_i)}, \tag{7.25}$$

where $\{\lambda_i^{(r)}\}_{i=1}^n$ denotes a trajectory of random draws from the sequence of auxiliary importance samplers m and R such trajectories are generated.

The idea of the EIS approach is to choose a sequence of samplers for $m(\lambda_i|\boldsymbol{\Lambda}_{i-1}, \boldsymbol{\phi}_i)$ exploiting the sample information on λ_i revealed by the observable data. As shown by Richard and Zhang (2007), the EIS principle is to choose the auxiliary parameters $\{\boldsymbol{\phi}_i\}_{i=1}^n$ in a way that provides a good match between $\Pi_{i=1}^n m(\lambda_i|\boldsymbol{\Lambda}_{i-1}, \boldsymbol{\phi}_i)$ and $\Pi_{i=1}^n f(y_i, \lambda_i|\mathbf{Y}_{i-1}, \boldsymbol{\Lambda}_{i-1}; \boldsymbol{\theta})$ in order to minimize the Monte Carlo sampling variance of $\hat{\mathcal{L}}_R(\mathbf{Y}; \boldsymbol{\theta})$. Richard and Zhang (2007) illustrate that the resulting high-dimensional minimization problem can be split up into solvable low-dimensional subproblems. This makes the approach tractable even for very high dimensions. The detailed EIS procedure is given by Hautsch (2008). See also Chap. 12 for a quite similar EIS procedure to estimate stochastic conditional intensity models.

An important advantage facilitating the computation of the function $f(\cdot)$ is that the time series recursion of the observation-driven components h_i, Φ_i and Ψ_i can be computed without the latent factor being known. As discussed in Sect. 7.2.1, this is because $\{h_i, \Phi_i, \Psi_i\}$ are driven based on innovations $\mathbf{z}_i$ that are observable given the history of $\{\xi_i, v_i, \rho_i\}$ and $\{h_i, \Phi_i, \Psi_i\}$. Then, h_i, Φ_i and Ψ_i can be computed in a first step according to the VARMA structure given by (7.19) to (7.22) and can be used in a second step to evaluate the sampler $\{m(\lambda_i|\boldsymbol{\Lambda}_{i-1}, \boldsymbol{\phi}_i\}_{i=1}^n$.

Filtered estimates of an arbitrary function of λ_i, $\vartheta(\lambda_i)$, given the observable information set up to t_{i-1} are given by

$$\mathbb{E}\left[\vartheta(\lambda_i)\,|\mathbf{Y}_{i-1}\right] = \frac{\int \vartheta(\lambda_i) p(\lambda_i|\mathbf{Y}_{i-1}, \boldsymbol{\Lambda}_{i-1}, \boldsymbol{\theta}) f(\mathbf{Y}_{i-1}, \boldsymbol{\Lambda}_{i-1}|\boldsymbol{\theta}) d\boldsymbol{\Lambda}_i}{\int f(\mathbf{Y}_{i-1}, \boldsymbol{\Lambda}_{i-1}|\boldsymbol{\theta}) d\boldsymbol{\Lambda}_{i-1}}. \tag{7.26}$$

The integral in the denominator corresponds to the marginal likelihood function of the first $i-1$ observations, $\mathcal{L}(\mathbf{Y}_{i-1};\boldsymbol{\theta})$, and can be evaluated on the basis of the sequence of auxiliary samplers $\{m(\lambda_j|\boldsymbol{\Lambda}_{j-1}, \hat{\boldsymbol{\phi}}_j^{i-1})\}_{j=1}^{i-1}$ where $\{\hat{\boldsymbol{\phi}}_j^{i-1}\}$ denotes the value of the EIS auxiliary parameters associated with the computation of $\mathcal{L}(\mathbf{Y}_{i-1};\boldsymbol{\theta})$ and $\boldsymbol{\theta}$ is set equal to its corresponding maximum likelihood estimate. Correspondingly, the numerator is computed by

$$\frac{1}{R}\sum_{r=1}^{R}\left\{\vartheta\left(\lambda_i^{(r)}(\boldsymbol{\theta})\right)\prod_{j=1}^{i-1}\left[\frac{f\left(y_j, \lambda_j^{(r)}(\hat{\boldsymbol{\phi}}_j^{i-1})|\mathbf{Y}_{j-1}, \boldsymbol{\Lambda}_{j-1}^{(r)}(\hat{\boldsymbol{\phi}}_{j-1}^{i-1}), \boldsymbol{\theta}\right)}{m\left(\lambda_j^{(r)}(\hat{\boldsymbol{\phi}}_j^{i-1})|\boldsymbol{\Lambda}_{j-1}^{(r)}(\hat{\boldsymbol{\phi}}_{j-1}^{i-1}), \hat{\boldsymbol{\phi}}_j^{i-1}\right)}\right]\right\}, \tag{7.27}$$

where $\{\lambda_j^{(r)}(\hat{\boldsymbol{\phi}}_j^{i-1})\}_{j=1}^{i-1}$ denotes a trajectory drawn from the sequence of importance samplers associated with $\mathcal{L}(\mathbf{Y}_{i-1};\boldsymbol{\theta})$, and $\lambda_i^{(r)}(\boldsymbol{\theta})$ is a random draw from the conditional density $p(\lambda_i|\mathbf{Y}_{i-1}, \boldsymbol{\Lambda}_{i-1}^{(r)}(\hat{\boldsymbol{\phi}}_{i-1}^{i-1}), \boldsymbol{\theta})$. The computation of the sequence of filtered estimates $\mathbb{E}\left[\vartheta(\lambda_i)\,|\mathbf{Y}_{i-1}\right]$, $i = 1, \ldots, n$, requires a re-run of the EIS algorithm for every i (from 1 to n). For more details, see Hautsch (2008). Then, the filtered residuals are given by

$$\hat{w}_i := \frac{\hat{\xi}_i}{\sqrt{\hat{h}_i \mathbb{E}\left[e^{\delta_1\lambda_i}\,|\mathbf{Y}_{i-1}\right]\hat{s}_{h,i}}},$$

$$\hat{u}_i := \frac{v_i}{\hat{\Phi}_i \mathbb{E}\left[e^{\delta_2\lambda_i}\,|\mathbf{Y}_{i-1}\right]\hat{s}_{v,i}},$$

$$\hat{\varepsilon}_i := \frac{\rho_i}{\hat{\Psi}_i \mathbb{E}\left[e^{\delta_3\lambda_i}\,|\mathbf{Y}_{i-1}\right]\hat{s}_{\rho,i}}.$$

7.2.3 Modelling Trading Processes

Tables 7.2 and 7.3 reproduces the estimates of Hautsch (2008) based on seasonally adjusted 5 min (ARMA(1,1)-pre-filtered) squared returns, average trade sizes and number of trades data from the NYSE stocks JP Morgan and IBM covering five months between 02/01/2001 and 31/05/2001. The underlying lag length is restricted to two, where $\mathbf{A}_2$ and $\mathbf{B}_2$ are diagonal matrices. The major findings are as follows:

Table 7.2 Maximum likelihood efficient importance sampling (ML-EIS) estimates of different parameterizations of SMEM specifications up to a lag order of $P = Q = 2$ models for the log return volatility, the average volume per trade and the number of trades per 5 min interval for the JP Morgan stock traded on the NYSE. Sample period from 02/01/01 to 31/05/01. Overnight observations are excluded. The models are re-initialized on each trading day. Standard errors are computed based on the inverse of the estimated Hessian. The ML-EIS estimates are computed using $R = 50$ Monte Carlo replications based on 5 EIS iterations.
Diagnostics: Log likelihood function (LL), Bayes Information Criterion (BIC), mean, standard deviation and Ljung–Box statistics of the filtered residuals (LB) and squared filtered residuals (LB2, only for the return process) as well as multivariate Ljung–Box statistic (MLB). The Ljung–Box statistics are computed based on 20 lags. Significance at the 1%, 5% and 10% levels are denoted by ***, ** and *, respectively. Results reproduced from Hautsch (2008)

	(1)	(2)	(3)	(4)	(5)
ω_1	1.996***	0.529***	−0.097***	0.167***	0.209
ω_2	−1.410***	−2.240***	−1.471***	−2.035***	−2.122***
ω_3	−0.016***	−0.225***	−0.337***	−0.008***	−0.005
α_0^{12}	0.859***	0.492***	−0.078***	0.023*	0.050***
α_0^{13}	0.910***	0.365***	−0.091***	0.022	−0.033
α_0^{23}	−0.882***	−1.158***	−0.953***	−0.941***	−0.958***
α_1^{11}	0.138***		0.097***	0.085***	0.072***
α_1^{12}	0.009			−0.004	−0.010
α_1^{13}	0.001***			0.000	0.001
α_1^{21}	−0.003			−0.022***	−0.029***
α_1^{22}	0.005***		0.005***	0.005***	0.010***
α_1^{23}	0.000**			0.000***	0.000**
α_1^{31}	0.550***			−0.037	0.096
α_1^{32}	0.388***			−0.054	−0.101**
α_1^{33}	0.216***		0.188***	0.212***	0.214***
α_2^{11}	0.117***			0.053**	0.049**
α_2^{22}	−0.002***			−0.001	−0.004**
α_2^{33}	−0.207***			−0.206***	−0.210***
β_1^{11}	0.051		0.972***	0.861***	0.789***
β_1^{12}	0.170***				−0.246***
β_1^{13}	0.000				−0.001
β_1^{21}	0.646***				0.066
β_1^{22}	0.708***		0.178***	0.115**	0.083
β_1^{23}	0.000				−0.001
β_1^{31}	1.166***				0.180**
β_1^{32}	0.527***				−0.427**
β_1^{33}	1.623***		0.836***	1.625***	1.655***
β_2^{11}	−0.068**			0.086	0.177
β_2^{22}	−0.017			0.025	−0.077
β_2^{33}	−0.625***			−0.626***	−0.657***
p_2	0.720***	0.863***	1.027***	0.854***	0.890***
m_2	6.947***	5.800***	4.917***	6.537***	5.869***
p_3	2.438***	2.414***	2.287***	2.454***	2.579***
m_3	2.373***	1.986***	2.620***	2.364***	2.173***

(continued)

Table 7.2 (continued)

	(1)	(2)	(3)	(4)	(5)
Latent component					
a	0.951***	0.907***	0.941***	0.930***	
δ_1	0.165***	0.339***	0.339***	0.350***	
δ_2	0.122***	0.176***	0.136***	0.132***	
δ_3	0.024***	0.009***	0.011***	0.015***	
General diagnostics					
LL	−18,306	−19,398	−18,201	−18,040	−18,009
BIC	−18,458	−19,461	−18,291	−18,184	−18,180
MLB	665.637***	12,670.643***	288.422***	171.290***	197.625***
Diagnostics for the return process					
Mean	−0.019	−0.011	−0.004	−0.005	−0.006
S.D.	0.999	1.026	1.048	1.067	1.056
LB	35.911**	31.627**	26.265	25.315	24.722
LB2	355.220***	35.564**	40.740***	6.513	11.903
Diagnostics for the volume process					
Mean	1.001	1.003	1.009	1.009	1.006
S.D.	0.600	0.609	0.598	0.597	0.592
LB	22.132	90.389***	46.434***	11.269	17.671
Diagnostics for the trading intensity process					
Mean	1.000	0.999	1.000	0.999	1.000
S.D.	0.273	0.306	0.276	0.273	0.273
LB	43.659***	6,337.353***	140.305***	50.553***	53.280***

First, it turns out that the latent common component is strongly autocorrelated with an autoregressive parameter being on average around $\hat{a} \approx 0.94$. Consequently, the latent factor seems to accommodate common long-run dependence, which is not captured by the observation-driven dynamics. Second, common shocks simultaneously increase all three trading components. As revealed by the parameters δ_1, δ_2 and δ_3, the joint factor influences primarily the volatility and trade size. Conversely, its impact on the trading intensity is relatively weak. These findings confirm the results by Xu and Wu (1999), Chan and Fong (2000), Huang and Masulis (2003) and Blume et al. (1994) documenting that trade size is obviously an important indicator for the quality of news. Consequently, a subordinated common (information) process is more strongly reflected in the average trade size rather than in the trading intensity.

Third, including the common latent factor induces a decline of the magnitude of the parameters α_0^{12} and α_0^{13}. This indicates that the *conditional* contemporaneous correlations between volatilities, volumes and trading intensities given the latent component are lower than the corresponding *unconditional* ones. Hence, a

Table 7.3 Maximum likelihood efficient importance sampling (ML-EIS) estimates of different parameterizations of SMEM specifications up to a lag order of $P = Q = 2$ models for the log return volatility, the average volume per trade and the number of trades per 5 min interval for the IBM stock traded on the NYSE. Sample period from 02/01/01 to 31/05/01. Overnight observations are excluded. The models are re-initialized on each trading day. Standard errors are computed based on the inverse of the estimated Hessian. The ML-EIS estimates are computed using $R = 50$ Monte Carlo replications based on 5 EIS iterations.
Diagnostics: Log likelihood function (LL), Bayes Information Criterion (BIC), mean, standard deviation and Ljung–Box statistics of the filtered residuals (LB) and squared filtered residuals (LB2, only for the return process) as well as multivariate Ljung–Box statistic (MLB). The Ljung–Box statistics are computed based on 20 lags. Significance at the 1%, 5% and 10% levels are denoted by ***, ** and *, respectively. Results reproduced from Hautsch (2008)

	(1)	(2)	(3)	(4)	(5)
ω_1	0.372***	0.504***	0.494***	0.316***	1.079***
ω_2	−0.784***	−1.420***	−1.831***	−1.307***	−1.403***
ω_3	−0.200***	−0.517***	−0.373***	−0.160***	−0.305***
α_0^{12}	0.792***	0.704***	0.596***	0.128***	0.140***
α_0^{13}	0.746***	0.182**	0.585***	0.099***	0.429***
α_0^{23}	−0.783***	−1.355***	−0.768***	−0.866***	−0.789***
α_1^{11}	0.183***		0.070***	0.068***	0.067***
α_1^{12}	0.025***			0.018**	0.015**
α_1^{13}	0.000			−0.001	0.002
α_1^{21}	−0.052***			−0.061***	−0.064***
α_1^{22}	0.021***		0.016***	0.012***	0.013***
α_1^{23}	0.001***			−0.001**	0.001**
α_1^{31}	−0.136***			−0.086**	0.015
α_1^{32}	0.344***			0.110**	0.136***
α_1^{33}	0.191***		0.153***	0.182***	0.151***
α_2^{11}	0.131***			0.086***	0.048**
α_2^{22}	−0.004***			0.001	0.001
α_2^{33}	−0.101***			−0.105***	−0.087***
β_1^{11}	0.580***		−0.228***	0.471***	0.596***
β_1^{12}	0.048***				−0.050
β_1^{13}	0.001				−0.007**
β_1^{21}	−0.108***				0.477***
β_1^{22}	0.866***		0.176***	0.400***	0.436***
β_1^{23}	0.000				−0.059***
β_1^{31}	−0.347***				0.114
β_1^{32}	0.701***				0.105
β_1^{33}	1.211***		0.900***	1.272***	1.122***
β_2^{11}	0.044			0.496***	0.414***
β_2^{22}	−0.009			−0.091**	−0.115**
β_2^{33}	−0.252***			−0.313***	−0.229**
p_2	0.847***	1.168***	0.965***	1.190***	1.115***
m_2	6.936***	4.466***	6.492***	4.953***	5.470***
p_3	2.303***	2.294***	2.156***	2.278***	2.172***
m_3	4.131***	3.499***	4.642***	4.255***	4.620***

(continued)

Table 7.3 (continued)

	(1)	(2)	(3)	(4)	(5)
Latent component					
a	0.942***	0.967***	0.940***	0.944***	
δ_1	0.154***	0.141***	0.263***	0.256***	
δ_2	0.146***	0.087***	0.133***	0.123***	
δ_3	0.044***	0.004***	0.012***	0.003**	
General diagnostics					
LL	−15,338	−16,959	−15,349	−15,106	−15,086
BIC	−15,490	−17,021	−15,439	−15,250	−15,257
MLB	240.460***	19,077.193***	833.269***	95.978***	76.669*
Diagnostics for the return process					
Mean	−0.009	−0.009	−0.006	0.000	0.000
S.D.	1.000	1.014	1.009	1.026	1.028
LB	17.720	17.105	19.767	23.732	24.735
LB2	154.973***	622.605***	313.360***	18.042	16.570
Diagnostics for the volume process					
Mean	0.999	1.003	1.001	1.005	1.004
S.D.	0.484	0.512	0.487	0.481	0.483
LB	47.007***	257.354***	54.532***	56.413***	54.164***
Diagnostics for the trading intensity process					
Mean	0.999	0.999	0.999	1.000	1.000
S.D.	0.216	0.246	0.217	0.216	0.216
LB	12.195	6,935.761 ***	84.642***	11.464	11.405

significant part of the contemporaneous relationships between conditional return variances and average trade sizes as well as trading intensities obviously stem from an underlying common component. Nonetheless, while the common factor comprises the positive dependence between return volatility and trade size to a large extent, it can only partly explain the positive correlation between trading intensities and volatilities. The parameter α_0^{23} is significantly negative and mainly unaffected by the inclusion of the latent component. This indicates that the (negative) contemporaneous relationship between trade size and trading intensity is *not* driven by a latent common component. Instead, according to Hautsch (2008), it might be rather explained by the common finding that high volumes are typically split over time leading to higher trading frequencies but also to smaller trade sizes.

Fourth, Panels (1)–(3) show the estimates of specifications omitting a common latent component and revealing significant cross-dependencies between volatilities, volumes and trading intensities. These dependencies, however, are clearly reduced as soon as a common latent component is taken into account (Panels (5)–(7)). Indeed, the size of cross-effects becomes close to zero and/or is insignificant. Likewise the non-diagonal elements in $\mathbf{B}_1$ shrink. As argued by Hautsch (2008), this finding indicates that most of the observed causalities between the individual variables are mainly due to the existence of a subordinated common (information) process jointly directing the individual components.

Moreover, the inclusion of the latent factor reduces the impact of the process-specific innovations ($\hat{\alpha}_1^{ii}$ and $\hat{\alpha}_2^{ii}$ for $i = 1, 2, 3$) and increases the persistence in the observation-driven dynamics. This finding is in accordance with the results shown for univariate SMEM processes in Sect. 6.4.2 and indicates that news seem to enter the model primarily through the latent factor reducing the impact of process-specific innovations.

Finally, the inclusion of a latent component leads to a reduction of the multivariate Ljung–Box statistic indicating that the common component successfully captures the multivariate dynamics and interdependencies between the individual processes. This is supported by a reduction of the Bayes information criterion indicating a better fit of the SVMEM compared to MEMs without a latent factor. Interestingly, the worst performance is observed for specification (4), where any observation-driven dynamics are omitted and only a parameter-driven dynamic is included. This indicates that a single common autoregressive component is obviously not sufficient to completely capture the dynamics of the multivariate system which confirms the findings by Andersen (1996) or Liesenfeld (1998). Ultimately, we can neither reject the parameter-driven dynamic nor the observation-driven dynamic confirming the basic idea of the proposed mixture model.

References

Andersen TG (1996) Return volatility and trading volume: An information flow interpretation of stochastic volatility. J Finance 51:169–204

Blume L, Easley D, O'Hara M (1994) Market statistics and technical analysis. J Finance 49(1):153–181

Bodnar T, Hautsch N (2011) Modeling time-varying covariances of trading processes: copula-based dynamic conditional correlation multiplicative error processes. Working Paper, Humboldt-Universität zu Berlin

Chan K, Fong W (2000) Trade size, order imbalance, and the volatility-volume relation. J Finan Econ 57:247–273

Cipollini F, Engle RF, Gallo GM (2007) Vector multiplicative error models: representation and inference. Working Paper, University of Florence

Engle RF (2002) Dynamic conditional correlation. J Bus Econ Stat 20:339–350

Engle RF, Hendry DF, Richard JF (1983) Exogeneity. Econometrica 51:277–304

Hautsch N (2008) Capturing common components in high-frequency financial time series: a multivariate stochastic multiplicative error model. J Econ Dyn Control 32:3978–4009

Hautsch N, Jeleskovic V (2008) High-frequency volatility and liquidity. In: Härdle W, Hautsch N, Overbeck L (ed) Applied quantitative finance. Springer, Berlin, Heidelberg

Hautsch N, Malec P, Schienle M (2010) Capturing the zero: a new class of zero-augmented distributions and multiplicative error processes. Discussion Paper 2010-055, Humboldt-Universität zu Berlin

Huang RD, Masulis RW (2003) Trading activity and stock price volatility: evidence from the London stock exchange. J Empir Financ 10:249–269

Liesenfeld R (1998) Dynamic bivariate mixture models: modeling the behavior of prices and trading volume. J Bus Econ Stat 16:101–109
Manganelli S (2005) Duration, volume and volatility impact of trades. J Finan Markets 8:377–399
Richard J-F, Zhang W (2007) Efficient high-dimensional importance sampling. J Econom 141:1385–1411
Xu XE, Wu C (1999) The intraday relation between return volatility, transactions, and volume. Int Rev Econ Fin 8:375–397

Chapter 8
Modelling High-Frequency Volatility

This chapter discusses different ways to estimate intraday volatility. Section 8.1 presents realized measures to estimate intraday quadratic variation. Here, we compactly illustrate fundamental approaches, such as the maximum likelihood estimator by Aït-Sahalia et al. (2005), the realized kernel estimator by Barndorff-Nielsen et al. (2008a) as well as the pre-averaging estimator by Jacod et al. (2009). We restrict ourselves to a rather intuitive discussion of the major principles behind these estimators. A more in-depth treatment, however, is beyond the scope of this book and we refer to the reader to the underlying literature. Moreover, we show applications of these estimators to estimate intraday variances based on different frequencies and discuss modelling and forecasting approaches. Section 8.2 deals with jump-robust and spot variance estimators. Section 8.3 illustrates intensity-based volatility estimators. Here, we discuss the estimation of trade-to-trade return variances based on high-frequency GARCH models as proposed by Engle (2000) and Ghysels and Jasiak (1998) and their relationships to spot variances. Section 8.4 shows how to use price durations to construct intensity-based volatility estimators.

8.1 Intraday Quadratic Variation Measures

Under the assumption of no arbitrage, price processes must follow a semi-martingale, see, e.g., Delbaen and Schachermayer (1994). Denote p_t as the asset's logarithmic price process and assume that it follows a continuous semi-martingale $p_t, t \geq 0$, of the form

$$p_t = p_0 + \int_0^t \mu_\tau d\tau + \int_0^t \sigma_\tau dW_\tau, \tag{8.1}$$

where W denotes a standard (one-dimensional) Wiener process, $(\mu_\tau)_{\tau \geq 0}$ is a finite variation càglàg drift process and $(\sigma_\tau)_{\tau \geq 0}$ is an adapted càdlàg volatility

N. Hautsch, *Econometrics of Financial High-Frequency Data*,
DOI 10.1007/978-3-642-21925-2_8,

process associated with the instantaneous conditional mean and volatility of the corresponding return.

To formally introduce the concept of quadratic variation, consider a stochastic process g_t, $t \geq 0$, and decompose the interval $[0,t]$ into intervals $s_0 < s_1 < \ldots < s_n = t$. Then,

$$V_n(0,t) := \sum_{i=1}^{n} |g_{s_i} - g_{s_{i-1}}| \tag{8.2}$$

is defined as the (absolute) *variation* of the stochastic process g on the partition $s_0, s_1, \ldots$ over the interval $[0,t]$. If $V_n(0,t) \overset{m.s.}{\to} V(0,t)$ for $n \to \infty$ with $V(0,t)$ denoting the stochastic limit (in mean square), g is referred to be of finite variation. In this case, the sum of absolute increments of the process on increasingly fine partitions for a fixed t is finite. This is true if the process has – loosely speaking – a sufficiently smooth sample path. Conversely, if the sample path is very "spiky", it might happen that the sum of process increments does not converge to a finite quantity but grows with $n \to \infty$.

Correspondingly, the *quadratic variation* (on the partition defined above) is given by

$$QV_n(0,t) := \sum_{i=1}^{n} (g_{s_i} - g_{s_{i-1}})^2 \tag{8.3}$$

with stochastic limit (in mean square) $QV(0,t)$. Using results from stochastic calculus, it can be shown that a finite variation process, i.e., $V(0,t) < \infty$, has a quadratic variation which stochastically converges to zero. Likewise it can be shown that a process with positive (finite) quadratic variation has infinite variation.

Consider the price process over the interval from 0 to t with continuously compounded return

$$r_{0,t} := p_t - p_0 = \int_0^t \mu_\tau d\tau + \int_0^t \sigma_\tau dW_\tau. \tag{8.4}$$

The quadratic variation of p over the interval $[0,t]$ is then given by

$$QV(0,t) = \int_0^t \sigma_\tau^2 d\tau \tag{8.5}$$

with

$$\sum_{j=1}^{n} (p_{j\Delta} - p_{(j-1)\Delta})^2 \overset{m.s.}{\to} QV(0,t) \quad \text{for } n \to \infty. \tag{8.6}$$

The reason for (8.5) is that the drift process μ_t has a quadratic variation of zero, while the quadratic variation of a Wiener process on $[0, t]$ equals t. Hence, μ_t does not affect the sample path variation of the return.

As the return process defined above is continuous, the quadratic variation equals the so-called *integrated variance*

$$IV(0,t) := \int_0^t \sigma_\tau^2 d\tau = QV(0,t). \tag{8.7}$$

This equality does not hold for processes containing a jump component, as, e.g., in jump-diffusion processes.

Loosely speaking, the quadratic variation of a process corresponds to the sum of its squared increments measured on infinitesimal intervals. Hence, it is a natural quantity reflecting the riskiness of an asset over a given time span.

The availability of high-frequency data opens up the possibility to observe the price path of an asset over arbitrarily small intervals and to generate estimators of the quadratic variation. For simplicity consider the (standardized) interval $[0, 1]$ corresponding to a day or any intraday interval. Then, the quadratic variation measured over $[0, 1]$ is naturally approximated by the so-called *realized variance*[1]

$$RV^n := \sum_{j=1}^{n} (p_{j\Delta} - p_{(j-1)\Delta})^2 := \sum_{j=1}^{n} r_{j\Delta,n}^2, \tag{8.8}$$

where n is the number of high-frequency intervals with length $\Delta = n^{-1}$ in which $[0, 1]$ is decomposed. Semimartingale theory ensures that the realized variance converges to $QV(0, 1)$ in probability if $n \to \infty$. These asymptotics are typically referred to as "in-fill" asymptotics. Hence, if it is possible to sample frequently enough, $QV(0, 1)$ can be estimated with minimal estimation error. This is also reflected in the asymptotic distribution (see Barndorff-Nielsen and Shephard 2002),

$$\sqrt{n}\frac{RV^n - QV}{\sqrt{2IQ}} \overset{a}{\sim} \mathcal{N}(0,1), \tag{8.9}$$

where $IV := IV(0,1)$ and $IQ := IQ(0,1) := \int_0^1 \sigma_\tau^4 d\tau$ defines the integrated quarticity which can be consistently estimated by the realized quarticity $RQ^n := \frac{1}{3}\sum_{j=1}^n r_{j\Delta,n}^4$.

The consistency of the realized volatility estimator builds on the assumption that prices behave according to the semi-martingale (8.1) and can be sampled arbitrarily frequently. In practice, however, the sampling frequency is inevitably limited by the actual quotation or transaction frequency. Moreover, as discussed

[1]In the literature it is often also called *realized volatility* though in a strict sense, the terminology "volatility" is typically used for σ_t rather than for σ_t^2.

in Chap. 4, transaction prices are subject to market microstructure effects, such as the discreteness of prices or the bid-ask bounce effect. Hence, instead of the semi-martingale process (8.1), it is more realistic to assume the observable price process to be given by

$$p_t = p_t^* + \varepsilon_t, \tag{8.10}$$

where p_t^* denotes the latent true price process following (8.1) and ε_t is a zero mean error term reflecting so-called market microstructure noise. If ε_t is assumed to be i.i.d. with $\mathbb{E}[\varepsilon_t^2] := \omega^2$, it is easily shown that the observed high-frequency returns $r_{j\Delta,n}$ follow an MA(1) process, see Sect. 8.1.1. Moreover, it is shown that

$$\mathbb{E}[RV^n] = IV(0,1) + 2n\omega^2. \tag{8.11}$$

Hence, the realized volatility estimator based on the observed price process is biased with bias term $2n\omega^2$, see, e.g., Hansen and Lunde (2006). For $n \to \infty$, RV_n diverges to infinity linearly in n. Thus, the estimate is dominated by market microstructure noise.

Consequently, sampling at a lower frequency, such as every 5, 10, 15 or 30 min, seems to alleviate the problem of market microstructure noise and thus is frequently applied in the literature. This so–called *sparse sampling*, however, comes at the cost of a less precise estimate of the actual volatility. To address the problem of not loosing too much information due to sparse sampling on the one hand and being not affected by market microstructure noise on the other hand, several competing estimators have been proposed in the recent literature. For an overview, see, e.g., Andersen et al. (2008).

Since it is beyond the scope of this book to provide an in-depth discussion of alternative quadratic variation estimators, we only briefly present three popular estimators accounting for the presence of noise. These are the maximum likelihood estimator proposed by Aït-Sahalia et al. (2005), the realized kernel estimator introduced by Barndorff-Nielsen et al. (2008a) and the pre-averaging estimator suggested by Jacod et al. (2009).

8.1.1 Maximum Likelihood Estimation

A relatively simple parametric estimator is proposed by Aït-Sahalia et al. (2005) and builds on the assumption of a high-frequency discrete (log) price process

$$p_{j\Delta} = p_{j\Delta}^* + \varepsilon_{j\Delta}, \quad \varepsilon_{j\Delta} \overset{i.i.d.}{\sim} (0, \omega^2), \tag{8.12}$$

where

$$r_{j\Delta,n}^* := p_{j\Delta}^* - p_{(j-1)\Delta}^* \overset{i.i.d.}{\sim} (0, \sigma^2 \Delta)$$

with $r_{j\Delta,n}^*$ being independent of $\varepsilon_{j\Delta}$.

Then, $r_{j\Delta,n} = r^*_{j\Delta,n} + \varepsilon_{j\Delta} - \varepsilon_{(j-1)\Delta}$ can be re-parameterized as a MA(1) process,

$$r_{j\Delta,n} = \mu_{j\Delta,n} + \eta\mu_{(j-1)\Delta,n}, \tag{8.13}$$

where $\mu_{j\Delta,n} \sim (0, \gamma^2)$. The parameters η and γ^2 can be identified by

$$\gamma^2(1+\eta^2) = \mathbb{V}[r_{j\Delta,n}] = \sigma^2\Delta + 2\omega^2,$$
$$\gamma^2\eta = \text{Cov}[r_{j\Delta,n}, r_{(j-1)\Delta,n}] = -\omega^2.$$

Consequently, the daily variance σ^2 as well as the microstructure noise variance can be estimated by

$$\hat{\sigma}^2 = \widehat{IV} = \Delta^{-1}\hat{\gamma}^2(1-\hat{\eta})^2, \tag{8.14}$$

$$\hat{\omega}^2 = -\hat{\gamma}^2\hat{\eta}, \tag{8.15}$$

where $\hat{\gamma}^2$ and $\hat{\eta}$ are maximum likelihood estimates based on a MA(1) process using high-frequency returns.

8.1.2 The Realized Kernel Estimator

While the maximum likelihood estimator above is consistent under the assumption of the given high-frequency price process, it is biased and inconsistent under more general price properties. Barndorff-Nielsen et al. (2008a) propose a realized kernel estimator which is consistent under more general conditions, such as, e.g., higher-order dependence or endogeneity in the noise process. The idea is to capture (potentially noise induced) serial correlations in trade-to-trade returns by a kernel. The *realized kernel estimator (RK)* proposed by Barndorff-Nielsen et al. (2008a) is defined by

$$K(\Delta) = \gamma_0(\Delta) + \sum_{h=1}^{H} k\left(\frac{h-1}{H}\right)\{\gamma_h(\Delta) + \gamma_{-h}(\Delta)\}, \tag{8.16}$$

where $\gamma_h(\Delta)$ denotes the hth realized autocovariance given by

$$\gamma_h(\Delta) = \sum_{j=1}^{n}(p_{i\Delta_n} - p_{(i-1)\Delta_n})(p_{(i-h)\Delta_n} - p_{(i-h-1)\Delta_n}) \tag{8.17}$$

with $h = -H, \ldots, -1, 0, 1, \ldots, H$ and $k(\cdot)$ denoting the kernel function depending on a bandwidth H. Barndorff-Nielsen et al. (2008a) suggest using the Tukey-Hanning$_2$ kernel with

$$k(x) = \sin^2\{\pi/2(1-x)^2\} \tag{8.18}$$

with optimal choice of the bandwidth, H, given by

$$H = c\zeta\sqrt{n}, \tag{8.19}$$

where $c = 5.74$ for the Tukey-Hanning$_2$ kernel and ζ is defined by

$$\zeta^2 = \omega^2 / \sqrt{\int_0^1 \sigma_u^4 du}.$$

The quantity $\int_0^1 \sigma_u^4 du$ can be estimated using the realized quarticity. To estimate ω^2, Barndorff-Nielsen et al. (2008b) suggest

$$\hat{\omega}^2 = \frac{1}{q}\sum_{i=1}^{q} \hat{\omega}_{(i)}^2,$$

with

$$\hat{\omega}_{(i)}^2 = \frac{RV_{(i)}^n}{2\tilde{n}_{(i)}}, \quad i = 1, \ldots, q,$$

and $RV_{(i)}^n$, $i = 1, \ldots, q$ are realized variance estimators $RV_{(i)}^n = \sum_{j=i}^n r_{j\Delta,n}^2$ sampling every $q = N/n$th trade using the first q trades per day as different starting points, N denoting the number of transactions in $[0, 1]$ and $\tilde{n}_{(i)}$ giving the number of non-zero returns that were used to compute $RV_{(i)}^n$. To robustify this estimator against serial dependence in the noise process, Barndorff-Nielsen et al. (2008b) propose using q such that every qth observation is, on average, 2 min apart. As discussed by the authors, this estimator is likely to be upward biased and therefore yields a rather conservative choice of the bandwidth. For more details, see Barndorff-Nielsen et al. (2008a,b).

8.1.3 The Pre-averaging Estimator

The principle behind the pre-averaging estimator proposed by Jacod et al. (2009) is – loosely speaking – to remove market microstructure noise by locally averaging high-frequency returns *before* squaring them and adding them up. The estimator is constructed by choosing a sequence k_n of integers satisfying

$$k_n \Delta^{1/2} = \theta + o(\Delta^{\frac{1}{4}}) \tag{8.20}$$

for some $\theta > 0$, and a non-zero real-valued function $g : [0, 1] \to \mathbb{R}$, which is continuous, piecewise continuously differentiable, and $g(0) = g(1) = 0$. A typical example of a function g, which we use in the empirical section, is given by $g(x) = \min\{x, (1-x)\}$ for $x \in (0,1)$. The pre-averaged returns are given as

$$\overline{Z}_i^n := \sum_{j=1}^{k_n} g\Big(\frac{j}{k_n}\Big) r_{(i+j)\Delta,n}, \qquad r_{i\Delta,n} = p_{i\Delta} - p_{(i-1)\Delta}. \tag{8.21}$$

Hence, $\overline{Z}_i^n$ corresponds to a weighted average of the increments $r_{i\Delta,n}$ in the local window $[i\Delta, (i+k_n)\Delta]$ which diminishes the influence of noise. The window size k_n is chosen to be of order $\Delta_n^{-1/2} = n^{1/2}$ leading to optimal convergence rates, see Jacod et al. (2009).

Then, the direct analogue to the realized variance estimator is given by

$$V(Z,n) = \sum_{i=0}^{[1/\Delta]-k_n} |\overline{Z}_i^n|^2 \tag{8.22}$$

yielding the pre-averaging estimator

$$C^n := \frac{\sqrt{\Delta}}{\theta\psi_2} V(Z,n) - \frac{\psi_1\Delta}{2\theta^2\psi_2} RV^n, \tag{8.23}$$

where $\psi_1 := \int_0^1 (g'(s))^2 ds = 1$ and $\psi_2 := \int_0^1 (g(s))^2 ds = 1/12$.

As shown by Jacod et al. (2009), we have

$$\frac{\Delta}{2} RV^n \approx \int_0^1 \omega_u^2 du + \frac{\Delta}{2} IV.$$

The error of this approximation is of order Δ and has expectation 0. Therefore, the statistic C^n actually estimates

$$\Big(1 - \frac{\psi_1^{k_n}\Delta}{2\theta^2\psi_2^{k_n}}\Big) IV.$$

In finite-samples we have to adjust for the true number of summands in $V(Z,n)$ inducing the adjustment term $[1/\Delta]/([1/\Delta]-k_n+1)$. This yields the finite-sample adjusted pre-averaging estimator

$$C_a^n = \left(1 - \frac{\psi_1^{k_n}\Delta}{2\theta^2\psi_2^{k_n}}\right)^{-1} \left(\frac{[1/\Delta]\sqrt{\Delta}}{([1/\Delta]-k_n+2)\theta\psi_2^{k_n}} V(Z,n) - \frac{\psi_1^{k_n}\Delta_n}{2\theta^2\psi_2^{k_n}} RV^n\right),$$

where $\psi_1^{k_n}$ and $\psi_2^{k_n}$ denote the finite-sample counterparts of ψ_1 and ψ_2 given by

$$\psi_1^{k_n} = k_n \sum_{j=1}^{k_n} \left(g\left(\frac{j+1}{k_n}\right) - g\left(\frac{j}{k_n}\right)\right)^2,$$

$$\psi_2^{k_n} = \frac{1}{k_n} \sum_{j=1}^{k_n-1} g^2\left(\frac{j}{k_n}\right).$$

Jacod et al. (2009) show that the estimator is consistent and asymptotically mixed normally distributed. It can be extended in various directions, for instance, by allowing for serially dependent noise and jumps in the underlying price process. Hautsch and Podolskij (2010) analyze the estimator's dependence of the pre-averaging parameter θ and suggest a data-driven MSE minimizing choice.

8.1.4 Empirical Evidence

Traditionally, realized variance estimators as discussed in the previous subsection are applied to daily (or even higher aggregated) data. Nevertheless, given the high trading frequency of blue chip stocks nowadays, there are sufficient high-frequency price (or mid-quote) observations to apply these estimators also to intraday data. As shown by Hautsch and Podolskij (2010), it is recommended to apply realized kernel and pre-averaging estimators using the highest possible sampling frequency. This is particularly important if the underlying interval over which the quadratic variation is measured shrinks. Figures 8.1–8.4 show estimates of the quadratic variation over 5 days based on intervals of 60, 30, 15 and 5 min, respectively. The realized kernel estimator is computed based on an optimal choice of H as illustrated in Sect. 8.1.2. The pre-averaging estimator is based on a fixed length of the pre-averaging interval according to $\theta = 0.2$. Both estimates are benchmarked with the squared return computed over the entire period and a plain realized variance estimator based on 2 min returns. To increase the efficiency of the realized variance estimator, it is sub-sampled, i.e., it corresponds to the average of all realized variance estimators sampled over 2-min grids but starting at different observations within the first 2-min interval.[2]

It is shown that squared returns yield quite noisy estimates, particularly when the underlying interval becomes small. The sub-sampled 2-min realized variance estimators are similar to the realized kernel and pre-averaging estimators. This finding indicates that at least for actively traded assets, (sub-sampled) realized variance estimators provide a powerful but simple way to estimate high-frequency

[2]All estimators are correspondingly scaled in order to account for "border effects" when the sampling grid does not exactly match the underlying sampling period.

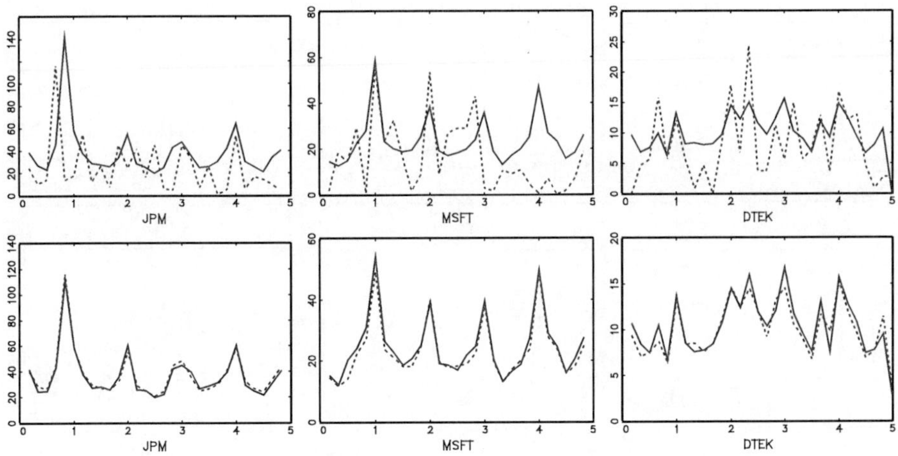

Fig. 8.1 Annualized quadratic variation estimates (in percent) of 60-min intervals for JP Morgan (NYSE), Microsoft (NASDAQ) and Deutsche Telekom (DTEK). Sample period: 01/06/09–05/06/09 for JP Morgan and Microsoft and 06/09/10–10/09/10 for Deutsche Telekom. *Upper panel*: 2-min sub-sampled realized variance estimator (*solid*) and squared 60-min returns (*dotted*). *Lower panel*: Realized kernel estimator (*solid*) and pre-averaging estimator (*dotted*)

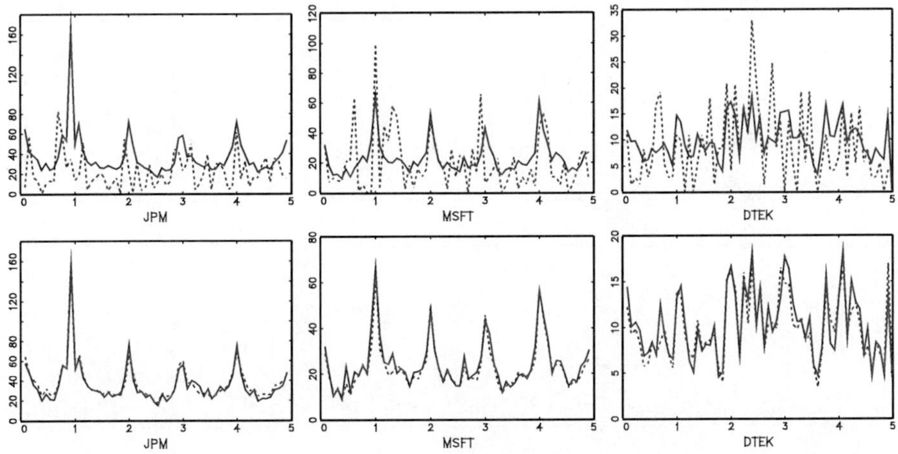

Fig. 8.2 Annualized quadratic variation estimates (in percent) of 30-min intervals for JP Morgan (NYSE), Microsoft (NASDAQ) and Deutsche Telekom (DTEK). Sample period: 01/06/09–05/06/09 for JP Morgan and Microsoft and 06/09/10–10/09/10 for Deutsche Telekom. *Upper panel*: 2-min sub-sampled realized variance estimator (*solid*) and squared 30-min returns (*dotted*). *Lower panel*: Realized kernel estimator (*solid*) and pre-averaging estimator (*dotted*)

variances with high precision without being affected by noise-induced biases. As expected, the estimators become more noisy and erratic if the underlying period shrinks. The time series of 60 min volatilities is relatively smooth and

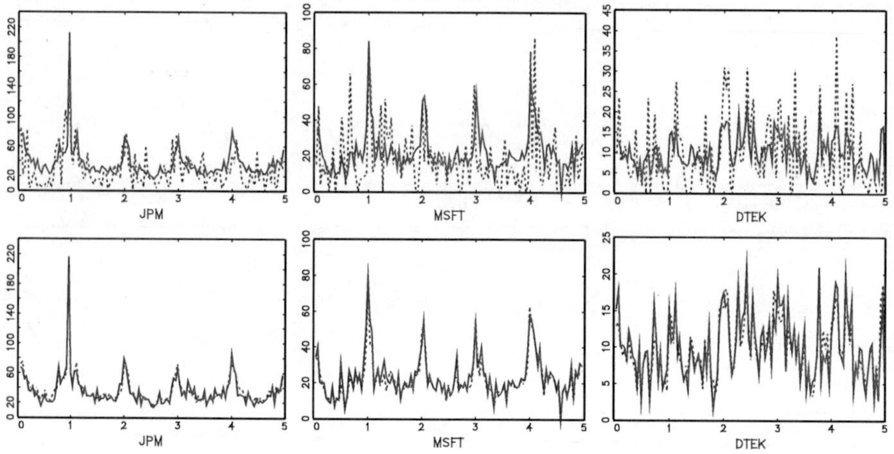

Fig. 8.3 Annualized quadratic variation estimates (in percent) of 15-min intervals for JP Morgan (NYSE), Microsoft (NASDAQ) and Deutsche Telekom (DTEK). Sample period: 01/06/09–05/06/09 for JP Morgan and Microsoft and 06/09/10–10/09/10 for Deutsche Telekom. *Upper panel*: 2-min sub-sampled realized variance estimator (*solid*) and squared 15-min returns (*dotted*). Lower panel: Realized kernel estimator (*solid*) and pre-averaging estimator (*dotted*)

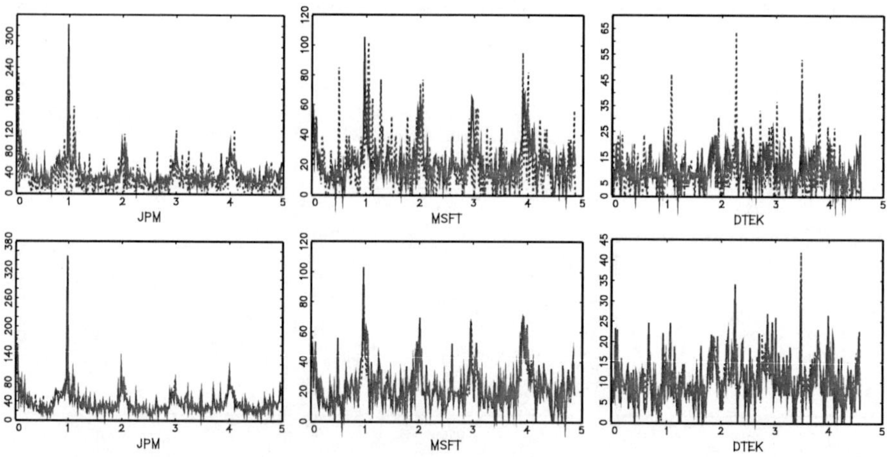

Fig. 8.4 Annualized quadratic variation estimates (in percent) of 5-min intervals for JP Morgan (NYSE), Microsoft (NASDAQ) and Deutsche Telekom (DTEK). Sample period: 01/06/09–05/06/09 for JP Morgan and Microsoft and 06/09/10–10/09/10 for Deutsche Telekom. *Upper panel*: 2-min sub-sampled realized variance estimator (*solid*) and squared 5-min returns (*dotted*). *Lower panel*: Realized kernel estimator (*solid*) and pre-averaging estimator (*dotted*)

reveals an obvious intraday seasonality pattern. In contrast, 5-min volatilities are based on less underlying data points and thus are less efficient and clearly more erratic. Moreover, the high-frequency estimators are naturally more affected by potential jumps. This is most evident for JPM revealing an obvious jump at the end of the first day. Overall, these results suggest that realized variance estimators are useful to estimate intraday variances as long as the interval of interest covers sufficiently many observations but reach their natural limits if the interval approaches transaction level. For these situations, spot variance estimators or trade-based volatility estimators as discussed in Sects. 8.2 and 8.4 are more useful.

8.1.5 Modelling and Forecasting Intraday Variances

The estimators as discussed in the previous subsection provide model-free ex post measures of the quadratic variation. To exploit this information for predictions of *future* volatilities, however, a time series model has to be employed. A typical feature of quadratic variation estimates on a daily level is their strong persistence as reflected by significantly positive and slowly decaying autocorrelation functions. To capture this behavior, Andersen et al. (2003) advocate using ARFIMA processes. However, on an intraday level, the statistical properties of realized variances differ in two respects. Firstly, the persistence in autocorrelations is lower. Secondly, intraday volatility is subject to significant seasonality patterns.

Figure 8.5 shows plots of the estimated intraday seasonality patterns of quadratic variation estimates covering 5 and 15 min. The underlying periodicities are estimated employing a flexible Fourier form as discussed in Chap. 5, i.e.,

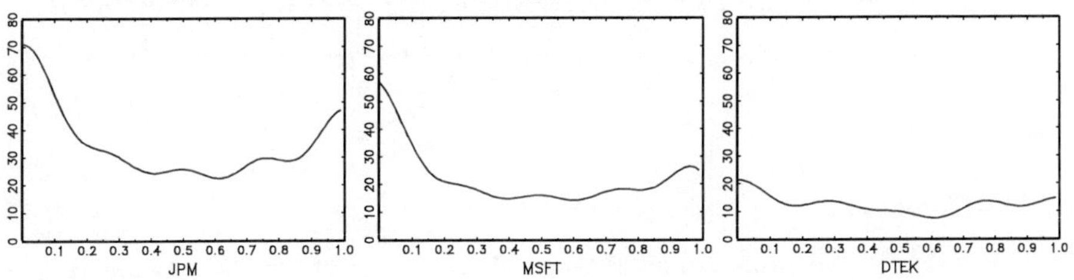

Fig. 8.5 Intraday seasonalities of pre-averaging estimates for JP Morgan (NYSE), Microsoft (NASDAQ) and Deutsche Telekom (XETRA). Sample period: June 2009 for JP Morgan and Microsoft and September 2010 for Deutsche Telekom. Seasonality estimated based on flexible Fourier form. *Top panel*: 15-min based estimates. *Bottom panel*: 5-min based estimates

$$RV_i = s_i \varepsilon_i,$$

$$s_i = \delta^s \cdot \bar{i} + \sum_{j=1}^{Q} \left\{ \delta^s_{c,j} \cos(\bar{i} \cdot 2\pi j) + \delta^s_{s,j} \sin(\bar{i} \cdot 2\pi j) \right\},$$

where ε_i is an i.i.d. mean one error term, RV_i denotes the corresponding volatility estimate associated with intraday interval i, s_i is the seasonality function and $\bar{i} \in [0, 1]$ denotes the standardized intraday time. We observe that the typical U-shaped intraday seasonality pattern is robustly estimated based on the different sampling intervals.

Figure 8.6 depicts the autocorrelation functions of the corresponding seasonally standardized volatility estimates. The ACFs are highly positive, but decay relatively quickly to zero. Such a dynamic behavior is straightforwardly captured by a standard ARMA process for logarithmic quadratic variation estimates

$$RV_i^a = RV_i s_i,$$

$$\ln RV_i^a = c + \sum_{j=1}^{P} \alpha \ln RV_{i-j}^a + \sum_{j=1}^{Q} \beta \varepsilon_{i-j} + \varepsilon_i,$$

where RV_i^a is the seasonally adjusted realized variance estimate and ε_i denotes a white noise error term. As discussed in Chap. 5, such a specification can be consistently and efficiently estimated in one step or, computationally easier but less efficient, in two steps. Though such a model is widely used in the given context, its

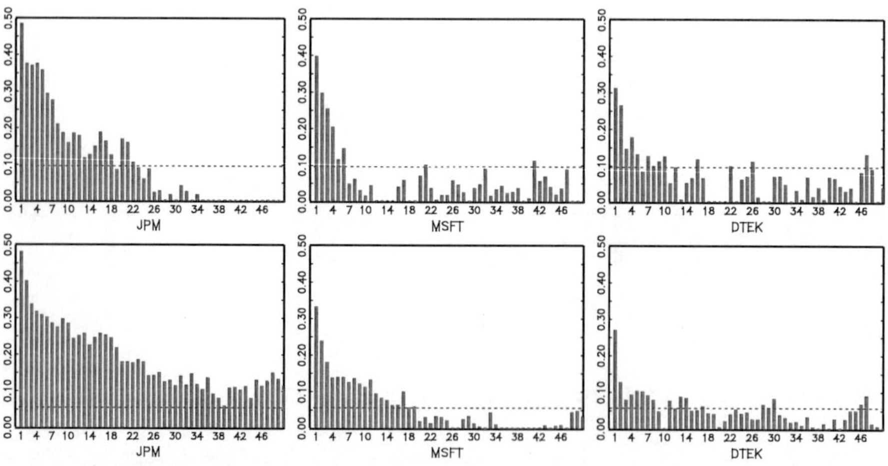

Fig. 8.6 Autocorrelations of seasonally adjusted pre-averaging estimates for JP Morgan (NYSE), Microsoft (NASDAQ) and Deutsche Telekom (XETRA). Sample period: June 2009 for JP Morgan and Microsoft and September 2010 for Deutsche Telekom. Seasonality estimated based on flexible Fourier form. *Top panel*: 15-min based estimates. *Bottom panel*: 5-min based estimates

main drawback is that estimates of log volatilities have to be transformed back to plain volatilities whenever one is interested in predictions of the non-transformed process. Alternatively, modelling RV directly can be straightforwardly performed in a MEM context as discussed in Chap. 5 and leading to

$$RV_i = s_i \Psi_i \varepsilon_i, \quad \varepsilon_i \sim Exp(1), \tag{8.24}$$

$$\Psi_i = \omega + \sum_{j=1}^{P} \alpha_{i-j} RV_{i-j}/s_{i-j} + \sum_{j=1}^{Q} \beta_{i-j} \Psi_{i-j}, \tag{8.25}$$

which can be estimated by QML. Obviously, this model is easily extended to a fractionally integrated process as illustrated in Chap. 6.

A more classical approach is to apply a GARCH process based on the corresponding seasonally adjusted intraday returns, i.e.,

$$r_i^a = \sqrt{r_i^2/s_i}\, sgn(r_i), \tag{8.26}$$

$$r_i^a = c + \varepsilon_i, \quad \varepsilon_i = u_i \sigma_i, \quad u_i \overset{i.i.d.}{\sim} (0,1), \tag{8.27}$$

$$\sigma_i^2 = \omega + \sum_{j=1}^{P} \alpha_j \varepsilon_{i-j}^2 + \sum_{j=1}^{Q} \beta_j \sigma_{i-j}^2, \tag{8.28}$$

where s_i is the (pre-estimated) intraday seasonality of *squared* returns and $sgn(r_i)$ denotes the sign of r_i.

However, as stressed by Hansen et al. (2011), GARCH models update only very slowly after sudden changes in volatility. The main reason for this slow responsiveness is that the innovation term ε_i^2 is a comparably noisy estimate of volatility. Consequently, it takes relatively long to reach a new variance level after a shock. Engle (2002) proposes augmenting a GARCH equation by the lagged realized variance. This idea has been put forward by Shepard and Sheppard (2009) and Hansen et al. (2011). Accordingly, by replacing ε_i^2 by RV_i we obtain

$$\sigma_i^2 = \omega + \sum_{j=1}^{P} \alpha_j RV_{i-j} + \sum_{j=1}^{Q} \beta_j \sigma_{i-j}^2.$$

In line with Hansen et al. (2011), we refer this specification to as a realized GARCH model.

Tables 8.1 and 8.2 show QML estimates of MEM(1,1), GARCH(1,1) and Realized GARCH(1,1) models for de-seasonalized 15-min and 5-min intraday volatilities for JPM, MSFT and DTEK. The underlying quadratic variations are estimated using the (finite-sample adjusted) pre-averaging estimator C_a^n. The intraday seasonalities are estimated using a flexible Fourier form as illustrated above. The highest persistence is revealed by the (realized) GARCH processes with

Table 8.1 QML estimates of MEM(1,1), GARCH(1,1) and Realized GARCH(1,1) models for 15-min de-seasonalized intraday volatilities for JP Morgan (NYSE), Microsoft (NASDAQ) and Deutsche Telekom (XETRA). Sample period: June 2009 for JP Morgan and Microsoft and September 2010 for Deutsche Telekom

par.	est.	p-val.	est.	p-val.	est.	p-val.
	JPM		MSFT		DTEK	
MEM estimates						
ω	0.084	0.036	0.210	0.000	0.180	0.034
α	0.347	0.000	0.289	0.000	0.222	0.000
β	0.572	0.000	0.501	0.000	0.599	0.000
GARCH estimates						
c	−0.030	0.770	0.036	0.187	−0.015	0.653
ω	0.097	0.027	0.095	0.178	0.345	0.002
α	0.073	0.002	0.036	0.076	0.107	0.005
β	0.830	0.000	0.869	0.000	0.550	0.000
Realized GARCH estimates						
c	−0.034	0.760	0.040	0.172	−0.017	0.658
ω	−0.001	0.531	0.887	0.000	0.189	0.023
α	0.170	0.004	0.228	0.028	0.325	0.000
β	0.831	0.000	−0.112	0.681	0.488	0.000

Table 8.2 QML estimates of MEM(1,1), GARCH(1,1) and Realized GARCH(1,1) models for 5-min de-seasonalized intraday volatilities for JP Morgan (NYSE), Microsoft (NASDAQ) and Deutsche Telekom (XETRA). Sample period: June 2009 for JP Morgan and Microsoft and September 2010 for Deutsche Telekom

par.	est.	p-val.	est.	p-val.	est.	p-val.
	JPM		MSFT		DTEK	
MEM estimates						
ω	0.079	0.003	0.194	0.004	0.198	0.048
α	0.293	0.000	0.210	0.000	0.176	0.000
β	0.629	0.000	0.596	0.000	0.627	0.000
GARCH estimates						
c	−0.008	0.629	0.028	0.117	−0.015	0.737
ω	0.022	0.004	0.058	0.061	0.191	0.029
α	0.045	0.000	0.047	0.003	0.110	0.000
β	0.934	0.000	0.895	0.000	0.700	0.000
Realized GARCH estimates						
c	−0.007	0.618	0.035	0.076	−0.019	0.782
ω	0.024	0.211	0.596	0.009	0.107	0.041
α	0.117	0.007	0.223	0.000	0.175	0.000
β	0.860	0.000	0.179	0.271	0.719	0.000

estimates of β close to one and α being comparably small. In the (plain) GARCH model, the innovation parameter α is significantly lower as in the realized GARCH and the multiplicative error models. This indicates that the past realized volatility

Table 8.3 Mincer–Zarnowitz forecasting regressions of 15-min de-seasonalized quadratic variations on corresponding one-step-ahead forecasts implied by MEM(1,1), GARCH(1,1) and Realized GARCH(1,1) models for JP Morgan (NYSE), Microsoft (NASDAQ) and Deutsche Telekom (XETRA). Sample period: June 2009 for JP Morgan and Microsoft and September 2010 for Deutsche Telekom

par.	est.	p-val.	R^2	est.	p-val.	R^2	est.	p-val.	R^2
	JPM			MSFT			DTEK		
MEM forecasts									
a	0.098	0.048		−0.019	0.585		−0.028	0.600	
b	0.895	0.000	0.302	1.019	0.000	0.191	1.027	0.000	0.137
GARCH forecasts									
a	0.022	0.405		−0.044	0.604		0.272	0.019	
b	0.977	0.000	0.175	1.042	0.000	0.066	0.726	0.000	0.052
Realized GARCH forecasts									
a	0.162	0.003		−0.741	1.000		0.186	0.017	
b	0.836	0.000	0.265	1.737	0.000	0.145	0.814	0.000	0.135

carries more information for the present conditional variance than the past squared return. This effect is also supported by the decline of the GARCH parameter β when moving from a GARCH specification to a realized GARCH model.[3] For a deeper discussion of these effects based on daily data, see Hansen et al. (2011).

A simple way to evaluate the goodness-of-fit and forecasting performance of volatility models is to apply forecasting regressions according to Mincer and Zarnowitz (1969) where actually observed quantities are regressed on their corresponding forecasts. As volatility is unobservable, it is replaced by an efficient quadratic variation estimate serving as benchmark. This is a common proceeding in the volatility literature, see, e.g., Andersen and Bollerslev (1998a), among others.

Tables 8.3–8.4 show the results of the corresponding forecasting regressions where the underlying pre-averaging based quadratic variation estimates are regressed on the corresponding MEM, GARCH and realized GARCH forecasts, i.e.,

$$C^n_{a,i} = a + b\widehat{QV}_i + \varepsilon_i, \quad \varepsilon_i \sim WN, \tag{8.29}$$

where $\widehat{QV}_i \in \{\hat{\Psi}_i, \hat{\sigma}^2_i\}$ in the context of the MEM or (realized) GARCH model, respectively.[4] Evaluating the estimates against the benchmark $a = 0$ (reflecting the forecast's unbiasedness), $b = 1$ (reflecting the forecast's efficiency) with

[3] Note that the p-values of the GARCH-RV model have to be interpreted with care since the process depends on pre-estimated regressors. Computing exact standard errors requires taking the sampling error of the pre-averaging estimator explicitly account. A deeper discussion of these effects is beyond the scope of this illustration.

[4] To reduce the potential effects of outliers, Pagan and Schwert (1990) suggest replacing the variables in (8.29) by their corresponding logarithmic transformations.

Table 8.4 Mincer–Zarnowitz forecasting regressions of 5-min de-seasonalized quadratic variations on corresponding one-step-ahead forecasts implied by MEM(1,1), GARCH(1,1) and Realized GARCH(1,1) models for JP Morgan (NYSE), Microsoft (NASDAQ) and Deutsche Telekom (XETRA). Sample period: June 2009 for JP Morgan and Microsoft and September 2010 for Deutsche Telekom

par.	est.	p-val.	R^2	est.	p-val.	R^2	est.	p-val.	R^2
	JPM			MSFT			DTEK		
MEM forecasts									
a	0.056	0.062		−0.086	0.902		0.010	0.455	
b	0.940	0.000	0.296	1.087	0.000	0.139	0.990	0.000	0.077
GARCH forecasts									
a	0.070	0.084		0.013	0.440		0.299	0.000	
b	0.924	0.000	0.170	0.990	0.000	0.080	0.700	0.000	0.048
Realized GARCH forecasts									
a	−0.047	0.871		−0.455	1.000		0.171	0.010	
b	1.041	0.000	0.282	1.458	0.000	0.125	0.829	0.000	0.075

a preferably high coefficient of determination R^2 (reflecting the correlatedness between realization and forecast) allows to compare the predictive performance of the competing models. We observe that MEM and realized GARCH specifications provide the highest R^2 and thus the highest correlation between the forecast and the realization. The similar performance of both approaches is expected as both specifications rely on the lagged realized volatility as innovation. Overall, the MEM approach yields the best forecasting performance which is not only reflected in the R^2 values but also in estimates of a being virtually zero and b being close to one. This is not necessarily true in the GARCH-type models which are partly biased and less efficient. Overall, these results support the usefulness of multiplicative error specifications for intraday volatility modelling and forecasting.

8.2 Spot Variances and Jumps

As illustrated above, for very high frequencies, the use of realized measures is limited by the availability of sufficient underlying transaction data. While for very liquid assets, quadratic variation estimates over short intervals are still more efficient than just using the squared returns measured over the corresponding interval, these differences vanish in case of less liquid stocks or if the length of the intervals shrinks to zero. Hence, if one is interested in estimating the volatility based on very high frequencies covering, e.g., only a couple of seconds, one is either left with the modelling of squared high-frequency returns or using smoothing techniques to exploit information of neighboring observations. The latter approach leads to an estimation of spot variances.

As in the given context, it is inevitable to account for the presence of jumps in the underlying price process, we modify (8.1) as follows:

$$p_t = p_0 + \int_0^t \mu_\tau d\tau + \int_0^t \sigma_\tau dW_\tau + \kappa_t N_t, \tag{8.30}$$

where κ_t denotes a random variable with mean $\mu_{\kappa,t}$ and variance $\sigma^2_{\kappa,t}$ and N_t is a counting process representing the number of jumps in the price path up to time t. Then, the quadratic variation over the interval $[0, t]$ is given by

$$QV(0,t) := \operatorname*{plim}_{\Delta \to 0} \sum_{j=1}^{\Delta^{-1}} (p_{j\Delta} - p_{(j-1)\Delta})^2 \tag{8.31}$$

$$= IV(0,t) + JV(0,t), \tag{8.32}$$

where

$$JV(0,t) := \sum_{j \le t} \kappa_j^2 \tag{8.33}$$

and $IV(0,t)$ given by (8.7).

In this case, the estimators discussed above are still consistent for $QV(0,t)$ but *not* for the integrated variance $IV(0,t)$. To consistently estimate $IV(0,t)$ in the absence of market microstructure noise, Barndorff-Nielsen and Shephard (2004) propose the realized bipower variation estimator given by

$$BPV^n = \frac{\pi}{2} \sum_{i=2}^{n} |p_{j\Delta} - p_{(j-1)\Delta}||p_{(j-1)\Delta} - p_{(j-2)\Delta}|. \tag{8.34}$$

A consistent estimator for $IV(0,t)$ in the presence of jumps and noise is given by Podolskij and Vetter (2009). Define the pre-averaged multipower variation as

$$V(Z, q_1, \ldots, q_l)^n := \sum_{i=0}^{n-lk_n+1} |\overline{Z}_i^n|^{q_1} |\overline{Z}_{i+k_n}^n|^{q_2} \cdots |\overline{Z}_{i+(l-1)k_n}^n|^{q_l}. \tag{8.35}$$

As shown by Podolskij and Vetter (2009), $V(Z, q_1, \ldots, q_l)^n$ is biased due to market microstructure noise. A bias-adjusted estimator is given by

$$BT^n := \frac{\sqrt{\Delta}}{\theta m_1^2 \psi_2} V(p, 1, 1)^n - \frac{\psi_1 \Delta}{2\theta^2 \psi_2} RV^n, \tag{8.36}$$

or, alternatively,

$$BT^n := \frac{\sqrt{\Delta}}{\theta m_1^2 \psi_2} V(p,1,1)^n - \frac{\psi_1}{\theta^2 \psi_2} \hat{\omega}^2, \tag{8.37}$$

with $BT^n \to IV$, $m_1 := \mathbb{E}[|\mathcal{N}(0,1)|]$ and ω^2 being consistently estimated by

$$\hat{\omega}^2 = -\frac{1}{N-1} \sum_{i=2}^{N} (p_{t_i} - p_{t_{i-1}})(p_{t_{i-1}} - p_{t_{i-2}}), \tag{8.38}$$

with p_{t_i} denoting the price at transaction i and N is the number of transactions within the interval $[0,1]$, see Oomen (2006).

Moreover, we have

$$BTV^n := \frac{\sqrt{\Delta}}{\theta \psi_2} (V(Z,2)^n - m_1^{-2} V(Z,1,1)^n) \to \sum_{j \le t} |\kappa_j|^2.$$

In finite samples, it is better to replace the constants ψ_1, ψ_2 by their empirical counterparts and to standardize the statistic $V(Z, q_1, \dots, q_l)^n$ by $[1/\Delta]/([1/\Delta] - lk_n + 2)$ (see Sect. 8.1) and to account for the true number of summands. For the empirical properties of this type of estimators, see Hautsch and Podolskij (2010).

The spot variance is defined as the derivative of the integrated variance IV, measured over a local window $t-h, t$, i.e.,

$$\sigma_t^2 := \lim_{h \to 0} \frac{IV(t-h,t)}{h} = \lim_{h \to 0} \frac{\int_{t-h}^{t} \sigma_\tau^2 d\tau}{h}. \tag{8.39}$$

Then, σ_t^2 can be estimated by

$$\hat{\sigma}_t^2 = \frac{n\widehat{IV}(t-h,t)}{h_n}, \tag{8.40}$$

where $\widehat{IV}(t-h,t)$ is a consistent estimator of $IV(t-h,t)$ and h_n is a local window depending on n with the property $h_n/n \to 0$ if $h_n \to \infty$.

As proposed by Bos et al. (2009), (8.37) can be adapted yielding a consistent estimate of σ_t^2 at the ith observation,

$$\hat{\sigma}_{t_i}^2 := \frac{\sqrt{\Delta}}{\theta m_1^2 \psi_2 h_n} \sum_{j=i-h_n}^{i-2k_n} |\overline{Z}_j^n||\overline{Z}_{j+k_n}^n| - \frac{\psi_1}{\theta^2 \psi_2} \hat{\omega}^2, \tag{8.41}$$

where $h_n = n^\beta$ with $\beta \in (0.5, 1]$ accounting for the fact that k_n is of order $n^{1/2}$. An obviously critical choice is the length of the local window h_n implying the usual trade-off between biasedness and efficiency of the estimator. Empirical evidence on the choice of h_n is provided by Bos et al. (2009). Underlying theory on nonparametric spot variance estimation is provided, among others, by Foster and Nelson (1996), Bandi and Reno (2009) or Kristensen (2010).

8.3 Trade-Based Volatility Measures

An alternative way of estimating spot volatility is to define volatility on a trade-to-trade basis and to construct estimators based on trade-to-trade returns. Engle (2000) defines returns per square root of time, $\tilde{r}_i := r_i/\sqrt{x_i}$ and suggests an ARMA-GARCH model given by

$$\begin{aligned}\tilde{r}_i &= c + \phi\tilde{r}_{i-1} + \theta\varepsilon_{i-1} + \varepsilon_i,\\ \varepsilon_i &= u_i\sqrt{h_i}, \quad u_i \sim (0,1),\\ h_i &= \omega_h + \alpha_h\varepsilon_{i-1}^2 + \beta_h h_{i-1}.\end{aligned} \tag{8.42}$$

According to this formulation, the variance per time unit, $h_i := \sigma_i^2/x_i$ with $\sigma_i^2 := \mathbb{V}[\varepsilon_i|\varepsilon_j, j \leq i-1]$ follows a GARCH(1,1) equation. This model is referred to as the simplest form of a so-called ultra-high-frequency GARCH (UHF-GARCH) which accounts for the irregular spacing of trades. The model builds on the assumption that the news from the last trade is measured as the square of the last price innovation per second, with the persistence of shocks being unaffected by trade durations. In this case, the coefficients are constant and the model can be estimated as a standard ARMA(1,1)-GARCH(1,l) model with the dependent variable defined as returns divided by the square root of trade-to-trade durations.

To allow for trade-to-trade durations influencing the volatility per time and assuming

$$\mathbb{E}[\varepsilon_i|\varepsilon_j, x_j, j \leq i-1; x_i] = 0, \tag{8.43}$$

the conditional variance of ε_i given the contemporaneous duration x_i, $\tilde{h}_i := \tilde{\sigma}_i^2/x_i$ with $\tilde{\sigma}_i^2 := \mathbb{V}[\varepsilon_i|\varepsilon_j, j \leq i-1, x_i]$, (8.42) can be specified as

$$\begin{aligned}\tilde{h}_i &= \omega_h + \alpha_h\varepsilon_{i-1}^2 + \beta_h\tilde{h}_{i-1} + \gamma_1 x_i^{-1} + \gamma_2\frac{x_i}{\Psi_i} + \gamma_3\Psi_i^{-1},\\ x_i &= \Psi_i\xi_i, \quad \xi_i \sim Exp(1),\\ \Psi_i &= \omega_x + \alpha_x x_{i-1} + \beta_x\Psi_{i-1}.\end{aligned} \tag{8.44}$$

As long as trade durations are assumed to be weakly exogenous for the volatility process, the model can be straightforwardly estimated by (Q)ML.

While Engle (2000) specifies the *variance per time unit*, Ghysels and Jasiak (1998) and Grammig and Wellner (2002) advocate GARCH processes for the *total variance*, $\sigma_i^2 = \mathbb{V}[\varepsilon_i|\varepsilon_j, x_j, j \leq i-1]$ in transaction time whose parameters are directed by an ACD process. Starting point is the assumption of a weak GARCH process according to Drost and Nijman (1993), where $c = \phi = \theta = 0$ and

$$P[\varepsilon_{i+1}^2|r_i, r_{i-1}, \ldots] = \sigma_{i+1}^2 = \omega_\sigma + \alpha_\sigma r_i^2 + \beta_\sigma \sigma_i^2, \tag{8.45}$$

where $P[r_{i+1}^2|r_i, r_{i-1}, \ldots]$ denotes the best linear predictor in terms of $(1, r_i, r_{i-1}, \ldots)$, i.e., $\mathbb{E}[(r_{i+1}^2 - P\{r_{i+1}^2|r_i, r_{i-1}, \ldots\})r_{i-n}^l] = 0$ for $i \geq 1$, $l = 0, 1, 2$ and $n = 0, 1, 2, \ldots$. Suppose that trade durations exactly correspond to the values predicted by the deterministic intraday component. Then, the corresponding GARCH process is observed based on "normal" business time and is referred to as a "normal duration GARCH process". However, in case of changes of the trading frequency, the normal duration GARCH process has to be aggregated or disaggregated. Drost and Werker (1996) provide a temporal aggregation formula for GARCH processes yielding a GARCH process with parameters depending on the underlying time between observations. In the given context of a weak GARCH process, this requires making the GARCH parameters dependent on the *expected* trade duration

$$\sigma_{i+1}^2 = \omega_{TA}(\Psi_{i+1}, \boldsymbol{\theta}_\sigma) + \alpha_{TA}(\Psi_{i+1}, \boldsymbol{\theta}_\sigma) r_i^2 + \beta_{TA}(\Psi_{i+1}, \boldsymbol{\theta}_\sigma)\sigma_i^2,$$

where "TA" stands for "temporal aggregation", $\boldsymbol{\theta}_\sigma := (\omega_\sigma, \alpha_\sigma, \beta_\sigma, k_\sigma)'$ with k_σ denoting the kurtosis of the underlying return distribution. Note that $\Psi_i := \mathbb{E}[x_i|\mathcal{F}_{i-1}]$ is deterministic given the information in t_{i-1}. Thus, the parameters $\boldsymbol{\theta}_{TA} := (\omega_{TA}, \alpha_{TA}, \beta_{TA})'$ are known in t_{i-1}. Drost and Werker's (1996) theorem on temporal aggregation yields the connection between $\boldsymbol{\theta}_\sigma$ and the parameters of the time aggregated process $\boldsymbol{\theta}_{TA}$:

Theorem 8.1 (Drost and Werker (1996)). *Assume $(y_i, i \in \mathbb{N})$ is a weak GARCH process with parameters $\boldsymbol{\theta}_\sigma = (\omega_{\sigma^2}, \alpha_\sigma, \beta_\sigma, k_\sigma)$, where k_σ is the kurtosis of y_i. Then, for each integer $m > 1$, the process $(\sum_{j=1}^{m-1} y_{i+j-1}, i \in m\mathbb{N})$ is a weak GARCH process with parameters*

$$\omega_m := \omega_{TA}(m, \theta_\sigma) = m\omega_\sigma \frac{1 - (\alpha_\sigma + \beta_\sigma)^m}{1 - (\alpha_\sigma + \beta_\sigma)},$$

$$\alpha_m := \alpha_{TA}(m, \theta_\sigma) = (\alpha_\sigma + \beta_\sigma)^m - \beta_\sigma,$$

$$\beta_m := \beta_{TA}(m, \theta_\sigma) \textit{ is the real solution of}$$

$$\frac{\beta_m}{1 + \beta_m^2} = \frac{a(\alpha_\sigma, \beta_\sigma, k_\sigma, m)(\alpha_\sigma + \beta_\sigma)^m - b(\alpha_\sigma, \beta_\sigma, m)}{a(\alpha_\sigma, \beta_\sigma, k_\sigma, m)(1 + (\alpha_\sigma + \beta_\sigma)^{2m}) - 2b(\alpha_\sigma, \beta_\sigma, m)},$$

$$k_\sigma := 3 + \frac{k_\sigma - 3}{m} + 6(k_\sigma - 1)$$
$$\times \frac{(m(1-\alpha_\sigma - \beta_\sigma)-1+(\alpha_\sigma+\beta_\sigma)^m)(\alpha_\sigma(1-(\alpha_\sigma+\beta_\sigma)^2)+\alpha_\sigma^2(\alpha_\sigma+\beta_\sigma))}{m^2(1-\alpha_\sigma-\beta_\sigma)(1-(\alpha_\sigma+\beta_\sigma)^2+\alpha_\sigma^2)},$$

where

$$b(\alpha_\sigma, \beta_\sigma, m) := (\alpha_\sigma(1-(\alpha_\sigma+\beta_\sigma)^2)+\alpha_\sigma^2(\alpha_\sigma+\beta_\sigma))\frac{1-(\alpha_\sigma+\beta_\sigma)^{2m}}{1-(\alpha_\sigma+\beta_\sigma)^2},$$

$$a(\alpha_\sigma, \beta_\sigma, k_\sigma, m) := m(1-\beta_\sigma)^2 + 2m(m-1)$$
$$\times \frac{(1-\alpha_\sigma-\beta_\sigma)^2(1-(\alpha_\sigma+\beta_\sigma)^2+\alpha_\sigma^2)}{(k_\sigma-1)(1-(\alpha_\sigma+\beta_\sigma)^2)}$$
$$+ 4\frac{(m(1-\alpha_\sigma-\beta_\sigma)-1+(\alpha_\sigma+\beta_\sigma)^m)}{1-(\alpha_\sigma+\beta_\sigma)^2},$$
$$\times \frac{(\alpha_\sigma(1-(\alpha_\sigma+\beta_\sigma)^2)+\alpha_\sigma^2(\alpha_\sigma+\beta_\sigma))}{1-(\alpha_\sigma+\beta_\sigma)^2}.$$

An important implication of this theorem is that $\boldsymbol{\theta}_{TA}(m, \boldsymbol{\theta}_{TA}(n, \boldsymbol{\theta})) = \boldsymbol{\theta}_{TA}(mn, \boldsymbol{\theta})$. Moreover, Drost and Werker (1996) show that the temporal aggregation results also hold if m and n are replaced by arbitrary real numbers. Then, the conditionally expected duration until the next trade Ψ_{i+1} (stemming from an ACD model) corresponds to the aggregation parameter m. Grammig and Wellner (2002) show how to simultaneously estimate the temporally aggregated GARCH process and the ACD process using GMM.

As illustrated by Meddahi et al. (2006), there are two main differences between Engle (2000) and the time-aggregated GARCH formulation of Ghysels and Jasiak (1998). Firstly, Engle (2000) considers a conditional variance given past information as well as *contemporaneous* trade durations while Ghysels and Jasiak (1998) condition on past information only. Under (8.43), we have

$$\sigma_i^2 := \mathbb{E}[\tilde{\sigma}_i^2 | \varepsilon_j, x_j, j \le i]. \tag{8.46}$$

Secondly, Engle's GARCH model for the variance per time units yields a time-varying parameter GARCH equation for the total variance process σ_i^2. Set for simplicity $\gamma_1 = \gamma_2 = \gamma_3 = 0$. Then, (8.44) can be written as

$$\tilde{\sigma}_i^2 = \omega x_i + \alpha \frac{x_i}{x_{i-1}} \varepsilon_{i-1}^2 + \beta \frac{x_i}{x_{i-1}} \tilde{\sigma}_{i-1}^2. \tag{8.47}$$

Using (8.46) and the definition of Ψ_i, we obtain

$$\sigma_i^2 = \omega\Psi_i + \alpha\frac{\Psi_i}{x_{i-1}}\varepsilon_{i-1}^2 + \beta\frac{\Psi_i}{x_{i-1}}\sigma_{i-1}^2 + \beta\frac{\Psi_i}{x_{i-1}}(\tilde{\sigma}_{i-1}^2 - \sigma_{i-1}^2), \tag{8.48}$$

which is obviously *not* a GARCH process for σ_i^2. Thus, Engle's (2000) high-frequency GARCH model is a time-varying parameter GARCH equation for $\tilde{h}_i$ but not for h_i.

Using an exact discretization of continuous time stochastic volatility processes observed at irregularly spaced times, Meddahi et al. (2006) show that it is appropriate to study the variance of ε_i given $\mathcal{F}_{i-1}^c = \sigma(\varepsilon_j, \nu_j, x_j, j \leq i-1; x_i)$ with ν_j denoting the spot variance of an underlying continuous time stochastic volatility model $d \ln p_t = \sqrt{\nu_t} dW_t$. They show that $\mathbb{V}[\varepsilon_i|\mathcal{F}_{i-1}^c]$ is well approximated by

$$\mathbb{V}[\varepsilon_i|\mathcal{F}_{i-1}^c] \approx \nu_{i-1}x_i, \tag{8.49}$$

which motivates studying the *conditional variance per time unit* $\mathbb{V}[\varepsilon_i|\mathcal{F}_{i-1}^c]/x_i$ as discrete time approximation of the underlying spot variance and supports the approach by Engle (2000).

However, Meddahi et al. (2006) also illustrate that in the given framework, $\mathbb{V}[\varepsilon_i|\mathcal{F}_{i-1}^c]/x_i$ follows an AR(1) process with time-varying parameters contradicting Engle (2000) and supporting the approach by Ghysels and Jasiak (1998). Hence, ideally combining the merits of both competing approaches would mean to (i) model the variance per time unit rather than the total variance, (ii) condition not only on the history of the process but also on the *contemporaneous* duration, and (iii) allow for time-varying parameters based on temporal aggregation.

Finally, note that the approaches discussed above ignore instantaneous simultaneity and causality effects between trade durations and stock price volatility. A simultaneous modelling of the price and duration process by explicitly accounting for the discreteness of price changes is proposed by Gerhard and Pohlmeier (2002). Renault and Werker (2011) study the simultaneous relationship between price volatility and trade durations in a continuous-time framework. They confirm the results of Gerhard and Pohlmeier (2002) and illustrate that a considerable proportion of intraday price volatility is caused by duration dynamics.

8.4 Volatility Measurement Using Price Durations

The use of price durations as defined in Chap. 4 opens up alternative ways of estimating volatility and price change risks. By definition they account for time structures in the price process and are of particular interest whenever an investor is able to determine his risk in terms of a certain price movement. On the basis of price durations, it is possible to estimate first passage times in the price process, i.e., the time until a certain price limit is exceeded. Equivalently, they allow for the quantification of the risk for a given price change within a particular time interval.

Price durations are naturally related to the price volatility of an asset. In order to clarify this relationship in more detail, it is worthwhile to consider the expected conditional volatility per time unit measured over the next trade duration x_{i+1}, i.e.,

$$\sigma^2(t_i) := \mathbb{E}\left[\frac{1}{x_{i+1}}(r_{i+1} - \bar{r}_{i+1})^2 \middle| \mathcal{F}_{t_i}\right], \quad i = 1, \ldots, n, \tag{8.50}$$

$$\text{with } r_i := \frac{p_i - p_{i-1}}{p_{i-1}},$$

$$\text{and } \bar{r}_i := \mathbb{E}[r_i | \mathcal{F}_{t_{i-1}}] := 0.$$

Here, r_i denotes the simple net return with a conditional mean assumed to be zero.[5] Note that (8.50) is quite imprecise with respect to the measure that is used within the integral. This aspect will be considered in more detail below.

In a GARCH framework, the event of interest is not the price at every transaction, but the price observed at certain points in time. Intra-day aggregates are typically used, e.g., on the basis of 5 or 10 min intervals.[6] Define in the following Δ as the length of the aggregation interval and $p(t)$ as the price that is valid at t. Then, a volatility measure based on *equi-distant* time intervals is obtained by

$$\sigma^2_{(r^\Delta)}(j\Delta) := \mathbb{E}\left[r^\Delta((j+1)\Delta)^2 | \mathcal{F}_{j\Delta}\right]\frac{1}{\Delta}, \quad j = 1, \ldots, n^\Delta, \tag{8.51}$$

$$\text{with } r^\Delta(t) := \frac{p(t) - p(t-\Delta)}{p(t-\Delta)}, \tag{8.52}$$

where $t = j\Delta$, $j = 1, \ldots, n^\Delta$, are the equi-distant time points associated with Δ and n^Δ denotes the sample size of Δ minute intervals. Thus, $\sigma^2_{(r^\Delta)}(t)$ is not based on the transaction process, but on certain (equi-distant) points in time and is typically estimated by standard GARCH-type models.[7] In general, volatility measurement based on equi-distant time intervals raises the question of an optimal aggregation level. A major problem is that the criteria of optimality for the determination of the aggregation level are not yet clear.

An alternative to an equi-distant interval framework is to specify a bivariate process, which accounts for the stochastic nature of both the process of price changes and the process of trade durations. These specifications allow for volatility estimates on the basis of a bivariate distribution. By defining the price change variable $d_i := p_i - p_{i-1}$, we have thus

$$\sigma^2_{(d,x)}(t_i) := \mathbb{E}\left[\frac{1}{x_{i+1}} \cdot d^2_{i+1} \middle| \mathcal{F}_{t_i}\right]\frac{1}{p_i^2}, \quad i = 1, \ldots, n. \tag{8.53}$$

---

[5]For the ease of exposition, no log returns are used. However, using log returns would not change the general proceeding.

[6]See, e.g., Andersen and Bollerslev (1998a,b) for further references.

[7]In this context, it is common to approximate $p(t)$ by the most recent observed price before t.

The limitation of such models to the analysis of price *changes* d_i is standard and not substantial since these models are usually estimated on the basis of transaction data. However, the process of price changes on the transaction level has a peculiar property which needs to be accounted for. Price changes take on only a few different values depending on the tick size and liquidity of the traded asset, see the empirical results in Chap. 3. For the implications of the tick size on the distributional and dynamical properties of the price process, see, e.g., Hautsch and Pohlmeier (2002). Most models concentrating on the bivariate process account explicitly for these market microstructure effects, like Russell and Engle (2005), Gerhard and Pohlmeier (2002), Liesenfeld and Pohlmeier (2003) or Rydberg and Shephard (2003). For more details, see Chap. 13.

A volatility estimator based on price durations is proposed by Gerhard and Hautsch (2002). They start with the assumption that a decision maker in need of a risk measure is able to express the size of a significant price change, dp. In this framework, the bivariate distribution of r_i and x_i is no longer of interest since r_i is reduced to the ratio of a constant and a conditionally deterministic variable. Define $\{t_i^{dp}\}_{i\in\{1,2,\ldots,n^{dp}\}}$ being the sequence of points of the thinned point process associated with price durations with respect to price change sizes of dp. Moreover, define $x_i^{dp} := t_i^{dp} - t_{i-1}^{dp}$ as the corresponding price duration. Then, we can formulate the conditional volatility per time measured over the spell of a price duration as

$$\sigma^2_{(x^{dp})}(t_i^{dp}) := \mathbb{E}\left[\frac{1}{x_{i+1}^{dp}}\middle|\mathcal{F}_{t_i^{dp}}\right]\left(\frac{dp}{p_{t_i^{dp}}}\right)^2 \tag{8.54}$$

$$= \sigma^{*2}_{(x^{dp})}(t_i^{dp}) \cdot \frac{1}{p^2_{t_i^{dp}}}, \quad i = 1,\ldots,n^{dp}, \tag{8.55}$$

where $\sigma^{*2}_{(x^{dp})}(t_i^{dp})$ stands for the conditional price change volatility from which the conditional volatility of returns can easily be recovered according to (8.55). Note that the estimation of $\sigma^2_{(x^{dp})}(t_i^{dp})$ necessitates the estimation of the conditional expectation $\mathbb{E}\left[\frac{1}{x_{i+1}^{dp}}\middle|\mathcal{F}_{t_i^{dp}}\right]$. This requires either specifying a stochastic process for $\frac{1}{x_i^{dp}}$ itself or, alternatively, computing the conditional distribution $\frac{1}{x_i^{dp}}$ using a transformation of the conditional distribution of x_i^{dp}. The latter procedure is typically quite cumbersome and requires the use of simulation techniques. For this reason, in Sect. 4.2.2, an approximation procedure based on a categorization approach is suggested.

Note that (8.55) is only defined at the particular points t_i^{dp} of the underlying thinned point process associated with dp-price durations. Thus, it does not provide a continuous-time volatility estimation. Alternatively, in line with Engle and Russell

(1998), an instantaneous volatility is defined as

$$\tilde{\sigma}^2(t) := \lim_{\Delta \downarrow 0} \frac{1}{\Delta} \mathbb{E}\left[\left(\frac{p(t+\Delta) - p(t)}{p(t)} \right)^2 \middle| \mathcal{F}_t \right]. \tag{8.56}$$

Equation (8.56) can be formulated in terms of the intensity function associated with the process of dp-price changes $\{t_i^{dp}\}_{i \in \{1,2,\ldots,n^{dp}\}}$. Thus,

$$\begin{aligned} \tilde{\sigma}^2_{(x^{dp})}(t) &:= \lim_{\Delta \downarrow 0} \frac{1}{\Delta} \mathbb{P}r\left[|p(t+\Delta) - p(t)| \geq dp \,|\mathcal{F}_t \right] \cdot \left[\frac{dp}{p(t)} \right]^2 \qquad (8.57)\\ &= \lim_{\Delta \downarrow 0} \frac{1}{\Delta} \mathbb{P}r\left[(N^{dp}(t+\Delta) - N^{dp}(t)) > 0 \,|\mathcal{F}_t \right] \cdot \left[\frac{dp}{p(t)} \right]^2 \\ &= \lambda^{dp}(t;\mathcal{F}_t) \cdot \left[\frac{dp}{p(t)} \right]^2, \end{aligned}$$

where $N^{dp}(t)$ denotes the counting process associated with cumulated absolute dp-price changes and $\lambda^{dp}(t;\mathcal{F}_t)$ the corresponding dp-price change intensity function. Expression (8.57) gives the expectation of the conditional dp-price change volatility per time in the next instant and thus, allows for a continuous picture of volatility over time.

Figures 8.7 and 8.8 show estimates of $\tilde{\sigma}^2_{(x^{dp})}(t)$ evaluated at the end of each price duration spell for JP Morgan, Microsoft and Deutsche Telekom. We choose different aggregation levels and estimate $\tilde{\sigma}^2_{(x^{dp})}(t)$ using an Exponential ACD(1,1) model with

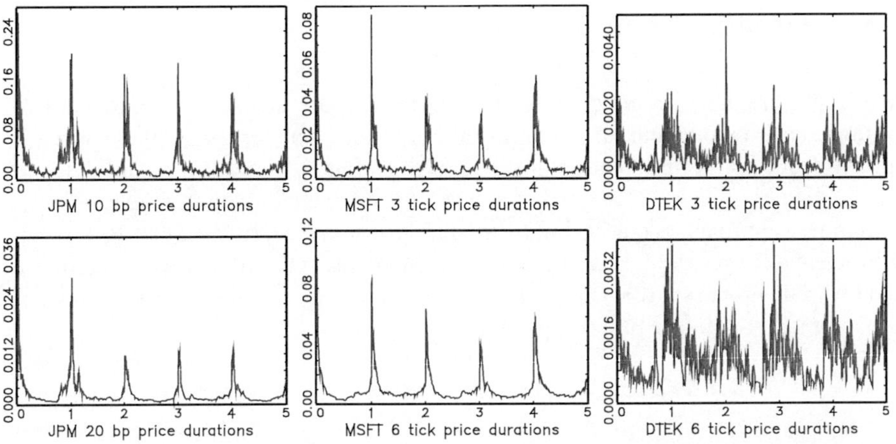

Fig. 8.7 Price duration based volatility estimates for JP Morgan (NYSE), Microsoft (NASDAQ) and Deutsche Telekom (XETRA). Sample period: 01/06/09–05/06/09 for JP Morgan and Microsoft and 06/09/10–10/09/10 for Deutsche Telekom. Estimates multiplied by $1e7$

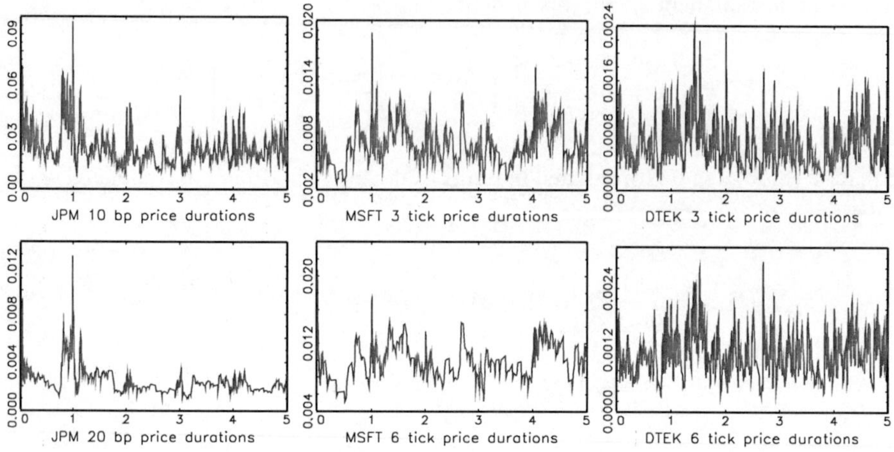

Fig. 8.8 Seasonally adjusted price duration based volatility estimates for JP Morgan (NYSE), Microsoft (NASDAQ) and Deutsche Telekom (XETRA). Sample period: 01/06/09–05/06/09 for JP Morgan and Microsoft and 06/09/10–10/09/10 for Deutsche Telekom. Estimates multiplied by $1e7$

multiplicative intraday seasonality function chosen as a flexible Fourier form (see Chap. 5). We observe that an increase in the aggregation level dp yields a natural "smoothing" over time since smaller price changes are filtered out. We also observe that these figures are strongly affected by intraday periodicities. Dividing by the corresponding seasonality function in Fig. 8.8 yields a more clear picture of intraday price intensities. Overall, we see an obvious similarity of the plots to the graphs shown in Sect. 8.1.

8.5 Modelling Quote Volatility

While the previous sections cover reduced-form (pure statistical) approaches to estimate and to model intraday volatility, this section discusses a structural approach arising from market microstructure literature. Define $r_{a,i} := \ln a_i - \ln a_{i-1}$ and $r_{b,i} := \ln b_i - \ln b_{i-1}$ as the log returns of ask and bid quotes, respectively. A common approach in market microstructure research (see, for instance, Roll 1984, Glosten and Harris 1988, Hasbrouck 1996 or Madhavan et al. 1997) is to decompose the bivariate process $\mathbf{r}_i := (r_{a,i}, r_{b,i})'$ into the sum of a common return component m_i and market-side-specific components $S_{a,i}$ and $S_{b,i}$,

$$\mathbf{r}_i = \mathbf{H}\boldsymbol{\xi}_i + \mathbf{B}'\mathbf{z}_i, \tag{8.58}$$

where

$$\mathbf{H} := \begin{pmatrix} 1 & 0 & 1 \\ 0 & 1 & 1 \end{pmatrix}$$

and $\boldsymbol{\xi}_i := (S_{a,i}, S_{b,i}, m_i)'$. Moreover, $\mathbf{B} := (\boldsymbol{\beta}_a, \boldsymbol{\beta}_b)$ is a $k \times 2$ parameter matrix associated with regressors $\mathbf{z}_i$ driving the individual components. The common component m_i captures the underlying (unobservable) "efficient return" driving both ask and bid returns. The quote-specific components $S_{a,i}$ and $S_{b,i}$ capture deviations between m_i and $\mathbf{r}_i$ and are associated with transaction costs driving the bid-ask spread. A similar framework for underlying quotes (in levels) is presented in Chap. 13. Then, m_i denotes the underlying efficient price, and ask and bid dynamics involve a cointegration relationship.

Hautsch et al. (2011) extend and apply this framework to jointly model quote and volatility dynamics. They refer $S_{a,i}$ and $S_{b,i}$ to as ask and bid noise returns, respectively, and model the dynamics of the unobservable return components in terms of a vector autoregressive (VAR) process of order one, i.e.,

$$\boldsymbol{\xi}_i = \boldsymbol{\mu} + \mathbf{F}\boldsymbol{\xi}_{i-1} + \varepsilon_i, \tag{8.59}$$

where $\boldsymbol{\mu} := (0, 0, c)'$ is a 3×1 vector, and $\mathbf{F}$ is a 3×3 matrix of the form

$$\mathbf{F} = \begin{pmatrix} \phi_a & 0 & 0 \\ 0 & \phi_b & 0 \\ 0 & 0 & \phi_m \end{pmatrix}.$$

According to traditional market microstructure theory, efficient returns should follow a white noise process implying ϕ_m and c to be zero. Conversely, ask and bid noise returns have a zero mean and should show mean-reverting behavior resulting in negative coefficients ϕ_a and ϕ_b. Hautsch et al. (2011) show that this notion is strongly supported by the data.

Hautsch et al. 2011 assume the 3×1 vector of innovations $\boldsymbol{\varepsilon}_i := (\varepsilon_{a,i}, \varepsilon_{b,i}, \varepsilon_{m,i})'$ to be conditionally normally distributed, i.e.,

$$\boldsymbol{\varepsilon}_i | \mathcal{F}_{i-1} \sim \mathcal{N}(\mathbf{0}, \boldsymbol{\Sigma}_i), \tag{8.60}$$

where $\boldsymbol{\Sigma}_i$ is specified as a diagonal matrix with $\boldsymbol{\Sigma}_i := \text{diag}(h_{a,i}, h_{b,i}, h_{m,i})$ and $\mathcal{F}_i$ denotes the information set up to i. Consequently, the off-diagonal elements are left unspecified. The components $h_{a,i}, h_{b,i}, h_{m,i}$ are referred to as (conditional) ask and bid noise variances as well as the efficient variance, respectively.

Hautsch et al. 2011 propose specifying the variance components in terms of a multiplicative error specification of the form

$$h_{j,i} = \sigma_{j,i}^2 s_{j,i} \exp\left(\gamma_j' \mathbf{w}_i\right), \quad j = \{a, b, m\}, \tag{8.61}$$

where $\mathbf{w}_i$ are regressors (multiplicatively) affecting the variance components with parameter γ_j. Moreover, $\sigma_{j,i}^2$ and $s_{j,i}$ denote components capturing volatility dynamics as well as deterministic time-of-day effects parameterized in terms of an EGARCH model and a flexible Fourier form, respectively,

$$\sigma_{j,i}^2 := \exp\left[\omega_j + \alpha_j \left(\frac{|\varepsilon_{j,i-1}|}{\sigma_{j,i-1}} - \sqrt{\frac{2}{\pi}}\right) + \beta_j \ln \sigma_{j,i-1}^2\right],$$

$$s_{j,i} := \exp\left[\sum_{q=1}^{Q} \psi_{j,s} \sin(2\pi q\tau) + \psi_{j,c} \cos(2\pi q\tau)\right],$$

with ω_j, α_j, and β_j are EGARCH parameters and ψ_j are Fourier form parameters. Finally, $\tau \in (0, 1)$ denotes the standardized time of the day.

Correspondingly, conditioning on past information and regressors, the conditional (co-)variances are given by

$$\sigma_{j,i}^2 = \mathbb{V}[r_{j,i} | \mathcal{F}_{i-1}, \mathbf{z}_{j,i}, \mathbf{w}_i] = h_{j,i} + h_{m,i}, \quad j \in \{a, b\}, \tag{8.62}$$

$$\sigma_{ab,i} = \mathrm{Cov}[r_{a,i}, r_{b,i} | \mathcal{F}_{i-1}, \mathbf{z}_{j,i}, \mathbf{w}_i] = h_{m,i}. \tag{8.63}$$

Similar to Engle and Patton (2004) (see also Chap. 13), parameterizing $r_{a,i}$ and $r_{b,i}$ directly yields a specification for changes in the log spread $s_i := \ln a_i - \ln b_i$ and the log mid-quote $mq_i := 0.5(\ln a_i + \ln b_i)$. Pre-multiplying (8.58) by the matrix $(1 : -1, 0.5 : 0.5)$ yields the reduced form

$$\Delta s_i = S_{a,i} - S_{b,i} + (\boldsymbol{\beta}_a - \boldsymbol{\beta}_b)'\mathbf{z}_i, \tag{8.64}$$

$$\Delta mq_i = c + 0.5(S_{a,i} + S_{b,i}) + m_i + 0.5(\boldsymbol{\beta}_a - \boldsymbol{\beta}_b)'\mathbf{z}_i, \tag{8.65}$$

where Δ denotes the first-difference operator. Correspondingly, the conditional variances of Δs_i and Δmq_i are given by

$$\sigma_{\Delta s,i}^2 = \mathbb{V}[\Delta s_i | \mathcal{F}_{i-1}, \mathbf{z}_i, \mathbf{w}_i] = h_{a,i} + h_{b,i}, \tag{8.66}$$

$$\sigma_{\Delta mq,i}^2 = \mathbb{V}[\Delta mq_i | \mathcal{F}_{i-1}, \mathbf{z}_i, \mathbf{w}_i] = 0.5(h_{a,i} + h_{b,i}) + h_{m,i}. \tag{8.67}$$

Hence, the conditional variance of spread changes equals the sum of the noise variances. As a result, if the noise variances are zero, the spread is constant, and quote returns and efficient returns coincide and correspond to the mid-quote return. Then, its conditional variance simply equals $h_{m,i}$.

Hautsch et al. (2011) estimate this model based on two steps. In the first stage, the model is estimated with constant variances using the Kalman filter as well as the parameters using the corresponding error prediction decomposition (see Harvey 1992). Using quasi-maximum likelihood arguments in an exponential family setting, the estimates are consistent, though not efficient, under distributional misspecification as long as the conditional means are correctly specified (see Gouriéroux et al. 1984). In the second step, Hautsch et al. (2011) estimate (univariate) EGARCH models based on the updated Kalman filter residuals, $e_{j,i} := \mathbb{E}[\varepsilon_{j,i}^* | \mathcal{F}_i]$, $j = \{a, b, m\}$. The model is applied to study the impact of macroeconomic news announcements on the individual quote and volatility components.

References

Andersen TG, Bollerslev T, Diebold F (2008) Parametric and Nonparametric Measurement of Volatility. In: Aït-Sahalia Y and Hansen L (ed) Handbook of Financial Econometrics. North Holland, Amsterdam

Andersen TG, Bollerslev T, Diebold F, Labys P (2003) Modeling and forecasting realized volatility. Econometrica 71(2):579–625

Aït-Sahalia Y, Mykland P, Zhang L (2005) How often to sample a continuous-time process in the presence of market microstructure noise. Rev Financ Stud pp. 351–416

Andersen TG, Bollerslev T (1998) Answering the skeptics: yes, standard volatility models do provide accurate forecasts Int Econ Rev 39(4):885–905

Bandi FM, Reno R (2009) Nonparametric stochastic volatility. Institute of Economic Research, Global COE Hi-Stat Discussion Paper gd08-035, Hitotsubashi University

Barndorff-Nielsen O, Hansen P, Lunde A, Shephard N (2008a) Designing realized kernels to measure the ex-post variation of equity prices in the presence of noise. Econometrica 76:1481–1536

Barndorff-Nielsen O, Hansen P, Lunde A, Shephard N (2008b) Realised kernels in practice: trades and quotes. Econom J 4:1–32

Barndorff-Nielsen O, Shephard N (2002) Econometric analysis of realized volatility and its use in estimating stochastic volatility models. J R Stat Soc Series B 64:253–280

Barndorff-Nielsen O, Shephard N (2004) Power and bipower variation with stochastic volatility and jumps. J Financ Econom 2(1):1–37

Bos CS, Janus P, Koopman SJ (2009) Spot variance path estimation and its application to high frequency jump testing. Tinbergen Institute Discussion Paper TI 2009-110/4, University of Oxford

Drost FC, Nijman TE (1993) Temporal aggregation of GARCH processes. Econometrica 61(2):909–927

Drost FC, Werker J (1996) Closing the GARCH gap: continuous time GARCH modeling. J Econom 74:31–57

Delbaen F, Schachermayer W (1994) A general version of the fundamental theorem of asset pricing. Mathematische Annalen 300:463–520

Engle RF (2000) The econometrics of ultra-high-frequency data. Econometrica 68(1):1–22

Engle RF (2002) New frontiers for ARCH models. J Appl Econom 17:425–446

Engle RF, Patton A (2004) Impact of trades in an error-correction model of quote prices. J Financ Markets 7:1–25

Engle RF, Russell JR (1998) Autoregressive conditional duration: a new model for irregularly spaced transaction data. Econometrica 66:1127–1162

Foster DP, Nelson DB (1996) Continuous record asymptotics for rolling sample variance estimators. Econometrica 64:139–174

Gerhard F, Hautsch N (2002) Volatility estimation on the basis of price intensities. J Empir Financ 9:57–89

Gerhard F, Pohlmeier W (2002) On the simultaneity of absolute price changes and transaction durations. Discussion paper, University of Konstanz

Ghysels E, Jasiak J (1998) GARCH for irregularly spaced financial data: The ACD-GARCH model. Stud Nonlinear Dynamics Econometrics 2:133–149

Glosten LR, Harris LE (1988) Estimating the components of the bid/ask spread. J Finan Econ 21:123–142

Gouriéroux C, Monfort A, Trognon A (1984) Pseudo maximum likehood methods: theory. Econometrica 52:681–700

Grammig J, Wellner M (2002) Modeling the interdependence of volatility and inter-transaction duration process. J Econom 106:369–400

Hansen PR, Huang Z, Shek HH (2011) Realized GARCH: A complete model of returns and realized measures of volatility. J Appl Econom forthcoming

Harvey AC (1989) Forecasting, structural time series models and the Kalman filter. Cambridge University Press, Cambridge

Hasbrouck J (1996) The dynamics of discrete bid and ask quotes. J Finance 6:2109–2142

Hautsch N, Hess D, Veredas D (2011) The impact of macroeconomic news on quote adjustments, noise, and informational volatility. J Bank Finance 35:2733–2746

Hautsch N, Podolskij M (2010) Pre-averaging based estimation of quadratic variation in the presence of noise and jumps: theory, implementation, and empirical evidence. Discussion Paper 2010/38, Collaborative Research Center 649 "Economic Risk", Humboldt-Universität zu Berlin

Hautsch N, Pohlmeier W (2002) Econometric analysis of financial transaction data: pitfalls and opportunities. Allgemeines Statistisches Archiv 86:5–30

Hansen PR, Lunde A (2006) Realized variance and market microstructure noise. J Bus Econ Stat 24(2):127–161

Jacod J, Li Y, Mykland P, Podolskij M, Vetter M (2009) Microstructure noise in the continuous case: the pre-averaging approach. Stochastic Process Appl 119:2249–2276

Kristensen D (2010) Nonparametric filtering of the realised spot volatility: a kernel-based approach. Econom Theory 26:60–93

Liesenfeld R, Pohlmeier W (2003) Ein dynamisches Hürdenmodell für Transaktionspreisveränderungen auf Finanzmärkten in Empirische Wirtschaftsforschung: Methoden und Anwendungen ed. by Franz W, Stadler M, H. J. Ramser. Mohr Siebeck

Madhavan A, Richardson M, Roomans M (1997) Why do security prices changes? a transaction-level analysis of NYSE stocks. Rev Financ Stud 10(4):1035–1064

Meddahi N, Renault E, Werker B (2006) GARCH and irregularly spaced data. Economic Letters 90:200–204

Mincer J, Zarnowitz V (1969) The evaluation of economic forecasts. In: Mincer J (ed) Economic forecasts and expectations. Columbia University Press, pp. 3–46

Oomen R (2006) Comment to realized variance and market microstructure noise. J Bus Econ Stat 24:195–202

Pagan AR, Schwert GW (1990) Alternative models for conditional stock volatility. J Econom 45:267–290

Podolskij M, Vetter M (2009) Bipower-type estimation in the noisy diffusion setting. Stochastic Process Appl 119:2803–2831

Renault E, Werker BJ (2011) Causality effects in return volatility measures with random times. Renault and Werker (2011), J of Econometrics, 160:272–279

Roll R (1984) A simple implicit measure of the effective bid-ask spread in an efficient market. J Finance 39:1127–1139

Russell JR, Engle RF (2005) A discrete-state continuous-time model of financial transactions prices and times: the autoregressive conditional multinomial-autoregressive conditional duration model. J Bus Econ Stat 23:166–180

Rydberg TH, Shephard N (2003) Dynamics of trade-by-trade price movements: decomposition and models. J Financ Econom 1:2–25

Shepard N, Sheppard K (2009) Realising the future: forecasting with high frequency based volatility (HEAVY) models. Department of Economics, Discussion Paper 438, University of Oxford

Chapter 9
Estimating Market Liquidity

Liquidity has been recognized as an important determinant of the efficient working of a market. Following the conventional definition of liquidity, an asset is considered as being liquid if it can be traded quickly, in large quantities and with little impact on the price.[1] According to this concept, the measurement of liquidity requires to account for three dimensions of the transaction process: time, volume and price. Kyle (1985) defines liquidity in terms of the tightness indicated by the bid-ask spread, the depth corresponding to the amount of one-sided volume that can be absorbed by the market without inducing a revision of the bid and ask quotes and resiliency, i.e., the time in which the market returns to its equilibrium.

The multi-dimensionality of the liquidity concept is also reflected in theoretical and empirical literature, where several strings can be divided: A wide range of the literature is related to the bid-ask spread as a measure of liquidity,[2] the price impact of volumes[3] and the analysis of market depth.[4]

In this chapter, we discuss different concepts of liquidity and econometric approaches to estimate liquidity demand, liquidity supply and market impact. Section 9.1 presents simple bid-ask spread and price impact measures. Section 9.2 discusses the application of volume durations to capture the time and volume dimension of liquidity. Here, we will particularly focus on excess volume durations as measures of one-sided trading intensity. Finally, Sect. 9.3 deals with dynamic models for limit order books.

---

[1]Classical references are, for example, Keynes (1930), Demsetz (1968), Black (1971) or Glosten and Harris (1988).

[2]See, for example, Conroy et al. (1990), Greene and Smart (1999), Bessembinder (2000) or Elyasiani et al. (2000).

[3]See, for example, Chan and Lakonishok (1995), Keim and Madhavan (1996) or Fleming and Remolona (1999).

[4]See, for example, Glosten (1994), Biais et al. (1995), Bangia et al. (2008) or Giot and Grammig (2006).

N. Hautsch, *Econometrics of Financial High-Frequency Data*,
DOI 10.1007/978-3-642-21925-2_9,

9.1 Simple Spread and Price Impact Measures

9.1.1 *Spread Measures*

A natural liquidity measure is the bid-ask spread reflecting the costs to cross the market. Besides the standard bid-ask spread

$$s_i := a_i - b_i, \tag{9.1}$$

with a_i and b_i denoting the ask and bid quotes at time t_i, also modified bid-ask spreads are often used, see, e.g., Goyenko et al. (2009). For instance, the *effective spread* is given by

$$s_{e,i} := 2|\ln p_i - \ln mq_i|, \tag{9.2}$$

where $mq_i = 0.5(a_i + b_i)$ denotes the mid-quote. The effective spread measures the (absolute) distance between the transaction price and the mid-quote and thus the *effective* transaction costs implied by a trade. If the effective spread is aggregated over time, we can compute a stock's dollar-volume-weighted average trading costs. Alternatively, the *realized spread* is given by

$$s_{r,i} := 2|\ln p_i - \ln mq_{i+j_x}|, \tag{9.3}$$

where j_x indicates the next trade which is at least x minutes apart. This measure gives the spread which is realized after x minutes and thus reflects the transaction costs which apply if the stock would be bought or sold back after this time. In applications, x is typically chosen as 5 min. If an identification of the trade direction is possible, the realized spread measure can be modified as

$$s_{r,i} := \begin{cases} 2(\ln p_i - \ln mq_{i+j_x}), & \text{if the } i\text{th trade is a buy,} \\ 2(\ln mq_{i+j_x} - \ln p_i), & \text{if the } i\text{th trade is a sell.} \end{cases} \tag{9.4}$$

Then, a simple measure of the permanent price impact of a trade in i is then given by the difference between the effective spread and the realized spread, i.e.,

$$s_{e,i} - s_{r,i} = 2|\ln mq_{i+j_x} - \ln mq_i|. \tag{9.5}$$

9.1.2 *Price Impact Measures*

Roll (1984) proposes a simple model of price dynamics if transaction costs are taken into account. Assume that midquotes follow a random walk process

$$mq_i = mq_{i-1} + \varepsilon_i, \tag{9.6}$$

with ε_i being white noise and prices are given by

$$p_i = mq_i + cy_i^b, \tag{9.7}$$

where

$$y_i^b = \begin{cases} 1 & \text{if } i \text{ is a buy,} \\ -1 & \text{if } i \text{ is a sell,} \end{cases}$$

and c reflects the effective trading costs corresponding to the half spread. Then, the model implies

$$\Delta p_i = mq_i + cy_i^b - (mq_{i-1} + cy_{i-1}^b) = c\Delta y_i^b + \varepsilon_i, \tag{9.8}$$

and thus $c = \sqrt{-\text{Cov}[\Delta p_i, \Delta p_{i-1}]}$. The effective transaction costs can be measured as the square root of the negative first order autocovariance of transaction price changes. Note that this measure is only valid as long as the first order autocovariance is negative. Generalizations of the Roll model are proposed by Hasbrouck (2007). Hasbrouck (2009) develops a Gibbs sampler to estimate transaction costs using daily data.

Amihud (2002) proposes an illiquidity measure given by

$$I = \overline{|r_d|/v_d}, \tag{9.9}$$

where d indexes trading days, r_d denotes daily log returns and v_d is the daily trading volume. Accordingly, the Amihud measure gives the average price change per trade size and thus reflects the average price impact of a transaction unit. Correspondingly,

$$L = I^{-1} = \overline{v_d/|r_d|}, \tag{9.10}$$

is a liquidity measure and is sometimes referred to as Amivest liquidity ratio (after a management firm that developed it).

Note that the price impact measures discussed above are "low-frequency" measures in the sense that they are computed based on aggregated data.

Corresponding high-frequency measures are traditionally obtained by price impact coefficients in regressions of trade-to-trade price changes on (signed) trading volume. Denote m_i as the expected value of the stock, conditional on the information set at i. Accordingly to Glosten and Harris (1988) we assume that m_i evolves as

$$m_i = m_{i-1} + \lambda q_i + \varepsilon_i,$$

where q_i denotes the signed trading volume at i and ε_i is a white noise process representing (non-predictable) public signals. The coefficient λ represents a market depth parameter reflecting the underlying demand or supply schedule. Then, if the price process follows (9.7), price changes are given by

$$\Delta p_i = \lambda q_i + c(y_i^b - y_{i-1}^b) + \varepsilon_i. \qquad (9.11)$$

As suggested by Brennan et al. (2009), this equation can be modified to allow for different price responses to buys and sells,

$$\Delta p_i = \alpha + \lambda_{buy}(q_i | q_i > 0) + \lambda_{sell}(q_i | q_i < 0) + c(y_i^b - y_{i-1}^b) + \varepsilon_i. \qquad (9.12)$$

Then, λ_{buy} and λ_{sell} are referred to as so-called "buy lambdas" and "sell lambdas", respectively, which reflect the trade-specific price impact and are easily estimated by OLS.

A more reduced-form approach to measure the price impact of a trade is to run the regression,

$$r_i = \lambda q_{i-1} + \varepsilon_i, \qquad (9.13)$$

and to estimate λ as a simple measure of the price impact. Modifications of these price impact regressions are proposed, e.g., by Hasbrouck (2009) by running the regression

$$r_{j\Delta} = \lambda_\Delta S_{j\Delta} + \varepsilon_{j\Delta}, \qquad (9.14)$$

where j indexes Δ-min intervals, m indexes all trades within the jth Δ-min interval, and $S_{j\Delta} = \sum_m sgn(v_{m,j})\sqrt{|v_{m,j}|}$ denotes the cumulative signed (dollar) volume over the jth interval. Then, λ_Δ reflects the derivative of the cost of demanding a certain amount of liquidity over Δ minutes. Typically, one chooses Δ as 5 min (see, e.g., Hasbrouck 2009 or Brennan et al. 2009).

Finally, note that all these measures require an identification of buyer and seller initiated trade. Hence, a noisy buy-sell identification can have a severe impact on the reliability of these measures. As already discussed in Chap. 3, these problems are particularly present if transactions are executed at prices within the bid-ask spread. For more refined (e.g., transaction cost based) measures, see, Harris (2003) or Hasbrouck (2007).

9.2 Volume Based Measures

9.2.1 The VNET Measure

As also reflected in Amihud's measure (see Sect. 9.1.2), a natural way to estimate the *realized* market depth is to relate the net trading volume to the corresponding price change over a fixed interval of time. However, as discussed by Engle and

Lange (2001), using a too small time interval can lead to measurement problems because the excess demand or the corresponding price change often can be zero. In contrast, using longer intervals reduces the ability of the measure to capture short-run dynamics that are of particular interest when the market is very active. For this reason, Engle and Lange (2001) propose the VNET measure, which measures the log net directional (buy or sell) volume over a price duration. Using a price duration as an underlying time interval avoids the aforementioned problems and links market volatility to market depth. Consider the sequence of points associated with dp-price changes, $\{t_i^{dp}\}$. Then, VNET is computed as

$$VNET_i := \ln \left| \sum_{j=N(t_{i-1}^{dp})}^{N(t_i^{dp})} y_j^b v_j \right|, \tag{9.15}$$

where $N(t)$ denotes the counting function associated with the transaction process. Hence, VNET measures the excess volume that can be traded before prices exceed a given threshold and therefore can be interpreted as the intensity in which excess demand flows into the market. An application of this concept in analyzing the depth of the market is found in Engle and Lange (2001).

Alternative liquidity measures can be constructed based on trading volumes reflecting (realized) liquidity demand. In this context, the cumulated trading volume, as empirically studied in Chap. 3, is a widely used measure and plays an important role for trading strategies, such as, for instance, VWAP strategies, see Brownlees et al. (2011). It is easily available based on transaction data, however, has the drawback that trading volume is also closely correlated to volatility, which can impede market liquidity.

A further category of liquidity proxies are intensity-based volume measures. Analyzing trade durations or, alternatively, the number of trades per time interval naturally reflect the trading frequency and liquidity demand in terms of trading opportunities. Alternatively, volume durations capture both the time and volume dimension of the intraday trading process. Though volume durations do not account for the price impact, they provide a reasonable measure of time costs of liquidity (see, e.g., Gouriéroux et al. 1999). Consider, for example, a trader who wants to execute a large order, but wants to avoid the costs of immediacy induced by a high bid-ask-spread. Then, she has the option of splitting her order and distributing the volume over time. Such a trader is interested in the expected time until the order is entirely executed. Accordingly, the expected volume duration yield the (time) costs of liquidity.

By defining volume durations not only based on the amount of volume shares, but also on the type of the corresponding transactions, it is possible to capture different components of the trading process. Hence, buy (sell) volume durations might be interpreted as the waiting time until a corresponding (unlimited) market order is executed. In this sense, forecasts of volume durations are associated with predictions of the absorptive capacities of the market. Alternatively, by the quantification of the time until a given volume on both market sides is traded, one obtains a liquidity measure that also accounts for the balance between the market sides. In this case, a market period is defined as liquid if unlimited market orders are executed quickly on both sides of the market.

9.2.2 Excess Volume Measures

9.2.2.1 Determinants of Excess Volume Durations

According to asymmetric information based market microstructure models as reviewed in Chap. 2, the presence of information on the market is associated with fast trading and one-sided volume confronting liquidity suppliers with risks due to inventory problems and adverse selection. Hautsch (2003) proposes to use excess volume durations as defined in Chap. 3 as natural indicators for the presence of information on the market. Excess volume durations measure the intensity of (one-sided) demand for liquidity and allow to account for the time and volume dimension of information-based trading. In this framework, small durations indicate a high demand of one-sided volume per time and reflect a high (il)liquidity risk. In contrast, long excess volume durations indicate either thin trading (long trade durations), or heavy, balanced trading. However, neither of the latter two cases confront the liquidity supplier with high liquidity risks. In this sense, excess volume durations serve as natural indicators for informed trading inducing adverse selection and inventory risks.

Hautsch (2003) models excess volume durations for several stocks traded at the NYSE. Table 9.1 reproduces part of the estimation results for IBM, JP Morgan and Philip Morris covering the period from 01/02/01 to 05/31/01. For each asset, two aggregation levels are chosen ensuring a satisfying number of observations per trading day and per stock leading to mean durations between approximately 4 and 10 min corresponding to approximately 40 and 130 observations per trading day, respectively. The excess volume duration series are seasonally adjusted by using a cubic spline regression based on 30 min nodes in a first step. The econometric framework is based on a Box–Cox ACD model (see Chap. 5) that allows for a parsimonious modelling of nonlinearities in the news impact function. Thus, the model is given by

$$(\Psi_i^{\delta_1} - 1)/\delta_1 = \omega + \sum_{j=1}^{P} \alpha_j \varepsilon_{i-j}^{\delta_2} + \sum_{j=1}^{Q} \beta_j (\Psi_{i-1}^{\delta_1} - 1)/\delta_1 + \mathbf{z}'_{i-1}\gamma. \tag{9.16}$$

Though this specification does not necessarily ensure positive values for Ψ_i and x_i, the actual estimates do not reveal any violations of this restriction. The ACD errors are assumed to follow a generalized F distribution as discussed in Chap. 5. The choice of the appropriate lag order is performed on the basis of the Bayes

Table 9.1 Estimates of Box–Cox-ACD(1,1) models with explanatory variables for excess volume durations. Based on the stocks IBM, JP Morgan and Philip Morris traded at the NYSE. Sample period from 01/02/01 to 05/31/01. QML standard errors in parantheses. Diagnostics: log likelihood function (LL), Bayes Information Criterion (BIC), LR-test for joint significance of all covariates ($\chi^2(6)$-statistic), as well as mean ($\bar{\hat{\varepsilon}}$), standard deviation (S.D.) and Ljung–Box statistic (LB) of ACD residuals. Reproduced from Hautsch (2003)

	IBM		JP Morgan		Philip Morris	
	$dv =$ 25,000	$dv =$ 50,000	$dv =$ 25,000	$dv =$ 50,000	$dv =$ 25,000	$dv =$ 50,000
ω	1.977	2.536	0.843	1.493	0.545	0.856
	(0.356)	(1.014)	(0.198)	(0.408)	(0.147)	(0.314)
α	0.292	0.339	0.151	0.163	0.137	0.154
	(0.041)	(0.070)	(0.027)	(0.054)	(0.024)	(0.047)
β	0.945	0.931	0.965	0.927	0.943	0.917
	(0.010)	(0.016)	(0.007)	(0.016)	(0.010)	(0.018)
δ_1	0.055	0.113	0.057	0.063	0.322	0.354
	(0.093)	(0.123)	(0.110)	(0.099)	(0.101)	(0.084)
δ_2	0.566	0.545	0.702	0.719	0.746	0.755
	(0.056)	(0.087)	(0.071)	(0.145)	(0.082)	(0.127)
a	1.615	2.206	0.981	1.001	1.352	0.857
	(0.207)	(0.214)	(0.113)	(0.138)	(0.053)	(0.115)
m	3.240	5.405	1.158	1.231	1.544	0.856
	(0.617)	(1.007)	(0.173)	(0.224)	(0.098)	(0.146)
η	0.050	–	0.142	0.086	–	0.187
	(0.027)	–	(0.059)	(0.061)	–	(0.082)
$\lvert dp_{i-1} \rvert$	−0.025	−0.127	−0.072	−0.140	−0.245	−0.314
	(0.018)	(0.039)	(0.058)	(0.069)	(0.124)	(0.139)
dp_{i-1}	0.023	−0.043	−0.101	−0.080	−0.204	−0.161
	(0.018)	(0.027)	(0.045)	(0.049)	(0.082)	(0.095)
dv_{i-1}	−0.213	−0.257	−0.091	−0.151	−0.063	−0.091
	(0.035)	(0.093)	(0.019)	(0.037)	(0.014)	(0.028)
$\overline{vol}_{i-1}$	0.007	−0.006	0.004	−0.016	−0.008	−0.023
	(0.004)	(0.009)	(0.005)	(0.011)	(0.007)	(0.012)
spd_{i-1}	−0.123	−0.094	−0.157	−0.223	−0.254	−0.362
	(0.033)	(0.043)	(0.042)	(0.089)	(0.060)	(0.113)
$dspd_{i-1}$	−0.027	−0.055	−0.259	−0.207	−0.615	−0.464
	(0.056)	(0.065)	(0.086)	(0.138)	(0.126)	(0.157)
obs	10,442	4,138	10,145	4,093	9,625	4,220
LL	−8,636	−3,382	−8,773	−3,386	−8,464	−3,563
BIC	−8,701	−3,436	−8,837	−3,444	−8,523	−3,621
$\chi^2(6)$	134.282	77.434	125.772	88.314	186.640	84.601
$\bar{\hat{\varepsilon}}$	1.005	1.013	1.005	1.017	1.008	1.016
S.D.	1.060	0.999	1.082	1.008	1.107	1.125
LB(20)	32.11	20.70	36.89	24.42	28.05	21.00

Information Criterion (BIC) leading to an ACD specification with a lag order of $P = Q = 1$ as the preferred model.

It is shown that the autocorrelation parameters are highly significant and reveal a strong serial dependence in the series of excess demand intensities. As indicated by the Ljung–Box statistics, nearly all estimated specifications are appropriate for capturing the dynamics of the duration process. In most cases, the Box–Cox parameter δ_1 is found to be between 0 and 0.3 rejecting both linear specifications ($\delta_1 = 1$) as well as logarithmic specifications ($\delta \to 0$). The estimates of δ_2 vary between 0.5 and 0.7 yielding evidence for a concave shape of the news impact curve. As indicated by the distribution parameters a, m and η, the high flexibility of the generalized F distribution is not necessarily required in all cases. In most cases, the estimates of η are close to zero indicating that a generalized gamma distribution seems to be sufficient.[5]

In order to capture the dependence between excess volume durations and past trading activities, Hautsch (2003) augments the model by six explanatory variables. Including the absolute midquote change $|dp_{i-1}|$, measured over the previous excess volume spell, yield insights whether the magnitude of quote adjustments during the past duration period has some predictability for the future liquidity demand intensity. Indeed, the absolute price change associated with a given excess volume is a proxy for the realized (absorbed) market depth. A significantly negative coefficient in most cases indicates that liquidity suppliers tend to adjust their quotes stronger when they expect high excess demands, and thus high liquidity risks. That is, the lower the market depth over the previous spell, the smaller the length of the subsequent excess volume duration.

Moreover, a highly significant coefficient associated with the signed price change (dp_{i-1}) measured over the previous duration period provides evidence in favor of asymmetric behavior of market participants. The sign of this coefficient is negative disclosing a higher (lower) excess demand intensity after positive (negative) price changes.

As a further explanatory variable the magnitude of the lagged (excess) volume is included. Note that a volume duration is measured as the time it takes a certain *minimum* excess volume to trade. Obviously, in the case of a large block trade, this threshold can be clearly exceeded. Since the extent to which the traded volume exceeds the threshold can contain important information for the appraisal of liquidity risk, the *effective* excess volume cumulated during the previous spell (dv_{i-1}) is included as a further regressor. A significantly negative impact of this variable in most regressions yields additional evidence for a strong clustering of excess volume intensities. To capture the fact that cumulative trading volumes are driven by both trade sizes and trading intensity, the average volume per transaction measured over the last duration spell, $\overline{vol}_{i-1}$, is included as a control variable but is shown to be insignificant. This might be induced by collinearity effects between dv_{i-1} and $\overline{vol}_{i-1}$ caused by the fact that large values of dv_{i-1} only occur when the

[5] In cases where $\eta \to 0$, the models are re-estimated using a generalized gamma distribution.

last trade in a duration period is a block trade. Then, the larger the quantity of this final transaction, the larger dv_{i-1} and therefore, the larger the average volume per trade, $\overline{vol}_{i-1}$.

Moreover, it is shown that the length of an excess volume duration is negatively correlated with the size of the spread posted at the beginning of the spell (spr_{i-1}). Hence, given the past excess volume intensity, the spread posted by the market maker indicates her assessment of the expected liquidity risk. Furthermore, the change of the spread during the previous spell, $dspr_{i-1} := spr_{i-1} - spr_{i-2}$, has a significantly negative impact on future excess volume durations. Thus, a widening of the bid-ask spread reflects an increase of the expected liquidity risk in the subsequent period.

9.2.2.2 Measuring Realized Market Depth

Similar to the VNET measure proposed by Engle and Lange (2001), the absolute price change measured over the corresponding duration period provides a measure of the realized market depth. The main difference between both concepts is the underlying time interval over which the price impact is measured. In case of excess volume durations, market depth is related to the intensity of the demand for (one-sided) liquidity and thus to liquidity supplier's inventory and adverse selection risk. Hence, they allow to analyze price adjustments in situations which are characterized by large imbalances between the buy and sell trading flow.

Table 9.2a reproduces part of Hautsch's (2003) results of linear regressions of the absolute price change, $|dp_i|$, measured over an excess volume duration episode on a set of regressors consisting mainly of the variables discussed in the previous subsection. In addition, Hautsch includes the contemporaneous excess volume duration, x_i, as well as the ACD residual $\hat{\varepsilon}_i = x_i/\hat{\Psi}_i$. Clearly, these variables are not weakly exogenous for $|dp_i|$. Nevertheless, one might argue that these regressors can be considered as being under the control of a market participant who influences the demand for liquidity, and thus the length of the underlying volume duration spell by her trading behavior. This specification enables to explore the determinants of the price impact given the length of the underlying spell. It allows to characterize the view of a trader who wants to place her volume in an optimal way and is interested in the relationship between the price impact and the time in which the volume is absorbed by the market.

An interesting question is whether the time in which a given excess volume is traded, plays an important role for the size of quote adjustments. Mostly highly significant coefficients associated with x_i indicate a positive correlation between the length of the time interval in which a certain excess volume is traded and the corresponding absolute price reaction. Obviously, a market maker has better chances of inferring from the trading flow when she is confronted with a more and thus uniform demand than in periods where she faces large single orders entering the market. Moreover, unexpectedly large (small) duration episodes, as measured by $\hat{\varepsilon}_i$, increase (decrease) the price impact even more. Hence, the market depth is

Table 9.2 Linear regressions for absolute price changes measured over excess volume durations. Dependent variable: $|dp_i|$. Based on the stocks IBM, JP Morgan and Philip Morris traded at the NYSE. Database extracted from the 2001 TAQ database, sample period from 01/02/01 to 05/31/01. Diagnostics: p-value of F-test for joint significance of explanatory variables (pv_F) and adjusted R-squared. HAC robust standard errors in parantheses. Reproduced from Hautsch (2003)

	IBM		JP Morgan		Philip Morris	
	$dv =$ 25,000	$dv =$ 50,000	$dv =$ 25,000	$dv =$ 50,000	$dv =$ 25,000	$dv =$ 50,000
$\lvert dp_{i-1}\rvert$	−0.000	0.097	0.062	0.046	0.083	0.085
	(0.000)	(0.018)	(0.014)	(0.020)	(0.011)	(0.017)
dv_{i-1}	−0.391	−0.066	−0.012	−0.010	−0.001	−0.012
	(0.379)	(0.026)	(0.002)	(0.008)	(0.001)	(0.005)
$\overline{vol}_{i-1}$	0.162	−0.008	−0.004	−0.005	−0.002	−0.000
	(0.182)	(0.005)	(0.001)	(0.003)	(0.000)	(0.001)
spr_{i-1}	−0.234	0.121	0.086	0.184	0.029	0.032
	(0.362)	(0.065)	(0.031)	(0.057)	(0.011)	(0.024)
$dspr_{i-1}$	0.184	−0.042	−0.039	−0.064	−0.012	−0.011
	(0.230)	(0.065)	(0.018)	(0.034)	(0.011)	(0.020)
$\hat{\varepsilon}_i$	−0.001	0.073	0.025	0.025	0.017	0.015
	(0.054)	(0.010)	(0.002)	(0.006)	(0.001)	(0.003)
x_i	0.064	0.022	0.007	0.025	0.001	0.008
	(0.057)	(0.009)	(0.002)	(0.006)	(0.001)	(0.002)
$const$	2.873	0.917	0.190	0.212	0.060	0.182
	(2.506)	(0.282)	(0.026)	(0.094)	(0.017)	(0.053)
obs	10,441	4,137	10,144	4,092	9,624	4,219
pv_F	0.000	0.000	0.000	0.000	0.000	0.000
$\bar{R}^2$	0.001	0.147	0.165	0.147	0.147	0.114

lower in periods of unexpectedly high excess demand intensity. These results seem to be counter-intuitive since one would expect that traders who trade large quantities quickly would have to bear additional "costs for immediacy". However, note that this empirical study does not consider the trading of block transactions. Such trades are typically executed on the upstairs market and are based on special agreements between the specialist and the trader (see, for example, Keim and Madhavan 1996) confronting the investor with additional transaction costs induced by higher bid-ask spreads. By explicitly focusing on floor trading at the NYSE, such effects are not captured and opposite tendencies seem to prevail. Hence, liquidity suppliers do not seem to draw inference from single large transactions, but from a more permanent one-sided trading flow. Accordingly, the higher the continuity of the one-sided trading flow, the higher the probability for the existence of information on the market. Obviously, single large trades are associated with individual liquidity traders, while a more continuous one-sided liquidity demand seems to be attributed to the existence of information.

9.3 Modelling Order Book Depth

The price impact measures discussed in the previous subsection reflect the *realized* price impact depending on the underlying (unobservable) supply and demand schedule in the market. However, limit order book data open up the possibility to estimate the *ex ante* price impact of a trade. This is particularly true if the complete bid and ask order curves are observable. Figure 9.1 illustrates a hypothetical limit order book and plots the cumulated pending ask and bid volume against the corresponding limit prices. Here, the steepness of the ask and bid curves are natural measures of the liquidity supply. The steeper the curves, the more liquidity supply is offered (i.e., sell offers on the buy (ask) side and buy offers on the sell (bid) side) and the smaller the price impact of a large order "walking up" or "walking down" the book. Given the objective to capture not only the volume around the best quotes but also pending quantities "behind" the market, i.e., deeper in the book, the underlying problem becomes high-dimensional. In some markets or for some stocks, depth is strongly concentrated closely to the market requiring to capture only a few levels. However, in less liquid markets, liquidity supply is dispersed over a wide range of price levels. Figure 9.2 shows snapshots of the order books of the National Australian Bank (NAB), BHP Billiton Limited (BHP), a resource company, Mount Isa Mines Limited (MIM), a base metal mining company and Woolworths (WOW) in the electronic trading system of the Australian Stock Exchange (ASX). Here, the order book for MIM reflects a situation where the market is very deep on the first order level but very thin behind the market. Here, large orders can be executed without price impact beyond the spread. However, as long as the first level is absorbed, trading costs might increase dramatically or – in the extreme case – the order cannot be completely executed because of a lack of offered liquidity. Conversely, the order

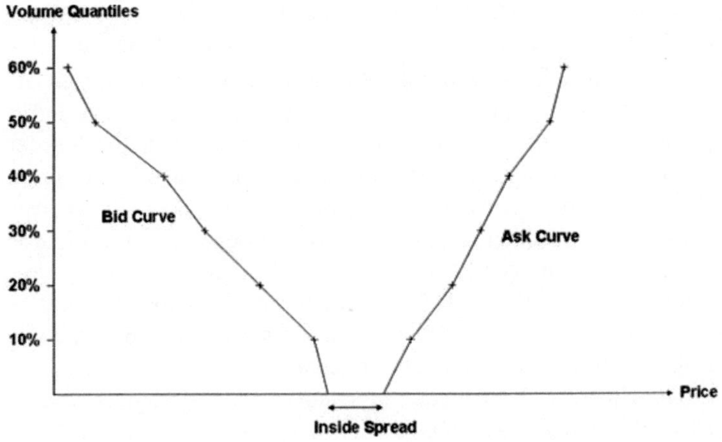

Fig. 9.1 Illustration of a limit order book. *Vertical axis*: Cumulated pending order volume. *Horizontal axis*: limit price

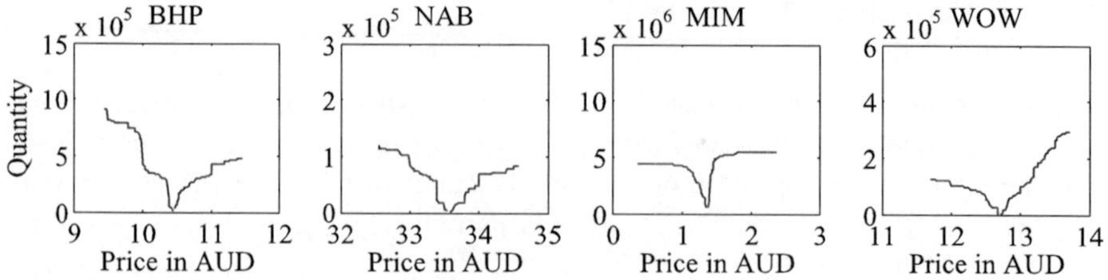

Fig. 9.2 Limit order books of the stocks BHP, NAB, MIM and WOW traded at the ASX on July 8, 2002 at 10:15

books of BHP and NAB reflect situations where the offered liquidity is spread over many levels. Since liquidity is less concentrated, it is much more likely that large orders walk up or down the book and thus face price impact costs.

9.3.1 A Cointegrated VAR Model for Quotes and Depth

In a limit order book market where liquidity and order activity is concentrated on a few levels close to the prevailing best ask and bid quotes, the dimensionality of the underlying process is still moderate. In this case, liquidity supply on the individual levels can be modelled using classical multivariate time series models. Hautsch and Huang (2009) propose a vector autoregressive model for best ask and bid quotes as well as several levels of depth. Let i index any order activity in the market and denote a_i and b_i as the best ask and bid quotes instantaneously after the ith order activity. Moreover, $v_i^{a,j}$ and $v_i^{b,j}$ for $j = 1, \ldots, k$, denote the depth on the jth best observed quote level on the ask and bid side, respectively.

Then, high-frequency dynamics in quotes and depths are captured in a K-dimensional vector of the logarithmic endogenous variables $\mathbf{y}_i := [\ln a_i, \ln b_i, \ln v_i^{a,1}, \ldots, \ln v_i^{a,k}, \ln v^{b,1}, \ldots, \ln v_i^{b,k}]'$ with $K = 2 + 2 \times k$. In Hautsch and Huang (2009), the quote levels associated with $v_i^{a,j}$ and $v_i^{b,j}$ are not observed on a *fixed* grid at and behind the best quotes. Hence, their price distance to a_i and b_i is not necessarily exactly $j-1$ ticks but might be higher if there are no limit orders on all possible price levels behind the market. This proceeding is justified for liquid assets where level j mostly corresponds to a distance of $j-1$ ticks to the corresponding best quote. Moreover, modelling log volumes instead of plain volumes reduces the impact of extraordinarily large volumes. This is also suggested by Potters and Bouchaud (2003) studying the statistical properties of the market impacts of trades. Moreover, using logs implies that changes in market depth can be interpreted as *relative* changes with respect to the current depth level.

Hautsch and Huang (2009) model log quotes and log depths in terms of a cointegrated VAR(P) model which is given in vector error correction (VEC) form

$$\Delta \mathbf{y}_i = \boldsymbol{\mu} + \boldsymbol{\alpha}\boldsymbol{\beta}'\mathbf{y}_{i-1} + \sum_{j=1}^{P-1} \boldsymbol{\Gamma}_j \Delta \mathbf{y}_{i-j} + \mathbf{u}_i, \tag{9.17}$$

where $\boldsymbol{\mu}$ is a $K \times 1$ vector of constants, $\boldsymbol{\alpha}$ and $\boldsymbol{\beta}$ denote the $K \times r$ loading and cointegrating matrices with $r < K$, and $\boldsymbol{\Gamma}_j$, $j = 1, \ldots, p-1$, is a $K \times K$ parameter matrix. The noise term $\mathbf{u}_i$ is assumed to be serially uncorrelated with zero mean and covariance $\boldsymbol{\Sigma}_u$. The model can be straightforwardly written in reduced VAR form,

$$\mathbf{y}_i = \boldsymbol{\mu} + \sum_{j=1}^{p} \mathbf{A}_j \mathbf{y}_{i-j} + \mathbf{u}_i, \tag{9.18}$$

where $\mathbf{A}_1 := \mathbf{I}_K + \boldsymbol{\alpha}\boldsymbol{\beta}' + \boldsymbol{\Gamma}_1$ with $\mathbf{I}_K$ denoting a $K \times K$ identity matrix, $\mathbf{A}_j := \boldsymbol{\Gamma}_j - \boldsymbol{\Gamma}_{j-1}$ with $1 < j < p$ and $\mathbf{A}_p := -\boldsymbol{\Gamma}_{p-1}$.

As the model is a specification for quote and depth *levels* rather than first differences, it is sufficiently general to accommodate both stationary and non-stationary behavior of the underlying variables and does not require a priori restrictions on the stationarity of order book depth. Using limit order book data from actively traded Euronext stocks in 2008, Hautsch and Huang (2009) show that order book depth can be actually strongly persistent on high frequencies and tend to reflect near-unit-root behavior. This general framework obviously also nests the case of stationary order book depth. Then, there is only one cointegration relationship between ask and bid quotes. In models involving only quote dynamics (see, e.g., Engle and Patton 2004) or spread dynamics (see, e.g., Lo and Sapp 2006), the error correction term $\boldsymbol{\beta}'\mathbf{y}_i$ is typically assumed to be equal to the spread implying a linear restriction $\mathbf{R}'\boldsymbol{\beta} = 0$ with $\mathbf{R}' = [1, 1, 0, \ldots, 0]$. For a discussion of the Engle and Patton (2004) model, see Chap. 13.

9.3.2 A Dynamic Nelson–Siegel Type Order Book Model

If market depth is less concentrated around the best quotes but more dispersed over several price levels, the dimensionality of the underlying process becomes high making VAR type models for individual depth levels rather cumbersome. In this case, dimension reduction techniques have to be applied. Proposing a suitable statistical model results in the problem of finding an appropriate way of reducing the high dimension without losing too much information on the spatial and dynamic structure of the process. Moreover, applicability of the model requires computational tractability as well as numerical stability.

A common way to reduce the dimensionality of multivariate processes is to apply a factor decomposition. The underlying idea is that the high-dimensional process is driven by only a few common factors which contain most underlying information. Factor models are often applied in the asset pricing literature to extract underlying common risk factors. In this spirit, a successful parametric factor model has been proposed, for instance, by Nelson and Siegel (1987) to model yield curves. Applying the Nelson–Siegel model to the order curve on one side of the market,[6] cumulative depth up to level j at period i is given by

$$v_i^j = \beta_{1i} + \beta_{2i}\left(\frac{1-e^{-\lambda_i j}}{\lambda_i j}\right) + \beta_{3i}\left(\frac{1-e^{-\lambda_i j}}{\lambda_i j} - e^{-\lambda_i j}\right). \tag{9.19}$$

The parameters β_{1i}, β_{2i} and β_{3i} can be interpreted as three latent dynamic factors with loadings 1, $(1 - e^{-\lambda_i j})/\lambda_i j$, and $\{(1 - e^{-\lambda_j j})/\lambda_i j\} - e^{-\lambda_i j}$, respectively. Then, β_{1i} represents a long-term factor whose loading is constant for all levels and thus reflects the (cross-sectional) average depth. With the loading of β_{2i} starting at one and decaying monotonically and quickly to zero, β_{2i} captures the overall slope of the order curve. Finally, β_{3i} is interpreted as a "curvature factor" with a loading starting at zero, increasing and decaying to zero in the limit. Finally, the parameter λ_i governs the speed of decay and drives the order schedule's curvature.

Then, denoting the factors in correspondence with the term structure literature by $L_i := \beta_{1i}$, $S_i := \beta_{2i}$ and $C_i := \beta_{3i}$, we can represent the model in state-space form

$$\mathbf{v}_i = \mathbf{A}\mathbf{f}_i + \boldsymbol{\varepsilon}_i, \tag{9.20}$$

where $\mathbf{f}_i = (L_i,\ S_i,\ C_i)'$ denotes the 3×1 vector of latent factors, $\mathbf{v}_i := \left(v_i^1, v_i^2, \ldots, v_i^J\right)'$ is the $J \times 1$ vector of level-dependent depth and

$$\mathbf{A} := \begin{pmatrix} 1 & \frac{1-e^{-\lambda\cdot 1}}{\lambda\cdot 1} & \frac{1-e^{-\lambda\cdot 1}}{\lambda\cdot 1} - e^{-\lambda\cdot 1} \\ 1 & \frac{1-e^{-\lambda_i\cdot 2}}{\lambda_i\cdot 2} & \frac{1-e^{-\lambda_i\cdot 2}}{\lambda_i\cdot 2} - e^{-\lambda_i\cdot 1} \\ \vdots & \vdots & \vdots \\ 1 & \frac{1-e^{-\lambda\cdot J}}{\lambda\cdot J} & \frac{1-e^{-\lambda\cdot J}}{\lambda\cdot J} - e^{-\lambda\cdot J} \end{pmatrix}$$

represents the $J \times 3$ matrix of factor loadings. Finally, for the $J \times 1$ vector of error terms ϵ_i one can assume

$$\boldsymbol{\varepsilon}_i := \left(\varepsilon_i^1,\ \varepsilon_i^2,\ \ldots, \varepsilon_i^J\right) \sim \text{i.i.d. } \mathcal{N}(\mathbf{0}, \boldsymbol{\Sigma})$$

with

$$\boldsymbol{\Sigma} = \operatorname{diag}\left\{\left(\sigma^{(1)}\right)^2, \left(\sigma^{(2)}\right)^2, \ldots, \left(\sigma^{(J)}\right)^2\right\}. \tag{9.21}$$

As suggested by Diebold and Li (2006), this model can be estimated for each observation i by regressing $\mathbf{v}_i$ on the matrix $\mathbf{A}$ using nonlinear least squares. If

[6] For ease of notation we omit the market-side specific index.

λ_i is fixed to a constant (as in typical term structure modelling, see, e.g., Diebold and Li 2006), the model can be even estimated by ordinary least squares.

Then, the dynamic behavior of limit order curves, captured by the latent factors $\mathbf{f}_i$, can be modelled using multivariate time series models. A natural choice is to model $\mathbf{f}_i$ as a VAR(Q) process,

$$\mathbf{f}_i = \boldsymbol{\mu} + \sum_{j=1}^{Q} \boldsymbol{\Phi}_j \mathbf{f}_{i-j} + \boldsymbol{\eta}_i, \tag{9.22}$$

where $\boldsymbol{\Phi}_j$ is a 3×3 parameter matrix, $\boldsymbol{\mu}$ denotes a 3×1 parameter vector, and the 3×1 zero mean error term vector $\boldsymbol{\eta}_i$ is assumed to be independent of $\boldsymbol{\varepsilon}_i$ with covariance matrix $\mathbf{H}$. Clearly, to allow for dynamic interdependencies between ask and bid order curves, it is suggested to model the dynamics of $\mathbf{f}_i^{ask}$ and $\mathbf{f}_i^{bid}$ in a joint six-dimensional VAR system.

The advantage of the Nelson–Siegel model is its flexibility while being very parsimonious and easy to estimate. Nevertheless, due its purely parametric form, its flexibility to capture any possible order book shape is naturally limited. Hence, more flexibility is provided by a model which captures the shapes of the order book curves in a fully nonparametric way.

Finally, note that the model does not necessarily ensure that order book curves are non-decreasing. This might cause problems for predictions of order book shapes in (extreme) areas where the book becomes very dispersed and data is sparse. In such a case, re-arrangement procedures for non-monotone estimates of monotone functions, as, e.g., proposed by Chernozhukov et al. (2009), might be used.

9.3.3 A Semiparametric Dynamic Factor Model

Härdle et al. (2009) propose modelling the order book curves using a dynamic semiparametric factor model (DSFM) as introduced by Park et al. (2009). The starting point is to assume that v_i^j follows an orthogonal L-factor model with $L \ll J$,

$$v_i^j = m_0 + \beta_{1i} m_1^j + \cdots + \beta_{Li} m_L^j + \varepsilon_i^j, \tag{9.23}$$

where m_0 is a constant, m_l^j is the j-level specific realization of a (time-invariant) factor loading with $m_l : \mathbb{R}^J \to \mathbb{R}$, β_{li} denotes the value of the corresponding factor at i and ε_i^j represents a white noise error term. Define $\mathbf{m} := (m_0, m_1 \ldots, m_L)'$ and $\mathbf{f}_i := (\beta_{0i}, \beta_{1i}, \ldots, \beta_{Li})'$ with $\beta_{0i} = 1$.[7] The DSFM is a generalization of the factor

[7]As both variables β_{li} and m_l are assumed to be unobservable, the definitions of factors and loadings are somewhat arbitrary. Here, we use a terminology which is consistent with the previous

model (9.23) and allows the factor loadings m_l to depend on explanatory variables, $\mathbf{z}_i^j$. Accordingly, it is given by

$$v_i^j = \sum_{l=0}^{L} \beta_{li} m_l(\mathbf{z}_i^j) + \varepsilon_i^j = \mathbf{f}_i' \mathbf{m}(\mathbf{z}_i^j) + \varepsilon_i^j, \tag{9.24}$$

where the processes $\mathbf{z}_i^j$, ε_i^j and $\mathbf{f}_i$ are assumed to be independent.

Park et al. (2009) propose estimating the unobservable factor loadings m_l using a series estimator. For $K \geq 1$, appropriate functions $\psi_k : [0,1]^d \to \mathbb{R}$, $k = 1, \ldots, K$, which are normalized such that $\int \psi_k^2(x)\,dx = 1$ holds, are selected. Park et al. (2009) select a tensor B-spline basis functions for ψ_k. Then, K denotes the number of knots used for the tensor B-spline functions and is interpretable as a bandwidth parameter. Accordingly, the loadings $\mathbf{m}$ are approximated by $\mathbf{B}\psi$, where $\mathbf{B} := (b_{lk}) \in \mathbb{R}^{(L+1)K}$ is a coefficient matrix, and $\psi := (\psi_1, \ldots, \psi_K)'$ denotes a vector of selected functions. Then, the first part in the right-hand side of (9.24) can be rewritten as

$$\mathbf{f}_i' \mathbf{m}(\mathbf{z}_i^j) = \sum_{l=0}^{L} \beta_{li} m_l(\mathbf{z}_i^j) = \sum_{l=0}^{L} \beta_{li} \sum_{k=1}^{K} b_{lk} \psi_k(\mathbf{z}_i^j) = \mathbf{f}_i' \mathbf{B} \psi(\mathbf{z}_i^j). \tag{9.25}$$

The coefficient matrix $\mathbf{B}$ and the factors $\mathbf{f}_i$ are estimated by least squares, i.e., $\hat{\mathbf{B}}$ and $\hat{\mathbf{f}}_i := \left(1, \hat{\phi}_{i,1}, \ldots, \hat{\phi}_{i,L}\right)'$ are defined as minimizers of the sum of squared residuals, $S(\mathbf{B}, \mathbf{f}_i)$,

$$\left(\hat{\mathbf{f}}_i, \hat{\mathbf{B}}\right) = \arg\min_{\mathbf{f}_i, \mathbf{B}} S(\mathbf{B}, \mathbf{f}_i) \tag{9.26}$$

$$= \arg\min_{\mathbf{f}_i, \mathbf{B}} \sum_{i=1}^{n} \sum_{j=1}^{J} \left\{ v_i^j - \mathbf{f}_i' \mathbf{B} \psi\left(\mathbf{z}_i^j\right) \right\}^2. \tag{9.27}$$

Park et al. (2009) suggest solving the minimization problem using a Newton–Raphson algorithm which is shown to converge to a solution at a geometric rate under weak regularity conditions. Moreover, they show that the difference between the estimated factors $\hat{\mathbf{f}}_i$ and the true factors $\mathbf{f}_i$ are asymptotically negligible. Consequently, it is justified to use in a second step multivariate time series specifications in order to model the dynamics of the factors.

The selection of the number of time-varying factors (L) and the number of knots K is performed by evaluating the proportion of explained variance (EV):

---

section and is commonly used in the literature of factor models. However, Härdle et al. (2009) use a converse terminology and define m_l as a factor and β_{li} as a factor loading.

$$EV(L) = 1 - \frac{\sum_{i=1}^{n}\sum_{j=1}^{J}\left\{v_i^j - \sum_{l=0}^{L}\hat{\beta}_{li}\widehat{m}_l\left(\mathbf{z}_i^j\right)\right\}^2}{\sum_{i=1}^{n}\sum_{j=1}^{J}\{v_i^j - \bar{v}\}^2}, \tag{9.28}$$

where $\bar{v}$ denotes the overall average.

Härdle et al. (2009) use this approach to model the limit order book curves of four stocks at the Australian Stock Exchange based on a 5 min frequency. They choose v_i^j as the seasonally adjusted order book depth, cumulated up to level j, observed at time interval i. Accordingly, as explanatory variables driving the loadings $\mathbf{m}$, they use the vector of relative distances between the corresponding limit price levels $j = 1, \ldots, J$, to the current mid-quote at the market. Hence, by using these so-called "relative order prices" instead of an absolute price grid, the order curves are interpreted as a relationship between the cumulated pending order volume and the order prices' relative distance to the current mid-quote level. In this context, Härdle et al. (2009) choose the knots underlying the tensor B-spline function such that they cover the complete range of realizations of the relative order prices.

Table 9.3 reproduces results from Härdle et al. (2009) and shows the model's explained variation for the stocks BHP, NAB, MIM and WOW traded at the ASX during July and August 2002. We observe that approximately 95% of the explained variation in order curves can be explained using two (time-varying) factors, whereas the marginal contribution of a potentially third factor is only very small. Consequently, a factor model involving one constant and two time-varying factors (similarly to the Nelson–Siegel approach illustrated above) is sufficient to capture the order book dynamics.

The goodness-of-fit of the model is illustrated by Fig. 9.3 giving snapshots of the estimated order curves on an arbitrary day. It turns out that the model captures the true order curves remarkably well. This is particularly true for price levels close to the best ask and bid quotes. Nevertheless, slight deviations are observed for price levels deeply in the book.

Figure 9.4 shows the estimates of the first and second factor loading $\widehat{m}_1$ and $\widehat{m}_2$ in dependence of the relative order price grids. The first factor obviously captures

Table 9.3 Explained variance of DSFM estimates of ask and bid order book curves depending on relative order prices based on different number of factors L and 20 B-spline knots. Based on trading at the ASX for the stocks, BHP, NAB, MIM and WOW, July 8 to August 16, 2002. Reproduced from Härdle et al. (2009)

L	BID				ASK			
	BHP	NAB	MIM	WOW	BHP	NAB	MIM	WOW
1	0 .925	0.934	0.990	0.916	0.916	0.909	0.946	0.938
2	0 .964	0.965	0.996	0.975	0.941	0.948	0.953	0.959
3	0 .971	0.976	0.996	0.981	0.941	0.961	0.949	0.964

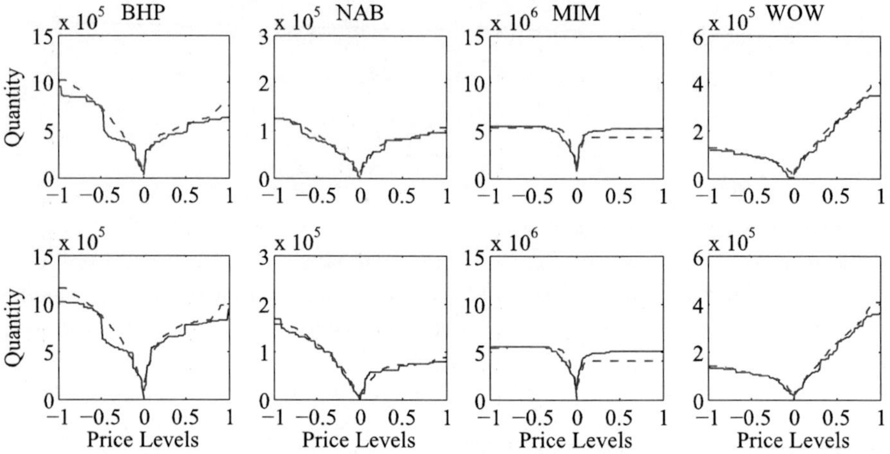

Fig. 9.3 Estimated (*dashed*) and observed (*solid*) limit order curves depending on relative order price levels for the stocks BHP, NAB, MIM and WOW traded at the ASX on July 8, 2002, 11:00 (*upper panel*) and 13:00 (*lower panel*). Reproduced from Härdle (2009)

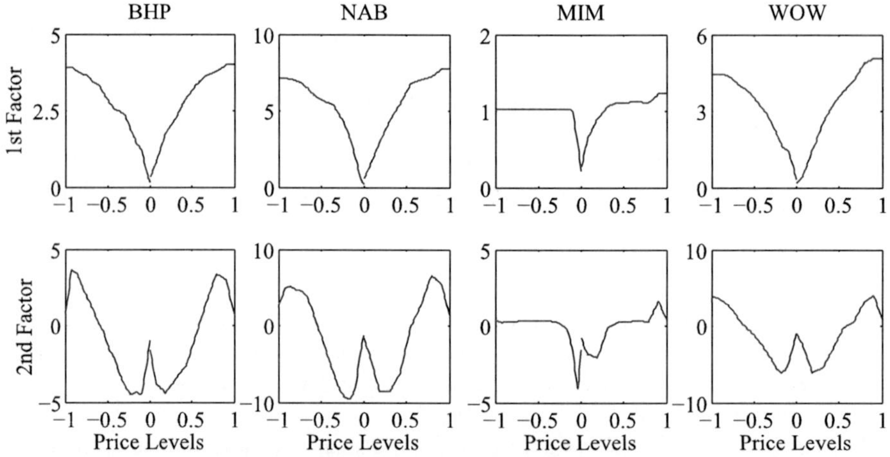

Fig. 9.4 Estimated first and second factor loadings of the limit order curves depending on relative order price levels for the stocks BHP, NAB, MIM and WOW traded at the ASX, July 8 to August 16. Reproduced from Härdle (2009)

the order book slope which is naturally associated with the average price impact on a particular side of the market. The second loading captures order curve fluctuations around the overall slope. Correspondingly, the second factor can be associated with a "curvature" factor in the spirit of Nelson and Siegel (1987). The shape of this loading reveals that the curve's curvature is particularly distinct for levels close to the best quotes and for levels very deep in the book where the curve seems to spread out. The shapes of the estimated loadings are remarkably similar for all stocks

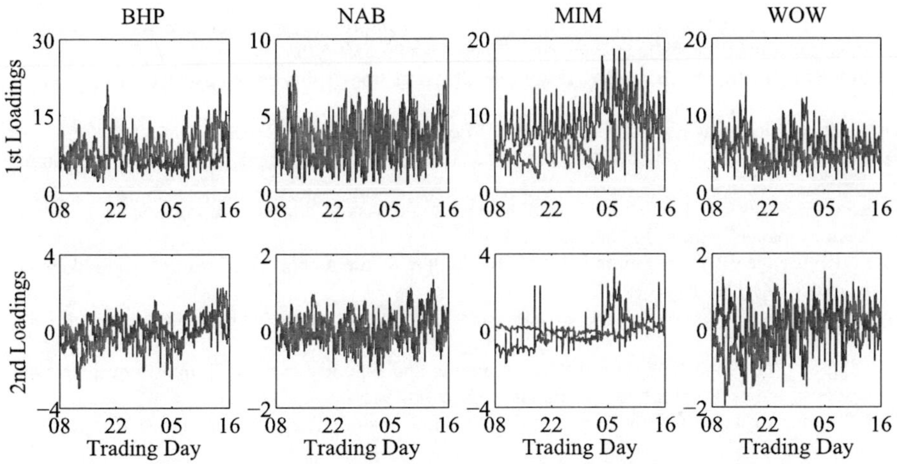

Fig. 9.5 Time series plots of the estimated first and second factors of the limit order curves depending on relative order price levels for the stocks BHP, NAB, MIM and WOW traded at the ASX, July 8 to August 16. Underlying aggregation level: 5 min. Reproduced from Härdle (2009)

except for MIM. As illustrated above, MIM is exceptional in the sense that order book depth is strongly concentrated at the best quotes.

The time series dynamics of the corresponding factors $\hat{\beta}_i^b$ and $\hat{\beta}_i^a$ are shown in Fig. 9.5. It is shown that the factors strongly vary over time reflecting time variations in the shape of the book. The series reveal clustering structures indicating a relatively high persistence in the processes. This result is not very surprising given the fact that order book inventories do not change too severely over short time horizons.

References

Amihud Y (2002) Illiquidity and stock returns: cross section and time-series effects. J Finan Markets 5:31–56

Bangia A, Diebold F, Schuermann T, Stroughair J (2008) Modeling liquidity risk, with implications for traditional market risk measurement and management. Discussion Paper FIN-99-062, New York University

Bessembinder H (2000) Tick size, spreads, and liquidity: an analysis of Nasdaq securities trading near ten dollars. J Financial Intermediation 9(3):213–239

Biais B, Hillion P, Spatt C (1995) An empirical analysis of the limit order book and the order flow in the Paris Bourse. J Finance 50:1655–1689

Black F (1971) Toward a fully automated exchange, part I, II. Financial Analysts J 27:29–34

Brennan MJ, Chordia T, Subrahmanyam A, Tong Q (2009) Sell-side liquidity and the cross-section of expected returns. Working Paper SSRN, http://ssrn.com/abstract=1396328

Brownlees C, Cipollini F, Gallo GM (2011) Intra-daily volume modeling and prediction for algorithmic trading. J Financ Econom 9:489–518

Chan LKC, Lakonishok J (1995) The behaviour of stock prices around institutional trades. J Finance 4:1147–1173

Chernozhukov V, Fernandez-Val I, Galichon A (2009) Improving point and interval estimates of monotone functions by rearrangement. Biometrika 96:559–575
Conroy RM, Harris RS, Benet BA (1990) The effects of stock splits on bid-ask spreads. J Finance 45:1285–1295
Demsetz H (1968) The cost of transacting. Q J Econ 82:33–53
Diebold F, Li C (2006) Forecasting the term structure of government bond yields. J Econom 130:337–364
Elyasiani E, Hauser S, Lauterbach B (2000) Market response to liquidity improvements: evidence from exchange listings. Financ Rev 35:1–14
Engle RF, Lange J (2001) Predicting VNET: a model of the dynamics of market depth. J Finan Markets 2(4):113–142
Engle RF, Patton A (2004) Impact of trades in an error-correction model of quote prices. J Financ Markets 7:1–25
Fleming M, Remolona E-M (1999) Price formation and liquidity in the U.S. treasury market: the response to public information. J Finance 54:1901–1915
Giot P, Grammig J (2006) How large is liquidity risk in an automated auction market? Empir Econ 30:867–887
Glosten LR (1994) Is the electronic open limit order book inevitable. J Finance 49:1127–1161
Glosten LR, Harris LE (1988) Estimating the components of the bid/ask spread. J Finan Econ 21:123–142
Gouriéroux C, Jasiak J, LeFol G (1999) Intra-day market activity. J Finan Markets 2:193–226
Goyenko RY, Holden CW, Trzcinka CA (2009) Do liquidity measures measure liquidity? J Finan Econ 92:153–181
Greene J, Smart S (1999) Liquidity provision and noise trading: evidence from the "investment dartboard" column. J Finance 54:1885–1899
Härdle WK, Hautsch N, Mihoci A (2009) Modelling and forecasting liquidity supply using semiparametric factor dynamics. Discussion Paper 2009-44, Collaborative Research Center 649 "Economic Risk", Humboldt-Universität zu Berlin
Harris L (2003) Trading & exchanges – market microstructure for practitioners. Oxford University Press, Oxford
Hasbrouck J (2007) Empirical market microstructure. Oxford University Press, Oxford
Hasbrouck J (2009) Trading costs and returns for U.S. equities: estimating effective costs from daily data. J Finance 64:1445–1477
Hautsch N (2003) Assessing the risk of liquidity suppliers on the basis of excess demand intensities. J Financ Econom 1:189–215
Hautsch N, Huang R (2009) The market impact of a limit order. Discussion Paper 2009/23, Collaborative Research Center 649 "Economic Risk", Humboldt-Universität zu Berlin
Keim DB, Madhavan A (1996) The upstairs market for large-block transactions: analysis and measurement of price effects. Rev Financ Stud 9:1–36
Keynes J (1930) Treatise on money. MacMillan
Kyle AS (1985) Continuous auctions and insider trading. Econometrica 53(6):1315–1335
Lo I, Sapp S (2006) A structural error-correction model of best prices and depths in the foreign exchange limit order market. Working paper, Bank of Canada
Nelson C, Siegel A (1987) Parsimonious modeling of yield curves. J Bus 60:473–489
Park B, Mammen E, Härdle W, Borak S (2009) Time series modelling with semiparametric factor dynamics. J Am Stat Assoc 104(485):284–298
Potters M, Bouchaud J (2003) More statistical properties of order books and price impact. Physica A 324:133–140
Roll R (1984) A simple implicit measure of the effective bid-ask spread in an efficient market. J Finance 39:1127–1139

Chapter 10
Semiparametric Dynamic Proportional Hazard Models

Proportional hazard (PH) models as introduced in Chap. 4 have a long history in labor economics and serve as a workhorse for the modelling of unemployment spells. Here, we discuss dynamic extensions which can be seen as the direct counterpart to the class of dynamic accelerated failure time (AFT) models to which the ACD model belongs to. As discussed in Chap. 4, a PH model can be estimated in different ways. One possibility is to adopt a fully parametric approach leading to a complete parameterization of the hazard function. Such a model is consistently estimated by maximum likelihood given that the chosen parameterization is correct. A further possibility is to refer to the results of Cox (1975) and to consistently estimate the parameter vector $\boldsymbol{\gamma}$ either by a partial likelihood approach or in a semiparametric way. In this framework, the model requires no specification of the baseline hazard $h_0(x)$ but of the integrated baseline hazard $H_0(x) := \int_0^x h_0(s)ds$. Then, $h_0(x)$ can be estimated semiparametrically or non-parametrically.[1]

In this chapter, we focus on the latter approach and consider a dynamic extension of *semiparametric* PH models. This class of models is introduced by Gerhard and Hautsch (2007) under the name *semiparametric autoregressive conditional proportional hazard (SACPH) model.* The basic idea is to specify a dynamic process directly for the *integrated* hazard function $H(x_i) := \int_0^{x_i} h(s)ds$. We illustrate that the categorization approach proposed by Han and Hausman (1990) is a valuable starting point for such a specification leading to a model that allows for a consistent estimation of the model parameters without requiring explicit parametric forms of the baseline hazard. Moreover, as in the non-dynamic case, discrete points of the baseline survivor function can be estimated simultaneously with the dynamic parameters.

A further strength of a dynamic model in terms of the integrated hazard function becomes evident in the context of censoring structures. Censoring occurs due to non-trading periods, when the exact end of a spell and, correspondingly, the

[1]See, for example, Han and Hausman (1990), Meyer (1990), Horowitz and Neumann (1987) or Horowitz (1996, 1999).

N. Hautsch, *Econometrics of Financial High-Frequency Data*,
DOI 10.1007/978-3-642-21925-2_10,

begin of the next spell, cannot be observed directly. Because of the relationship between the integrated hazard and the conditional survivor function (see Chap. 4), an autoregressive model for the integrated hazard rate is a natural way to account for censoring mechanisms in a dynamic framework.

In Sect. 10.1, we discuss general challenges when specifying dynamics in semi-parametric PH models and motivate the idea behind the SACPH model. Section 10.2 introduces the SACPH model and illustrates ML estimation. Theoretical properties are discussed in Sect. 10.3. Here, we focus on the derivation of the theoretical ACF, as well as on the effects of the discretization approach on the estimation quality of the model. Section 10.4 considers extensions of the basic SACPH model, where we discuss regime-switching baseline hazard functions, unobserved heterogeneity and censoring effects. Diagnostic tests for the model are given in Sect. 10.5. Section 10.6 illustrates the application of the SACPH approach for modelling price change volatilities based on censored price durations.

10.1 Dynamic Integrated Hazard Processes

Recall the definition of the standard PH model, as given by (4.32) in Chap. 4,

$$h(x_i; \mathbf{z}_{i-1}) = h_0(x_i) \exp(-\mathbf{z}_{i-1}' \boldsymbol{\gamma}).$$

We formulate the PH model in terms of a log-linear representation of the baseline hazard function, as illustrated in (4.37),

$$\ln H_0(x_i) = \phi(\mathcal{F}_{i-1}; \boldsymbol{\theta}) + \varepsilon_i^*, \tag{10.1}$$

where $\phi(\mathcal{F}_{i-1}; \boldsymbol{\theta})$ denotes a possibly nonlinear function depending of past durations, marks and a parameter vector $\boldsymbol{\theta}$. The error ε_i^* follows per construction an i.i.d. standard extreme value distribution. Hence, we obtain a regression model based on a standard extreme value distributed error term as a natural starting point for a dynamic extension.

In order to simplify the following discussion, define

$$x_i^* := \ln H_0(x_i), \tag{10.2}$$

where $H_0(x_i)$ denotes the integrated baseline hazard function as defined in Chap. 4. In the case of a non-specified baseline hazard $h_0(x_i)$, the transformation from x_i to x_i^* is unknown, and thus, x_i^* is interpreted as a latent variable. Then, the PH model can be represented in more general form as a latent variable model

$$x_i^* = \phi(\mathcal{F}_{i-1}; \boldsymbol{\theta}) + \varepsilon_i^*. \tag{10.3}$$

Using the terminology of Cox (1981), there are two ways to incorporate a dynamic in a latent variable model. On the one hand, *observation driven* models are characterized by a conditional mean function which is measurable with respect to some *observable* information set. On the other hand, in *parameter driven* models, the conditional mean function is measurable with respect to some *unobservable* information set $\mathcal{F}_i^* = \sigma(x_i^*, x_{i-1}^*, \ldots, x_1^*, \mathbf{z}_i, \mathbf{z}_{i-1}, \ldots, \mathbf{z}_1)$. At first glance, the distinction seems point-less since (10.2) is a one-to-one mapping from x_i to x_i^* and thus, the information sets coincide. However, the importance of the distinction will become clear once both approaches have been outlined in greater detail.

An observation driven dynamic model of x_i^* can be depicted as in (10.3). The estimation of this type of model uses the partial likelihood procedure proposed by Cox (1975), which has been shown by Oakes and Cui (1994) to be available, even in the dynamic case. Therefore, the two-step estimation of the parameters $\boldsymbol{\theta}$ and the baseline intensity h_0 is still possible. A simple example would be to specify $\phi(\cdot)$ in terms of lagged durations (see, e.g., Hautsch 1999). However, it turns out, that the dynamic properties of such models are non-trivial and that in most applications, AR type structures are not sufficient to capture the persistence in financial duration processes. The inclusion of a MA term is not easy in the given context and requires to build the dynamic on variables which are unobservable leading directly to a parameter driven model.

Hence, an alternative is to specify the model dynamics directly in terms of the log integrated baseline hazard function leading to a *parameter driven* dynamic PH model of the form

$$x_i^* = \phi(\mathcal{F}_{t_{i-1}}^*; \boldsymbol{\theta}) + \varepsilon_i^*. \tag{10.4}$$

In this context, two main problems have to be resolved. First, since this specification involves dynamics in terms of the log integrated hazard function, the partial likelihood approach proposed by Cox (1975) is not available. This becomes obvious, if one considers an AR(1) process for x_i^* as a special case of (10.4),

$$x_i^* = \alpha x_{i-1}^* + \varepsilon_i^* = \alpha \ln H_0(x_{i-1}) + \varepsilon_i^*. \tag{10.5}$$

Then, x_i^* is a function of the baseline hazard h_0 and the latter is left unspecified. Hence, a separate estimation of α and h_0 is not possible. Second, a further challenge is that x_i^* itself is not observable directly. As dynamics attached to a latent variable have to be integrated out, maximum likelihood estimation involves an n-fold integral, see, e.g., Chap. 6.

To circumvent these problems, Gerhard and Hautsch (2007) propose a model which is based on the categorization framework as discussed in Han and Hausman (1990). This approach is a valuable alternative to the partial likelihood approach and allows for a simultaneous estimation of discrete points of the baseline survivor function and of the parameter vector $\boldsymbol{\gamma}$. A further advantage of a categorization approach is that x_i^* can be observed through a threshold function. This property is exploited by specifying a model based on an observation driven dynamic which does not necessitate the use of extensive simulation methods but allows for straightforward ML estimation. As it will be demonstrated in more detail in the next subsection, the SACPH model embodies characteristics of both observation driven specifications as well as parameter driven models.

10.2 The Semiparametric ACPH Model

By adopting the discretization approach as discussed in Sect. 4.2.2.1, we recall the definition of μ_k^* as the value of the latent variable x_i^* at the boundary $\bar{x}_k$, (4.40),

$$\mu_k^* := \ln H_0(\bar{x}_k), \qquad k = 1, \ldots, K-1.$$

Define in the following

$$x_i^d := k \cdot \mathbb{1}_{\{\bar{x}_{k-1} < x_i \leq \bar{x}_k\}} \tag{10.6}$$

as an ordered integer variable indicating the observed category. Moreover, let $\mathcal{F}_i^d := \sigma(x_i^d, x_{i-1}^d, \ldots, x_1^d, \mathbf{z}_i, \mathbf{z}_{i-1}, \ldots, \mathbf{z}_1)$ be the information set generated by the *categorized* durations. This is a standard approach in the analysis of grouped durations, see, e.g., Thompson (1977), Prentice and Gloeckler (1978), Meyer (1990), Kiefer (1988), Han and Hausman (1990), Sueyoshi (1995) or Romeo (1999). Clearly, the discretization approach implies some loss of information since $\mathcal{F}_i^d \subset \mathcal{F}_i$.

To avoid the computational challenges of a pure parameter driven dynamic, Gerhard and Hautsch (2007) suggest specifying the dynamic on the basis of conditional expectations of the error ε_i^*,

$$e_i := \mathbb{E}\left[\varepsilon_i^* | \mathcal{F}_i^d\right]. \tag{10.7}$$

The conditional expectation e_i relates to the concept of generalized errors, see Gouriéroux et al. (1987), or Bayesian errors, see Albert and Chib (1995).

The resulting SACPH(P,Q) model is given by

$$x_i^* = \phi_i + \varepsilon_i^*, \tag{10.8}$$

where $\phi_i := \phi(\mathcal{F}_{i-1}; \boldsymbol{\theta})$ is defined through a recursion, conditioned on an initial ϕ_0,

$$\phi_i = \sum_{j=1}^{P} \alpha_j (\phi_{i-j} + e_{i-j}) + \sum_{j=1}^{Q} \beta_j e_{i-j}. \tag{10.9}$$

The computation of ϕ_i allows us to use a conditioning approach which exploits the observation of ϕ_{i-1}, and thus prevents us from computing high-dimensional integrals. The SACPH model is built on an ARMA structure based on the conditional expectation of the latent variable given the *observable* categorized durations. This specification poses a dynamic extension of the approach of Han and Hausman (1990) since the autoregressive structure is built on values of a transformation of h_0 that is assumed to be constant during the particular categories. The covariance stationarity conditions for the SACPH(P,Q) model correspond to the stationarity conditions of a standard ARMA model and are given by $\sum_j \alpha_j < 1$.

The SACPH model can be rewritten in terms of the hazard function as

$$h(x_i; \mathcal{F}_{i-1}^d) = h_0(x_i) \exp(-\phi_{i-1}). \tag{10.10}$$

The dynamic structure of the SACPH model given by (10.8) and (10.9) relies on a recursive updating structure. For illustration, consider a special case of the SACPH model when h_0 is known. In this case, the transformation from x_i to x_i^* is replaced by a measurable one-to-one function, so that $\mathcal{F}_i^d = \mathcal{F}_i^* = \mathcal{F}_i$. Then, $e_i = \varepsilon_i^*$ and $\phi_i = \mathbb{E}[x_i^* | \mathcal{F}_{i-1}^*] - \mathbb{E}[\varepsilon_i^*]$, and thus, the SACPH model coincides with a standard ARMA process of the log integrated hazard function. Assume in the following a Weibull specification for the baseline hazard with $\lambda_0(x) = ax^{a-1}$. Then, we obtain a Weibull ACPH(1,1) (WACPH(1,1)) model of the form

$$a \ln x_i = \phi_i + \varepsilon_i^*, \tag{10.11}$$

where $\phi_i = \alpha(\phi_{i-1} + \varepsilon_{i-1}^*) + \beta\varepsilon_{i-1}^*$. The WACPH(1,1) model can be written as an ARMA model for log durations based on a standard extreme value distribution, i.e.,

$$\ln x_i = \alpha \ln x_{i-1} + \frac{\beta}{a}\varepsilon_{i-1}^* + \frac{\varepsilon_i^*}{a}. \tag{10.12}$$

However, in the general case where h_0 is unknown, the SACPH model does *not* exactly correspond to an ARMA model for log durations. In this case, the parameters α_j and β_j can be interpreted only as approximations to the parameters $\tilde{\alpha}_j$ and $\tilde{\beta}_j$ in the ARMA model of the form

$$x_i^* = \sum_{j=1}^{P} \tilde{\alpha}_j x_{i-1}^* + \sum_{j=1}^{Q} \tilde{\beta}_j \varepsilon_{i-j}^* + \varepsilon_i^*. \tag{10.13}$$

The exactness of this approximation obviously increases with the fineness of the chosen categorization. In the limiting case for $K \to \infty$, h_0 is quasi-observable, and the SACPH model converges to (10.13) with parameters $\alpha_j = \tilde{\alpha}_j$ and $\beta_j = \tilde{\beta}_j$.

The proposed dynamic has the advantage that a computationally simple maximum likelihood estimator of the dynamic parameters α and β and the parameters of the baseline hazard approximation μ^* is directly available.

The computation of the log likelihood function requires to compute the generalized errors e_i by

$$e_i := \mathbb{E}\left[\varepsilon_i^* | \mathcal{F}_i^d\right] = \mathbb{E}\left[\varepsilon_i^* | x_i^d, \phi_i\right]$$
$$= \begin{cases} \dfrac{\kappa(-\infty, \nu_{i,1})}{1 - S_{\varepsilon^*}(\nu_{i,1})} & \text{if } \bar{x}_i = 1, \\ \dfrac{\kappa(\nu_{i,k-1}, \nu_{i,k})}{S_{\varepsilon^*}(\nu_{i,k-1}) - S_{\varepsilon^*}(\nu_{i,k})} & \text{if } \bar{x}_i \in \{2, \ldots, K-1\}, \\ \dfrac{\kappa(\nu_{i,K-1}, \infty)}{S_{\varepsilon^*}(\nu_{i,K-1})} & \text{if } \bar{x}_i = K, \end{cases} \tag{10.14}$$

where $\nu_{i,k} := \mu_k^* - \phi_i$, $\kappa(s_1, s_2) := \int_{s_1}^{s_2} u f_{\varepsilon^*}(u) du$ and $f_{\varepsilon^*}(\cdot)$ and $S_{\varepsilon^*}(\cdot)$ denote the p.d.f. and survivor function of the standard extreme value distribution, respectively.[2]

Since the observation driven dynamic enables us to use the standard prediction error decomposition, the likelihood is evaluated in a straightforward iterative fashion: The function ϕ_i is initialized with its unconditional expectation $\phi_0 := \mathbb{E}[\phi_i]$. Then, based on the recursion, (10.9), as well as the definition of the generalized errors, (10.14), the likelihood contribution of observation i given the observation rule (10.6) is computed as

$$\mathbb{P}r\left[x_i^d = k \,|\mathcal{F}_{i-1}^d\right] = \begin{cases} 1 - S_{\varepsilon^*}(\mu_1^* - \phi_i) & \text{if } x_i^d = 1, \\ S_{\varepsilon^*}(\mu_1^* - \phi_i) - S_{\varepsilon^*}(\mu_2^* - \phi_i) & \text{if } x_i^d = 2, \\ \vdots & \\ S_{\varepsilon^*}(\mu_{K-1}^* - \phi_i) & \text{if } x_i^d = K. \end{cases} \tag{10.15}$$

By denoting the matrix of the underlying data as $\mathbf{Y}$, the log likelihood function is given by

$$\ln \mathcal{L}(\mathbf{Y}; \boldsymbol{\theta}) = \sum_{i=1}^{n} \sum_{k=1}^{K} \mathbb{1}_{\{x_i^d = k\}} \ln \mathbb{P}r\left[x_i^d = k \,|\mathcal{F}_{i-1}^d\right]. \tag{10.16}$$

As illustrated in Chap. 4, the baseline survivor function and (discrete) baseline hazard function is computed based on (4.41) to (4.43).

10.3 Properties of the Semiparametric ACPH Model

10.3.1 Autocorrelation Structure

Since the dynamics of the model are incorporated in a *latent* structure, an important question is how the autoregressive parameters can be interpreted in terms of the

---

[2]For an extended discussion of generalized errors in the context of non-dynamic models, see Gouriéroux et al. (1987).

observable durations. The main challenge is that no closed form expression for the generalized errors and the p.d.f of the latent variable x_i^* can be given, so that one needs to resort to numerical methods to evaluate the model's ACF.

Gerhard and Hautsch (2007) conduct a simulation study and simulate SACPH processes, (10.8) and (10.9), based on exogenously given categories which are associated with predetermined quantiles of the unconditional distribution of x_i^*. The resulting empirical ACF of the SACPH model, (10.8)-(10.9), as well as of the ARMA process in (10.13) is computed for $\alpha = \tilde{\alpha}$ and $\beta = \tilde{\beta}$. This process corresponds to the limiting case of the SACPH process when $K \rightarrow \infty$, i.e., when h_0 is known, and can be interpreted as benchmark process. Moreover, the ACF of the resulting *observable* durations implied by WSACPH models with shape parameter $a = 0.5$ is computed.

Figures 10.1 and 10.2 reproduce the individual autocorrelation patterns computed by Gerhard and Hautsch (2007) for SACPH(1,0) dynamics with parameters $\alpha = 0.5$ and $\alpha = 0.9$, respectively, based on $n = 100{,}000$ drawings. Each process is simulated under three different groupings for the durations. The first grouping is based on only two categories, in particular below and above the 0.5-quantile of x_i^*. Correspondingly, the other categorizations are based on the 0.25-, 0.5- and

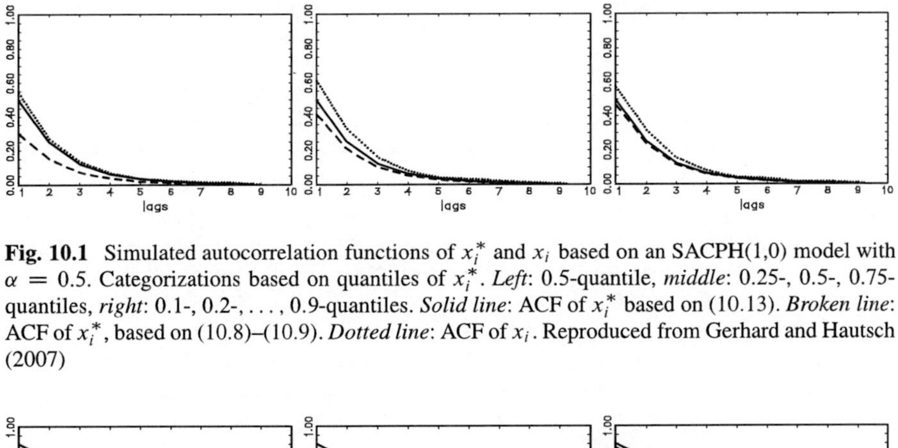

Fig. 10.1 Simulated autocorrelation functions of x_i^* and x_i based on an SACPH(1,0) model with $\alpha = 0.5$. Categorizations based on quantiles of x_i^*. *Left*: 0.5-quantile, *middle*: 0.25-, 0.5-, 0.75-quantiles, *right*: 0.1-, 0.2-, ..., 0.9-quantiles. *Solid line*: ACF of x_i^* based on (10.13). *Broken line*: ACF of x_i^*, based on (10.8)–(10.9). *Dotted line*: ACF of x_i. Reproduced from Gerhard and Hautsch (2007)

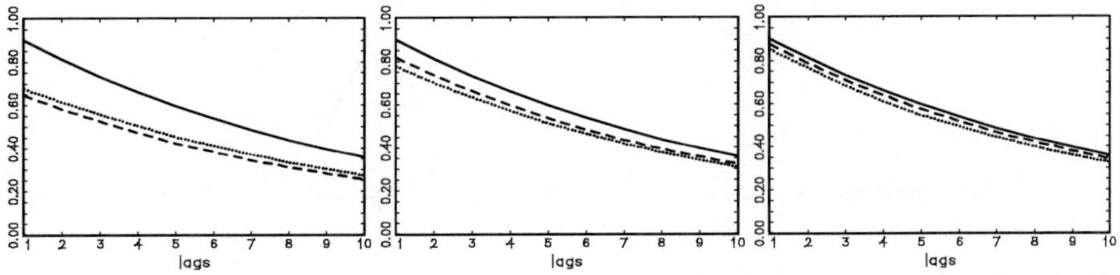

Fig. 10.2 Simulated autocorrelation functions of x_i^* and x_i based on an SACPH(1,0) model with $\alpha = 0.9$. Categorizations based on quantiles of x_i^*. *Left*: 0.5-quantile, *middle*: 0.25-, 0.5-, 0.75-quantiles, *right*: 0.1-, 0.2-, ..., 0.9-quantiles. *Solid line*: ACF of x_i^* based on (10.13). *Broken line*: ACF of x_i^*, based on (10.8)–(10.9). *Dotted line*: ACF of x_i. Reproduced from Gerhard and Hautsch (2007)

0.75-quantiles, as well as the 0.1-, 0.2-,. . .,0.9-quantiles. The former categorization is the worst possible approximation of the true baseline hazard h_0 while the latter covers a more realistic case using a moderate number of thresholds. We observe a close relationship between the different autocorrelation functions which is rather driven by the chosen categorization than by the strength of the serial dependence. While for the two-category-grouping clear differences in the autocorrelation functions are observed, quite similar shapes are revealed for the finer categorizations. It is shown that the ACF implied by the SACPH model converges towards the ACF of the pure ARMA model for the log integrated hazard, (10.13), when the categorization becomes finer. Moreover, there is a quite close relationship between the ACF of the latent variable x_i^* and the observable variable x_i.

These results are confirmed based on SACPH(1,1) processes (Figs. 10.3 and 10.4). For a sufficiently fine categorization, the autocorrelation functions of the latent and the observable processes are quite similar. Hence, it can be concluded that the ACF implied by the estimated ARMA coefficients of an SACPH model is a good proxy for the ACF of a pure ARMA process for $\ln H_0(x_i)$ and for the ACF of the observed durations.

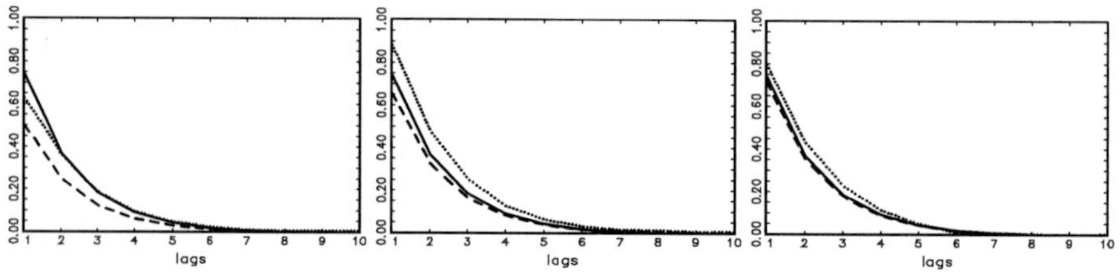

Fig. 10.3 Simulated autocorrelation functions of x_i^* and x_i based on an SACPH(1,1) model with $\alpha = 0.5$ and $\beta = 0.7$. Categorizations based on quantiles of x_i^*. *Left*: 0.5-quantile, *middle*: 0.25-, 0.5-, 0.75-quantiles, *right*: 0.1-, 0.2-, . . . , 0.9-quantiles. *Solid line*: ACF of x_i^* based on (10.13). *Broken line*: ACF of x_i^*, based on (10.8)-(10.9). *Dotted line*: ACF of x_i. Reproduced from Gerhard and Hautsch (2007)

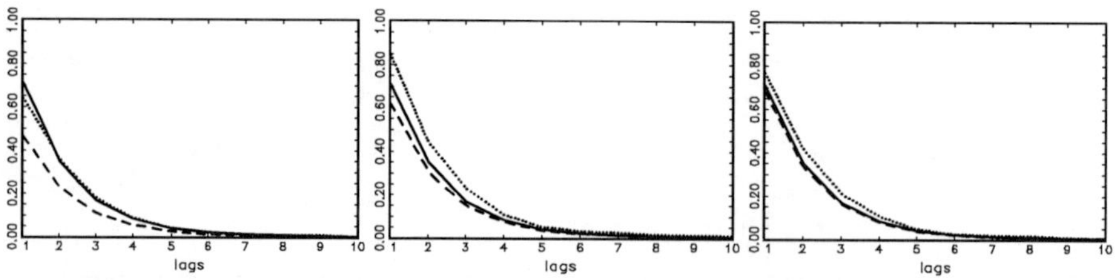

Fig. 10.4 Simulated autocorrelation functions of x_i^* and x_i based on an SACPH(1,1) model with $\alpha = 0.5$ and $\beta = 0.5$. Categorizations based on quantiles of x_i^*. *Left*: 0.5-quantile, *middle*: 0.25-, 0.5-, 0.75-quantiles, *right*: 0.1-, 0.2-, . . . , 0.9-quantiles. *Solid line*: ACF of x_i^* based on (10.13). *Broken line*: ACF of x_i^*, based on (10.8)-(10.9). *Dotted line*: ACF of x_i. Reproduced from Gerhard and Hautsch (2007)

10.3.2 Estimation Quality

Gerhard and Hautsch (2007) perform a Monte Carlo study to evaluate the small sample bias incurred by the discretization approach for different sample sizes and two different categorizations based on $K = 2$ and $K = 10$ categories. The categories are chosen in accordance with the 0.5-quantile, as well as the 0.1-, ..., 0.9-quantile of x_i^*. The two-category-model is replicated for two sample sizes $n = 50$ and $n = 500$. The model with more thresholds is only estimated for a small sample size $n = 50$. This set-up allows us to compare the improvement achieved by increasing the number of observations vs. the benefit of a better approximation of the baseline hazard function. Parameter estimations are based on the SACPH(1,0) and SACPH(0,1) model. Since the focus is on the bias of the dynamic parameters, the threshold parameters are fixed to their true values. A range of parameter values for α and β are covered in the simulations, concisely, $\alpha, \beta \in \mathcal{Q} = \{-0.9, -0.8, \ldots, 0.8, 0.9\}$ providing $n_i^{MC} = 1{,}000$, $i \in \mathcal{Q}$, replications for each value. The errors ε_i^* are drawn from the standard extreme value distribution as in the assumed DGP. Overall results for all $n^{MC} = 19{,}000$ replications are reported in Table 10.1. It provides descriptive statistics of the difference between the true parameters and the corresponding estimates, $\alpha^{(i)} - \hat{\alpha}^{(i)}$, and $\beta^{(i)} - \hat{\beta}^{(i)}$, for $i = 1, \ldots, n^{MC}$. Though we aggregate over all parameter values, the small sample properties match the expectation build from asymptotic theory, i.e., the variance decreases over an increasing sample size. The results indicate that even a moderately sized sample of 50 observations is sufficient to obtain reasonable results. Particularly for the SACPH(1,0) model, the asymptotic properties seem to hold quite nicely. To gain more insight into the consequences the discretization grid of the durations bears for the estimation, the results of the Monte Carlo experiment are scrutinized with respect to the parameters of the model, α and β. Simulation results for each of the 19 considered values of the true parameter in the DGP are illustrated in the Box plots reported in Figs. 10.5 through 10.7. Overall, the results indicate that the bias incurred for an SACPH(1,0) based on $K = 2$ categories is reduced considerably once a higher parameterized model based on $K = 10$ categories is employed. Furthermore, for a reasonable sample size ($n = 500$), even for the two-category-model the performance of the estimator is quite satisfying over all parameter values considered.

Table 10.1 Results of a Monte Carlo study based on SACPH(P,Q) models with K categories and n observations. Shown diagnostics: bias, mean squared error (MSE) and mean absolute error (MAE)

P	Q	K	n	bias	MSE	MAE
1	0	2	50	−0.006	0.029	0.118
1	0	10	50	0.004	0.009	0.073
1	0	2	500	0.005	0.002	0.037
0	1	2	50	−0.007	0.074	0.202
0	1	10	50	−0.005	0.018	0.093
0	1	2	500	0.015	0.020	0.095

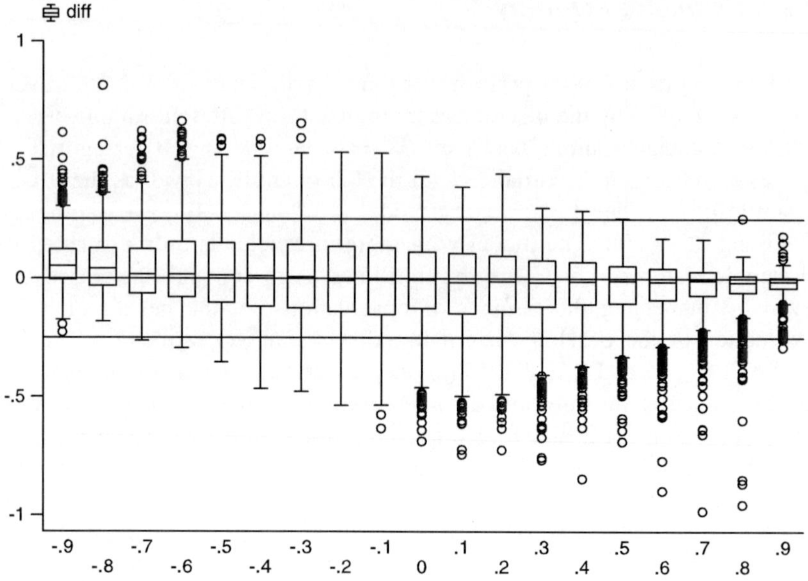

Fig. 10.5 Box plots of $\alpha^{(i)} - \hat{\alpha}^{(i)}$ for 19 values of the parameter $\alpha^{(i)}$ in a Monte Carlo study. SACPH(1,0) model, $K = 2, n = 50$. Reproduced from Gerhard and Hautsch (2007)

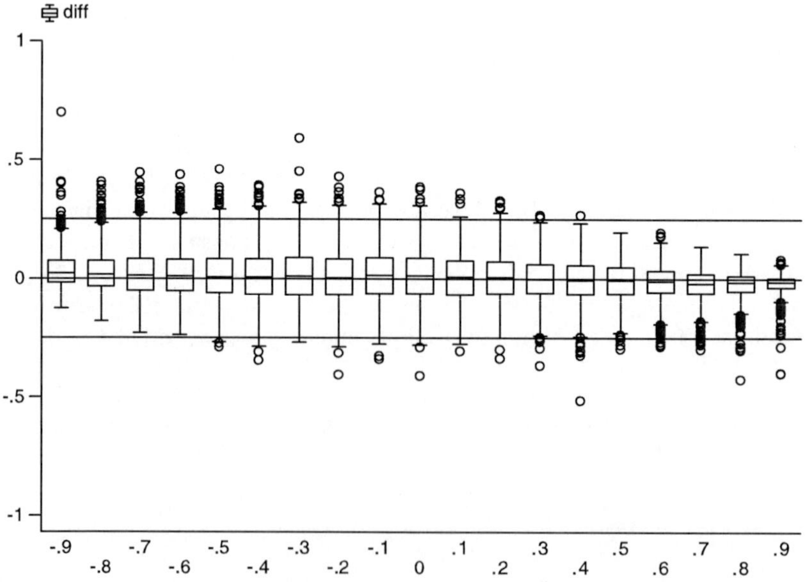

Fig. 10.6 Box plots of $\alpha^{(i)} - \hat{\alpha}^{(i)}$ for 19 values of the parameter $\alpha^{(i)}$ in a Monte Carlo study. SACPH(1,0) model, $K = 10, n = 50$. Reproduced from Gerhard and Hautsch (2007)

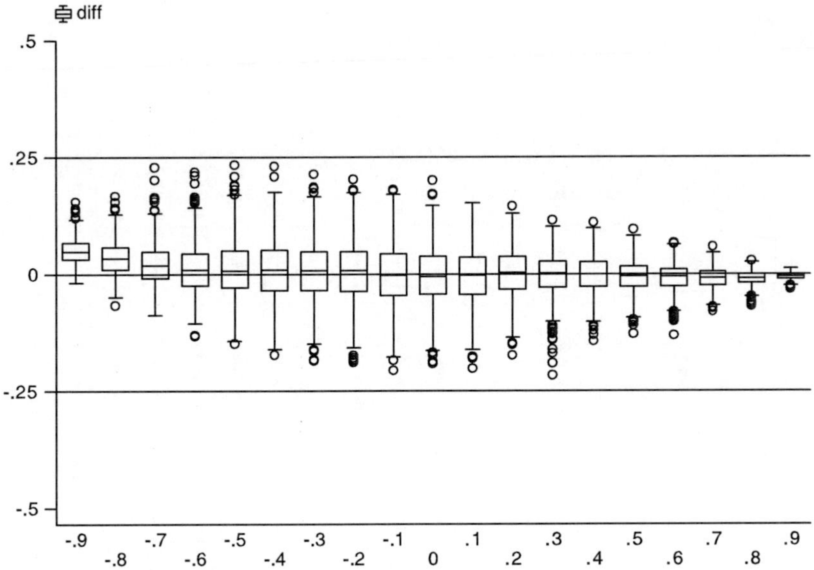

Fig. 10.7 Box plots of $\alpha^{(i)} - \hat{\alpha}^{(i)}$ for 19 values of the parameter $\alpha^{(i)}$ in a Monte Carlo study. SACPH(1,0) model, $K = 2$, $n = 500$. Reproduced from Gerhard and Hautsch (2007)

Figures 10.8–10.10 reproduce the corresponding results for SACPH(0,1) models. Though qualitatively similar, it is evident from the study that the SACPH(0,1) model performs worse than the corresponding SACPH(1,0) model. After an increase in the number of categories from $K = 2$ to $K = 10$, the approximation reaches about the quality of the SACPH(1,0) process with $K = 2$ categories, except for the parameter value $\beta = 0.9$. The reason for this can be found in the differing ACF of an AR(1) and a MA(1) process. The relatively bad performance of the SACPH(0,1) process for parameters β with a large absolute value is due to the flattening out of the ACF towards the limits of the invertible region.

10.4 Extended SACPH Models

10.4.1 Regime-Switching Baseline Hazard Functions

The standard PH model underlies the assumption that, for any two sets of explanatory variables $\mathbf{z}_1$ and $\mathbf{z}_2$, the hazard functions are related by

$$h(x; \mathbf{z}_1) \propto h(x; \mathbf{z}_2).$$

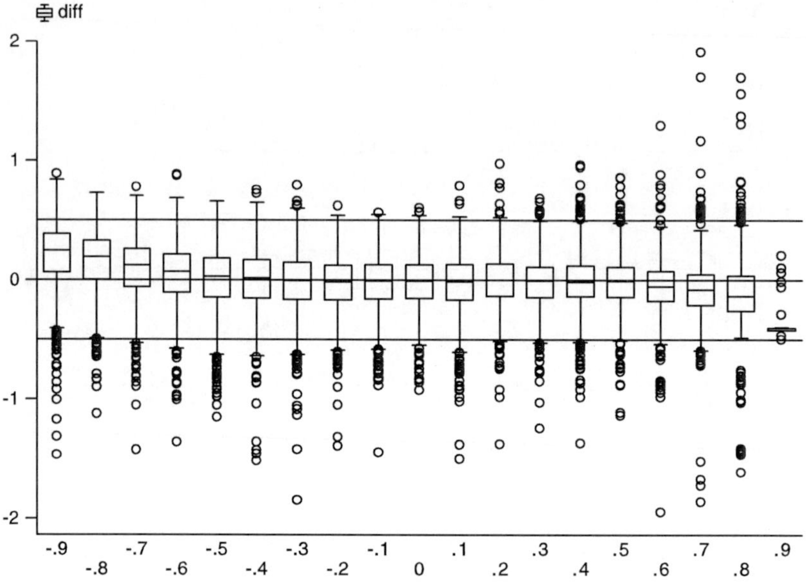

Fig. 10.8 Box plots of $\beta^{(i)} - \hat{\beta}^{(i)}$ for 19 values of the parameter $\beta^{(i)}$ in a Monte Carlo study. SACPH(0,1) model, $K = 2, n = 50$. Reproduced from Gerhard and Hautsch (2007)

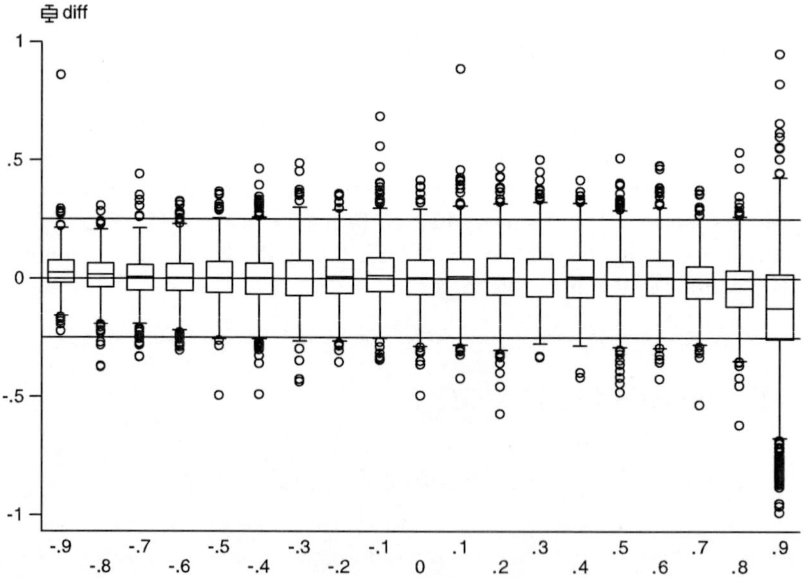

Fig. 10.9 Box plots of $\beta^{(i)} - \hat{\beta}^{(i)}$ for 19 values of the parameter $\beta^{(i)}$ in a Monte Carlo study. SACPH(0,1) model, $K = 10, n = 50$. Reproduced from Gerhard and Hautsch (2007)

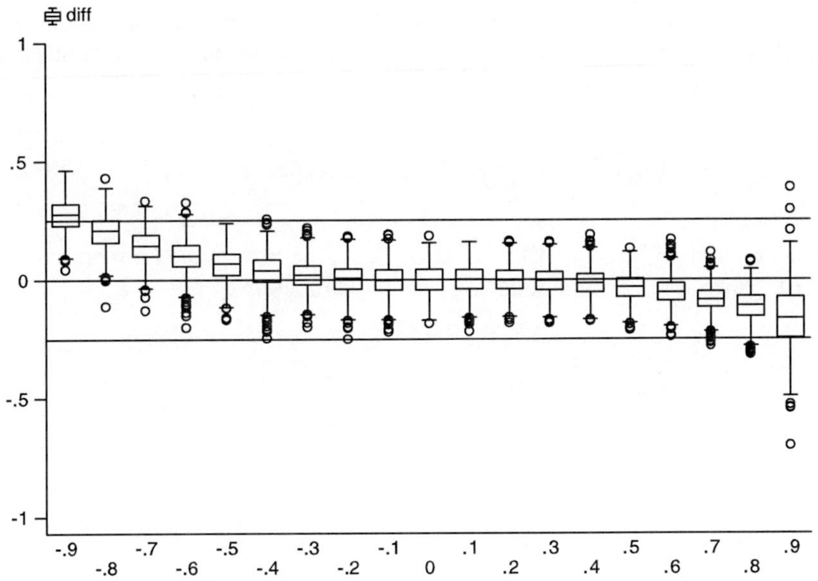

Fig. 10.10 Box plots of $\beta^{(i)} - \hat{\beta}^{(i)}$ for 19 values of the parameter $\beta^{(i)}$ in a Monte Carlo study. SACPH(0,1) model, $K = 2$, $n = 500$. Reproduced from Gerhard and Hautsch (2007)

To obtain more flexibility and to relax the proportionality assumption, we stratify the data set and define regime-dependent baseline hazard functions $h_{0,m}(x)$, $m = 1, \ldots, M$. Assume a state defining integer variable $M_i = 1, \ldots, M$, which is weakly exogenous for x_i and determines the functional relationship between x_i^* and the baseline hazard $h_{0,m}(x)$. Then, the transformation from x_i to x_i^* is state-dependent and is given by

$$x_i^* = \sum_{m=1}^{M} \mathbb{1}_{\{M_i=m\}} \ln H_{0,m}(x_i), \tag{10.17}$$

where $H_{0,m}(x_i)$ denotes the regime-dependent integrated baseline hazard. The assumption of a model with M distinct baseline hazard functions translates to M sets of distinct threshold parameters $\mu_{k,m}^*$, where $k = 1, \ldots, K-1$ and $m = 1, \ldots, M$. Correspondingly, we obtain M baseline survivor functions evaluated at the $K-1$ thresholds

$$S_{0,m}(\bar{x}_k) = \exp(-\exp(\mu_{k,m}^*)), \quad k = 1, \ldots, K-1, \quad m = 1, \ldots, M. \tag{10.18}$$

Consequently, the generalized residuals are also state-dependent, i.e.,

$$e_i := \mathbb{E}\left[\varepsilon_i^* \,|\, \mathcal{F}_i^d\right] = \sum_{m=1}^{M} \mathbb{1}_{\{R_i=m\}} e_{i,m},$$

where $e_{i,m}$ is computed according to (10.14) based on the corresponding set of threshold parameters $\mu^*_{k,m}$. Hence, a SACPH model with state-dependent baseline hazard functions is defined as

$$h(x_i; \mathcal{F}^d_{i-1}) = \sum_{m=1}^{M} h_{0,m}(x_i) \mathbb{1}_{\{M_i=m\}} \exp(-\phi_{i-1}). \tag{10.19}$$

The calculation of the log likelihood is based on the procedure proposed in Sect. 10.2, therefore, we obtain the log likelihood function by

$$\ln \mathcal{L}(\mathbf{Y}; \boldsymbol{\theta}) = \sum_{i=1}^{n} \sum_{k=1}^{K} \sum_{m=1}^{M} \mathbb{1}_{\{x_i^d=k\}} \mathbb{1}_{\{M_i=m\}} \ln \mathbb{P}r\left[x_i^d = k \,|M_i = m; \mathcal{F}^d_{i-1}\right]. \tag{10.20}$$

10.4.2 Censoring

A major advantage of the SACPH model is that it can easily accommodate for censoring structures. Define in the following $x_i^{d,l}$ and $x_i^{d,u}$ as the discretized counterparts to the duration boundaries x_i^l and x_i^u as given in Sect. 4.3.1. Hence, $x_i^d \in [x_i^{d,l}, x_i^{d,u}]$, where $x_i^{d,l}$ and $x_i^{d,u}$ are computed corresponding to (4.51) by accounting for the observation rule, (10.6). Then, in the case of a censored observation i, the corresponding log likelihood contribution in (10.20) is replaced by

$$\mathbb{P}r\left[x_i^l \le x_i^d \le x_i^u \,|\mathcal{F}^d_{i-1}, c_{i-1}, c_i, c_{i+1}\right] = S_{\varepsilon^*}(\mu^{*l} - \phi_i) - S_{\varepsilon^*}(\mu^{*u} - \phi_i), \tag{10.21}$$

where

$$\begin{cases} \mu^{*l} = \mu^*_k & \text{if} \quad x_i^{d,l} = k+1, \\ \mu^{*u} = \mu^*_k & \text{if} \quad x_i^{d,u} = k. \end{cases}$$

The derivation of the generalized residuals needs to be slightly modified. In the case of censoring, the conditional expectation of ε_i^* is computed as

$$\begin{aligned} e_i &= \mathbb{E}\left[\varepsilon_i^* | x_i^d, \phi_i, c_{i-1}, c_i, c_{i+1}\right] = \mathbb{E}\left[\varepsilon_i^* | x_i^{d,l}, x_i^{d,u}, \phi_i\right] \\ &= \frac{\kappa(\nu_i^l, \nu_i^u)}{S_{\varepsilon^*}(\nu_i^u) - S_{\varepsilon^*}(\nu_i^l)}, \end{aligned} \tag{10.22}$$

where $\nu_i^l := \mu^{*l} - \phi_i$ and $\nu_i^u := \mu^{*u} - \phi_i$.

10.4.3 Unobserved Heterogeneity

A further advantage of the SACPH model is that it is readily extended to allow for unobservable heterogeneity. In duration literature, it is well known that ignoring unobserved heterogeneity can lead to biased estimates of the baseline hazard function.[3] Following Han and Hausman (1990), unobserved heterogeneity effects can be captured by a random variable which enters the hazard function multiplicatively leading to a mixed SACPH model. Lancaster (1997) illustrates that the inclusion of a heterogeneity variable can capture errors in the regressors. In financial duration data, unobservable heterogeneity effects can be driven by different groups of traders or different states of the market.

The standard procedure to account for unobserved heterogeneity in the SACPH model is to introduce an i.i.d. random variable v_i in the specification (10.10) to obtain

$$\lambda(x_i; \mathcal{F}^d_{i-1}, v_i) = h_0(x_i) \cdot v_i \cdot \exp(-\phi_{i-1}). \tag{10.23}$$

Assume for the random variable v_i a Gamma distribution with mean one and variance η^{-1}, which is standard for this type of mixture models, see, e.g., Lancaster (1997) or Sect. 5.3.2. Then, the survivor function of the compounded model is obtained by integrating out v_i

$$S(x_i; \mathcal{F}^d_{i-1}) = \left[1 + \eta^{-1} \exp(-\phi_i) H_0(x_i)\right]^{-\eta}. \tag{10.24}$$

Note that this is identical to the survivor function of a BurrII(η) distribution under appropriate parameterization (see appendix).

The latter gives rise to an analogue model based on the discretization approach outlined in Sect. 10.2. By augmenting the log-linear model of the integrated baseline hazard by the compounder, we obtain an extended ACPH(P,Q) model based on the modified latent process

$$x_i^* = \ln(\eta) + \phi_i + \varepsilon_i^*, \tag{10.25}$$

where the error term ε_i^* follows in this case a BurrII(η) distribution with density function

$$f_{\varepsilon^*}(s) = \frac{\eta \exp(s)}{[1 + \exp(s)]^{\eta+1}}. \tag{10.26}$$

It is easily shown that

$$\lim_{\eta \to \infty} \left[1 + \eta^{-1} \phi_i H_0(x_i)\right]^{-\eta} = \exp(-H_0(x_i)\phi_i),$$

[3] See, e.g., Lancaster (1979) or Heckmann and Singer (1984) among others.

i.e., for $\eta^{-1} = \mathbb{V}(v_i) \to 0$, the BurrII($\eta$) distribution converges to the standard extreme value distribution. Hence, if no unobservable heterogeneity effects exist, the model coincides with the basic SACPH model.

The estimation procedure is similar to the procedure described in Sect. 10.2. The difference is that the model is now based on a BurrII(η) distribution. Apart from an obvious adjustment to the generalized errors, the relationship between the estimated thresholds and the estimation of the distribution function of the error term, as given in (4.41) for the standard extreme value distribution, is slightly modified to

$$S_0(\bar{x}_k) = \frac{1}{\left[1 + \exp(\mu_k^* - \ln(\eta))\right]^{\eta}}, \qquad k = 1, \ldots, K-1. \tag{10.27}$$

10.5 Testing the SACPH Model

An obvious way to test for correct specification of the SACPH model is to evaluate the properties of the series of the estimated log integrated hazard $\hat{\varepsilon}_i^* = \ln \hat{H}(x_i)$ which should be i.i.d. standard extreme value or BurrII(η) distributed, respectively. However, the difficulty is that we cannot estimate ε_i^* but only its conditional expectation $\hat{e}_i = \hat{\mathbb{E}}[\varepsilon_i^* | \mathcal{F}_i^d]$. Thus, the SACPH model has to be evaluated by comparing the distributional and dynamical properties of $\hat{e}_i$ with their theoretical counterparts.

The theoretical mean of e_i is straightforwardly computed as

$$\mathbb{E}[e_i] = \mathbb{E}[\mathbb{E}[\varepsilon_i^* | x_i^d, \phi_i]] = \mathbb{E}[\varepsilon_i^*]. \tag{10.28}$$

However, the computation of higher order moments of e_i is a difficult task. The reason is that the categorization is based on x_i^*, and thus the category boundaries for ε_i^*, $\nu_{i,k} = \mu_i^* - \phi_i$, are time-varying and depend itself on lags of e_i. Therefore, the derivation of theoretical moments can only be performed on the basis of the *estimated* model dynamics, and thus, they are of limited value for powerful diagnostic checks of the model. Hence, only upper limits for the moments in the limiting case $K \to \infty$ can be given. In this case, $e_i = \varepsilon_i^*$, and thus, the moments of e_i correspond to the moments of the standard extreme value or BurrII(η) distribution, respectively.

Gerhard and Hautsch (2007) propose evaluating the *dynamic* properties of the $\hat{e}_i$ series based on a test for serial dependence according to Gouriéroux et al. (1985). The test is based on the direct relationship between the score of the observable and the latent model. By accounting for unobserved heterogeneity (see Sect. 10.4.3), the latent model is written as

$$x_i^* = \ln(\eta) + \phi_i + u_i, \tag{10.29}$$

$$u_i = \tilde{\alpha}_j u_{i-j} + \varepsilon_i^*, \tag{10.30}$$

where ε_i^* is i.i.d. BurrII(η) distributed and j denotes the tested lag. Then, the null hypothesis is $H_0 : \tilde{\alpha}_j = 0$. The test is based on the score of the observable model. Following Gouriéroux et al. (1985), the observable score is equal to the conditional expectation of the latent score, given the observable categorized variable, i. e.,

$$\frac{\partial \ln \mathcal{L}(\mathbf{Y};\boldsymbol{\theta})}{\partial \boldsymbol{\theta}} = \mathbb{E}\left[\left. \frac{\partial \ln \mathcal{L}^*(\mathbf{L}^*;\boldsymbol{\theta})}{\partial \boldsymbol{\theta}} \right| \mathcal{F}_i^d \right], \tag{10.31}$$

where $\ln \mathcal{L}^*(\cdot)$ denotes the log likelihood function of the *latent* model and $\mathbf{L}^*$ denotes the $n \times 1$ vector of the *latent* realizations x_i^*. Under the assumption of a BurrII(η) distribution for ε_i^*, the log likelihood function of the latent model is given by

$$\begin{aligned}
\ln \mathcal{L}^*(\mathbf{L}^*;\boldsymbol{\theta}) &= \sum_{i=j+1}^{n} \ln f_{\varepsilon^*}(u_i - \tilde{\alpha}_j u_{i-j}) \\
&= \sum_{i=j+1}^{n} \left[\ln(\eta) + \tilde{\alpha}_j u_{i-j} - u_i - (\eta+1)\ln\left[1 + \exp(\tilde{\alpha}_j u_{i-j} - u_i)\right]\right].
\end{aligned}$$

Under the null, the score with respect to $\tilde{\alpha}_j$, $s(\tilde{\alpha_j})$ is given by

$$\begin{aligned}
s(\tilde{\alpha}_j) &= \mathbb{E}\left[\left. \frac{\partial \ln \mathcal{L}^*(\mathbf{L}^*;\boldsymbol{\theta})}{\partial \tilde{\alpha}_j} \right| \mathcal{F}_i^d \right] \\
&= \sum_{i=j+1}^{n} \mathbb{E}\left[e_{i-j}^* \middle| \mathcal{F}_i^d \right] \left[1 - (\eta+1)\mathbb{E}\left[\left. \frac{\exp(\varepsilon_i^*)}{1+\exp(\varepsilon_i^*)} \right| \mathcal{F}_i^d \right]\right] \\
&= \sum_{i=j+1}^{n} e_{i-j} \left[1 - (\eta+1)\tilde{e}_i\right],
\end{aligned} \tag{10.32}$$

where

$$\tilde{e}_i := \mathbb{E}\left[\left. \frac{\exp(\varepsilon_i^*)}{1+\exp(\varepsilon_i^*)} \right| \mathcal{F}_i^d \right]. \tag{10.33}$$

Then,

$$s(\hat{\tilde{\alpha}}_j) = \sum_{i=j+1}^{n} \hat{e}_{i-j} \left[1 - (\hat{\eta}+1)\hat{\tilde{e}}_i\right]. \tag{10.34}$$

Under the null, the expectation of $\hat{\tilde{e}}_i$ is given by $\mathbb{E}[\hat{\tilde{e}}_i] = (\eta+1)^{-1}$, and thus, $\mathbb{E}[s(\hat{\tilde{\alpha}}_j)] = 0$. Exploiting the asymptotic normality of the score, i.e.,

$$\frac{1}{\sqrt{n}} s(\tilde{\alpha}_j) \xrightarrow{d} \mathcal{N}\left(0, \operatorname{plim} \frac{1}{n} \sum_{i=j+1}^{n} e_{i-j}^2 \, [1 - (\eta + 1)\tilde{e}_i]^2 \right),$$

a χ^2-statistic for the null hypothesis $H_0 : \tilde{\alpha}_j = 0$ is obtained by

$$\Upsilon^{(j)} = \frac{\left[\sum_{i=j+1}^{n} \hat{e}_{i-j} \left[1 - (\hat{\eta} + 1)\hat{\tilde{e}}_i\right]\right]^2}{\sum_{i=j+1}^{n} \hat{e}_{i-j}^2 \left[1 - (\hat{\eta} + 1)\hat{\tilde{e}}_i\right]^2} \overset{a}{\sim} \chi^2(1). \tag{10.35}$$

Correspondingly, for the standard extreme value case, the test modifies to

$$\Upsilon^{(j)} = \frac{\left[\sum_{i=j+1}^{n} \hat{e}_{i-j} \left(\hat{\tilde{e}}_i - 1\right)\right]^2}{\sum_{i=j+1}^{n} \hat{e}_{i-j}^2 \left[\hat{\tilde{e}}_i - 1\right]^2} \overset{a}{\sim} \chi^2(1) \tag{10.36}$$

with

$$\tilde{e}_i := \mathbb{E}\left[\exp(\varepsilon_i^*) | \mathcal{F}_i^d\right]. \tag{10.37}$$

10.6 Estimating Volatility Using the SACPH Model

Gerhard and Hautsch (2007) apply the SACPH model for volatility estimation on the basis of price durations. They use data originating from Bund future trading. The Bund future is one of the most actively traded future contracts in Europe. The contract is written on a synthetic long-term German government bond. Prices are denoted in basis points of the face value (so-called "ticks") corresponding to 10 EURO. As the minimum tick size is comparably large, price movements in Bund future trading are very discrete. As shown, e.g., by Hautsch and Pohlmeier (2002), trade-to-trade price changes are most likely zero or just one tick. As such price discreteness challenges volatility estimators based on continuous distributions, Gerhard and Hautsch (2007) rely on the concept of price durations. Due to the flexibility of the SACPH model to capture also censored price durations, arising, e.g., from non-trading periods, it is particularly suitable to study volatility dynamics also on higher aggregation levels. To illustrate this feature, Gerhard and Hautsch (2007) choose aggregation levels associated with large cumulated price movements that might last over several trading days.

10.6.1 Data and the Generation of Price Events

The underlying data stems from Bund future trading at the London International Financial Futures and Options Exchange (LIFFE) which has been taken over by Euronext 2002 and in meantime belongs to NYSE Euronext. The sample covers transaction data on 11 contracts and 816 trading days between 04/05/94 and 06/30/97 with a total of about $n \approx 2 \cdot 10^6$ transactions. Though the data is relatively old, it still reveals very similar features as also found for more recent Bund future data (though the underlying trading frequency arguably has increased). Particularly for the effects discussed in the sequel, the data is still quite representative.

Gerhard and Hautsch generate price durations x_i^{dp} as described in Sect. 3.2.2 using the aggregation level $dp \in \{10, 15, 20\}$ ticks, corresponding to 7,491, 3,560, and 2,166 observations. Price change events can also be caused by news occurring during non-trading periods inducing censoring structures. Since the resulting price event is observable at the earliest at the beginning of the next trading day, it is not possible to identify whether the price movement is caused by overnight news or by recent information. To overcome this problem, Gerhard and Hautsch (2007) consider the first price event occurring within the first 15 min of a trading day as censored, i.e., this price change is assumed to be driven by events occurring during the non-trading period. In contrast, price events observed after the first 15 min of a trading day are assumed to be driven by recent information. For these observations, the duration is measured as the time since the last observation of the previous trading day. Figure 10.11 illustrates the identification rule graphically. It shows the arrival times of three transactions occurring on two subsequent trading days. According to this rule, the spell between A and B is assumed to be censored and has a minimum length of 15 min and a maximum length of 15 h and 30 min. In contrast, the duration episode between A and C can be measured exactly as 16 h.

Based on the sample used, there are 600, 457, and 318 censored durations for the 10, 15 and 20 tick aggregation level, respectively. Gerhard and Hautsch choose a categorization which ensures satisfactory frequencies in each category. For this reason, the categorization differs for the individual aggregation levels. Table 10.2 shows the distribution of the categorized durations based on the finest grouping which is used for the 10 tick level.

A peculiar feature of Bund future trading data is the occurrence of periodicities occurring not only within a trading day but also over the lifetime of a contract

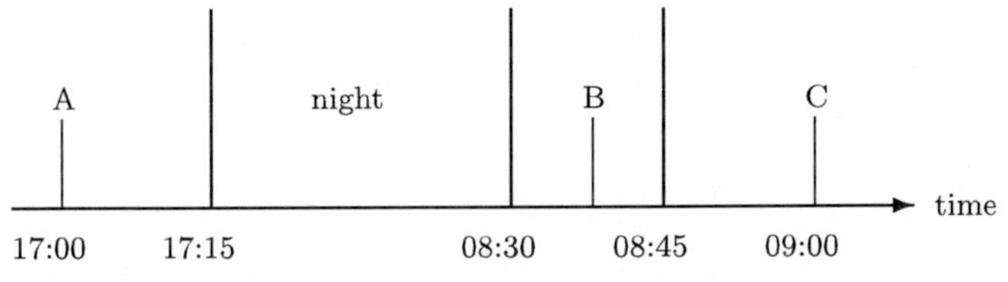

Fig. 10.11 Identification of censored durations in the Bund future trading

Table 10.2 Distribution of categorized price durations based on 10, 15 and 20 tick price changes. Based on BUND futures trading at LIFFE. Means and standard deviations (only for non-censored durations) in minutes. Censored observations are stated by their corresponding lower boundaries. Reproduced from Gerhard and Hautsch (2007)

	10 Ticks				15 Ticks				20 Ticks			
Categories	Non-Censored			Cens.	Non-Censored			Cens.	Non-Censored			Cens.
	Obs	Mean	Std.Dv.	Obs	Obs	Mean	Std.Dv.	Obs	Obs	Mean	Std.Dv.	Obs
[0′, 5′]	670	2.37	1.48	68	125	2.35	1.57	21	34	1.98	1.48	7
(5’, 10’]	618	7.46	1.45	52	117	7.76	1.43	19	26	1.64	5.21	4
(10’, 20’]	1,071	14.72	2.84	96	237	14.60	2.87	45	84	14.98	2.92	23
(20’, 40’]	1,345	29.06	5.77	139	399	29.81	5.66	79	149	31.09	5.79	35
(40’, 1 h]	762	49.42	5.75	85	336	49.11	5.78	52	148	50.00	5.90	27
(1 h, 2 h]	974	84.46	16.81	79	489	43.92	16.64	85	243	87.33	17.46	55
(2 h, 3 h]	375	145.91	17.05	32	251	84.90	17.78	36	134	149.25	16.55	27
(3 h, 4 h]	214	207.81	17.09	5	146	146.65	17.23	15	85	209.46	17.03	13
(4 h, 6 h]	249	293.42	36.50	12	240	204.80	35.66	23	158	293.29	34.62	15
(6 h, 8 h]	89	407.12	33.21	21	138	294.75	34.35	30	130	406.71	37.31	30
(8 h, 24 h]	311	1,089.20	177.29	11	339	406.28	222.35	52	294	1,145.20	218.72	79
(24 h, 36 h]	59	1,621.92	168.52	0	83	1,110.03	148.70	0	116	1,627.54	155.37	3
(36 h, 48 h]	36	2,708.46	174.35	0	40	2,731.32	160.93	0	51	2,694.04	170.45	0
(48 h, ∞)	118	5,050.50	2,397.13	0	163	4,841.94	2,931.29	0	196	4,875.33	2,809.73	0
Total	6,891	321.62	912.70	600	3,103	539.95	1,309.20	457	1,848	968.74	1,739.66	318

(so-called "time-to-maturity seasonalities"). Both types of seasonality effects are captured by a Fourier series approximation $s(t) = s(\delta^s, \bar{t}, Q)$ of order $Q = 6$ (see (5.34) in Chap. 5). In case of intraday effects, periodicities are quantified based on the normalized intraday time $\bar{t} \in [0, 1]$ given by

$$\bar{t} = \frac{\text{seconds since 7:30}}{\text{seconds between 7:30 and 16:15}}. \tag{10.38}$$

Accordingly, for time-to-maturity seasonalities

$$\bar{t} = \frac{\text{days to maturity}}{150} \tag{10.39}$$

is used. Observations with values of more than 150 days are captured by a dummy variable. Including the seasonality variables in a static way, the resulting SACPH model is given by

$$x_i^* = \phi_i + s(t_{i-1}) + \varepsilon_i^*,$$

$$\phi_i = \sum_{j=1}^{P} \alpha_j (\phi_{i-j} + e_{i-j}) + \sum_{j=1}^{Q} \beta_j e_{i-j}.$$

To account for the fact that the durations are categorized, Gerhard and Hautsch (2007) reformulate the conditional price change volatility derived in Chap. 8 as

$$\sigma^{*2}_{(x^{dp})}(t_i^{dp}) = (dp)^2 \sum_{k=1}^{K-1} \mathbb{P}r\left[x_{i+1}^{d,dp} = k \,\middle|\, \mathcal{F}_{t_i}^d; s(t_i)\right]$$
$$\cdot \mathbb{E}\left[\frac{1}{x_{i+1}^{d,dp}} \,\middle|\, x_{i+1}^{d,dp} = k; \mathcal{F}_{t_i}^d, s(t_i)\right], \tag{10.40}$$

where $x_i^{d,dp}$ denotes the categorized dp-price duration. Moreover, in the context of grouped durations, the following approximation is assumed to hold:

$$\mathbb{E}\left[\frac{1}{x_{i+1}^{d,dp}} \,\middle|\, x_{i+1}^{d,dp} = k; \mathcal{F}_{t_i}^d, s(t_i)\right] \approx \mathbb{E}\left[\frac{1}{x_{i+1}^{d,dp}} \,\middle|\, x_{i+1}^{d,dp} = k\right].$$

Then, an obvious sample estimator for the second factor in the weighted sum of (10.40) is obtained. This, however, does not exclude the conditioning information, but merely expresses the fact that all information contained in the regressors enters exclusively the first factor $\mathbb{P}r\left[x_{i+1}^{d,dp} = k \,\middle|\, \mathcal{F}_{t_i}^d; s(t_i)\right]$ in (10.40) for which an estimator is available by adopting (10.15).

10.6.2 Empirical Findings

Table 10.3 reproduces the estimation results from Gerhard and Hautsch (2007) for SACPH specifications without any seasonality variables (Panel A), with only intraday seasonalities (Panel B) and including all seasonality variables (Panel C) based on 15 tick price durations. It is shown that the goodness-of-fit is significantly improved by the inclusion of seasonality variables. The autoregressive parameters are highly significant indicating strong serial dependence. Hence, even on the highest aggregation level, price durations are strongly clustered. Applying the test

Table 10.3 ML estimates of SACPH(1,2) models for Bund future price durations using 15 tick price changes. Standard errors computed based on OPG estimates. Diagnostics: Log Likelihood (LL), Bayes Information Criterion (BIC), mean ($\bar{\hat{e}}_i$) and standard deviation (S.D.) of SACPH residuals. Reproduced from Gerhard and Hautsch (2007)

	A		B		C	
	est.	S.E.	est.	S.E.	est.	S.E.
Thresholds						
ν_2^* ($\bar{x}_2 = 10'$)	−3.990	0.182	−3.770	0.208	−4.422	0.246
ν_3^* ($\bar{x}_3 = 20'$)	−3.255	0.179	−3.042	0.205	−3.673	0.243
ν_4^* ($\bar{x}_4 = 40'$)	−2.543	0.178	−2.332	0.203	−2.949	0.241
ν_5^* ($\bar{x}_5 = 1\,\text{h}$)	−2.117	0.177	−1.900	0.202	−2.514	0.241
ν_6^* ($\bar{x}_6 = 2\,\text{h}$)	−1.595	0.175	−1.354	0.200	−1.963	0.239
ν_7^* ($\bar{x}_7 = 3\,\text{h}$)	−1.344	0.175	−1.080	0.200	−1.685	0.239
ν_8^* ($\bar{x}_8 = 4\,\text{h}$)	−1.198	0.174	−0.919	0.199	−1.520	0.238
ν_9^* ($\bar{x}_9 = 6\,\text{h}$)	−0.942	0.173	−0.639	0.198	−1.232	0.237
ν_{10}^* ($\bar{x}_{10} = 8\,\text{h}$)	−0.777	0.172	−0.456	0.198	−1.042	0.237
ν_{11}^* ($\bar{x}_{11} = 24\,\text{h}$)	−0.178	0.169	0.200	0.194	−0.329	0.235
ν_{12}^* ($\bar{x}_{12} = 36\,\text{h}$)	−0.009	0.168	0.376	0.194	−0.127	0.235
ν_{13}^* ($\bar{x}_{13} = 48\,\text{h}$)	0.084	0.168	0.473	0.193	−0.014	0.234
Intraday seasonalities						
δ_1^s			0.814	0.141	0.853	0.153
$\delta_{s,1}^s$			0.461	0.054	0.456	0.056
$\delta_{s,2}^s$			−0.068	0.038	−0.066	0.040
$\delta_{s,3}^s$			−0.020	0.034	−0.036	0.037
$\delta_{s,4}^s$			0.023	0.033	0.029	0.035
$\delta_{s,5}^s$			−0.057	0.031	−0.072	0.033
$\delta_{s,6}^s$			−0.034	0.029	−0.022	0.031
$\delta_{c,1}^s$			0.191	0.035	0.228	0.035
$\delta_{c,2}^s$			0.169	0.031	0.189	0.032
$\delta_{c,3}^s$			−0.007	0.029	0.002	0.030
$\delta_{c,4}^s$			0.042	0.030	0.042	0.031
$\delta_{c,5}^s$			0.068	0.029	0.063	0.031
$\delta_{c,6}^s$			−0.044	0.027	−0.042	0.028

(continued)

Table 10.3 (continued)

	A		B		C	
	est.	S.E.	est.	S.E.	est.	S.E.
Seasonalities over the future's maturity						
$\mathbf{1}_{>150}$					0.418	0.271
δ_1^{*s}					−1.455	0.276
$\delta_{s,1}^{*s}$					−0.038	0.104
$\delta_{s,2}^{*s}$					0.233	0.063
$\delta_{s,3}^{*s}$					−0.048	0.049
$\delta_{s,4}^{*s}$					0.080	0.040
$\delta_{s,5}^{*s}$					0.070	0.034
$\delta_{s,6}^{*s}$					0.026	0.032
$\delta_{c,1}^{*s}$					0.508	0.055
$\delta_{c,2}^{*s}$					0.008	0.044
$\delta_{c,3}^{*s}$					−0.046	0.038
$\delta_{c,4}^{*s}$					0.018	0.036
$\delta_{c,5}^{*s}$					−0.055	0.035
$\delta_{c,6}^{*s}$					0.014	0.030
Dynamic parameters						
α_1	0.962	0.006	0.961	0.006	0.978	0.005
β_1	−0.788	0.014	−0.779	0.013	−0.850	0.014
β_2	−0.084	0.013	−0.081	0.013	−0.085	0.013
Diagnostics						
Obs	3,559		3,559		3,559	
LL	−7,480		−7,331		−7,208	
BIC	−7,541		−7,445		−7,380	
$\bar{\hat{e}}_i$	−0.561		−0.561		−0.557	
S.D.	1.166		1.159		1.169	

on serial correlation as illustrated in Sect. 10.5, Gerhard and Hautsch (2007) show that the null hypothesis of remaining serial correlation in generalized residuals is clearly rejected. Hence, the SACPH model does a good job in capturing the serial dependence in the data.

Table 10.4 reproduces the results for the fully parameterized model (specification C) for all three aggregation levels using a gamma compounded SACPH model accounting also for unobservable heterogeneity effects. It is shown that the heterogeneity variance increases with the aggregation level. For 20 tick price changes, the heterogeneity variance takes on a value of 1.156 corresponding to a specification which is close to an ordered logit model.[4]

[4]Recall that the BurrII(1) distribution coincides with the logistic distribution.

Table 10.4 ML estimates of gamma compounded SACPH(1,2) models for Bund future price durations using 10, 15 and 20 tick price changes. Standard errors computed based on OPG estimates. Diagnostics: Log Likelihood (LL), Bayes Information Criterion (BIC), mean ($\bar{\hat{e}}_i$) and standard deviation (S.D.) of SACPH residuals. Reproduced from Gerhard and Hautsch (2007)

	$dp = 10$		$dp = 15$		$dp = 20$	
	est.	S.E.	est.	S.E.	est.	S.E.
Thresholds						
μ_1^* ($\bar{x}_1 = 5'$)	−6.034	0.301				
ν_2^* ($\bar{x}_2 = 10'$)	−5.194	0.303	−4.590	0.439		
ν_3^* ($\bar{x}_3 = 20'$)	−4.300	0.307	−3.756	0.441	−2.344	0.565
ν_4^* ($\bar{x}_4 = 40'$)	−3.429	0.312	−2.900	0.446	−1.481	0.565
ν_5^* ($\bar{x}_5 = 1\,\text{h}$)	−2.933	0.317	−2.339	0.452	−0.910	0.567
ν_6^* ($\bar{x}_6 = 2\,\text{h}$)	−2.164	0.328	−1.546	0.465	−0.146	0.574
ν_7^* ($\bar{x}_7 = 3\,\text{h}$)	−1.772	0.335	−1.114	0.475	0.241	0.579
ν_8^* ($\bar{x}_8 = 4\,\text{h}$)	−1.506	0.340	−0.847	0.481	0.486	0.583
ν_9^* ($\bar{x}_9 = 6\,\text{h}$)	−1.119	0.350	−0.357	0.495	0.949	0.592
ν_{10}^* ($\bar{x}_{10} = 8\,\text{h}$)	−0.940	0.356	−0.012	0.506	1.371	0.602
ν_{11}^* ($\bar{x}_{11} = 24\,\text{h}$)	0.133	0.387	1.446	0.560	2.832	0.641
ν_{12}^* ($\bar{x}_{12} = 36\,\text{h}$)	0.451	0.397	1.923	0.581	3.523	0.666
ν_{13}^* ($\bar{x}_{13} = 48\,\text{h}$)	0.698	0.404	2.209	0.595	3.916	0.680
Intraday seasonalities						
δ_1^s	1.407	0.136	1.855	0.247	1.374	0.324
$\delta_{s,1}^s$	0.896	0.056	1.021	0.095	0.624	0.125
$\delta_{s,2}^s$	0.080	0.037	0.050	0.062	−0.138	0.085
$\delta_{s,3}^s$	0.054	0.033	−0.003	0.055	−0.149	0.077
$\delta_{s,4}^s$	0.114	0.030	0.092	0.053	−0.001	0.075
$\delta_{s,5}^s$	−0.077	0.030	−0.145	0.051	−0.162	0.072
$\delta_{s,6}^s$	−0.001	0.028	−0.049	0.049	−0.079	0.068
$\delta_{c,1}^s$	−0.050	0.035	0.264	0.055	0.356	0.074
$\delta_{c,2}^s$	0.170	0.030	0.365	0.052	0.280	0.069
$\delta_{c,3}^s$	0.043	0.029	0.086	0.050	−0.098	0.068
$\delta_{c,4}^s$	0.150	0.030	0.178	0.051	0.113	0.068
$\delta_{c,5}^s$	0.157	0.029	0.143	0.049	0.000	0.067
$\delta_{c,6}^s$	−0.022	0.026	−0.050	0.046	0.040	0.064
Seasonalities over the future's maturity						
$\mathbf{1}_{>150}$	0.353	0.328	0.902	0.374	1.398	0.467
δ_1^{*s}	−3.264	0.384	−2.742	0.465	−2.250	0.569
$\delta_{s,1}^{*s}$	−0.181	0.141	−0.084	0.181	0.180	0.202
$\delta_{s,2}^{*s}$	0.291	0.094	0.396	0.107	0.490	0.119
$\delta_{s,3}^{*s}$	−0.070	0.076	−0.043	0.083	0.031	0.094
$\delta_{s,4}^{*s}$	0.113	0.067	0.171	0.071	0.166	0.084
$\delta_{s,5}^{*s}$	0.041	0.062	0.066	0.064	0.107	0.078
$\delta_{s,6}^{*s}$	−0.015	0.056	−0.014	0.057	0.042	0.071
$\delta_{c,1}^{*s}$	0.981	0.087	1.010	0.122	1.005	0.116

(continued)

Table 10.4 (continued)

	$dp = 10$		$dp = 15$		$dp = 20$	
	est.	S.E.	est.	S.E.	est.	S.E.
$\delta^{*s}_{c,2}$	−0.056	0.073	−0.038	0.080	−0.084	0.080
$\delta^{*s}_{c,3}$	−0.055	0.067	−0.112	0.067	−0.198	0.076
$\delta^{*s}_{c,4}$	0.040	0.061	0.021	0.062	0.002	0.072
$\delta^{*s}_{c,5}$	−0.071	0.056	−0.133	0.058	−0.133	0.069
$\delta^{*s}_{c,6}$	0.076	0.053	0.002	0.054	0.134	0.067
Heterogeneity variance						
$\mathbb{V}[v_i]$	0.576	0.047	0.897	0.074	1.156	0.095
Dynamic parameters						
α_1	0.913	0.009	0.982	0.004	0.993	0.002
β_1	−0.680	0.015	−0.828	0.018	−0.882	0.022
β_2	−0.055	0.012	−0.097	0.017	−0.067	0.022
Diagnostics						
Obs	7,490		3,559		2,165	
LL	−14,597		−7,141		−4,304	
BIC	−14,793		−7,317		−4,465	
$\bar{\hat{e}}_i$	−0.812		−0.176		0.245	
S.D.	1.476		1.662		1.809	

Figure 10.12 from Gerhard and Hautsch (2007) gives the resulting pattern of $\sigma^{*2}_{(x^{dp})}(t_i^{dp})$ in dependence of the time to maturity for 20 tick price changes. The intraday seasonality coefficients are fixed to a value associated with 14:00 GMT while the dynamic variables are set to their unconditional expectations. It is shown that volatility is highest within a time horizon of about 90 days to maturity. This is probably caused by the roll-over from the previous contract to the front month contract inducing a price discovery process. This period seems to be finished after around 80 days to maturity leading to a stabilization of the volatility pattern at a relatively constant level. The corresponding volatility pattern based on 10 tick price changes (Fig. 10.13) reveals a similar picture.

Figures 10.14 and 10.15 show the corresponding intraday seasonality patterns for 10 tick and 20 tick price durations evaluated based on a fixed time to maturity of 30 days. We observe a slight volatility spike at the beginning of the trading day which is well explained by the price finding process at the opening, where the price information conveyed by U.S. traders needs to be scrutinized. Then, volatility declines leading to the well known lunch time effect around 11:00 GMT. Volatility increases again around 13:20 GMT which obviously corresponds to the opening of the CBOT. After that period, volatility drops continually, interrupted by an additional spike before 15:00 GMT which is obviously associated with the opening of the NYSE. Overall, during the afternoon, volatility remains on a higher level than in the morning.

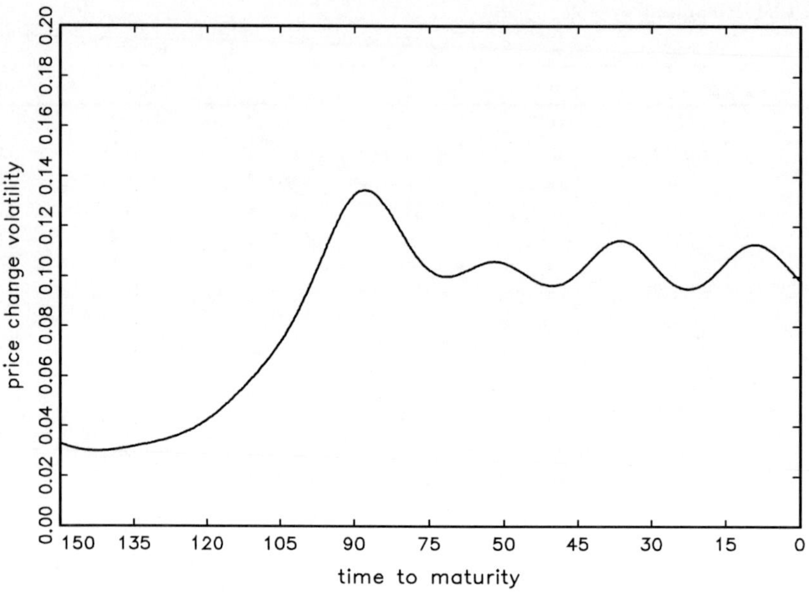

Fig. 10.12 Price change volatility $\sigma^{*2}_{(x^{dp})}(t_i^{dp})$ vs. time to maturity for $dp = 20$. 14:00 GMT, Bund future trading, LIFFE. Based on estimates by Gerhard and Hautsch (2007)

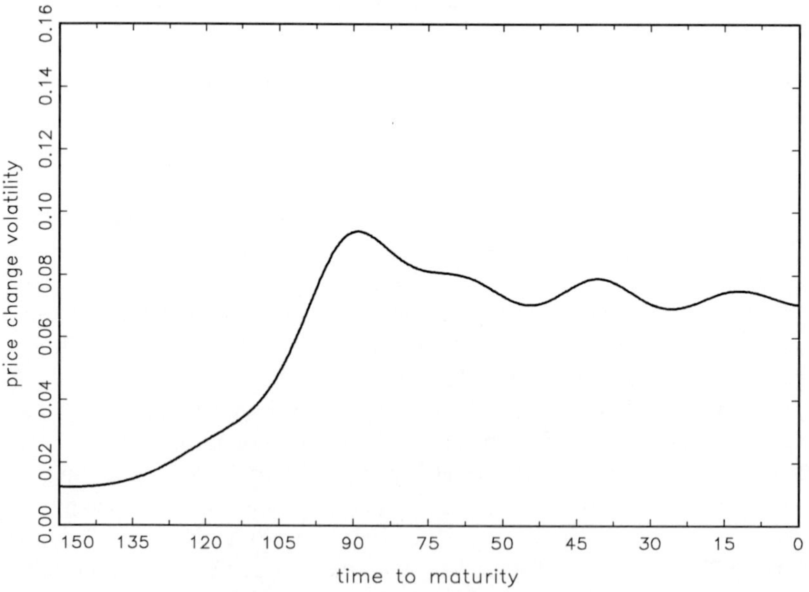

Fig. 10.13 Price change volatility $\sigma^{*2}_{(x^{dp})}(t_i^{dp})$ vs. time to maturity for $dp = 10$. 14:00 GMT, Bund future trading, LIFFE. Based on estimates by Gerhard and Hautsch (2007)

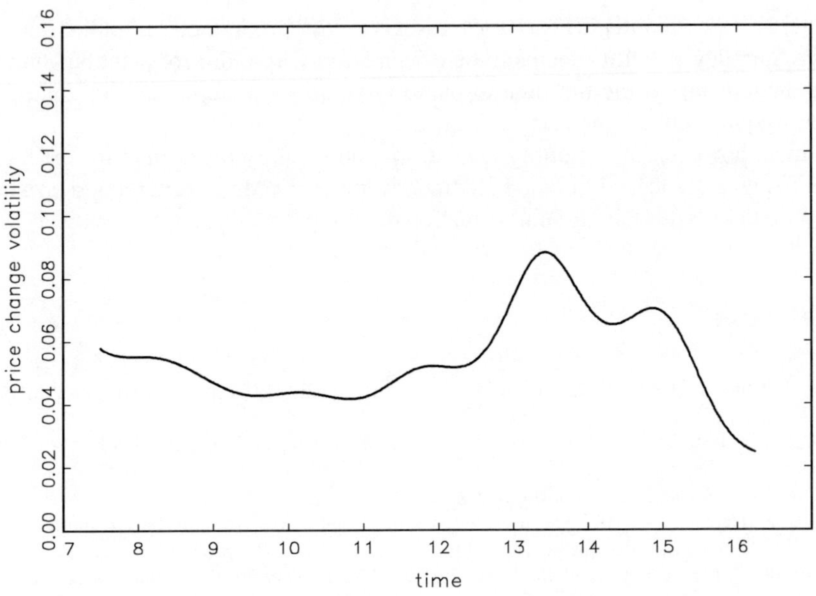

Fig. 10.14 Price change volatility $\sigma^{*2}_{(x^{dp})}(t_i^{dp})$ vs. intraday time $dp = 10$. 30 days to maturity, Bund future trading, LIFFE. Based on estimates by Gerhard and Hautsch (2007)

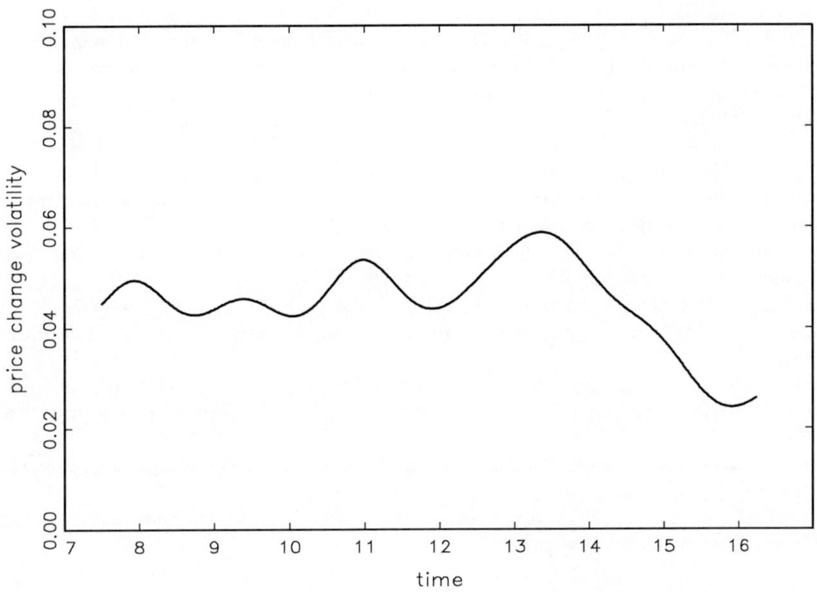

Fig. 10.15 Price change volatility $\sigma^{*2}_{(x^{dp})}(t_i^{dp})$ vs. intraday time for $dp = 20$. 30 days to maturity, Bund future trading, LIFFE. Based on estimates by Gerhard and Hautsch (2007)

The 20 tick volatility (Fig. 10.15) reveals a less pronounced intraday pattern. Hence, intraday volatility seems to be driven mainly by more frequent but smaller price movements, while the time-to-maturity volatility patterns are dominated by lower frequent but larger price changes. The CBOT opening effect can still be identified, however, the volatility level drops significantly faster than for the 5 tick price changes. Hence, additional U.S. traders obviously do not generate price events above 20 ticks, after the initial information dissemination.

References

Albert J, Chib S (1995) Bayesian residual analysis for binary response regression models. Biometrika 82:747–759

Cox DR (1981) Statistical analysis of time series: some recent developments. Scand J Stat 8:93–115

Cox DR (1975) Partial likelihood. Biometrika 62:269

Gerhard F, Hautsch N (2007) A dynamic semiparametric proportional hazard model. Stud Nonlinear Dynamics Econometrics 11, http://www.bepress.com/snde/vol11/iss2/art1

Gouriéroux C, Monfort A, Renault E, Trognon A (1987) Generalised residuals. J Econom 34: 5–32

Gouriéroux C, Monfort A, Trognon A (1985) A general approach to serial correlation. Econom Theory pp. 315–340

Han A, Hausman JA (1990) Flexible parametric estimation of duration and competing risk models. J Appl Econom 5:1–28

Hautsch N (1999) Analyzing the time between trades with a gamma compounded hazard model. An Application to LIFFE Bund Future Transctions Discussion Paper 99/03, Center of Finance and Econometrics, University of Konstanz, Konstanz

Hautsch N, Pohlmeier W (2002) Econometric analysis of financial transaction data: pitfalls and opportunities. Allgemeines Statistisches Archiv 86:5–30

Heckmann JJ, Singer B (1984) Econometrics duration analysis. J Econom 24:63–132

Horowitz JL (1996) Semiparametric estimation of a regression model with an unknown transformation of the dependent variable. Econometrica 64:103–137

Horowitz JL, Neumann GR (1987) Semiparametric estimation of employment duration models. Econometric Review 6(1):5–40

Kiefer NM (1988) Economic duration data and hazard functions. J Econ Lit 26:646–679

Lancaster T (1979) Econometric methods for the duration of unemployment. Econometrica 47(4):939–956

Lancaster T (1997) The econometric analysis of transition data. Cambridge University Press

Meyer BD (1990) Unemployment insurance and unemployment spells. Econometrica 58(4): 757–782

Oakes D, Cui L (1994) On semiparametric inference for modulated renewal processes. Biometrika 81:83–90

Prentice RL, Gloeckler LA (1978) Regression analysis of grouped survival data with application to breast cancer data. Biometrics 34:57–67

Romeo CJ (1999) Conducting inference in semiparametric duration models and inequality restrictions on the shape of the hazard implied by job search theory. J Appl Econom 14: 587–605

Sueyoshi GT (1995) A class of binary response models for grouped duration data. J Appl Econom 10:411–431

Thompson WA (1977) On the treatment of grouped durations in life studies. Biometrics 33: 463–470

Chapter 11
Univariate Dynamic Intensity Models

This chapter presents dynamic parameterizations of the intensity function. We model the intensity in continuous time which allows to update the intensity process whenever required. This is in contrast to Chaps. 5, 6 and 10 discussing discrete-time processes which are only updated at discrete points in time. Moreover, instead of specifying a model for durations (Chaps. 5 and 6) or the integrated (baseline) hazard function (Chap. 10), we discuss dynamic models which are built directly on the intensity. As illustrated in this and the next chapter, such a framework yields a valuable approach to account for time-varying covariates and multivariate structures.

We focus on two general ways of introducing dynamic structures in intensity processes. The first possibility is to parameterize the intensity function in terms of an autoregressive structure, which is updated at each occurrence of a new point. Following this idea, Russell (1999) proposes a dynamic extension of a parametric proportional intensity (PI) model that he calls *autoregressive conditional intensity* (ACI) model. This model will be presented in Sect. 11.1. Generalizations of this framework are presented in Sect. 11.2. Here, we will discuss long memory ACI models, generalizations allowing for accelerated failure time structures and thus nesting special cases of the ACD model as well as component ACI models.

A valuable alternative to an autoregressive intensity process is a so-called *self-exciting* intensity process where the intensity is driven by a function of the backward recurrence time to all previous points. A particular type of linear self-exciting processes is introduced by Hawkes (1971) where the intensity is governed by the sum of negative exponential functions of the time to all previous events. As shown in Sect. 11.3, such a process naturally accounts for events which are clustered in time and is well suited to model the evolution of market activity and trading intensities.

N. Hautsch, *Econometrics of Financial High-Frequency Data*,
DOI 10.1007/978-3-642-21925-2_11,

11.1 The Autoregressive Conditional Intensity Model

Assume the existence of a time-varying covariate process which occurs at discrete points $t_1^0, t_2^0, \ldots, t_{n^0}^0$.[1] Then, $N^0(t)$ and $\breve{N}^0(t)$ denote the corresponding right-continuous and left-continuous counting processes, respectively, associated with the arrival times of the covariate process z_i^0. Moreover, let $\{\tilde{t}_i\}$ be the pooled process of *all* points t_i and t_i^0.

A straightforward way of specifying an autoregressive intensity process is to parameterize the intensity function in terms of a time series model. Then, $\lambda(t; \mathcal{F}_t)$ follows a dynamic process that is updated whenever a new point occurs. As proposed by Russell (1999), the intensity is driven by three components: one component $\Phi(t)$ capturing the dynamic structure, a baseline intensity component $\lambda_0(t)$ (Russell calls it backward recurrence time function), as well as a deterministic function of time (e.g., periodicity component) $s(t)$. By imposing a multiplicative structure, the ACI model is obtained by a dynamic extension of a (parametric) proportional intensity (PI) model and is given by

$$\lambda(t; \mathcal{F}_t) = \Phi(t)\lambda_0(t)s(t). \tag{11.1}$$

By including both time-invariant and time-varying covariates in static form and specifying $\Phi(t)$ in logarithmic form (to ensure non-negativity), $\Phi(t)$ is given by

$$\Phi(t) = \exp\left(\tilde{\Phi}_{\breve{N}(t)+1} + \mathbf{z}'_{\breve{N}(t)}\boldsymbol{\gamma} + \mathbf{z}^{0\prime}_{\breve{N}^0(t)}\boldsymbol{\vartheta}\right), \tag{11.2}$$

$$\tilde{\Phi}_i = \omega + \sum_{j=1}^{P} \alpha_j \tilde{\varepsilon}_{i-j} + \sum_{j=1}^{Q} \beta_j \tilde{\Phi}_{i-j}, \tag{11.3}$$

where $\boldsymbol{\gamma}$ and $\boldsymbol{\vartheta}$ denote coefficient vectors and the innovation term ε_i can specified either as

$$\tilde{\varepsilon}_i := 1 - \varepsilon_i = 1 - \Lambda(t_{i-1}, t_i) \tag{11.4}$$

or, alternatively, as

$$\tilde{\varepsilon}_i := -0.5772 - \ln \varepsilon_i = -0.5772 - \ln \Lambda(t_{i-1}, t_i) \tag{11.5}$$

with $\varepsilon_i := \Lambda(t_{i-1}, t_i) := \int_{t_{i-1}}^{t_i} \lambda(s; \mathcal{F}_s) ds$ denoting the integrated intensity function. Note that $\Phi(t)$ is a left-continuous function that only changes during a spell due to the evolution of the time-varying covariate process. Hence, in absence of time-varying covariates, $\Phi(t)$ remains constant between t_{i-1} and t_i, i.e.,

[1] Such a process might be associated, for instance, with the arrival of new (limit) orders in the market.

$$\Phi(t) = \Phi(t_i) \qquad \text{for } t_{i-1} < t \le t_i.$$

In that case, $\Phi(t_i)$ is known instantaneously after the occurrence of t_{i-1} and does not change until t_i. Then, $\lambda(t)$ changes between t_{i-1} and t_i only as a deterministic function of time (according to $\lambda_0(t)$ and $s(t)$). Under the conditions given in Theorem 4.2 in Chap. 4, the integrated intensity follows an i.i.d. standard exponential process, hence, the model innovations $\tilde{\varepsilon}_i$ are i.i.d. exponential variates in (11.4) or extreme value variates in (11.5) that are centered by their unconditional mean and enter the model negatively. This implies a positive value of α whenever the intensity is positively autocorrelated.

Note that in contrast to the SACPH model presented in Chap. 10, the baseline intensity function $\lambda_0(t)$ is fully parameterized. Russell (1999) suggests to specify $\lambda_0(t) = \lambda_0(x(t))$ according to a standard Weibull parameterization

$$\lambda_0(t) = \exp(\omega)x(t)^{a-1}, \qquad a > 0, \tag{11.6}$$

where a value of a larger (smaller) than one is associated with an upward (downward) sloping intensity function, i.e., "positive" or respectively "negative" duration dependence. Alternatively, a standard Burr type baseline intensity function is obtained by

$$\lambda_0(t) = \exp(\omega)\frac{x(t)^{a-1}}{1 + \kappa x(t)^a}, \qquad a > 0,\ \eta \ge 0, \tag{11.7}$$

which allows for non-monotonous hazard shapes.

Note that $\tilde{\Phi}_i$ follows an ARMA type dynamic that is updated by i.i.d. zero mean innovations. Thus, $\mathbb{E}[\tilde{\Phi}_i] = 0$ and the weak stationarity of $\tilde{\Phi}_i$ is ensured if the roots of the lag polynomial based on the persistence parameters $\beta_1, \ldots, \beta_Q$ lie inside the unit circle.

To show the relationship to the ACD model, recall (5.8) from Chap. 5 that the intensity representation of an ACD model is given by

$$\lambda(t; \mathcal{F}_t) = h_\varepsilon\left(\frac{x(t)}{\Psi_{\breve{N}(t)+1}}\right)\frac{1}{\Psi_{\breve{N}(t)+1}},$$

where $\Psi_i := \mathbb{E}[x_i|\mathcal{F}_{i-1}]$ denotes the conditionally expected duration mean and h_ε denotes the (baseline) hazard function induced by the assumed distribution of ε_i, $i = 1, \ldots, n$, with $\varepsilon_i := x_i\Psi_i^{-1}$ being i.i.d. We observe two major differences to the ACD framework: Firstly, in contrast to Φ_i, Ψ_i is a discrete-time function which is only updated at the end of each spell. Secondly, the baseline hazard h_ε does not only depend on $x(t)$ but on an "accelerated" time scale $x(t)/\Psi_{\breve{N}(t)+1}$. Hence, for both models to coincide we have to rule out the possibility of time-varying covariates and have to impose specific assumptions on the form of the baseline intensity. For simplicity, set $s(t) = 1$ and assume $\vartheta = 0$. Then, $\lambda_0(t) = 1$ implies $\Psi_i = \Phi_i^{-1}$ with

$$
\begin{aligned}
\Psi_i &= \exp\left(\tilde{\Psi}_i - \mathbf{z}_{i-1}'\boldsymbol{\gamma}\right), \\
\tilde{\Psi}_i &= -\omega + \alpha(x_{i-1}/\Psi_{i-1} - 1) + \beta(\tilde{\Psi}_{i-1} + \omega) \\
&= \tilde{\omega} + \alpha x_{i-1}/\Psi_{i-1} + \beta\tilde{\Psi}_{i-1},
\end{aligned}
\tag{11.8}
$$

corresponding to a Log ACD (type II) model with $\tilde{\omega} := (\beta - 1)\omega - \alpha$ (see Chap. 5). Note that a constant baseline intensity function is equivalent to the assumption of conditionally exponentially distributed durations. A similar relationship can be derived for a Weibull parameterization of the baseline intensity. Hence, the fact that AFT and PI models coincide in case of Weibull or exponential distributions carries over to the coincidence between ACI and Log-ACD models (as long as there are no time-varying covariates). Building on the close relationship between ACI and ACD models in case of conditionally exponentially distributed durations and exploiting the inverse relationship between intensity and conditionally expected durations, Hamilton and Jorda (2002) propose a so-called autoregressive conditional hazard (ACH) model of the form

$$
\lambda(t;\mathcal{F}_t) = \frac{1}{\Psi_{\breve{N}(t)+1} + \mathbf{z}_{\breve{N}^0(t)}^{0\prime}\boldsymbol{\vartheta}},
\tag{11.9}
$$

corresponding to a particular type of ACI model.

The explicit derivation of theoretical moments of the intensity function is complicated by the fact that conditional expectations of $\lambda(t;\mathcal{F}_t)$ typically cannot be computed analytically. This is because the relationship between the intensity function at some point t_i and the expected time until the next point t_{i+1} generally cannot be expressed in closed form. In general, the computation of the conditional expected arrival time of the next point, $\mathbb{E}[t_i | \mathcal{F}_{t_{i-1}}]$, is performed by exploiting the relationship $\varepsilon_i := \Lambda(t_{i-1}, t_i) \sim Exp(1)$. However, in case of a Weibull parameterization of $\lambda_0(t)$, (11.6), $\Lambda(t_{i-1}, t_i)$ can be actually expressed in closed form. Then, under the assumption of no (time-varying) covariate arrival during the current spell, $\mathbb{E}[t_i | \mathcal{F}_{t_{i-1}}]$ is computed as

$$
\mathbb{E}[t_i | \mathcal{F}_{t_{i-1}}] = t_{i-1} + \mathbb{E}\left[\left[\frac{\varepsilon_i a}{\Phi(t_i)\exp(\omega)}\right]^{1/a} \middle| \mathcal{F}_{t_{i-1}}\right].
\tag{11.10}
$$

Correspondingly, the conditionally expected intensity at the next point t_i is derived as

$$
\begin{aligned}
\mathbb{E}[\lambda(t_i) | \mathcal{F}_{t_{i-1}}] &= \Phi(t_i)\exp(\omega)\mathbb{E}\left[(t_i - t_{i-1})^{a-1} \middle| \mathcal{F}_{t_{i-1}}\right] \\
&= \Phi(t_i)\exp(\omega)\mathbb{E}\left[\left[\frac{\varepsilon_i a}{\Phi(t_i)\exp(\omega)}\right]^{\frac{a-1}{a}} \middle| \mathcal{F}_{t_{i-1}}\right].
\end{aligned}
\tag{11.11}
$$

Hence, these expressions are computed based on the conditional mean of transformations of an exponential variate. The calculation of autocovariances of the intensity function and of the corresponding durations x_i has to be performed numerically on the basis of simulation procedures. Figures 11.1–11.3 show simulated ACI(1,1) processes based on different parameter settings with a constant baseline intensity function $\lambda_0(t) = \exp(\omega)$. The figures depict the ACF of $\lambda(t_i; \mathcal{F}_{t_i})$ (measured at the points $t_1, t_2, \ldots$), as well as the ACF of the resulting duration process. The parameterizations reflect different persistence levels. The processes shown in Figs. 11.1 and 11.2 are based on a value of $\beta = 0.97$ and thus are close to the non-stationary region. Both processes imply significantly autocorrelated durations coming along with an exponentially decay of the ACF. In particular, the ACF of the durations in Fig. 11.1 shows a shape that is quite typical for financial duration series. Obviously, the persistence of the duration processes strongly depends on the value of β, while the strength of the serial dependence is driven by the innovation parameter α. For example, for $\alpha = 0.05$ and $\beta = 0.7$ (Fig. 11.3), the ACF of the duration series declines sharply and quickly

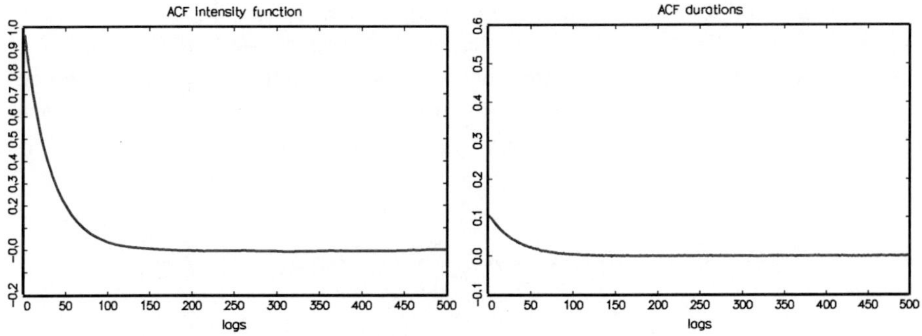

Fig. 11.1 ACF of univariate ACI(1,1) processes. *Left*: ACF of $\lambda(t_i; \mathcal{F}_{t_i})$. *Right*: ACF of x_i. Based on 5,000,000 drawings. $\omega = 0, \alpha = 0.05, \beta = 0.97$

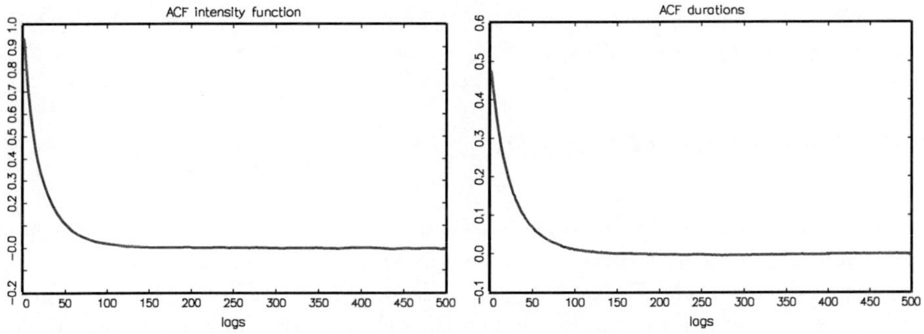

Fig. 11.2 ACF of univariate ACI(1,1) processes. *Left*: ACF of $\lambda(t_i; \mathcal{F}_{t_i})$. *Right*: ACF of x_i. Based on 5,000,000 drawings. $\omega = 0, \alpha = 0.2, \beta = 0.97$

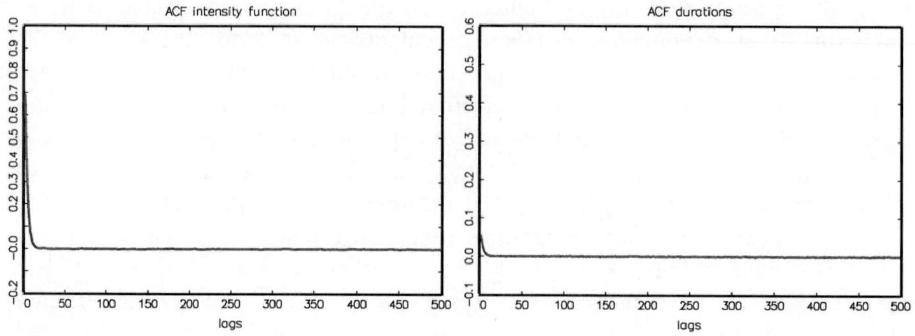

Fig. 11.3 ACF of univariate ACI(1,1) processes. *Left*: ACF of $\lambda(t_i; \mathcal{F}_{t_i})$. *Right*: ACF of x_i. Based on 5,000,000 drawings. $\omega = 0, \alpha = 0.05, \beta = 0.7$

tends towards zero for higher lags. Thus, autocorrelation patterns of intensities that resemble the autocorrelation structures of typical financial duration series require values of β quite close to one.

The log likelihood function of the ACI model is obtained by adapting (4.24) in an obvious way. For example, by assuming a Weibull type baseline intensity function, (11.6), the integrated intensity function can be computed analytically, and the log likelihood function is obtained based on piecewise analytical integrations of the intensity function over all points $\tilde{t}_j$ with $t_{i-1} < \tilde{t}_j \leq t_i$, hence

$$\ln \mathcal{L}(\mathbf{Y}; \boldsymbol{\theta}) = \sum_{i=1}^{n} \left\{ - \sum_{j} \left[\Phi(\tilde{t}_j) s(\tilde{t}_j) \exp(\omega) \frac{1}{a} (\tilde{t}_{j+1} - \tilde{t}_j)^a \right] \right.$$

$$\left. + \ln \left(\Phi(t_i) s(t_i) \exp(\omega) x_i^{a-1} \right) \right\}. \qquad (11.12)$$

However, in the case of more sophisticated parameterizations of $\lambda_0(t)$, $\Lambda(t_{i-1}, t_i)$ has to be computed numerically.

11.2 Generalized ACI Models

In this section, we discuss several extensions of the basic ACI model. Section 11.2.1 introduces an ACI specification which captures long memory dynamics in the intensity. Section 11.2.2 presents a generalized ACI specification which includes the case of accelerated failure time dynamics and nests special cases of the ACD model. In Sect. 11.2.3, we briefly illustrate how to extend the ACI model to allow also for low-frequency (e.g., daily) dynamics. Finally, Sect. 11.2.4 gives empirical illustrations of the individual specifications.

11.2.1 Long-Memory ACI Models

Empirical applications of the ACI model (see Sects. 11.2.4 or 12.2.1) show that financial intensity processes reveal a strong persistence and often indicate the presence of long range dependence.

A simple extension of the basic model, which is easy to estimate but allows for higher persistence, is to specify a two-component ACI model given by

$$\tilde{\Phi}_i = w\tilde{\Phi}_{i,1} + (1-w)\tilde{\Phi}_{i,2}, \tag{11.13}$$

$$\tilde{\Phi}_{i,1} = \alpha_1\tilde{\varepsilon}_{i-1} + \tilde{\Phi}_{i-1,1}, \tag{11.14}$$

$$\tilde{\Phi}_{2,i} = \alpha_2\tilde{\varepsilon}_{i-1} + \beta\tilde{\Phi}_{i-1,2}, \tag{11.15}$$

where $\tilde{\Phi}_{i,1}$ and $\tilde{\Phi}_{i,2}$ denote two intensity components leading to a weighted sum of an integrated process and a weakly stationary process. This model is the counterpart to the two-component ACD model specified by Engle (2000) building on the work of Ding and Granger (1996) (see Sect. 6.3).

Alternatively, ACI dynamics can be straightforwardly extended to allow for long range dependence. Similar to the case of ACD models as discussed in Sect. 6.3, $\tilde{\Phi}_i$ might be parameterized in terms of an infinite series representation according to Koulikov (2003), i.e.,

$$\tilde{\Phi}_i = \omega + \alpha\sum_{j=1}^{\infty}\theta_{j-1}\tilde{\varepsilon}_{i-j}, \quad \theta_j := \sum_{k=0}^{j}\beta^k\theta^*_{j-k}, \tag{11.16}$$

where

$$\theta^*_j := \frac{\Gamma(d+j)}{\Gamma(d)\Gamma(1+j)} \quad \forall\, j \geq 0,$$

are the coefficients of the expansion of $(1-z)^{-d}$ with $|\beta| < 1$. This yields a long memory ACI specification

$$\tilde{\Phi}_i = \omega + \alpha(1-\beta L)^{-1}(1-L)^{-d}\tilde{\varepsilon}_{i-1} \tag{11.17}$$

with $(1-L)^{-d}\tilde{\varepsilon}_{i-1} = \sum_{j=1}^{\infty}\theta_{j-1}\tilde{\varepsilon}_{i-j}$. The model obviously nests the basic ACI specification for $d = 0$ and reveals (stationary) long memory dynamics for $d \in (0, 0.5)$. For a deeper discussion of these types of long memory specifications, see Sect. 6.3.

11.2.2 An AFT-Type ACI Model

As discussed in Sect. 11.1, the ACI and ACD model only coincide in case of Weibull parameterizations of the baseline intensity $\lambda_0(t)$. The reason is that under such a parameterization, AFT and PI specifications coincide since accelerating factors in the baseline intensity can be alternatively written as proportional factors. Therefore, to bridge the gap between the ACI and ACD model and to obtain a more general specification which nests both types of structures, the ACI model has to be extended to allow for AFT effects. Accordingly, a generalized ACI specification nesting both AFT and PI models is given by specifying $\lambda_0(t)$ as

$$\lambda_0(t) = \lambda_0\left(\eta(t)\right), \tag{11.18}$$

where

$$\eta(t) := x(t) \cdot \left[\Phi_{\breve{N}(t)+1} s(t)\right]^{\delta}, \tag{11.19}$$

corresponding to the time elapsed since the last event scaled by $\Phi_{\breve{N}(t)+1}$ and $s(t)$. Hence, $\eta(t)$ can be interpreted as a transformation of the time scale on which the baseline intensity $\lambda_0(\cdot)$ is defined. If $\delta > 0$, ACI dynamics and seasonality effects accelerate the time until the next event, whereas for $\delta < 0$, the time scale is decelerated. By assuming that the ACI dynamics are given by (11.17), setting for convenience $s(t) = 1$, and ruling out time-varying covariates ($\boldsymbol{\vartheta} = 0$), we can distinguish between the following special cases:

1. If the baseline intensity $\lambda_0(\cdot)$ is non-specified, and $\delta = 1$, $\omega = \alpha = \beta = 0$, the model corresponds to the standard (non-dynamic) AFT model (see, e.g., Kalbfleisch and Prentice 1980) as given by

   $$\lambda(t;\mathcal{F}_t) = \lambda(t) = \lambda_0\left[x(t)\exp(\mathbf{z}_{\breve{N}}(t)'\boldsymbol{\gamma})\right]\exp(\mathbf{z}_{\breve{N}}(t)'\boldsymbol{\gamma}).$$

   As discussed in Chap. 4, it can be alternatively written as a log-linear model in terms of the durations $x_i := t_i - t_{i-1}$,

   $$\ln x_i = -\mathbf{z}_{i-1}'\boldsymbol{\gamma} + \xi_i, \quad i = 1,\ldots,n,$$

   where ξ_i is an error term following a non-specified continuous distribution.
2. If the baseline intensity $\lambda_0(\cdot)$ is non-specified, and $\delta = \omega = \alpha = \beta = 0$, the model corresponds to the well-known class of (non-dynamic) semiparametric PH models as discussed in Chap. 4 and given by

   $$\lambda(t;\mathcal{F}_t) = \lambda(t) = \lambda_0(x(t))\exp(\mathbf{z}_{\breve{N}}(t)'\boldsymbol{\gamma}).$$

3. If $\lambda_0(\cdot) = 1$, we have $\Phi_i = \Psi_i^{-1}$ with

   $$\Psi_i = \exp\left(\tilde{\Psi}_i - \mathbf{z}_{i-1}'\boldsymbol{\gamma}\right),$$
   $$\tilde{\Psi}_i = -\omega + \alpha(1-\beta L)^{-1}(1-L)^{-d}(u_{i-1}),$$

which corresponds to a long memory Log-ACD model based on centered standardized durations $u_i := x_i/\Psi_i - 1$ as innovations.

4. If $\lambda_0(\eta(t)) = p \cdot \eta(t)^{p-1}(1 + \kappa\eta(t)^p)^{-1}$ and $\delta = 1$, then

$$\Phi_i = \Psi_i^{-1} \left[\frac{\kappa^{1+1/p}\Gamma(1+1/\kappa)}{\Gamma(1+1/p)\Gamma(\kappa^{-1} - 1/a)} \right] =: \Upsilon_i,$$

$$\Upsilon_i = \exp\left(\tilde{\Upsilon}_i - \mathbf{z}'_{i-1}\boldsymbol{\gamma}\right),$$

$$\tilde{\Upsilon}_i = -\omega - \alpha(1-\beta L)^{-1}(1-L)^{-d}\tilde{\varepsilon}_{i-1},$$

corresponding to a special type of long memory ACD model with the (Burr parameterized) centered integrated intensity as innovation term.

An obvious extension is to allow for component-specific acceleration effects by re-formulating (11.19) as

$$\eta(t) := x(t) \cdot \Phi_{\breve{N}(t)+1}^{\delta_\Phi} s(t)^{\delta_s}, \tag{11.20}$$

where δ_Φ and δ_s are specific acceleration parameters separately affecting the individual components. This specification allows to identify whether a possible acceleration/decelaration effect is mainly driven by seasonality effects or intra-day dynamics. For $\delta_\Phi = \delta_s = \delta$, the model collapses to the basic specification (11.19).

11.2.3 A Component ACI Model

Similarly to the component MEM discussed in Sect. 6.4.3, the (generalized) ACI model can be extended to capture also daily dynamics. Denote $\tau(t)$ as an integer variable indexing the current trading day observed at time t. Thus $\tau(t)$ is constant during a day and jumps only from day to day. Furthermore, define $t^{\dagger}_{\tau(t)}$ and $t^{\ddagger}_{\tau(t)}$ as the time of the opening and closure of the trading day $\tau(t)$, respectively. Then, the ACI model might be re-specified as

$$\lambda(t; \mathcal{F}_t) = \lambda_0(\eta(t))\, \Psi(t) s(t) \varphi_{\tau(t)}, \tag{11.21}$$

where $\eta(t)$ is defined as in Sect. 11.2.2 and $\varphi_{\tau(t)}$ is a function varying only on a daily level. It is given by

$$\varphi_{\tau(t)} = a\zeta(t) + b\varphi_{\tau(t)-1}, \tag{11.22}$$

where the innovation term $\zeta(t)$ is specified as

$$\zeta(t) = 1 - \frac{\Lambda\left(t^{\dagger}_{\tau(t)-1}, t^{\ddagger}_{\tau(t)-1}\right)}{N\left(t^{\ddagger}_{\tau(t)-1}\right) - N\left(t^{\dagger}_{\tau(t)-1}\right)} \tag{11.23}$$

exploiting the basic results of the martingale theory of point processes. As shown in Chap. 4, $(N(t) - N(s)) - \Lambda(t,s)$ is a mean zero martingale. Consequently, $1 - \Lambda(t,s)/(N(t) - N(s))$ and thus $\zeta(t)$ are mean zero martingales as well.

Hence, both types of dynamics Φ_i and $\varphi_{\tau(t)}$ are driven by functions of the lagged integrated intensity. In order to distinctly disentangle both types of dynamics, Φ_i should be re-initialized at the beginning of each day. Then, intensity dynamics might reveal long memory dynamics within individual trading days and short memory dynamics across trading days.

11.2.4 Empirical Application

Table 11.1 shows estimates of different generalized ACI specifications based on JPM 10bp price durations for June 2009. In summary, the most general specification is given by

$$\begin{aligned}
\lambda(t;\mathcal{F}_t) &= \lambda_0\left(\eta(t)\right)\Psi(t)s(t)\varphi_{\tau(t)},\\
\Psi(t) &= \exp\left(\tilde{\Phi}_{\breve{N}(t)+1}\right)\\
\tilde{\Phi}_i &= \omega + \alpha(1-\beta L)^{-1}(1-L)^{-d}\tilde{\varepsilon}_{i-1}\\
&\quad + \varsigma(1-\beta L)^{-1}(1-L)^{-d}(|\tilde{\varepsilon}_{i-1}| - \mathbb{E}[|\tilde{\varepsilon}_{i-1}|]),\\
\varphi_{\tau(t)} &= a\zeta(t) + b\varphi_{\tau(t)-1}.\\
\lambda_0(\eta(t)) &= \exp(\omega)\frac{\eta(t)^{a-1}}{1+\kappa\eta(t)^a},\\
\eta(t) &= x(t)\cdot\Phi^{\delta_\Phi}_{\breve{N}(t)+1}\varphi^{\delta_\varphi}_{\tau(t)}s(t)^{\delta_s},\\
\zeta(t) &= 1 - \frac{\Lambda\left(t^{\dagger}_{\tau(t)-1}, t^{\ddagger}_{\tau(t)-1}\right)}{N\left(t^{\ddagger}_{\tau(t)-1}\right) - N\left(t^{\dagger}_{\tau(t)-1}\right)},\\
s(t) &= 1 + \sum_{j=1}^{Q}\delta_j(t_i - \tau_j)\mathbb{1}_{\{t_i>\tau_j\}}.
\end{aligned}$$

Table 11.1 Maximum likelihood estimates of univariate generalized ACI models. Based on 10bp-price durations of the JPM stock traded at the New York Stock Exchange, June 2009. 28,077 observations. The spline function is based on 6 equally spaced nodes between 9:30 a.m. and 4:00 p.m. Standard errors are based on the outer product of gradients

	A		B		C		D		E	
	est.	p-v.	est.	p-v.	est.	p-v.	est.	p-v.	est.	p-v.
Baseline intensity parameters										
ω	1.587	0.000	1.306	0.000	1.260	0.000	2.071	0.000	2.071	0.000
p	1.124	0.000	1.125	0.000	1.123	0.000	1.327	0.000	1.327	0.000
κ	0.160	0.000	0.158	0.000	0.153	0.000	2.512	0.000	2.512	0.000
Acceleration parameters										
δ_Φ							1.061	0.000	1.061	0.000
δ_s							1.017	0.000	1.061	
δ_φ							0.960	0.000	1.061	
Dynamic parameters										
α	0.087	0.000	0.089	0.000	0.151	0.000	0.172	0.000	0.172	0.000
ς	0.035	0.000	0.039	0.000	0.069	0.000	0.084	0.000	0.084	0.000
β	0.988	0.000	0.982	0.000	0.076	0.000	0.082	0.002	0.082	0.002
a			0.104	0.000	0.090	0.000	0.089	0.000	0.089	0.000
b			0.335	0.000	0.489	0.000	0.462	0.000	0.462	0.000
d					0.338	0.000	0.343	0.000		
Seasonality parameters										
ν_1	−1.836	0.000	−1.874	0.000	−1.907	0.000	−1.917	0.000	−1.917	0.000
ν_2	1.555	0.000	1.611	0.000	1.670	0.000	1.646	0.000	1.646	0.000
ν_3	0.195	0.020	0.173	0.027	0.151	0.018	0.183	0.002	0.183	0.002
ν_4	0.160	0.016	0.164	0.009	0.156	0.003	0.141	0.002	0.141	0.002
ν_5	0.068	0.379	0.057	0.433	0.011	0.851	0.017	0.727	0.017	0.727
ν_6	0.922	0.000	0.891	0.000	0.864	0.000	0.678	0.000	0.678	0.000
LL	−18,080		−18,034		−17,956		−17,569		−17,569	
BIC	−18,141		−18,106		−18,032		−17,661		−17,661	
Residual diagnostics										
Mean of e_i	1.012		1.012		1.017		1.009		1.009	
S.D. of e_i	1.088		1.087		1.093		0.989		0.989	
LB(20) of e_i	95.670		85.461		26.572		23.923		23.923	

Note: In specification E, the restriction $\delta := \delta^\Phi = \delta^\varphi = \delta^s$ is imposed

Hence, the model allows for long memory dynamics, a daily dynamic component as well as acceleration effects of both intradaily and daily dynamics. Moreover, the long memory specification of $\tilde{\Phi}_i$ allows for asymmetric effects of past innovations $\tilde{\varepsilon}_i$ on the current intensity. It is specified in analogy with Nelson's (1991) EGARCH specification and allows for a kinked news impact curve governed by the parameter ς.

Panel A shows the results of a basic ACI specification (with asymmetric news impact) which is subsequently extended by an inter-day component (Panel B), long memory dynamics (Panel C) and acceleration effects (Panels D and E). Intraday periodicities are captured according to (5.33) using $Q = 6$ knots. The estimates of ς reflect the presence of asymmetric news impact effects. Hence, in periods where the price intensity, i.e., the instantaneous volatility, is higher than expected, we observe a stronger impact on the expected intensity than in periods where the intensity is lower than expected. Moreover, we observe clear evidence for positive serial dependencies in the inter-day component (Panel B). Hence, confirming the results by Brownlees et al. (2011), the (average) daily level of price intensities is time-varying and positively autocorrelated. Nevertheless, even though the specification leads to a better goodness-of-fit of the model, it is not sufficient to fully capture the dynamic properties of the process. The inclusion of a long memory component is necessary to achieve a significant improvement of the model's dynamic properties as indicated by the residual diagnostics and the BIC. We find a value of d ranging around 0.33 indicating the presence of long range dependence with the process being covariance stationary. Furthermore, clear evidence for acceleration effects (Panel D) are found. The estimated acceleration parameters support the idea of an accelerated specification in contrast to a proportional model. We observe that the individual parameters are relatively similar and are close to one yielding specification F as the most parsimonious one. Finally, though a Burr parameterization allows for quite flexible shapes of the baseline intensity, the residual diagnostics still reveal significant overdispersion and thus a limited ability of the model to fully capture the distributional properties in the data. Actually, it is shown that this overdispersion is mainly caused by very long price durations which are not easily captured by a standard parametric form and requires even more flexible baseline intensities.

11.3 Hawkes Processes

A valuable alternative to an autoregressive specification is to parameterize the intensity function in terms of a self-exciting process. In general form, such a process can be written as

$$\begin{aligned} \lambda(t;\mathcal{F}_t) =& \varphi\left(\mu(t) + \int_{-\infty}^{t} w(t-s)dN(s)\right) \\ =& \varphi\left(\mu(t) + \sum_{i\geq 1} \mathbb{1}_{\{t_i \leq t\}} w(t-t_i)\right), \end{aligned} \tag{11.24}$$

where φ is a possibly nonlinear function, $\mu(t)$ is a constant, and $w(\cdot)$ is a non-negative weighting function. The processes are referred to as *Hawkes processes* as Hawkes (1971) was among the first who systematically studied this class of processes. Hawkes processes serve as epidemic models since the occurrence of a number of events increases the probability for further events. In natural sciences, they play an important role in modelling the emission of particles from a radiating body or in forecasting seismic events (see, for example, Vere-Jones 1970, Vere-Jones and Ozaki 1982 or Ogata 1988). Bowsher (2007) is among the first applications using these types of processes to model financial data.

In self-exciting processes, the intensity is driven by a weighted non-increasing function of the backward recurrence time to all previous points. Accordingly, the intensity is high whenever we have observed many events in recent periods. Such a specification naturally captures event clustering and thus positive autocorrelations in event durations. If $\varphi(\lambda) \neq \lambda$, we obtain the class of non-linear Hawkes processes which are studied, e.g., by Brémaud and Massoulié (1996). For instance, choosing $\varphi(\lambda) = \exp(\lambda)$ allows to preserve the non-negativity of the process without additional parameter constraints.

Note that the Hawkes model allows to estimate dependencies in the intensity process without imposing parametric time series structures. The dependence is rather specified in terms of the elapsed time since past events. Hence, the marginal contribution of previous events on the current intensity is independent of the number of intervening events.

In case of $\varphi(\lambda) = \lambda$, we obtain the class of linear Hawkes processes as studied by Hawkes (1971). In this case, the intensity is a linear function of the sum of the time distance to all past points. The strength of the impact of past events is driven by the function w which might be economically interpreted as an impulse response function. Hawkes (1971) suggests parameterizing $w(t)$ as

$$w(t) = \sum_{j=1}^{P} \alpha_j \exp(-\beta_j t), \tag{11.25}$$

where $\alpha_j \geq 0$, $\beta_j \geq 0$ for $j = 1, \ldots, P$, are parameters and P determines the (exogenously chosen) order of the process. The parameter α_j determines the scale, whereas β_j determines the time decay of the influence of past points of the process. Accordingly, the response of a previous event on the intensity function in t decays exponentially and is driven by the parameter β_j. For $P > 1$, the Hawkes(P) model is based on the superposition of differently parameterized exponentially decaying weighted sums of the backward recurrence time to all previous points. In this case, identification requires some constraint on the parameters β_j, e.g., $\beta_1 > \ldots > \beta_P$. As shown by Hawkes and Oakes (1974), a Hawkes process can be represented (and generated) by clusters of Poisson processes. This principle can be exploited to simulate a Hawkes process. For more details, see, Møller and Rasmussen (2005) or Daley and Vere-Jones (2005).

The function $\mu(t)$ can be specified in terms of covariates $\mathbf{z}_{\breve{N}(t)}$ and $\mathbf{z}^0_{\breve{N}^0(t)}$ and a seasonality function $s(t)$, thus

$$\mu(t) = \omega + s(t) + \mathbf{z}'_{\check{N}(t)}\boldsymbol{\gamma} + \mathbf{z}^{0\prime}_{\check{N}^0(t)}\boldsymbol{\vartheta}, \tag{11.26}$$

which, however, does not ensure the non-negativity of $\mu(t)$. Alternatively, $\mu(t)$ might be specified in logarithmic form. Seasonality functions might enter the model also multiplicatively. Then, the model is given by

$$\lambda(t;\mathcal{F}_t) = s(t)\left\{\mu(t) + \sum_{j=1}^{P}\sum_{i=1}^{\check{N}(t)} \alpha_j \exp(-\beta_j(t-t_i))\right\}. \tag{11.27}$$

Obviously, alternative functional forms for the decay function are possible. For instance, an alternative parameterization is

$$w(t) = \frac{H}{(t+\kappa)^p}, \tag{11.28}$$

with parameters H, κ, and $p > 1$ featuring a hyperbolic decay. Such weight functions are typically applied in seismology and allow to capture long range dependence. Since financial intensity processes also tend to reveal long memory behavior, this specification may be interesting in financial applications as well.

However, the choice of an exponential decay simplifies the derivation of the theoretical properties of the model. More details can be found in Hawkes (1971), Hawkes and Oakes (1974) or Ogata and Akaike (1982). As shown by Hawkes (1971), stationarity of the process requires $0 < \int_0^\infty w(s)ds < 1$ which is ensured by $\sum_{j=1}^P \alpha_j/\beta_j < 1$. For the special case $P = 1$ and $\mu(t) = \mu$, the unconditional mean of $\lambda(t;\mathcal{F}_t)$ is given by

$$\mathbb{E}[\lambda(t_i)] = \frac{\mu}{1 - \int_0^\infty \alpha \exp(-\beta u)du} = \frac{\mu\beta}{\beta - \alpha}. \tag{11.29}$$

As in the case of ACI processes, the computation of the conditional expectation of the next point of the process, $\mathbb{E}[t_i \mid \mathcal{F}_{t_{i-1}}]$, cannot be expressed in closed form. Therefore, t_i is computed as the solution of

$$\begin{aligned}\varepsilon_i &= \Lambda(t_{i-1}, t_i)\\ &= \int_{t_{i-1}}^{t_i} \mu(s)ds - \sum_{j=1}^{P}\sum_{k=1}^{i-1} \frac{\alpha_j}{\beta_j}\left\{\exp(-\beta_j(t_i - t_k)) - \exp(-\beta_j(t_{i-1} - t_k))\right\},\end{aligned} \tag{11.30}$$

where ε_i denotes an i.i.d. standard exponential variate. The solution of (11.30) leads to a nonlinear function of an exponentially distributed random variable, $t_i = g_1(\varepsilon_i;\mathcal{F}_{t_{i-1}})$. Its conditional mean can be calculated using the law of iterated expectations (under the assumption of no time-varying covariates) as

$$\mathbb{E}\left[t_i \,|\, \mathcal{F}_{t_{i-1}}\right] = \mathbb{E}\left[\mathbb{E}\left[g_1(\varepsilon_i; \mathcal{F}_{t_{i-1}}) | \, \varepsilon_i\right] | \, \mathcal{F}_{t_{i-1}}\right]. \tag{11.31}$$

Then, the conditional expectation of $\lambda(t_i)$ given the information set at t_{i-1} is computed as

$$\mathbb{E}\left[\lambda(t_i) | \, \mathcal{F}_{t_{i-1}}\right] = \mu(t_{i-1}) + \sum_{j=1}^{P} \sum_{k=1}^{i-1} \alpha_j \mathbb{E}\left[\exp(-\beta_j t_i) \big| \, \mathcal{F}_{t_{i-1}}\right] \exp(\beta_j t_k). \tag{11.32}$$

This expressions requires to compute the conditional expectation of a function of the time until the next point. However, this conditional expectation typically cannot be expressed in closed form and requires to use simulation methods.

Figures 11.4–11.6 show the autocorrelation functions of the intensities $\lambda(t_i; \mathcal{F}_{t_i})$ and of the resulting durations $x_i = t_i - t_{i-1}$, evaluated at each event, for the intensities following Hawkes processes with different parameterizations. It is shown that the persistence of the intensity process and the resulting duration process strongly depends on the ratio α/β. Figure 11.4 is based on a parameterization

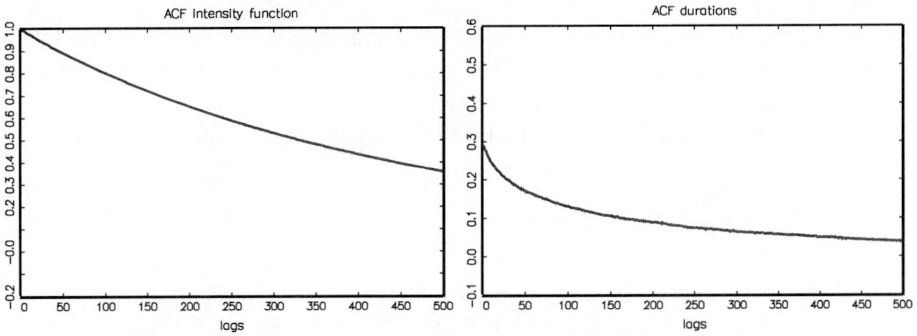

Fig. 11.4 ACF of univariate Hawkes(1) processes. *Left*: ACF of $\lambda(t_i; \mathcal{F}_{t_i})$. *Right*: ACF of x_i. Based on 5,000,000 drawings. $\omega = 0.2$, $\alpha = 0.2$, $\beta = 0.21$

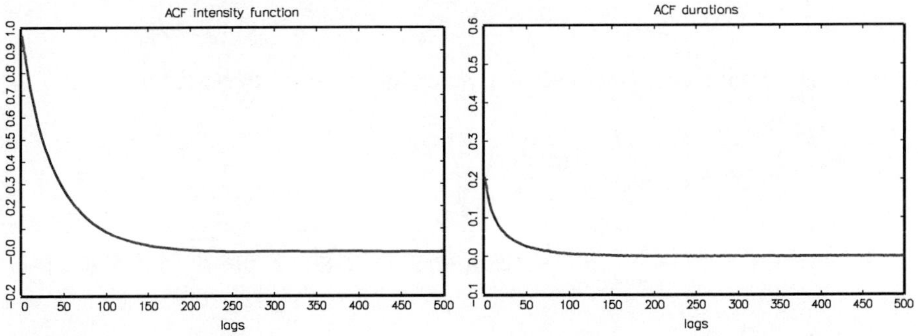

Fig. 11.5 ACF of univariate Hawkes(1) processes. *Left*: ACF of $\lambda(t_i; \mathcal{F}_{t_i})$. *Right*: ACF of x_i. Based on 5,000,000 drawings. $\omega = 0.2$, $\alpha = 0.2$, $\beta = 0.25$

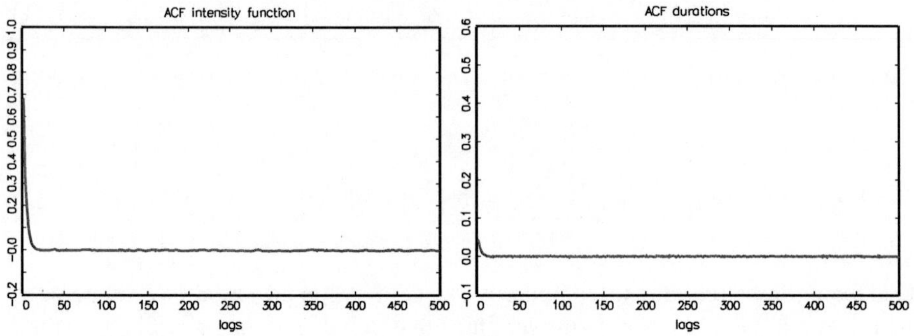

Fig. 11.6 ACF of univariate Hawkes(1) processes. *Left*: ACF of $\lambda(t_i; \mathcal{F}_{t_i})$. *Right*: ACF of x_i. Based on 5,000,000 drawings. $\omega = 0.2$, $\alpha = 0.2$, $\beta = 0.5$

leading to a quite persistent intensity process. The ACF of the implied duration process shows a pattern that is very realistic for financial duration series (compare to the empirical autocorrelation functions shown in Chap. 3). Hence, the Hawkes process allows for persistent duration processes implying a slowly decaying ACF. It turns out that the persistence in the implied duration series is generally higher than for ACI processes. Figure 11.5 shows a less persistent process associated with autocorrelation functions that clearly decay faster than in Fig. 11.4. In contrast, Fig. 11.6 reveals only very weak serial dependencies.

The Hawkes(P) model is estimated by ML using the log likelihood function

$$\begin{aligned}\ln \mathcal{L}(\mathbf{Y}; \boldsymbol{\theta}) = \sum_{i=1}^{n} \Bigg\{ & - \int_{t_{i-1}}^{t_i} \mu(s) ds \\ & - \sum_{j=1}^{P} \sum_{k=1}^{i-1} \frac{\alpha_j}{\beta_j} \left[\exp(-\beta_j (t_i - t_k)) - \exp(-\beta_j (t_{i-1} - t_k)) \right] \\ & + \ln \left[\mu(t_i) + \sum_{j=1}^{P} \sum_{k=1}^{i-1} \alpha_j \exp(-\beta_j (t_i - t_k)) \right] \Bigg\}, \end{aligned} \tag{11.33}$$

which can be computed on the basis of a recursion and can be re-written as

$$\begin{aligned}\ln \mathcal{L}(\mathbf{Y}; \boldsymbol{\theta}) = \sum_{i=1}^{n} \Bigg\{ & - \int_{t_{i-1}}^{t_i} \mu(s) ds \\ & - \sum_{j=1}^{P} \sum_{k=1}^{i-1} \frac{\alpha_j}{\beta_j} \left[\exp(-\beta_j (t_i - t_k)) - \exp(-\beta_j (t_{i-1} - t_k)) \right] \\ & + \ln \left[\mu(t_i) + \sum_{j=1}^{P} \alpha_j A_i^j \right] \Bigg\}, \end{aligned} \tag{11.34}$$

where

$$A_i^j := \sum_{k=1}^{i-1} \exp(-\beta_j(t_i - t_k)) = \exp(-\beta_j(t_i - t_{i-1}))(1 + A_{i-1}^j). \quad (11.35)$$

An important advantage of the Hawkes parameterization is that $\Lambda(t_i, t_{i-1})$, and thus the log likelihood function, can be computed in closed form and require no numerical integration. Nevertheless, the occurrence of time-varying covariates requires integrating the function $\mu(s)$ piecewise over all intervening points $\{\tilde{t}_j\}$ with $t_{i-1} < \tilde{t}_j \leq t_i$.

References

Bowsher CG (2007) Modelling security markets in continuous time: intensity based, multivariate point process models. J Econom 141:876–912
Brémaud P, Massoulié L (1996) Stability of nonlinear Hawkes processes. Ann Probab 24: 1563–1588
Brownlees C, Cipollini F, Gallo GM (2011) Intra-daily volume modeling and prediction for algorithmic trading. J Financ Econom 9:489–518
Daley D, Vere-Jones D (2005) An introduction to the theory of point processes. Volume I: Elementary theory and methods. Springer, New York
Ding Z, Granger CWJ (1996) Modeling volatility persistence of speculative returns: a new approach. J Econom 73:185–215
Engle RF (2000) The econometrics of ultra-high-frequency data. Econometrica 68(1):1–22
Hamilton JD, Jorda O (2002) A model of the federal funds rate target. J Polit Econ 110:1135
Hawkes AG (1971) Spectra of some self-exciting and mutually exciting point processes. Biometrika 58:83–90
Hawkes AG, Oakes D (1974) A cluster process representation of a self-exciting process. J Appl Probab 11:493–503
Kalbfleisch JD, Prentice RL (1980) The statistical analysis of failure time data, Wiley, New York
Koulikov D (2003) Modeling sequences of long memory non-negative covariance stationary random variables. Discussion Paper 156, CAF
Møller J, Rasmussen J (2005) Perfect simulation of Hawkes processes. Adv Appl Probab 37: 629–646
Nelson D (1991) Conditional heteroskedasticity in asset returns: a new approach. J Econom 43:227–251
Ogata Y (1988) Statistical models for earthquake occurrences and residual analysis for point processes. J Am Stat Assoc 83:9–27
Ogata Y, Akaike H (1982) On linear intensity models for mixed Doubly stochastic poisson and self-exciting point processes. J R Stat Soc Series B 44:102–107
Russell JR (1999) Econometric modeling of multivariate irregularly-spaced high-frequency data. Working Paper, University of Chicago
Vere-Jones D (1970) Stochastic models for earthquake occurrence. J R Stat Soc Series B 32:1–62
Vere-Jones D, Ozaki T (1982) Some examples of statistical inference applied to earthquake data. Ann Inst Stat Math 34:189–207

Chapter 12
Multivariate Dynamic Intensity Models

This chapter presents multivariate extensions of dynamic intensity models. Section 12.1 considers multivariate autoregressive conditional intensity (ACI) models which have been originally proposed by Russell (1999). Applications of this framework are presented in Sect. 12.2. We illustrate how a multivariate ACI model is used to estimate simultaneous buy and sell trading intensities and to model order aggressiveness in an open limit order book market. Section 12.3 presents multivariate Hawkes processes, discusses their statistical properties and shows in an empirical illustration how Hawkes processes can be employed to estimate multivariate price intensities. Finally, Sect. 12.4 discusses Stochastic Conditional Intensity (SCI) processes as introduced by Bauwens and Hautsch (2006). These models are based on the assumption that the conditional intensity, given the (observable) history of the process, is not deterministic but stochastic and follows a dynamic process. Similarly to the Stochastic Multiplicative Error Model presented in Chap. 6, the underlying idea is to introduce a latent dynamic factor which drives commonalities in intensities. We discuss the model's properties, statistical inference and present an empirical application to the analysis of commonalities in intraday volatility.

12.1 Multivariate ACI Models

Recall the setup and notation of the univariate ACI model introduced in Chap. 11. Then, a multivariate ACI model is a straightforward multivariate extension and is based on the specification of the vector of K intensity functions

$$\boldsymbol{\lambda}(t;\mathcal{F}_t) := \begin{bmatrix} \lambda^1(t;\mathcal{F}_t) \\ \lambda^2(t;\mathcal{F}_t) \\ \vdots \\ \lambda^K(t;\mathcal{F}_t) \end{bmatrix}, \qquad (12.1)$$

N. Hautsch, *Econometrics of Financial High-Frequency Data*,
DOI 10.1007/978-3-642-21925-2_12,

where each component is parameterized as

$$\lambda^k(t;\mathcal{F}_t) = \Psi^k(t)\lambda_0^k(t)s^k(t), \qquad k = 1,\ldots,K, \tag{12.2}$$

where $\Psi^k(t)$, $\lambda_0^k(t)$ and $s^k(t)$ are the corresponding k-type dynamic component, baseline intensity function and a deterministic function of time. Russell (1999) proposes specifying $\Psi^k(t)$ as[1]

$$\Psi^k(t) = \exp\left(\tilde{\Psi}^k_{\breve{N}(t)+1} + \mathbf{z}'_{\breve{N}(t)}\boldsymbol{\gamma}^k + \mathbf{z}^{0\prime}_{\breve{N}^0(t)}\boldsymbol{\vartheta}^k\right), \tag{12.3}$$

where the vector $\tilde{\Psi}_i := \left(\tilde{\Psi}_i^1, \tilde{\Psi}_i^2, \ldots, \tilde{\Psi}_i^K\right)'$ is parameterized in terms of a VARMA type specification, given by

$$\tilde{\Psi}_i = \left(\mathbf{A}^k\breve{\varepsilon}_{i-1} + \mathbf{B}^k\tilde{\Psi}_{i-1}\right)y^k_{i-1}, \tag{12.4}$$

with $\mathbf{A}^k = \{\alpha_j^k\}$ being a $K \times 1$ vector associated with the innovation term $\breve{\varepsilon}_{i-1}$, $\mathbf{B}^k = \{\beta_{ij}\}$ being a $K \times K$ matrix of persistence parameters and y_i^k defined as an indicator variable taking the value one if the ith event is of type k and being zero otherwise. Hence, each of the processes $\tilde{\Phi}_i^k$, $k = 1,\ldots,K$, correspond to a univariate autoregressive process with regime-switching dynamics in dependence of the type of the most recent point. The individual processes are linked together since they are jointly updated by ε_i at each point of the pooled process.

As proposed by Russell (1999), ε_i can be specified in terms of the integrated intensity associated with the type of the most current point leading to

$$\breve{\varepsilon}_i = \sum_{k=1}^{K}\left(1 - \Lambda^k(t_{i-1}^k, t_i^k)\right)y_i^k \tag{12.5}$$

or

$$\breve{\varepsilon}_i = \sum_{k=1}^{K}\left(-0.5772 - \ln\Lambda^k(t_{i-1}^k, t_i^k)\right)y_i^k, \tag{12.6}$$

where

$$\Lambda^k(t_{i-1}^k, t_i^k) = \sum_j \int_{\tilde{t}_j^k}^{\tilde{t}_{j+1}^k} \lambda^k(s;\mathcal{F}_s)ds, \tag{12.7}$$

Then, at each event t_i, all K processes are updated by the realization of the integrated intensity with respect to the most recent process, where the impact of the innovation

[1]For ease of illustration, we restrict our analysis to a lag order of one. The extension to higher order specifications is straightforward.

on the K processes can be different and also varies with the type of the most recent point. As shown in Chap. 4, under correct specification of $\lambda(t; \mathcal{F}_t)$, the integrated intensity functions $\{\Lambda(t_{i-1}^k, t_i^k)\}_{\{i=1,\ldots,n^k\}}$ are i.i.d. standard exponential variates. Consequently, ε_i corresponds to a (random) mixture of zero mean i.i.d. variates and thus is itself a zero mean i.i.d. random variable. As a result, stability conditions depend on the eigenvalues of the matrices $\mathbf{B}^k$. Because of the regime-switching nature of the persistence matrix, the derivation of stationarity conditions is difficult. However, a sufficient (but not necessary) condition is that the eigenvalues of the matrices $\mathbf{B}^k$ for all $k = 1, \ldots, K$ lie inside the unit circle.

Alternatively, Bowsher (2007) suggests specifying the innovation in terms of the integrated intensity of the pooled process, i.e.

$$\tilde{\varepsilon}_i = 1 - \Lambda(t_{i-1}, t_i), \tag{12.8}$$

where $\Lambda(t_{i-1}, t_i) = \sum_{k=1}^{K} \Lambda^k(t_{i-1}, t_i)$ denotes the integrated intensity of the pooled process computed between the two most recent points. Consequently, $\tilde{\varepsilon}_i$ is also a mean zero i.i.d. innovation term.

Alternatively, we can exploit the fact that $N(t) - \int_0^t \lambda(s)ds$ is a martingale difference (see Chap. 4). Consequently, $(N^k(t_i) - N^k(t_{i-1})) - \int_{t_{i-1}}^{t_i} \lambda^k(s)ds$ is a martingale difference as well. Then, an alternative specification of ACI dynamics is obtained by

$$\mathbf{\Phi}_i = \mathbf{A}\tilde{\boldsymbol{\varepsilon}}_{i-1} + \mathbf{B}\mathbf{\Psi}_{i-1}, \tag{12.9}$$

where both $\mathbf{A}$ and $\mathbf{B}$ are $(K \times K)$ parameter matrices and

$$\tilde{\boldsymbol{\varepsilon}}_i := \left(\tilde{\varepsilon}_i^1, \tilde{\varepsilon}_i^2, \ldots, \tilde{\varepsilon}_i^K\right)'$$

with

$$\tilde{\varepsilon}_i^j := (N^j(t_i) - N^j(t_{i-1})) - \Lambda^j(t_{i-1}, t_i) = y_i^j - \Lambda^j(t_{i-1}, t_i),$$

for $k = 1, \ldots, K$.

The baseline intensity functions are extensions of the univariate case. The most simple way is to assume that $\lambda_0^k(t)$ is constant, but depends on the type of the event that occurred most recently, i.e.,

$$\lambda_0^k(t) = \exp(\omega_r^k)\, y_{\breve{N}(t)}^k, \quad r = 1, \ldots, K, \quad k = 1, \ldots, K. \tag{12.10}$$

Alternatively, the baseline intensity function may be specified as a product of Weibulll hazard rates, i.e.,

$$\lambda_0^k(t) = \exp(\omega^k) \prod_{r=1}^{K} x^r(t)^{a_r^k - 1}, \; a_r^k > 0, \tag{12.11}$$

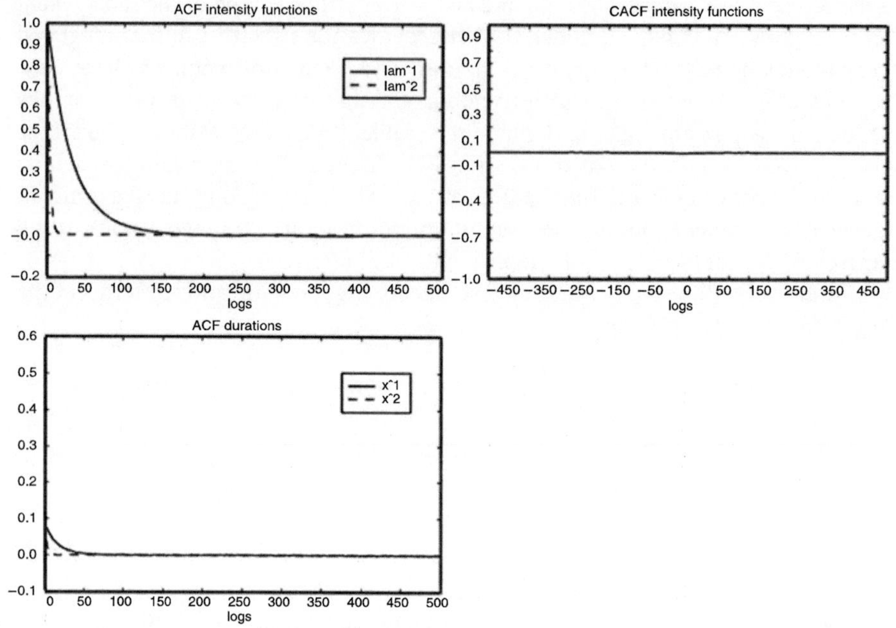

Fig. 12.1 Simulated bivariate ACI(1,1) processes. *Top*: ACF and CACF of $\lambda^{1,2}(t_i; \mathcal{F}_{t_i})$. *Bottom*: ACF of $x_i^{1,2}$. $\omega^1 = \omega^2 = 0$, $\alpha_1^1 = \alpha_2^2 = 0.05$, $\alpha_2^1 = \alpha_1^2 = 0$, $\beta_{11} = 0.97$, $\beta_{12} = \beta_{21} = 0$, $\beta_{22} = 0.7$. Based on 5,000,000 drawings

or of Burr hazard rates,

$$\lambda_0^k(t) = \exp(\omega^k) \prod_{r=1}^{K} \frac{x^r(t)^{a_r^k - 1}}{1 + \kappa_r^k x^r(t)^{a_r^k}}, \quad a_r^k > 0, \kappa_r^k \geq 0. \tag{12.12}$$

A special case occurs when the kth process depends only on its own backward recurrence time, which corresponds to $a_r^k = 1$ and $\kappa_r^k = 0$, $\forall\ r \neq s$.

The computations of $\mathbb{E}\left[t_i^k \middle| \mathcal{F}_{t_{i-1}} \right]$ and $\mathbb{E}\left[\lambda^k(t_i) \middle| \mathcal{F}_{t_{i-1}} \right]$ are complicated by the fact that in the multivariate case, $\lambda_0^k(t)$ depends on the backward recurrence times of *all* K processes. In that case, there exists no closed form solution for the integrated intensity,[2] and thus, $\Lambda^k(t_{i-1}^k, t_i^k)$ as well as the conditional expectations of t_i^k and $\lambda^k(t_i)$ have to be computed numerically.

Figures 12.1–12.3 show the autocorrelation functions (ACFs) of bivariate simulated ACI(1,1) processes. In order to illustrate the dynamic interactions between the two processes, the cross-autocorrelation functions (CAFCs) are shown as

[2] With the exception of the case of a constant baseline intensity function.

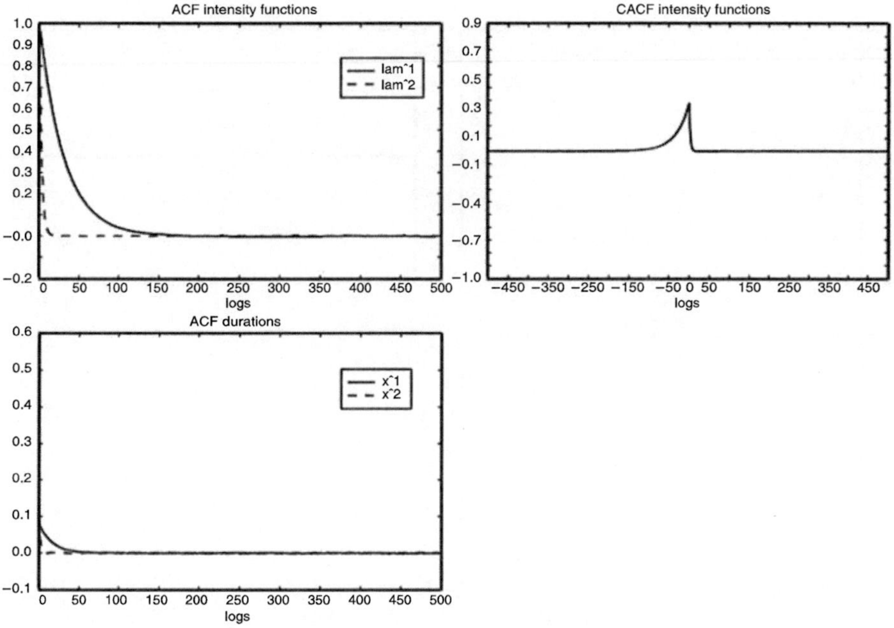

Fig. 12.2 Simulated bivariate ACI(1,1) processes. *Top*: ACF and CACF of $\lambda^{1,2}(t_i; \mathcal{F}_{t_i})$. *Bottom*: ACF of $x_i^{1,2}$. $\omega^1 = \omega^2 = 0$, $\alpha_1^1 = \alpha_2^1 = \alpha_2^2 = 0.05$, $\alpha_1^2 = 0$, $\beta_{11} = 0.97$, $\beta_{12} = \beta_{21} = 0$, $\beta_{22} = 0.7$. Based on 5,000,000 drawings

well.[3] All simulations are based on constant baseline intensity functions implying no interdependence between the processes. Figure 12.1 is based on a completely diagonal ACI specification. Thus, the processes do not interact, which is also revealed by the CACF. In contrast, the process in Fig. 12.2 implies an asymmetric interdependence ($\alpha_1^2 = 0$) between the processes, while Fig. 12.3 is associated with (symmetric) interactions in both directions ($\alpha_2^1 = \alpha_1^2 = 0.05$). The CACF shows an increase around zero, which is the more pronounced the stronger the interactions between the processes. The asymmetry of the dip results from the fact that the process-specific serial dependencies are of different strengths. Thus, the higher persistence of $\lambda^1(t_i; \mathcal{F}_{t_i})$ implies higher values of $\text{Corr}\left[\lambda^1(t_i; \mathcal{F}_{t_i}), \lambda^2(t_{i-j}; \mathcal{F}_{t_{i-j}})\right]$ for $j < 0$ than for $j > 0$.

The log likelihood function of the multivariate ACI model based on data $\mathbf{Y}$ and parameters $\boldsymbol{\theta}$ is computed as

$$\ln \mathcal{L}(\mathbf{Y}; \boldsymbol{\theta}) = \sum_{k=1}^{K} \sum_{i=1}^{n} \left\{ -\Lambda^k(t_{i-1}, t_i) + y_i^k \ln \lambda^k(t_i; \mathcal{F}_{t_i}) \right\}, \tag{12.13}$$

where $\Lambda^k(t_{i-1}, t_i)$ is calculated according to (12.7).

[3]The graphs of the CACF depict the plot of $\text{Corr}\left[\lambda^1(t_i; \mathcal{F}_{t_i}), \lambda^2(t_{i-j}; \mathcal{F}_{t_{i-j}})\right]$ vs. j.

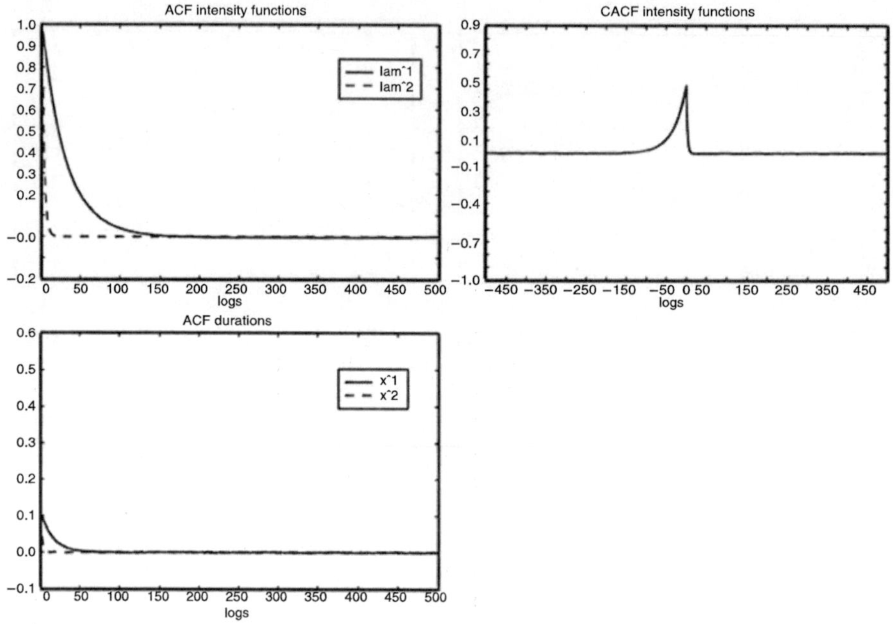

Fig. 12.3 Simulated bivariate ACI(1,1) processes. *Top*: ACF and CACF of $\lambda^{1,2}(t_i; \mathcal{F}_{t_i})$. *Bottom*: ACF of $x_i^{1,2}$. $\omega^1 = \omega^2 = 0$, $\alpha_1^1 = \alpha_2^1 = \alpha_1^2 = \alpha_2^2 = 0.05$, $\beta_{11} = 0.97$, $\beta_{12} = \beta_{21} = 0$, $\beta_{22} = 0.7$. Based on 5,000,000 drawings

12.2 Applications of Multivariate ACI Models

12.2.1 Estimating Simultaneous Buy/Sell Intensities

Hall and Hautsch (2007) apply bivariate ACI specifications to model the simultaneous buy and sell trade arrival process at the Australian Stock Exchange in dependence of the state of the underlying order book. For such an application, a multivariate intensity model is particularly attractive as it avoids any aggregation and allows to process all limit order book activities.

Exploiting information from fully re-constructed limit order books, Hall and Hautsch (2007) construct various variables allowing to test hypotheses on the trading behavior of market participants. As summarized in Table 12.1, four groups of explanatory variables are used. The first type of variables captures the effect of overnight returns and consists of the log ratio between the previous day's closing price and the current opening price (OCP) as well as its interaction with two time dummies ($OCP1$ and $OCP2$) which allow to investigate whether the influence of overnight returns changes over a trading day.

The second category consists of variables that are observed only when a market order is executed, such as the difference between the posted log limit price and the

Table 12.1 Definition of explanatory variables used by Hall and Hautsch (2007)

Variables observed at each transaction	
$TRVB$	Traded buy volume
$TRVS$	Traded sell volume
Time-varying covariates associated with the limit order process	
$AVOL$	Total volume on the ask queue
$BVOL$	Total volume on the bid queue
D_VOL	AVOL-BVOL
A_x	Price associated with the $x\%$ quantile of the cumulated volume on the ask queue
B_x	Price associated with the $x\%$ quantile of the cumulated volume on the bid queue
MQ	Midquote $((A_0 + B_0)/2)$
$DFF1$	$DFF1 = \|B_{10} - mq\| - \|A_{10} - mq\|$
$DFF2$	$DFF2 = \|B_{20} - B_{10}\| - \|A_{20} - A_{10}\|$
$DFF4$	$DFF4 = \|B_{40} - B_{20}\| - \|A_{40} - A_{20}\|$
$DFF6$	$DFF6 = \|B_{60} - B_{40}\| - \|A_{60} - A_{40}\|$
$DFF9$	$DFF9 = \|B_{90} - B_{60}\| - \|A_{90} - A_{60}\|$
$SPRD$	Bid-ask spread
DMQ	Midquote-change $DMQ = MQ_i - MQ_{i-1}$
$QASK$	1: if limit order is an ask, 0: if limit order is a bid
$QVOLA$	Quoted volume if limit order is an ask, (0 for bids)
$QVOLB$	Quoted volume if limit order is a bid, (0 for asks)

prevailing log ask price in case of a buy or log bid price in case of a sell (DQP) as a measure for the investor's implicit marginal valuation of the asset. Moreover, the change of the best ask/bid log quote due to a transaction is included (DP). To control for potential non-linearities, DP is interacted with a dummy variable indicating whenever best ask and bid prices are shifted by more than one tick ($DP1$). Moreover, the signed log traded quantity at the most recent transaction (TRV) as well as its interaction with a dummy variable indicating volumes greater than the median volume ($TRV1$) and the signed difference between the quoted and the traded log volume as a measure for the non-executed quantity of a market order (DV) are used.

The third group of variables captures the most recent state of the order book as well as characteristics of the most recent limit order arrivals, such as the signed log volume of a limit order set below (above), at or above (below) the most current best ask (bid) price (LO_{-1}, LO_0, LO_{+1}) and the signed log volume of a cancellation above or below the current best ask/bid ($CAN0$, $CAN1$). Moreover, the relative price differential between limit prices associated with specific volume quantiles (relative to the current mid-quote price) ($ASL\cdot$, $BSL\cdot$) capture the piecewise slopes of the ask and bid curves. Further order book characteristics are the log aggregated trading volume on the bid and ask queues (AV and BV) reflecting the overall current aggregated demand and supply in the market and the relative spread between the current best bid and best ask price (SPR) relative to the midquote as a measure for the tightness of the market.

The fourth group of variables are indicators for market movements during the last 30 min. They contain the signed order flow computed as the cumulated change of the pending log ask volume in excess of the log bid volume during the last 30 min (OFL), the corresponding log midquote return (DMQ) as well as the changes of the ask and bid slopes during the last 30 min ($DASL$ and $DBSL$). Finally, the order flow volatility measured by the square root of the cumulated squared (de-meaned) differences between the pending log ask and bid volume ($OFLV$) during the last 30 min, and the midquote volatility, measured by the square root of the cumulated squared de-meaned log midquote changes during the last 30 min ($VOLA$), reflect the current uncertainty in the market.

Table 12.2 reproduces a part of the estimation results from Hall and Hautsch (2007) for bivariate ACI models for the ASX stocks BHP Billiton Limited (BHP), the National Australian Bank (NAB) and Telstra (TLS), during the period July to August 2002. Hall and Hautsch (2007) estimate bivariate ACI(1,1) specifications for buy (B) and sell (S) trades represented by $k \in \{B, S\}$. The dynamics are specified by (12.4) with $\mathbf{B}^b = \mathbf{B}^s = \mathbf{B}$ and innovations given by (12.5). The specifications include time-varying covariates which are updated whenever the order book is changed due to the arrival of new orders or changes of pending orders. All regressors enter the model in lagged form. The baseline intensity function is specified in terms of a Burr parameterization as given in (12.12) yielding the best goodness-of-fit in terms of the BIC for specifications excluding spill-over effects, i.e., $p_B^S = p_S^B = 1$ and $\kappa_B^S = \kappa_S^B = 0$. The seasonality function is assumed to be common for both processes and is specified as a linear spline function based on six equally-spaced nodes between 10:30 and 16:00.

For all stocks, the intensity processes are very persistent with persistence parameters close to unity and comparably small innovation coefficients. Moreover, there is empirical evidence for mostly positive interdependencies between the buy and sell side. The residual diagnostics are based on the estimated process-specific integrated intensities $\hat{\Lambda}_i^k := \hat{\Lambda}^k(t_{i-1}^k, t_i^k), k \in \{B, S\}$ which should be i.i.d. standard exponentially distributed in case of correct specification (see also Chap. 4). The relatively low values of the Ljung–Box statistics based on the ACI residuals indicate that the specifications widely capture the dynamics in the data. Testing for excess dispersion according to Engle and Russell's (1998) test as discussed in Sect. 4.1.7 reveals remaining slight over-dispersion in the residuals. This shows that even the flexible Burr distribution cannot fully capture the distributional properties of the data. This is a well-known result for trade durations, see also Bauwens et al. (2004) and the corresponding discussion in Chap. 11.

By denoting the buy and sell intensities as $\lambda^B(t; \mathcal{F}_t)$ and $\lambda^S(t; \mathcal{F}_t)$, respectively, Hall and Hautsch (2007) define the so-called *net buy pressure* formally as

$$\Delta^B(t) := \ln \lambda^B(t; \mathcal{F}_t) - \ln \lambda^S(t; \mathcal{F}_t). \tag{12.14}$$

Table 12.2 Estimates of bivariate ACI(1,1) model for the buy (B) and sell (S) intensity with covariates evaluated at each order arrival. Based on all market orders of NAB, BHP and TLS traded at ASX from July 8 to August 30, 2002. Standard errors are computed based on the outer product of gradients. The time scale is standardized by the average duration between two trades. The time series are re-initialized at each trading day. Significance at the 1%, 5% and 10% levels are denoted by ***, ** and *, respectively. Reproduced from Hall and Hautsch (2007)

	Ask side				Bid side		
	NAB	BHP	TLS		NAB	BHP	TLS
ACI parameters							
ω^B	−1.14***	−0.45***	−0.33	ω^S	−0.64***	−0.40***	−0.51***
p_B^B	0.86***	0.89***	0.87***	p_S^S	0.87***	0.85***	0.88***
κ_B^B	0.11***	−3.15***	0.00	κ_S^S	0.11***	−3.36***	0.00
α_B^B	0.12***	0.04***	0.10***	α_S^B	0.05***	0.00	−0.02***
α_B^S	0.04***	0.00	−0.02***	α_S^S	0.09***	0.07***	0.05***
β_{BB}	0.86***	0.99***	0.93***	β_{SB}	0.01	0.01***	0.04***
β_{BS}	−0.05***	0.02***	0.02***	β_{SS}	0.97***	0.97***	0.98***
Impact of overnight price changes							
$OCP1$	−0.56	0.11	0.78	$OCP1$	0.69	−0.02	0.56
$OCP2$	−0.48	0.29	1.00	$OCP2$	0.67	0.00	0.59
Impact of the most recent transaction							
DQP	−0.06	0.00	0.01	DQP	0.17*	0.07**	0.00
DP	−9.60***	−6.18***	−3.34***	DP	8.70***	7.92***	3.39***
$DP1$	2.37***	1.42***	−0.24	$DP1$	−1.42**	−0.07	0.27
TRV	0.17***	0.07***	0.02	TRV	−0.18***	−0.20***	−0.04*
$TRV1$	0.07***	0.15***	0.15***	$TRV1$	−0.04	−0.12***	−0.17***
DV	−0.34***	0.53***	0.75***	DV	0.43***	−0.30***	0.24
Impact of current limit order arrivals							
LO_{-1}	0.26***	0.12***	0.07***	LO_{-1}	−0.30***	−0.17***	−0.10***
LO_0	0.03	0.03*	−0.01	LO_0	−0.01	−0.02	0.03*
LO_{+1}	0.06**	0.05***	0.05**	LO_{+1}	−0.02	−0.05**	0.00
$CAN0$	−0.09*	0.09**	0.13***	$CAN0$	0.06	−0.07	−0.20***
$CAN1$	0.06	0.02	−0.02	$CAN1$	−0.02	0.04	−0.06
Impact of the current state of the market							
AV	1.17	−0.81	0.32	AV	0.25	0.29	0.41
BV	0.28	0.44	−0.32	BV	−0.04	−0.90	−0.39
$BSL02$	−3.43***	−4.00***	−4.84***	$BSL02$	1.32***	4.27***	7.45***
$ASL02$	2.04***	3.42***	7.46***	$ASL02$	−1.18***	−3.39***	−3.05***
$BSL05$	−0.57	−0.71***	−3.36***	$BSL05$	1.71***	1.08***	4.84***
$ASL05$	1.49***	0.69**	5.41***	$ASL05$	−0.77**	−0.06	−2.32***
$BSL10$	0.34	−0.20	−1.42***	$BSL10$	0.40*	−0.12	2.56***
$ASL10$	0.10	−0.11	3.64***	$ASL10$	0.31	0.49**	−0.88**
$BSL20$	0.27	−0.09	1.50**	$BSL20$	0.24	0.87***	1.00*
$ASL20$	0.46*	−0.12	0.65	$ASL20$	0.31	0.00	0.13

(continued)

Table 12.2 (continued)

	Ask side				Bid side		
	NAB	BHP	TLS		NAB	BHP	TLS
$BSL50$	0.28	−0.55	−0.49	$BSL50$	0.73***	0.81**	0.59
$ASL50$	−0.16	−0.15	−0.01	$ASL50$	0.60	−0.15	−0.54
$BSL90$	0.20	−0.33	−0.46	$BSL90$	−0.17	0.72	−0.23
$ASL90$	1.89***	−0.02	0.39	$ASL90$	0.97	−0.11	0.62
SPR	−10.32***	−6.50***	−3.01***	SPR	−8.91***	−5.38***	−2.67***
Impact of the market activity during last 30 min							
$VOLA$	0.29	0.15	0.42	$VOLA$	0.24	0.04	0.08
DMQ	−2.25	−2.26	−1.31	DMQ	0.14	1.75	1.81
$DASL$	−0.76**	0.21	0.23	$DASL$	−0.22	−0.44	0.60
$DBSL$	0.56	0.11	0.25	$DBSL$	−0.43	−0.07	−0.87**
OFL	−0.12	0.02	−0.21	OFL	−0.14	0.15	0.05
$OFLV$	0.08***	0.04***	0.04***	$OFLV$	0.07***	0.07***	0.02**
Seasonality parameters							
ν_1	−1.37***	−0.60***	−0.88***	ν_4	1.53***	1.74***	1.43***
ν_2	1.37***	0.28	0.86***	ν_5	−0.43***	0.03	−0.05
ν_3	−0.80***	−0.65***	−0.83***	ν_6	0.17	0.69***	0.63***
Diagnostics							
n	34,296	44,604	30,878				
n^0	84,164	107,568	73,682				
LL	−50,243	−67,475	−48,269				
BIC	−50,713	−67,957	−48,724				
	Residuals ask side				Residuals sell side		
Mean of $\hat{\Lambda}_i^k$	1.00	0.99	1.00	1.00	1.00	1.00	
S.D. of $\hat{\Lambda}_i^k$	1.03	1.01	1.04	1.03	1.04	1.04	
LB(20) of $\hat{\Lambda}_i^k$	57.15***	78.64***	27.14	27.50	34.52**	63.82***	
Exc. disp.	3.25***	0.86	3.71***	2.70***	3.58	3.13***	

Diagnostics: Log Likelihood (LL), Bayes Information Criterion (BIC) and diagnostics (mean, standard deviation, Ljung–Box statistic and excess dispersion test) of the ACI residuals $\hat{\Lambda}_i^k$, $k \in \{B, S\}$. n: number of trades, n^0: number of total observations (including all limit order activities).

Then, the marginal change of $\Delta^B(t)$ induced by a change of $\mathbf{z}_{M^0(t)}$ is computed as

$$\frac{\partial \Delta^B(t)}{\partial \mathbf{z}_{M^0(t)}} = \boldsymbol{\gamma}^B - \boldsymbol{\gamma}^S. \tag{12.15}$$

The corresponding estimates of $\boldsymbol{\gamma}^B - \boldsymbol{\gamma}^S$ are reported in Table 12.3, where the standard errors are computed by applying the Delta method using the estimated covariance matrix of the parameter estimates shown in Table 12.2.

Table 12.3 Estimates of the difference $\gamma^B - \gamma^S$ based on the estimates reported in Table 12.2. Standard errors are computed by applying the Delta method. Significance at the 1%, 5% and 10% levels are denoted by ***, ** and *, respectively. Reproduced from Hall and Hautsch (2007)

	NAB	BHP	TLS
Impact of overnight price changes			
$OCP1$	−1.25	0.13	0.22
$OCP2$	−1.15	0.29	0.41
Impact of the most recent transaction			
DQP	−0.23*	−0.07*	0.00
DP	−18.30***	−14.10***	−6.74***
$DP1$	3.79***	1.49**	−0.51
TRV	0.35***	0.26***	0.06**
$TRV1$	0.10***	0.27***	0.33***
DV	−0.77***	0.83***	0.51*
Impact of the current state of the market			
AV	0.92	−1.10	−0.09
BV	0.32	1.34	0.07
$BSL02$	−4.76***	−8.27***	−12.29***
$ASL02$	3.21***	6.81***	10.51***
$BSL05$	−2.28***	−1.79***	−8.20***
$ASL05$	2.26***	0.75	7.74***
$BSL10$	−0.06	−0.08	−3.98***
$ASL10$	−0.21	−0.60**	4.52***
$BSL20$	0.03	−0.95***	0.50
$ASL20$	0.15	−0.12	0.52
$BSL50$	−0.45	−1.35***	−1.08**
$ASL50$	−0.76	0.00	0.52
$BSL90$	0.37	−1.05	−0.23
$ASL90$	0.91	0.10	−0.23
SPR	−1.41***	−1.11**	−0.34
Impact of the market activity during last 30 min			
$VOLA$	0.05	0.11	0.33
DMQ	−2.40	−4.01***	−3.12
$DASL$	−0.55	0.65	−0.37
$DBSL$	0.99	0.18	1.12**
OFL	0.02	−0.13	−0.26
$OFLV$	0.00	−0.03	0.01
Impact of current limit order arrivals			
LO_{-1}	0.56***	0.29***	0.17***
LO_0	0.04	0.04*	−0.03
LO_{+1}	0.08**	0.10***	0.05*
$CAN0$	−0.15**	0.16***	0.33***
$CAN1$	0.08	−0.02	0.04

The major findings reported by Hall and Hautsch (2007) are as follows: First, the distance of a limit price to the prevailing best ask or bid quote does not significantly affect the overall trading intensity and the resulting net buy pressure. However, the buy (sell) intensity is decreasing (increasing) after an upward (downward) movement of the best ask (bid) quote as a result of an aggressive market order. Then, traders obviously become reluctant to post a buy (sell) market order as they face worse terms of trade. Secondly, the buy (sell) trading intensity increases with the volume of a buy (sell) order and decreases with the volume of a sell (buy) order. As indicated by the coefficient of $TRV1$, the informational value of transaction volumes seems to increase with the order size. Third, limit orders posted in the spread significantly increase the trading intensity on the opposite side of the market. Hence, traders tend to take the liquidity offered by aggressive limit orders. Conversely, they refrain from trading when the offered liquidity on top of the book vanishes. Fourth, an increase of the ask (bid) slope decreases (increases) the buy intensity and increases (decreases) the sell intensity. Likewise the net buy pressure significantly increases (decreases) when the steepness of the slope of the ask (bid) reaction curve declines. Hall and Hautsch (2007) argue that this finding supports the notion that standing limit prices provide information about traders' upper tail expectations confirming the idea of crowding out effects according to Parlour (1998). Hence, if the market is very deep on a particular side, the execution probability of additionally posted limit orders (entering at the end of the queue) shrinks. As a consequence, traders directly post a market order. Fifth, the bid-ask spread is negatively correlated with the overall trading intensity. The results indicate also a negative relationship between the bid-ask spread and the resulting buy-sell pressure. Sixth, as long as the current state of the limit order book is taken into account, recent market activities have only limited impact on trading intensities. This finding suggests that the current state of the book carries sufficient information to which investors particularly pay attention. Finally, it is shown that trading activity on both sides of the market is significantly higher in periods of high order flow volatility confirming the hypothesis that differences in traders' expectations increase the intensity in market order trading.

Overall, these results show that the limit order book has a significant impact on the buy and sell trading intensity. Traders seem to monitor the book and strategically post orders in order to reduce trading costs and to increase execution probabilities. Moreover, the results by Hall and Hautsch (2007) indicate that market participants seem to trade not only due to liquidity reasons but also tend to exploit information on liquidity supplier's price expectations revealed by the book.

12.2.2 Modelling Order Aggressiveness

Hall and Hautsch (2006) employ a six-dimensional ACI process to study traders' order aggressiveness in the open limit order book market of the ASX. They apply a modification of the order aggressiveness categorization scheme proposed by

Table 12.4 Classification of order aggressiveness at the ASX as used by Hall and Hautsch (2006)

Aggressive buy order	Quoted volume exceeds first ask level
Normal buy order	Quoted volume does not exceed first ask level
Most aggressive ask order	Limit price undercuts current best ask
Aggressive ask order	Limit price is the current best ask
Normal ask order	Limit price is above current best ask
Canceled ask order	Cancellation of a limit ask order
Aggressive sell order	Quoted volume exceeds the first bid level
Normal sell order	Quoted volume does not exceed first bid level
Most aggressive bid order	Limit price overbids current best bid
Aggressive bid order	Limit price is at current best bid
Normal bid order	Limit price is below current best bid
Canceled bid order	Cancellation of a limit bid order

Biais et al. (1995) that classifies orders according to their implied price impact and their position in the order book. Hall and Hautsch adapt this scheme to the ASX and define a "normal" market order as a buy or sell order whose volume can be fully matched with pending limit orders. Ask and bid limit orders are classified according to the distance between the posted limit price and the current best bid and ask price. Accordingly, "most aggressive" limit orders are orders whose price undercuts or overbids the current best ask or bid limit price, respectively, "aggressive" limit orders are orders placed directly in the current first level of the ask or bid queue, and "normal" limit orders are orders entering the higher levels of the order book. Finally, canceled limit orders are regarded as the least aggressive orders. The corresponding classification is shown in Table 12.4.

To thin the point process, Hall and Hautsch (2006) suggest focusing only on most aggressive market orders, limit orders and cancellations on both sides of the market yielding a six-dimensional point process. On top of this classification scheme, only those orders are selected whose volumes are substantially larger than the average order volume. This is motivated by the notion that order aggressiveness is naturally linked to the size of the posted volume. Actually, for larger volumes (greater or equal than the 75%-quantile), the economic trade-off between the costs of immediacy and the pick-off risk is much more relevant than for small orders.

Table 12.5 reproduces parts of the estimation results from Hall and Hautsch (2006) based on a six-dimensional ACI specification similar to that presented in the previous section. To reduce the number of parameters, the persistence matrix **B** is specified as diagonal matrix. Moreover, there are no time-varying covariates as regressors are treated as constant during each spell. To capture the state of the market, Hall and Hautsch specify covariates using the log aggregated volumes pending on the ask and bid queues, $AV = \ln(avol)$ and $BV = \ln(bvol)$, the (signed) cumulative change in the logarithmic aggregated ask volume (DAV), the logarithmic aggregated bid volume (DBV) as well as the mid-quote (MQ) process during the past 5 min. Moreover, the current volatility (VL), measured by the average squared mid-quote changes during the past 5 min as well as the current

Table 12.5 Maximum likelihood estimates of six-dimensional ACI(1,1) models for intensity processes of (1) aggressive buy orders, (2) aggressive sell orders, (3) aggressive ask limit orders, (4) aggressive bid limit orders, (5) aggressive cancellations of ask orders, (6) aggressive cancellations of bid orders. Backward recurrence functions are specified in terms of individual univariate Weibull parameterizations. The persistence vectors $\mathbf{A}^k$ are fully parameterized, whereas $\mathbf{B}$ is parameterized as diagonal matrix. Three spline functions are specified for market orders (ν^{12}), limit orders (ν^{34}) and cancellations (ν^{56}) based on one hour nodes between 10 a.m. and 4 p.m. All covariates except VOL are scaled by 10. Standard errors are computed based on OPG estimates. The time series are re-initialized at each trading day. Significance at the 1%, 5% and 10% levels are denoted by ***, ** and *, respectively. Reproduced from Hall and Hautsch (2006)

	BHP	NAB	TLS		BHP	NAB	TLS
	Constants and backward recurrence parameters						
ω^1	−1.059***	−0.336***	−1.085***	p^1	0.844***	0.797***	0.832***
ω^2	−0.301***	−0.285***	−1.207***	p^2	0.845***	0.778***	0.836***
ω^3	−0.793***	−0.782***	−0.722***	p^3	0.834***	0.905***	0.825***
ω^4	−0.497***	−1.348***	−0.420*	p^4	0.818***	0.862***	0.700***
ω^5	−1.352***	−1.397***	−1.025***	p^5	0.742***	0.749***	0.653***
ω^6	−1.704***	−1.442***	−1.363***	p^6	0.762***	0.741***	0.671***
	Innovation parameters						
α_1^1	0.115***	0.114***	0.235***	α_2^1	0.003	0.029**	0.046
α_1^2	−0.010	0.013	0.179***	α_2^2	0.120***	0.115***	0.244***
α_1^3	0.052***	0.038***	0.160	α_2^3	−0.010	−0.009	−0.007
α_1^4	−0.033**	−0.009	0.076	α_2^4	0.018	0.042***	0.263*
α_1^5	0.031**	0.037**	0.074**	α_2^5	0.046***	0.049***	−0.030
α_1^6	0.044***	0.027*	0.049**	α_2^6	0.010	0.042***	0.030*
α_3^1	0.042***	−0.002	0.085	α_4^1	0.028**	0.053***	−0.079*
α_3^2	0.026*	−0.007	−0.159**	α_4^2	0.064***	0.048***	0.046
α_3^3	0.097***	0.073***	0.438***	α_4^3	0.012	0.008	−0.393**
α_3^4	0.076***	−0.004	0.010	α_4^4	0.116***	0.083***	−0.042
α_3^5	0.051***	0.034**	0.063	α_4^5	0.031*	0.056**	0.017
α_3^6	0.056***	0.032**	−0.038	α_4^6	0.022	0.010	−0.014
α_5^1	−0.008	−0.017	−0.019	α_6^1	0.002	−0.054**	0.127***
α_5^2	0.023	0.013	0.185***	α_6^2	0.017	−0.024	−0.056***
α_5^3	−0.007	0.017	0.186*	α_6^3	0.000	0.014	−0.451**
α_5^4	0.058*	0.015	0.129	α_6^4	0.001	−0.002	−0.088
α_5^5	0.002	0.097***	0.068**	α_6^5	0.005	0.030	0.029
α_5^6	0.094***	0.002	0.030	α_6^6	0.067***	0.119***	0.019
	Persistence parameters						
β^{11}	0.980***	0.961***	0.922***	β^{44}	0.950***	0.984***	−0.251***
β^{22}	0.969***	0.968***	0.820***	β^{55}	0.980***	0.980***	0.948***
β^{33}	0.955***	0.991***	0.481***	β^{66}	0.979***	0.982***	−0.973***
	Seasonality parameters						
$\nu_{11:00}^{12}$	−0.673***	−0.646***	8.226***	$\nu_{11:00}^{34}$	0.160	−0.334	3.955**
$\nu_{12:00}^{12}$	0.138	0.266	−14.730***	$\nu_{12:00}^{34}$	−1.159**	0.183	−7.780**
$\nu_{13:00}^{12}$	−0.981***	−1.341***	2.870*	$\nu_{13:00}^{34}$	−0.438***	−1.551***	3.113***

(continued)

Table 12.5 (continued)

	BHP	NAB	TLS		BHP	NAB	TLS
$\nu^{12}_{14:00}$	3.161***	3.553***	10.182***	$\nu^{34}_{14:00}$	2.956	3.548***	5.563**
$\nu^{12}_{15:00}$	−0.286	−0.521*	−6.032**	$\nu^{34}_{15:00}$	0.025***	−1.032**	−4.674
$\nu^{12}_{16:00}$	0.468	0.516	6.824**	$\nu^{34}_{16:00}$	−0.237	0.652	1.557
$\nu^{56}_{11:00}$	−0.572*	−1.387***	−0.264	$\nu^{56}_{14:00}$	3.755***	2.814***	4.863***
$\nu^{56}_{12:00}$	0.266	1.555***	0.359	$\nu^{56}_{15:00}$	−1.313***	−1.044***	−2.817***
$\nu^{56}_{13:00}$	−1.531***	−1.605***	−2.265***	$\nu^{56}_{16:00}$	1.071*	0.799*	3.798***
	Explanatory variables						
AD^1	−7.287***	−2.318***	−18.057***	BD^1	3.974***	1.903***	8.315***
AD^2	4.400***	1.340***	4.035***	BD^2	−7.144***	−1.247***	−16.624***
AD^3	−7.708***	−2.072***	−28.527***	BD^3	0.148	−0.185	4.385***
AD^4	0.709*	−0.302	2.784**	BD^4	−8.672***	−2.400***	−26.667***
AD^5	4.822***	6.074***	2.902***	BD^5	1.516***	−0.048	1.510**
AD^6	0.584	0.177	−1.085	BD^6	6.410***	6.437***	4.674***
AV^1	3.108***	1.579***	10.325***	DAV^1	−0.649**	−0.411**	−2.109***
AV^2	0.477	−0.006	1.283	DAV^2	−0.698**	−0.161	−1.309***
AV^3	8.480***	5.378***	23.421***	DAV^3	0.172	−0.456*	−0.662
AV^4	11.203***	−10.071***	−5.250***	DAV^4	−0.910**	−0.232	−0.938***
AV^5	−0.602	0.635	5.198***	DAV^5	1.341***	1.811***	−0.390
AV^6	−3.195***	−7.523***	−4.753***	DAV^6	0.363	−0.968***	0.524*
BV^1	0.379	−1.429**	−2.643*	DBV^1	−0.440**	−0.629***	−1.498***
BV^2	1.869***	−0.271	10.025***	DBV^2	−0.290	−0.549***	−1.588***
BV^3	−4.407***	−5.105***	−6.776***	DBV^3	−0.447*	−0.228	−0.991*
BV^4	15.485***	11.177***	21.712***	DBV^4	0.539**	−0.167	−0.105
BV^5	−4.124***	−5.123***	−8.142***	DBV^5	−1.576***	−0.289	0.112
BV^6	−1.756**	2.774***	2.399**	DBV^6	0.543*	1.046***	−0.980***
MQ^1	−1.849***	−0.305***	−0.925**	VL^1	3.773***	0.242***	0.124
MQ^2	1.551***	0.248***	1.836***	VL^2	4.790***	0.232***	0.073
MQ^3	−0.257*	−0.010	−0.770*	VL^3	3.936***	0.042	0.055
MQ^4	0.467***	−0.016	0.911*	VL^4	3.921***	0.144***	−0.090
MQ^5	−1.098***	−0.096	0.608	VL^5	5.712***	0.266***	0.055
MQ^6	1.788***	−0.062	0.440	VL^6	5.848***	0.273***	0.069
SP^1	−1.925***	−0.386***	−1.594				
SP^2	−1.699***	−0.297***	−3.223*				
SP^3	0.639***	0.199***	1.661***				
SP^4	0.549***	0.173***	1.753***				
SP^5	−0.857***	0.044	−1.218**				
SP^6	−0.969***	0.124***	−0.887*				
	Diagnostics						
Obs	9316	10463	3102				
LL	−20145	−23836	−6343				
BIC	−20721	−24419	−6850				

(continued)

Table 12.5 (continued)

	BHP	NAB	TLS	BHP	NAB	TLS
	Aggressive buy orders			Aggressive sell orders		
Mean of $\hat{\varepsilon}_i$	1.014	1.001	0.952	0.994	1.000	1.042
S.D. of $\hat{\varepsilon}_i$	1.159	1.067	1.190	1.139	1.058	1.241
LB(20) of $\hat{\varepsilon}_i$	13.039	18.503	14.846	14.905	28.017	26.972
Exc. disp.	5.379***	2.533**	3.369***	5.106***	2.295**	4.501***
	Aggressive ask limit orders			Aggressive bid limit orders		
Mean of $\hat{\varepsilon}_i$	1.020	0.999	0.978	1.015	0.997	1.032
S.D. of $\hat{\varepsilon}_i$	1.095	1.058	1.088	1.091	1.035	1.083
LB(20) of $\hat{\varepsilon}_i$	10.806	22.876	29.364*	20.823	13.765	10.760
Exc. disp.	2.740***	1.796*	1.009	2.714***	0.985	1.151
	Aggressive ask cancellations			Aggressive scellations		
Mean of $\hat{\Lambda}_i^k$	0.991	1.004	1.029	1.006	1.023	1.005
S.D. of $\hat{\Lambda}_i^k$	0.975	0.963	0.975	0.932	0.972	0.905
LB(20) of $\hat{\Lambda}_i^k$	17.313	20.241	31.632**	12.733	27.965	37.679***
Exc. disp.	0.549	0.744	0.464	1.368	0.578	1.701*

Diagnostics: Log Likelihood (LL), Bayes Information Criterion (BIC) and diagnostics (mean, standard deviation, Ljung–Box statistics and excess dispersion test) of ACI residuals $\hat{\Lambda}_i^k$.

bid-ask spread (SP) are included. Finally, by denoting $p_{0.05,a}$ ($p_{0.05,b}$) as the limit price associated with the 5% level of cumulated ask (bid) depth and define mq as the midquote, then $AD = \ln[0.05 \cdot avol/(p_{0.05,a} - mq)]$ and $BD = \ln[0.05 \cdot bvol/(mq - p_{0.05,b})]$ define the (log) ratio between the current 5% volume quantile and the corresponding price impact.

As in Sect. 12.2.1, dynamic spill-overs between the individual intensity processes are primarily positive. This is true for dependencies between the two sides of the market as well as across the different aggressiveness categories. For instance, a higher market order activity also increases the intensity of order cancellations. Summarizing the results by Hall and Hautsch (2006), one observes that an increase of the depth on the ask side increases the aggressiveness in sell market order trading, decreases it in sell limit order trading and increases it in ask cancellations. The converse is true for the bid side and confirms the idea that high depth on one side of the market induces a crowding out from limit orders to market orders on the opposite side of the market (see Parlour 1998). This is also confirmed by a significantly negative relationship between the depth on a particular side of the market and traders' preference to post aggressive market orders on that side. This result, however, also supports the notion that traders might exploit information from the book to infer on price expectations. Hence, greater ask (bid) depth might indicate that a relatively higher proportion of volume is to be sold (bought) at a comparably low (high) price.

Moreover, Hall and Hautsch observe that an increase in one-sided cumulated depth during the past 5 min decreases the intensity of market and limit orders on both sides of the market, and simultaneously increases the cancellation intensity on the same side. Hence, after periods in which significant one-sided volume has been accumulated in the queues, mean reversion effects seem to cause a reduction of the overall order flow and an increase in traders' incentive to remove pending orders.

Finally, it is shown that recent downward (upward) movements of mid-quotes have a significantly negative (positive) effect on traders' aggressiveness on the buy (sell) side. Obviously, a movement of the mid-quote accompanied by the absorption of a substantial part of one-sided depth induces higher trading costs on that side making traders more reluctant to post further aggressive orders.

12.3 Multivariate Hawkes Processes

12.3.1 Statistical Properties

The univariate Hawkes model, as discussed in Sect. 11.3, is readily extended to the multivariate case. Hence, in a K-dimensional Hawkes process, the intensity function associated with the kth process is given by

$$\lambda^k(t;\mathcal{F}_t) = \mu^k(t) + \sum_{l=1}^{K}\sum_{j=1}^{P}\sum_{m=1}^{\breve{N}^l(t)} \alpha_j^{kl} \exp(-\beta_j^{kl}(t - t_m^l)), \tag{12.16}$$

with $\alpha_j^{kl} \geq 0$, $\beta_j^{kl} \geq 0$. Thus, in the multivariate case, $\lambda^k(t;\mathcal{F}_t)$ depends not only on the backward recurrence time to all k-type points, but also on the backward recurrence time to all other points of the pooled process.

The conditional moments of λ^k as well as the conditional expectation of the arrival time of the next point of the process, $\mathbb{E}\left[t_i^k \,\middle|\, \mathcal{F}_{t_{i-1}}\right]$ cannot be expressed in closed form and are computed in a way similar to that shown in Sect. 11.3. Hawkes (1971) provides parameter restrictions for α^{kl} and β^{kl} under which the multivariate process given by (12.16) is weakly stationary. In general these conditions cannot be explicitly expressed in closed form and have to be verified by numerical methods.[4]

Figures 12.4–12.6 show the autocorrelation and cross-autocorrelation functions implied by simulated bivariate Hawkes(1) processes. In Fig. 12.4, it is assumed that $\alpha^{12} = \alpha^{21} = 0$, i.e., both processes are not interdependent. This property is reflected by the plot of the CACF, which is zero. Moreover, we observe distinct differences in the autocorrelation patterns of the intensity function, as well as in the resulting duration processes which are caused by the different persistence of the processes. In Fig. 12.5, an asymmetric interdependence is assumed ($\alpha^{21} = 0$). Here, it is illustrated that even interdependencies in only one direction lead to strong contemporaneous and nearly symmetric CACF patterns. This is obviously

[4] For more details, see Hawkes (1971).

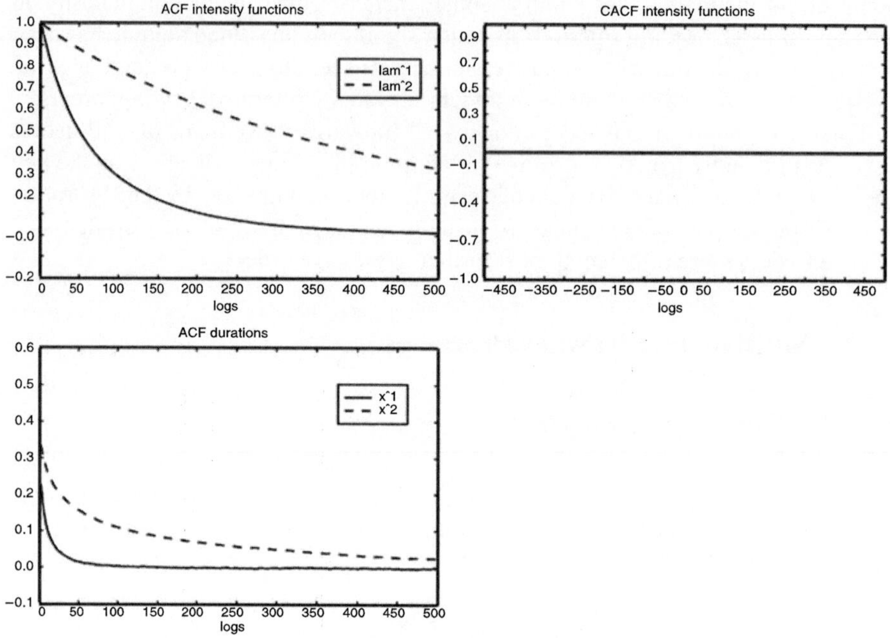

Fig. 12.4 Simulated bivariate Hawkes(1) processes. *Top*: ACF and CACF of $\lambda^{1,2}(t_i; \mathcal{F}_{t_i})$. *Bottom*: ACF of $x_i^{1,2}$. $\omega^1 = \omega^2 = \ln(0.1)$, $\alpha^{11} = \alpha^{22} = 0.2$, $\beta^{11} = 0.25$, $\beta^{22} = 0.21$, $\alpha^{12} = \alpha^{21} = 0$. Based on 5,000,000 drawings

due to the strong serial dependence of the underlying single processes. Figure 12.6 shows the resulting plots of a bivariate Hawkes process where both single processes are (symmetrically) interdependent. Despite the different parameterizations of the individual processes, the resulting intensity functions and autocorrelation functions are quite similar which is caused by the strong interdependence of both processes.

Since the parameters associated with $\lambda^k(t; \mathcal{F}_t)$ are variation free, the log likelihood function of the complete model can be computed as the sum of the log likelihood contributions of each single process $k = 1, \ldots K$. Therefore,

$$\ln \mathcal{L}(\mathbf{Y}; \boldsymbol{\theta}) = \sum_{k=1}^{K} \sum_{i=1}^{n} \Bigg\{ - \int_{t_{i-1}}^{t_i} \mu^k(u) du - \sum_{l=1}^{K} \sum_{j=1}^{P} \sum_{m=1}^{\check{N}^l(t_i)} \frac{\alpha_j^{kl}}{\beta_j^{kl}} \left\{ \exp(-\beta_j^{kl}(t_i - t_m^l)) - \exp(-\beta_j^{kl}(t_{i-1} - t_m^l)) \right\} + \ln \left[\mu^k(t_i) + \sum_{l=1}^{K} \sum_{j=1}^{P} \sum_{m=1}^{\check{N}^l(t_i)} \alpha_j^{kl} A_i^{j,kl} \right] \Bigg\}, \quad (12.17)$$

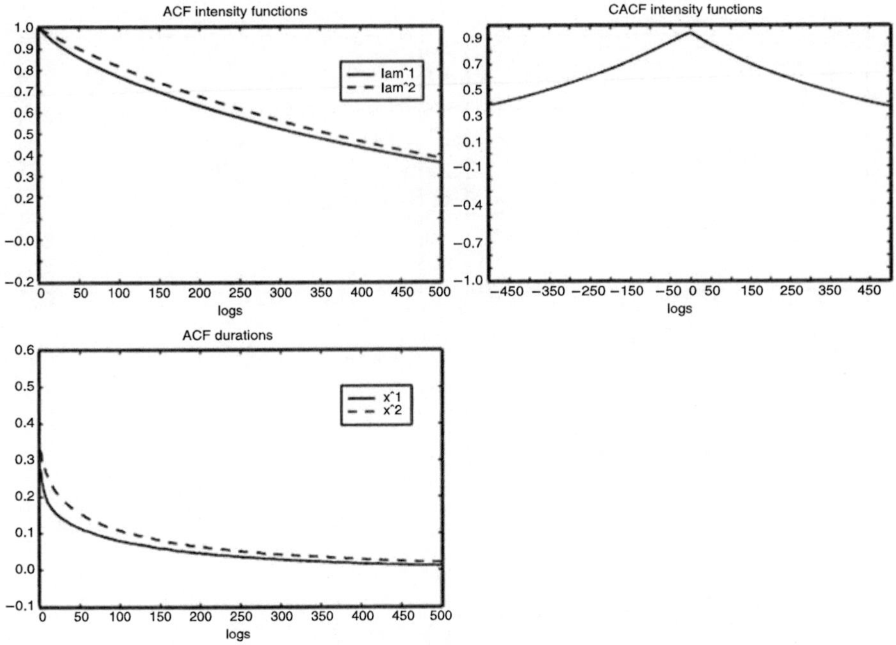

Fig. 12.5 Simulated bivariate Hawkes(1) processes. *Top*: ACF and CACF of $\lambda^{1,2}(t_i; \mathcal{F}_{t_i})$. *Bottom*: ACF of $x_i^{1,2}$. $\omega^1 = \omega^2 = \ln(0.1)$, $\alpha^{11} = \alpha^{22} = 0.2$, $\beta^{11} = 0.4$, $\beta^{22} = 0.21$, $\alpha^{12} = 0.1$, $\alpha^{21} = 0$, $\beta^{12} = 0.3$. Based on 5,000,000 drawings

where

$$A_i^{j,kl} = \sum_{m=1}^{\check{N}^l(t_i)} \exp(-\beta_j^{kl}(t_i - t_m^l))(y_{i-1}^l + A_{i-1}^{j,kl}), \tag{12.18}$$

and y_i^l takes the value one if event i is of type l and zero otherwise. Thus, a multivariate Hawkes model can be estimated by separately maximizing the log likelihood components of the individual processes. This property makes the model attractive for the analysis of high-dimensional point processes.

12.3.2 Estimating Multivariate Price Intensities

This section illustrates the application of a Hawkes process to multivariate volatility estimation based on price intensities. The study uses price durations generated from a sample consisting of five NYSE stocks during the period 01/02/01 to 02/28/01: Boeing, Coca-Cola, General Electric, Home Depot and Philip Morris.

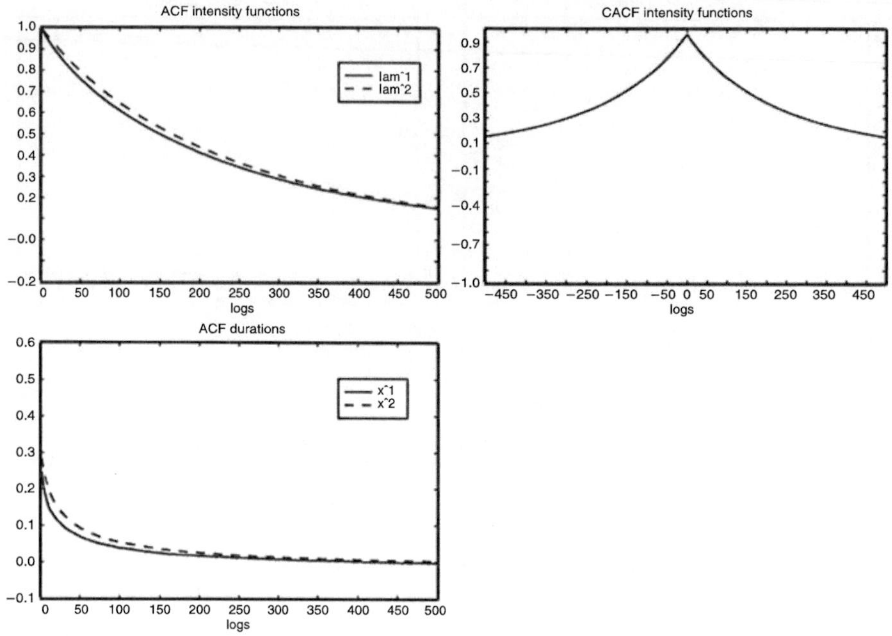

Fig. 12.6 Simulated bivariate Hawkes(1) processes. *Top*: ACF and CACF of $\lambda^{1,2}(t_i; \mathcal{F}_{t_i})$. *Bottom*: ACF of $x_i^{1,2}$. $\omega^1 = \omega^2 = \ln(0.1)$, $\alpha^{11} = \alpha^{22} = 0.2$, $\beta^{11} = 0.4$, $\beta^{22} = 0.25$, $\alpha^{12} = \alpha^{21} = 0.1$, $\beta^{12} = 0.3$, $\beta^{12} = 0.6$. Based on 5,000,000 drawings

Table 12.6 Descriptive statistics and Ljung–Box statistics of $0.20 midquote change durations for the Boeing, Coca-Cola, General Electric, Home Depot and Philip Morris stocks traded at the NYSE. Data extracted from the TAQ database, sample period from 01/02/01 to 02/28/01

	Boeing	Coca-Cola	General Electric	Home Depot	Philip Morris
Obs	16,236	12,678	17,637	18,789	11,152
Mean	57.252	72.757	52.872	49.351	83.158
S.D.	69.222	99.296	66.507	65.094	129.665
LB(20)	2,974	2,248	7,695	3,824	3,442

Overnight spells are removed. Descriptive statistics of durations in seconds.

The generation of price durations is performed according to the procedure described in Sect. 3.2.2. The price durations are computed based on midquote changes of the size $0.20. As shown by the summary statistics in Table 12.6, average price durations are between 45 and 80 s, and thus are associated with an intraday volatility measured at a very high frequency.[5]

The multivariate $0.20 price change intensity is modelled using a five-dimensional Hawkes process. According to the BIC, a lag order of $P=1$ is chosen.

[5]Since the aggregation level is quite low, we neglect overnight effects.

To capture intraday periodicities, the Hawkes component for each process multiplicatively interacts with a linear spline function based on the nodes 9:30, 10:00, 11:00, 12:00, 13:00, 14:00 and 15:00. The model is estimated by maximizing the log likelihood function given in (12.17) and exploiting the property that the log likelihood components associated with the individual processes can be maximized separately. For numerical reasons, the individual series are standardized by the average duration mean of the pooled process. This is, however, only a matter of scaling and does not change the order of the processes.

Table 12.7 gives the estimation results of the five-dimensional Hawkes model. The entries in the table represent the coefficients α^{kl} and β^{kl} associated with the

Table 12.7 ML estimates of a five-dimensional Hawkes(1) model for \$0.20-price durations of the Boeing (B), Coca-Cola (C), General Motors (G), Home Depot (H) and Philip Morris (P) stock traded at the NYSE. Data from the TAQ database, sample period 01/02/01 to 02/28/01. Standard errors based on OPG estimates

	Boeing		Coca-Cola		GM		Home Depot		PM	
	est.	S.E.	est.	S.E.	est.	S.E.	est.	S.E.	est.	S.E.
ω	0.106	0.010	0.060	0.006	0.066	0.010	0.108	0.080	0.026	0.005
$\alpha^{\cdot B}$	0.021	0.001	0.011	0.000	0.023	0.001	0.006	0.056	0.013	0.000
$\alpha^{\cdot C}$	0.015	0.001	0.031	0.002	0.028	0.002	0.006	0.062	0.009	0.000
$\alpha^{\cdot G}$	0.017	0.006	0.015	0.004	0.021	0.006	0.012	0.796	0.003	0.001
$\alpha^{\cdot H}$	0.012	0.005	0.005	0.001	0.020	0.008	0.038	1.005	0.009	0.005
$\alpha^{\cdot P}$	0.004	0.002	0.004	0.001	0.011	0.003	0.009	0.314	0.038	0.011
$\beta^{\cdot B}$	0.016	0.009	0.210	0.110	0.176	0.070	0.554	0.386	0.419	0.152
$\beta^{\cdot C}$	0.136	0.069	0.034	0.044	0.211	0.068	0.142	1.492	0.182	0.092
$\beta^{\cdot G}$	0.303	0.190	0.188	0.326	0.013	0.005	0.108	1.998	0.019	0.013
$\beta^{\cdot H}$	0.209	0.266	0.646	0.437	0.182	0.106	0.043	0.455	0.229	0.110
$\beta^{\cdot P}$	0.121	0.190	0.729	0.579	0.123	0.089	0.101	0.550	0.034	0.020
s_1	−2.780	0.194	−1.893	0.265	−3.210	0.145	−1.898	0.210	−2.396	0.243
s_2	2.684	0.232	1.957	0.325	3.146	0.174	1.888	0.265	2.483	0.289
s_3	0.093	0.110	−0.197	0.162	0.076	0.091	−0.135	0.134	−0.292	0.149
s_4	−0.025	0.106	0.103	0.161	−0.003	0.088	0.093	0.131	0.341	0.153
s_5	0.185	0.110	0.019	0.164	0.031	0.092	0.231	0.134	−0.154	0.157
s_6	−0.272	0.110	0.251	0.168	0.000	0.093	−0.199	0.133	−0.047	0.157
s_7	0.407	0.123	−0.655	0.187	0.061	0.102	0.075	0.147	0.209	0.171
Obs					76,492					
LL					−1,92,325					
					Diagnostics of $\hat{\Lambda}^k$					
Mean	0.99		0.99		0.99		0.99		0.99	
S.D.	0.98		1.02		0.99		0.98		1.01	
LB	42.95	0.00	26.06	0.16	104.15	0.00	39.25	0.00	48.32	0.00
Disp.	0.94	0.34	2.12	0.03	0.09	0.92	1.90	0.05	1.48	0.13

Diagnostics: Log Likelihood (LL) and diagnostics (mean, standard deviation, Ljung–Box statistic with respect to 20 lags as well as excess dispersion test (the latter two statistics inclusive p-values)) of residuals $\hat{\Lambda}^k$, $k = 1, \ldots, 5$.

impact of the l-type backward recurrence times on the kth series. We observe highly significant estimates in most cases. The significance of the interaction coefficients ($l \neq k$) illustrates the existence of strong contemporaneous correlations and cross-autocorrelations between the individual series. The estimated seasonalities indicate the well known inverse U-shape intraday pattern and are very similar for all series.[6] Diagnostics are computed based on $\hat{\Lambda}_i^k := \hat{\Lambda}^k(t_{i-1}^k, t_i^k)$. The test against excess dispersion provides satisfying results since the null hypothesis of no excess dispersion is not rejected in most cases. In contrast, the relatively high Ljung–Box statistics associated with the individual series $\hat{\Lambda}_i^k$ indicate that the dynamics are not completely captured by the model. These results show that particularly in the multivariate case, there exists a clear trade-off between a satisfying goodness-of-fit and model parsimony.

To illustrate the estimated dynamics graphically, the resulting autocorrelation and cross-autocorrelation functions of the estimated intensity functions are plotted in Fig. 12.7. Strong interdependencies between the single price intensity series yield empirical evidence for significant co-movements in price change volatility.[7] The contemporaneous correlation is around 0.6 for all series accompanied by highly persistent cross-autocorrelations in both directions. Hence, volatility shocks in one series lead to persistent spill-over effects in the other series causing strong multidimensional clustering structures. The shape of the individual (cross-)autocorrelation functions is quite similar and symmetric for all series and reveal no clear lead-lag structure. These results can be readily explained in the light of a common information flow that jointly influences the intraday volatility of individual stocks.

12.4 Stochastic Conditional Intensity Processes

12.4.1 Model Structure

An important property of Hawkes and ACI processes, as discussed in the previous sections, is that the conditional intensity function, given the history of the process, is completely deterministic. This is due to the fact that in both frameworks, the intensity process is parameterized in terms of the history of *observable* factors. However, the assumption that the intensity function is completely explained by the observable process history is at least questionable.

The so-called stochastic conditional intensity (SCI) model introduced by Bauwens and Hautsch (2006) captures time-varying unobservable heterogeneity in terms of a joint latent component. The SCI model is based on the assumption

---

[6]For sake of brevity, the corresponding plots are not shown.

[7]Recall the close relationship between the price intensity $\lambda^{dp}(t;\mathcal{F}_t)$ and the corresponding instantaneous volatility $\tilde{\sigma}^2_{(x^{dp})}(t)$ as given in (8.57).

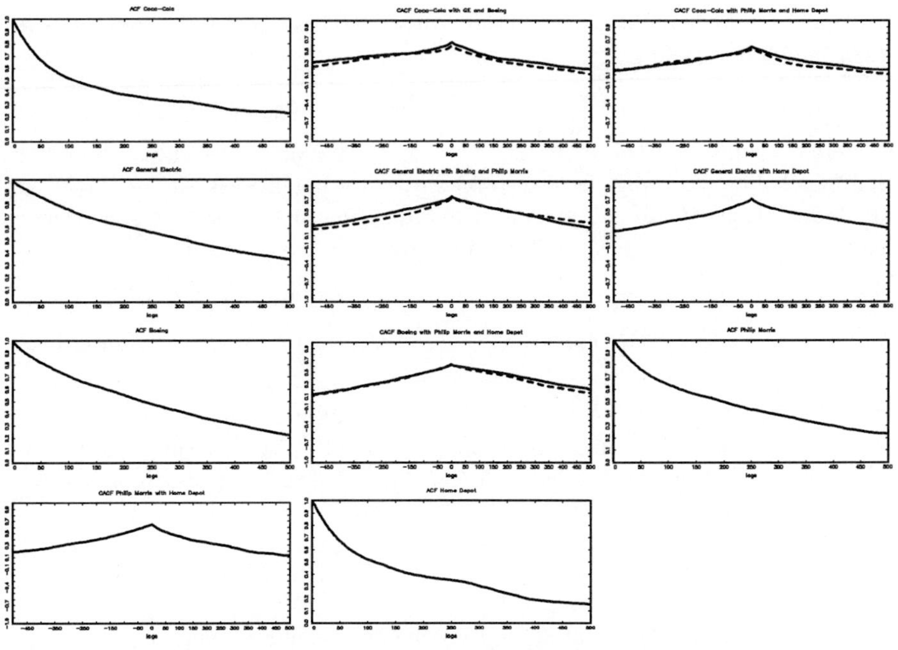

Fig. 12.7 Autocorrelation and cross-autocorrelation functions of the estimated intensity functions based on a five-dimensional Hawkes(1) model for \$0.20-price durations of the Boeing (B), Coca-Cola (C), General Electric (G), Home Depot (H) and Philip Morris (P) stock traded at the NYSE. Sample period 01/02/01 to 02/28/01. *First row*: *left*: ACF of Coca-Cola, *middle*: CACF C-G (*solid line*) and C-B (*broken line*), *right*: CACF C-P (*solid line*) and C-H (*broken line*). *Second row*: *left*: ACF of General Electric, *middle*: CACF G-B (*solid line*) and G-P (*broken line*), *right*: CACF G-H. *Third row*: *left*: ACF of Boeing, *middle*: CACF of B-P (*solid line*) and B-H (*broken line*), *right*: ACF of Philip Morris. *Fourth row*: *left*: CACF P-H, *middle*: ACF of Home Depot

that the conditional intensity function, given the process history, consists of two components: a univariate latent one and an observation driven one. In this sense, the SCI model combines features of parameter driven models and observation driven models and mimics the structure of the Stochastic MEM discussed in Chap. 6. The observable component can be specified univariately or multivariately. In the latter case, the latent factor corresponds to a common component that captures the impact of a general factor influencing all individual point processes and interacts with their observable component. Two limit cases emerge naturally: one when the latent factor is irrelevant and the intensity is driven by the dynamics of the specific components and the other when the specific components are not relevant and the latent factor completely dominates. In intermediate cases, the model dynamics are driven by the interaction between the latent dynamic factor which is updated by latent innovations and the observable dynamic component, which is updated by process-specific innovations. In this sense, the model can be interpreted as a dynamic extension of a doubly stochastic Poisson process (see, e.g., Grandell 1976 or Cox and Isham 1980).

The basic SCI model is based on a decomposition of the intensity function into a latent and an observable component. In this context, we define the information set $\mathcal{F}_t$ more explicitly as $\mathcal{F}_t := \sigma(\mathcal{F}_t^o \cup \mathcal{F}_t^*)$, consisting of an observable conditioning set $\mathcal{F}_t^o$ including the complete *observable* history of the process up to t (inclusive possible explanatory variables) and an *unobservable* history $\mathcal{F}_t^*$ of some latent factor $\lambda^*_{\breve{N}(t)+1}$. Then, the basic SCI model is given by

$$\lambda^k(t;\mathcal{F}_t) = \lambda^{o,k}(t;\mathcal{F}_t^o)\left(\lambda^*_{\breve{N}(t)+1}\right)^{\delta^{k*}}, \tag{12.19}$$

where $\lambda^{o,k}(t;\mathcal{F}_t^o)$ denotes a k-type conditionally deterministic intensity component based on the observable elements included in $\mathcal{F}_t^o$ and δ^{k*} drives the process-specific influence of λ_i^*. As in the Stochastic MEM, the latent factor is conditionally log-normally distributed and follows an AR(1) process,

$$\ln \lambda_i^* = a^* \ln \lambda_{i-1}^* + u_i^*, \qquad u_i^* \sim \text{i.i.d. } \mathcal{N}(0,1). \tag{12.20}$$

The latent factor is indexed by the left-continuous counting function, i.e., it does not change during a spell. In particular, it is assumed that λ_i^* has left-continuous sample paths with right-hand limits which means that it is updated instantaneously after the occurrence of t_{i-1} and remains constant until (and inclusive) t_i. In order to obtain a valid intensity process, it is assumed that the latent innovations u_i^* are independent from the series of integrated intensities $\varepsilon_i := \Lambda(t_{i-1}, t_i)$ which are i.i.d. standard exponentially distributed (under correct model specification). An important prerequisite for weak stationarity of the SCI model is weak stationarity of the latent component which is fulfilled for $|a^*| < 1$.[8] The latent factor dynamics can be extended to ARMA(P,Q) parameterizations, however, often, a simple AR(1) dynamic is shown to be sufficient in capturing latent dynamics. Notice that a constant is omitted since the observation driven component $\lambda^o(t;\mathcal{F}_t^o)$ encloses a constant that would be not identified otherwise.

By defining $\lambda_i^{k*} := \delta^{k*} \ln \lambda_i^*$ as the latent factor influencing the k-type component, it is easy to see that

$$\lambda_i^{k*} = a^* \lambda_{i-1}^{k*} + \delta^{k*} u_i^*.$$

Hence, δ^{k*} corresponds to a scaling factor that scales the latent component influencing the $k-$type intensity process. This flexibility ensures that the impact of a

[8]Of course, under the normality assumption, the latent factor is even strictly stationary.

latent shock u_i^* on the individual processes can differ and is driven by the parameter δ^{k*}.[9]

This approach can be extended by specifying δ^{k*} in a time-varying fashion. Then, the process-specific scaling factor can change over time in order to allow for conditional heteroscedasticity. An important source of heteroscedasticity could be intraday seasonality associated with deterministic fluctuations of the overall information and activity flow. For example, it could be due to institutional settings, like the opening of other related markets. Hence, a reasonable specification is to index δ^{k*} by the counting function and parameterize it in terms of a linear spline function

$$\delta_i^{k*} = s_0^* \left(1 + \sum_{m=1}^{M} \nu_{0,m}^{k*} \mathbb{1}_{\{\tau(t_i) \geq \bar{\tau}_m\}} (\tau(t_i) - \bar{\tau}_m) \right), \tag{12.21}$$

where $\tau(t)$ denotes the calendar time at t, $\bar{\tau}_m$, $m = 1, \ldots, M-1$, denote the exogenously given (calendar) time points and s_0^* and $\nu_{0,m}^{k*}$ are the corresponding coefficients of the spline function.

A further generalization of the SCI model is to allow for regime-switching latent dynamics. Then, a more flexible SCI model is obtained by specifying the autoregressive parameter in dependence of the length of the previous spell. Such a specification is in line with a threshold model (see Chap. 6) and is obtained by

$$\ln \lambda_i^* = a_m^* \mathbb{1}_{\{\bar{x}_{m-1} < x_{\breve{N}(t)} \leq \bar{x}_m\}} \ln \lambda_{i-1}^* + u_i^*, \quad m = 1, \ldots, M-1, \tag{12.22}$$

where $\bar{x}_m$ denotes the exogenously given thresholds (with $\bar{x}_0$:=0), and a_m^* are regime-dependent latent autoregressive parameters.

The observation-driven component $\lambda^{o,k}(t; \mathcal{F}_t^o)$ is specified in terms of an ACI model. For $k = 1$, parameterizing $\lambda^{0,1}(t; \mathcal{F}_t^o) = \lambda^0(t; \mathcal{F}_t^o)$ univariately by (11.1) through (11.5) results in an univariate SCI(P,Q) model. In this framework, the innovation term is specified either in plain or in logarithmic form, i.e., by

$$\breve{\varepsilon}_i = 1 - \Lambda^o(t_{i-1}, t_i) \qquad \text{or} \tag{12.23}$$

$$\breve{\varepsilon}_i = -0.5772 - \ln \Lambda^o(t_{i-1}, t_i), \tag{12.24}$$

where

$$\Lambda^o(t_{i-1}, t_i) := \sum_j \int_{\tilde{t}_j}^{\tilde{t}_j+1} \lambda^o(s; \mathcal{F}_s^o) ds = \sum_j \int_{\tilde{t}_j}^{\tilde{t}_j+1} \frac{\lambda(s; \mathcal{F}_s) ds}{\lambda_i^{*\delta^*}} \tag{12.25}$$

[9]Note that δ^{k*} can be even negative. Hence, theoretically it is possible that the latent component simultaneously increases one component while decreasing the other component.

with $j: t_{i-1} \leq \tilde{t}_j \leq t_i$ and $\sum_j \int_{\tilde{t}_j}^{\tilde{t}_{j+1}} \lambda(s;\mathcal{F}_s)ds = \varepsilon_i \sim$ i.i.d. $Exp(1)$. Hence, the innovation of an univariate SCI process is based on the integral over the *observable* intensity component $\lambda^o(t;\mathcal{F}_t^o)$ which equals an i.i.d. standard exponential variate that is standardized by a stationary log-normal random variable. Of course, $\Lambda^o(t_{i-1},t_i)$ is not anymore i.i.d. exponentially distributed then. Nevertheless, it can be centered by 1 or -0.5772, respectively, in order to make it comparable to the innovation of the pure ACI model as benchmark case. Because $\Lambda^o(t_{i-1},t_i)$, and thus the SCI innovation depends on lags of λ_i^*, the component $\lambda^o(t;\mathcal{F}_t^o)$ is directly affected by lags of the latent factor. Hence, λ_i^* influences the intensity $\lambda(t;\mathcal{F}_t)$ not only contemporaneously (according to (12.19)), but also through its lags.

Due to the log-linear structure of both components $\lambda^o(t;\mathcal{F}_t^o)$ and λ_i^*, the model dynamics are characterized by a direct interaction between the latent dynamic factor, which is updated by latent innovations u_i^* and the observable dynamic component, which is updated by the innovations $\breve{\varepsilon}_i$. By excluding covariates and assuming for simplicity $\lambda_0(t) = 1$ and $s(t) = 1$, the intensity function for the SCI(1,1) model is rewritten in logarithmic form as

$$\begin{aligned}\ln\lambda(t;\mathcal{F}_t) &= \Psi(t) + \delta^*\ln\lambda^*_{\breve{N}(t)+1}\\ &= \alpha\breve{\varepsilon}_{\breve{N}(t)} + \beta\Psi(t_{\breve{N}(t)}) + a^*\delta^*\ln\lambda^*_{\breve{N}(t)} + \delta^* u^*_{\breve{N}(t)}\\ &= \alpha\breve{\varepsilon}_{\breve{N}(t)} + \delta^*(a^*-\beta)\ln\lambda^*_{\breve{N}(t)} + \delta^* u^*_{\breve{N}(t)} + \beta(\ln\lambda(t_{\breve{N}(t)};\mathcal{F}_{t_{\breve{N}(t)}}).\end{aligned} \tag{12.26}$$

Hence, the SCI model can be represented as an ARMA model for the log intensity function that is augmented by a further dynamic component.

By specifying $\lambda^o(t;\mathcal{F}_t^o)$ multivariately, as described in Sect. 12.1, a multivariate SCI model is obtained. In this case, λ_i^* is updated at every point of the *pooled* process. The innovation term is specified in terms of innovations of multivariate SCI models

$$\breve{\varepsilon}_{\breve{N}(t)} = 1 - \Lambda^{o,k}\left(t^k_{\breve{N}^k(t)-1}, t^k_{\breve{N}^k(t)}\right) y^k_{\breve{N}(t)} \quad \text{or} \tag{12.27}$$

$$\breve{\varepsilon}_{\breve{N}(t)} = -0.5772 - \ln\Lambda^{o,k}\left(t^k_{\breve{N}^k(t)-1}, t^k_{\breve{N}^k(t)}\right) y^k_{\breve{N}(t)}, \tag{12.28}$$

where

$$\Lambda^{o,k}(t^k_{i-1}, t^k_i) := \sum_j \int_{\tilde{t}_j}^{\tilde{t}_{j+1}} \lambda^{o,k}(s;\mathcal{F}_s^o)ds = \sum_j \int_{\tilde{t}_j}^{\tilde{t}_{j+1}} \frac{\lambda^s(s;\mathcal{F}_s)ds}{\lambda_j^{*\delta^{k*}}} \tag{12.29}$$

with $j: t^k_{i-1} \leq \tilde{t}_j \leq t^k_i$ and $\sum_j \int_{\tilde{t}_j}^{\tilde{t}_{j+1}} \lambda^k(s;\mathcal{F}_s)ds = \varepsilon_i \sim$ i.i.d. $Exp(1)$. Hence, in the multivariate setting, $\Lambda^{o,k}(t^k_{i-1}, t^k_i)$ corresponds to an i.i.d. exponential variate that is piecewise standardized by a stationary log-normal random variable.

12.4.2 Probabilistic Properties of the SCI Model

The computation of conditional moments of $\lambda(t; \mathcal{F}_t)$ given the *observable* information set is difficult since the latent variable has to be integrated out. Bauwens and Hautsch (2006) numerically compute conditional moments given $\mathcal{F}^o_{t_{i-1}}$ for arbitrary parameterizations of $\lambda^o(t; \mathcal{F}^o_t)$. Let $\mathbf{y}_i$ a row of the matrix of *observable* variables $\mathbf{Y}$ and correspondingly, we define $\mathbf{Y}_i = \{\mathbf{y}_j\}_{j=1}^i$. Accordingly, $\boldsymbol{\Lambda}^*_i = \{\lambda^*_j\}_{j=1}^i$ defines the sequence of *latent* variables until t_i. Furthermore, let $f(\mathbf{Y}_i, \boldsymbol{\Lambda}^*_i|\boldsymbol{\theta})$ denote the joint density function of $\mathbf{Y}_i$ and $\boldsymbol{\Lambda}^*_i$, and $p(\lambda^*_i|\mathbf{Y}_i, \boldsymbol{\Lambda}^*_i, \boldsymbol{\theta})$ the conditional density of λ^*_i given $\mathbf{Y}_i$ and $\boldsymbol{\Lambda}^*_i$. Then, the conditional expectation of an arbitrary function of λ^*_i, $\vartheta(\lambda^*_i)$ given the *observable* information set up to t_{i-1} can be computed as

$$\mathbb{E}\left[\vartheta(\lambda^*_i)\left|\mathcal{F}^o_{t_{i-1}}\right.\right] = \frac{\int \vartheta(\lambda^*_i) p(\lambda^*_i|\mathbf{Y}_{i-1}, \boldsymbol{\Lambda}^*_{i-1}, \boldsymbol{\theta}) f(\mathbf{Y}_{i-1}, \boldsymbol{\Lambda}^*_{i-1}|\boldsymbol{\theta}) d\boldsymbol{\Lambda}^*_i}{\int f(\mathbf{Y}_{i-1}, \boldsymbol{\Lambda}^*_{i-1}|\boldsymbol{\theta}) d\boldsymbol{\Lambda}^*_{i-1}}. \quad (12.30)$$

The integrals in this ratio cannot be computed analytically, but can be simulated numerically, e.g., by efficient importance sampling. The computations of the conditional expectations $\mathbb{E}\left[t_i\left|\mathcal{F}^o_{t_{i-1}}\right.\right]$ and $\mathbb{E}\left[\lambda(t_i)\left|\mathcal{F}^o_{t_{i-1}}\right.\right]$ can be performed by exploiting the exponential distribution of the integrated intensity $\varepsilon_i := \Lambda(t_{i-1}, t_i)$. Then, by conditioning on predetermined values of ε_i and λ^*_{i-1} and applying the law of iterated expectations, $\mathbb{E}\left[t_i\left|\mathcal{F}^o_{t_{i-1}}\right.\right]$ is computed as[10]

$$\mathbb{E}\left[t_i\left|\mathcal{F}^o_{t_{i-1}}\right.\right] = \mathbb{E}\left[g_1(\cdot)\left|\mathcal{F}^o_{t_{i-1}}\right.\right] \quad (12.31)$$

$$g_1(\cdot) = \mathbb{E}\left[t_i\left|\lambda^*_{i-1}; \mathcal{F}^o_{t_{i-1}}\right.\right] = \mathbb{E}\left[g_2(\cdot)\left|\lambda^*_{i-1}; \mathcal{F}^o_{t_{i-1}}\right.\right], \quad (12.32)$$

where $t_i = g_2(\varepsilon_i; \mathcal{F}^o_{t_{i-1}}, \lambda^*_{i-1})$ is determined by solving the equation $\Lambda(t_i, t_{i-1}) = \varepsilon_i$ for given values of ε_i, λ^*_{i-1} and $\mathcal{F}^o_{t_{i-1}}$. The complexity of the expression for t_i depends on the parameterization of $\lambda^o(t; \mathcal{F}^o_t)$. However, as discussed in Chap. 11, at least for the multivariate case, closed-form solutions for $g_2(\cdot)$ do not exist. After computing the conditional expectation of t_i given λ^*_{i-1} (see (12.32)), the next step is to integrate over the latent variable according to (12.31) based on the integrals in (12.30). The computation of $\mathbb{E}\left[\lambda(t_i)\right| \mathcal{F}_{t_{i-1}}]$ is performed similarly.

Figures 12.8–12.12 show the (cross-)autocorrelation functions of the individual intensity components and the corresponding duration processes based on simulations performed by Bauwens and Hautsch (2006). Here, bivariate SCI(1,1) processes are simulated using logarithmic innovations according to (12.28) and constant baseline intensity functions. Figures 12.8 and 12.9 are based on SCI specifications implying no interactions between the observation driven components $\lambda^{o,k}(t, \mathcal{F}^o_t)$ but being driven by a highly autocorrelated latent factor. As expected, it turns out that the impact of the latent factor strongly depends on the magnitude of

[10]For ease of notation, we neglect the existence of time-varying covariates.

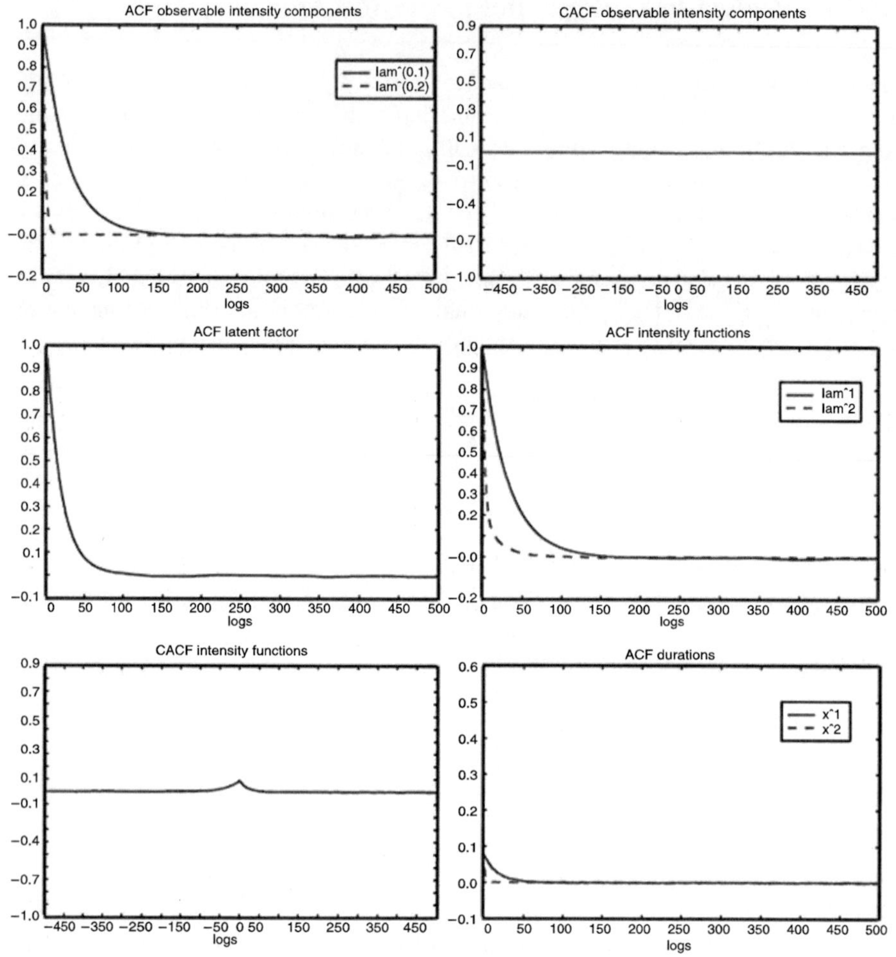

Fig. 12.8 Simulated bivariate SCI(1,1) processes. *Upper*: ACF and CACF of $\lambda^{o,1}(t_i;\mathcal{F}^o_{t_i})$ and $\lambda^{o,2}(t_i;\mathcal{F}^o_{t_i})$. *Middle*: ACF of λ^*_i, as well as of $\lambda^1(t_i;\mathcal{F}_{t_i})$ and $\lambda^2(t_i;\mathcal{F}_{t_i})$. *Lower*: CACF of $\lambda^1(t_i;\mathcal{F}_{t_i})$ vs. $\lambda^2(t_i;\mathcal{F}_{t_i})$, as well as ACF of x^1_i and x^2_i. $\omega^1 = \omega^2 = 0$, $\alpha^1_1 = \alpha^2_2 = 0.05$, $\alpha^1_2 = \alpha^2_1 = 0$, $\beta_{11} = 0.97$, $\beta_{12} = \beta_{21} = 0$, $\beta_{22} = 0.7$, $a^* = 0.95$, $\delta^{1*} = \delta^{2*} = 0.01$. Based on 5,000,000 drawings. Reproduced from Bauwens and Hautsch (2006)

the latent variances. In case of $\delta^{1*} = \delta^{2*} = 0.01$ (Fig. 12.8), only a very weak cross-autocorrelation between $\lambda^1(t;\mathcal{F}_t)$ and $\lambda^2(t;\mathcal{F}_t)$ can be identified. In contrast, in Fig. 12.9, the impact of the latent factor is clearly stronger. Here, it causes also slight contemporaneous correlations between the *observable* components $\lambda^{o,1}(t_i;\mathcal{F}^o_{t_i})$ and $\lambda^{o,2}(t_i;\mathcal{F}^o_{t_i})$. This is due to the fact that λ^*_i influences the intensity components not only contemporaneously but also through the lagged innovations $\breve{\varepsilon}_i$. It is shown that the latent dynamics dominate the dynamics of the processes $\lambda^1(t_i;\mathcal{F}_{t_i})$ and $\lambda^2(t_i;\mathcal{F}_{t_i})$, as well as of x^1_i and x^2_i, leading to quite similar autocorrelation functions.

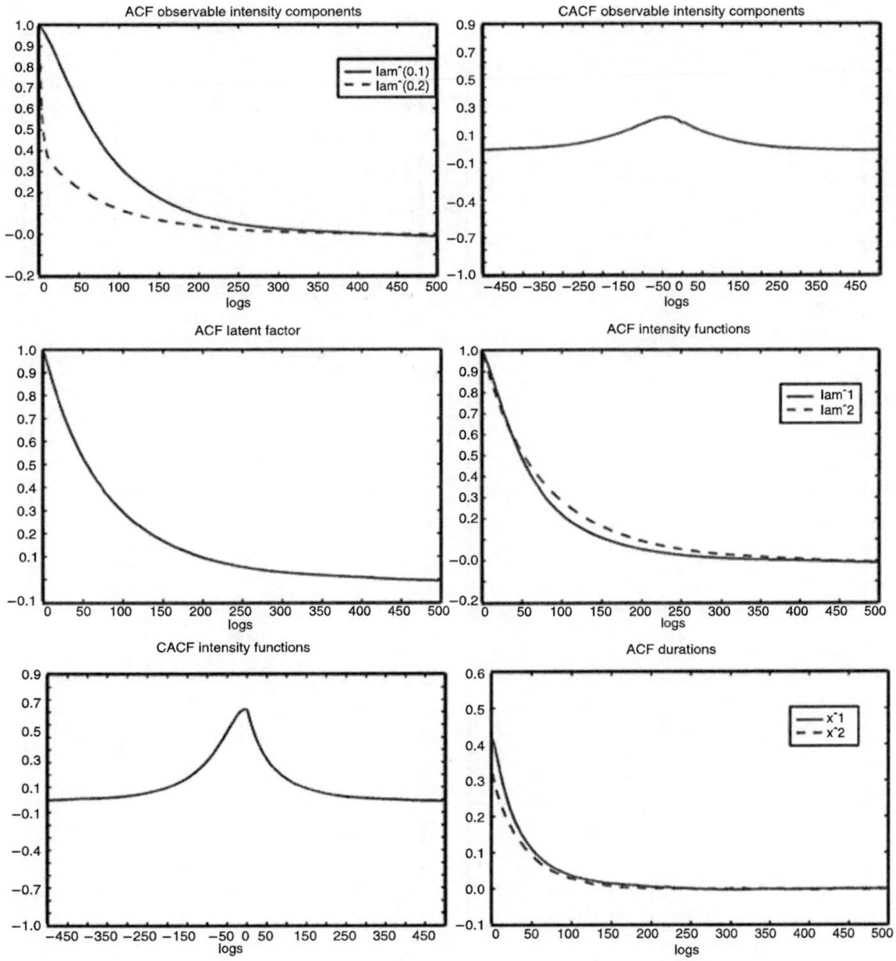

Fig. 12.9 Simulated bivariate SCI(1,1) processes. *Upper*: ACF and CACF of $\lambda^{o,1}(t_i;\mathcal{F}^o_{t_i})$ and $\lambda^{o,2}(t_i;\mathcal{F}^o_{t_i})$. *Middle*: ACF of λ^*_i, as well as of $\lambda^1(t_i;\mathcal{F}_{t_i})$ and $\lambda^2(t_i;\mathcal{F}_{t_i})$. *Lower*: CACF of $\lambda^1(t_i;\mathcal{F}_{t_i})$ vs. $\lambda^2(t_i;\mathcal{F}_{t_i})$, as well as ACF of x^1_i and x^2_i. $\omega^1 = \omega^2 = 0$, $\alpha^1_1 = \alpha^2_2 = 0.05$, $\alpha^1_2 = \alpha^2_1 = 0$, $\beta_{11} = 0.97$, $\beta_{12} = \beta_{21} = 0$, $\beta_{22} = 0.7$, $a^* = 0.99$, $\delta^{1*} = \delta^{2*} = 0.1$. Based on 5,000,000 drawings. Reproduced from Bauwens and Hautsch (2006)

Moreover, a clear increase of the autocorrelations in the duration processes is observed.

Figure 12.10 shows the corresponding plots of symmetrically interdependent ACI components ($\alpha^1_2 = \alpha^2_1 = 0.05$). Here, the latent variable triggers the contemporaneous correlation between the two processes and drives the autocorrelation functions of the individual intensity components towards higher similarity. While the CACF of $\lambda^{o,1}(t_i;\mathcal{F}^o_{t_i})$ vs. $\lambda^{o,2}(t_i;\mathcal{F}^o_{t_i})$ reveals the well-known asymmetric shape as discussed in Sect. 12.1, the joint latent factor causes a more symmetric shape of

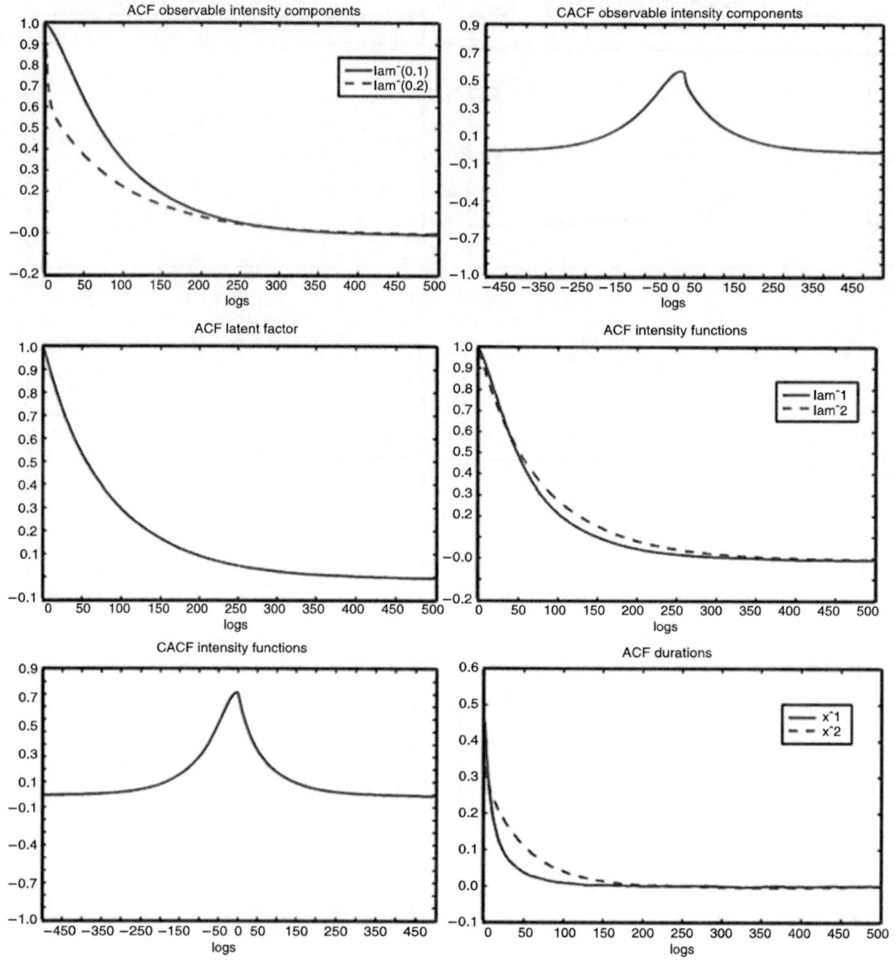

Fig. 12.10 Simulated bivariate SCI(1,1) processes. *Upper*: ACF and CACF of $\lambda^{o,1}(t_i;\mathcal{F}^o_{t_i})$ and $\lambda^{o,2}(t_i;\mathcal{F}^o_{t_i})$. *Middle*: ACF of λ^*_i, as well as of $\lambda^1(t_i;\mathcal{F}_{t_i})$ and $\lambda^2(t_i;\mathcal{F}_{t_i})$. *Lower*: CACF of $\lambda^1(t_i;\mathcal{F}_{t_i})$ vs. $\lambda^2(t_i;\mathcal{F}_{t_i})$, as well as ACF of x^1_i and x^2_i. $\omega^1 = \omega^2 = 0$, $\alpha^1_1 = \alpha^1_2 = \alpha^2_1 = \alpha^2_2 = 0.05$, $\beta_{11} = 0.97$, $\beta_{12} = \beta_{21} = 0$, $\beta_{22} = 0.7$, $a^* = 0.99$, $\delta^{1*} = \delta^{2*} = 0.1$. Based on 5,000,000 drawings. Reproduced from Bauwens and Hautsch (2006)

the CACF between $\lambda^1(t_i;\mathcal{F}_{t_i})$ and $\lambda^2(t_i;\mathcal{F}_{t_i})$. The DGP associated with Fig. 12.11 is based on relatively weak dynamics in the individual intensity components $\lambda^{o,1}(t_i;\mathcal{F}^o_{t_i})$ and $\lambda^{o,2}(t_i;\mathcal{F}^o_{t_i})$ while the impact of the latent factor is comparably strong. Here, λ^*_i completely dominates the dynamics of the joint system. It causes strong and similar autocorrelations in the components $\lambda^{o,1}(t_i;\mathcal{F}^o_{t_i})$ and $\lambda^{o,2}(t_i;\mathcal{F}^o_{t_i})$ as well as in $\lambda^1(t_i;\mathcal{F}_{t_i})$ and $\lambda^2(t_i;\mathcal{F}_{t_i})$. Moreover, its impact on the CACF is clearly stronger than in the cases outlined above. In fact, the contemporaneous correlation between $\lambda^1(t_i;\mathcal{F}_{t_i})$ and $\lambda^2(t_i;\mathcal{F}_{t_i})$ is nearly one. Nevertheless, the CACF dies

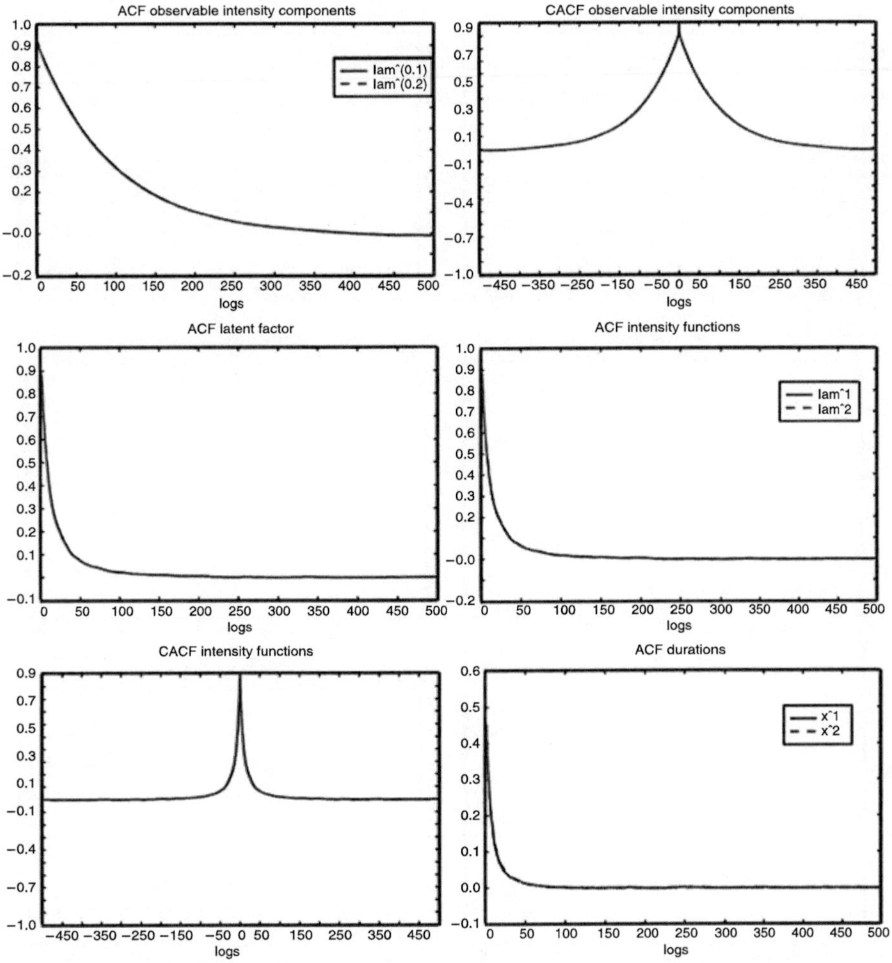

Fig. 12.11 Simulated bivariate SCI(1,1) processes. *Upper*: ACF and CACF of $\lambda^{o,1}(t_i; \mathcal{F}^o_{t_i})$ and $\lambda^{o,2}(t_i; \mathcal{F}^o_{t_i})$. *Middle*: ACF of λ^*_i, as well as of $\lambda^1(t_i; \mathcal{F}_{t_i})$ and $\lambda^2(t_i; \mathcal{F}_{t_i})$. *Lower*: CACF of $\lambda^1(t_i; \mathcal{F}_{t_i})$ vs. $\lambda^2(t_i; \mathcal{F}_{t_i})$, as well as ACF of x^1_i and x^2_i. $\omega^1 = \omega^2 = 0$, $\alpha^1_1 = \alpha^1_2 = \alpha^2_1 = \alpha^2_2 = 0.05$, $\beta_{11} = \beta_{22} = 0.2$, $\beta_{12} = \beta_{21} = 0$, $a^* = 0.95$, $\delta^{1*} = \delta^{2*} = 0.5$. Based on 5,000,000 drawings. Reproduced from Bauwens and Hautsch (2006)

out quite quickly which is due to the AR(1) structure in the latent process. The parameterization underlying Fig. 12.12 resembles the specification in Fig. 12.10. Here, the scaling factors are chosen as $\delta^{1*} = 0.1$ and $\delta^{2*} = -0.1$. It is shown that the latent component influences $\lambda^1(t_i; \mathcal{F}_{t_i})$ positively while influencing $\lambda^2(t_i; \mathcal{F}_{t_i})$ negatively which causes clear negative cross-autocorrelations between $\lambda^1(t_i; \mathcal{F}_{t_i})$ and $\lambda^2(t_i; \mathcal{F}_{t_i})$ as well as a flattening of the CACF between $\lambda^{o,1}(t_i; \mathcal{F}^o_{t_i})$ and $\lambda^{o,2}(t_i; \mathcal{F}^o_{t_i})$ compared to Fig. 12.10.

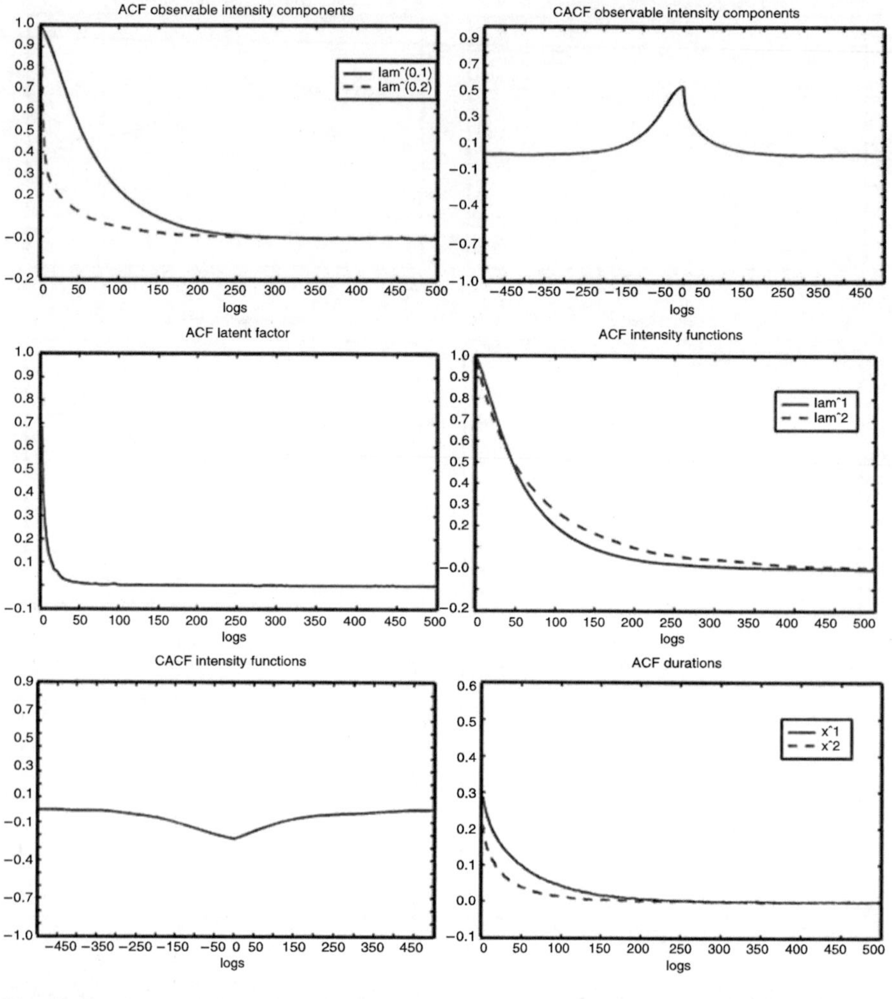

Fig. 12.12 Simulated bivariate SCI(1,1) processes. *Upper*: ACF and CACF of $\lambda^{o,1}(t_i;\mathcal{F}^o_{t_i})$ and $\lambda^{o,2}(t_i;\mathcal{F}^o_{t_i})$. *Middle*: ACF of λ^*_i, as well as of $\lambda^1(t_i;\mathcal{F}_{t_i})$ and $\lambda^2(t_i;\mathcal{F}_{t_i})$. *Lower*: CACF of $\lambda^1(t_i;\mathcal{F}_{t_i})$ vs. $\lambda^2(t_i;\mathcal{F}_{t_i})$, as well as ACF of x^1_i and x^2_i. $\omega^1 = \omega^2 = 0$, $\alpha^1_1 = \alpha^2_1 = \alpha^1_2 = \alpha^2_2 = 0.05$, $\beta_{11} = 0.97$, $\beta_{12} = \beta_{21} = 0$, $\beta_{22} = 0.7$, $a^* = 0.99$, $\delta^{1*} = 0.1$, $\delta^{2*} = -0.1$. Based on 5,000,000 drawings. Reproduced from Bauwens and Hautsch (2006)

12.4.3 Statistical Inference

If the latent variables $\boldsymbol{\Lambda}^*_n := \{\ln\lambda^*_i\}^n_{i=1}$ were observable, the likelihood function would be given by

$$\mathcal{L}(\mathbf{Y},\boldsymbol{\Lambda}_n^*;\boldsymbol{\theta}) = \prod_{i=1}^{n}\prod_{k=1}^{K}\exp(-\Lambda^k(t_{i-1},t_i))\left[\lambda^{o,k}(t_i;\mathcal{F}_{t_i}^o)\lambda_i^{*\delta^{k*}}\right]^{y_i^k}, \qquad (12.33)$$

where

$$\Lambda^k(t_{i-1},t_i) = \lambda_i^{*\delta^{k*}}\;\Lambda^{o,k}(t_{i-1},t_i). \qquad (12.34)$$

As for stochastic MEMs discussed in Chap. 7, the main computationally difficulty is that the latent process is not observable, hence the conditional likelihood function must be integrated with respect to $\boldsymbol{\Lambda}_n^*$ using the assumed distribution of the latter. Accordingly, the unconditional or integrated likelihood function is given by

$$\mathcal{L}(\mathbf{Y};\boldsymbol{\theta}) = \int\prod_{i=1}^{n}\prod_{k=1}^{K}\lambda_i^{*\delta^{k*}y_i^k}\exp(-\lambda_i^{*\delta^{k*}}\;\Lambda^{o,k}(t_{i-1},t_i))\left[\lambda^{o,k}(t_i;\mathcal{F}_{t_i}^o)\right]^{y_i^k}$$

$$\times\frac{1}{\sqrt{2\pi}}\exp\left[-\frac{(\ln\lambda_i^*-m_i^*)^2}{2}\right]d\boldsymbol{\Lambda}_n^*. \qquad (12.35)$$

$$= \int\prod_{i=1}^{n}f(\mathbf{y}_i,\ln\lambda_i^*|\mathbf{Y}_{i-1},\boldsymbol{\Lambda}_{i-1}^*,\boldsymbol{\theta})d\boldsymbol{\Lambda}_n^*, \qquad (12.36)$$

where the last equality clearly defines the function $f(\cdot)$ and $m_i^* := \mathbb{E}[\ln\lambda_i^*|\mathcal{F}_{i-1}^*]$ denotes the conditional mean of the latent component.

The computation of the n-dimensional integral in (12.35) must be done numerically. Bauwens and Hautsch (2006) propose using efficient importance sampling as discussed in Chap. 7. In order to implement the EIS algorithm, the integral (12.36) is rewritten as

$$\mathcal{L}(\mathbf{Y};\boldsymbol{\theta}) = \int\prod_{i=1}^{n}\frac{f(\mathbf{y}_i,\ln\lambda_i^*|\mathbf{Y}_{i-1},\boldsymbol{\Lambda}_{i-1}^*,\boldsymbol{\theta})}{m(\ln\lambda_i^*|\boldsymbol{\Lambda}_{i-1}^*,\phi_i)}\prod_{i=1}^{n}m(\ln\lambda_i^*|\boldsymbol{\Lambda}_{i-1}^*,\phi_i)d\boldsymbol{\Lambda}_n^*, \qquad (12.37)$$

and is approximated by

$$\mathcal{L}(\mathbf{Y};\boldsymbol{\theta}) \approx \hat{\mathcal{L}}_R(\mathbf{Y};\boldsymbol{\theta}) = \frac{1}{R}\sum_{r=1}^{R}\prod_{i=1}^{n}\frac{f(\mathbf{y}_i,\ln\lambda_i^{*(r)}|\mathbf{Y}_{i-1},\boldsymbol{\Lambda}_{i-1}^{*(r)},\boldsymbol{\theta})}{m(\ln\lambda_i^{*(r)}|\boldsymbol{\Lambda}_{i-1}^{*(r)},\phi_i)}, \qquad (12.38)$$

where $\{\ln\lambda_i^{*(r)}\}_{i=1}^n$ denotes a trajectory of random draws from the sequence of auxiliary importance samplers $\{m(\ln\lambda_i^*|\boldsymbol{\Lambda}_{i-1}^*,\phi_i)\}_{i=1}^n$, and R is the number of generated trajectories.

Let $k(\boldsymbol{\Lambda}_i^*,\phi_i)$ denote a density kernel for $m(\ln\lambda_i^*|\boldsymbol{\Lambda}_{i-1}^*,\phi_i)$, given by

$$k(\boldsymbol{\Lambda}_i^*,\phi_i) = m(\ln\lambda_i^*|\boldsymbol{\Lambda}_{i-1}^*,\phi_i)\chi(\boldsymbol{\Lambda}_{i-1}^*,\phi_i), \qquad (12.39)$$

where the integrating constant is given by

$$\chi(\boldsymbol{\Lambda}_{i-1}^{*}, \phi_i) = \int k(\boldsymbol{\Lambda}_i^{*}, \phi_i) d \ln \lambda_i^{*}. \tag{12.40}$$

As discussed in Sect. 7.2, the implementation of EIS requires the selection of a class of density kernels $k(\cdot)$ for the auxiliary sampler $m(\cdot)$ that provide a good approximation to the product $f(\mathbf{y}_i, \ln \lambda_i^* | \mathbf{Y}_{i-1}, \boldsymbol{\Lambda}_{i-1}^*, \boldsymbol{\theta})\ \chi(\boldsymbol{\Lambda}_{i-1}^*, \phi_i)$. A convenient and efficient possibility is to use a parametric extension of the direct samplers implied by the assumed distribution of $\ln \lambda_i^*$ and to approximate the function

$$\prod_{k=1}^{K} \lambda_i^{*\delta^{k*} y_i^k} \exp(-\lambda_i^{*\delta^{k*}} \Lambda^{o,k}(t_{i-1}, t_i)), \tag{12.41}$$

that appears in (12.35) by the normal density kernel

$$\zeta(\ln \lambda_i^*) = \exp(\phi_{1,i} \ln \lambda_i^* + \phi_{2,i} (\ln \lambda_i^*)^2). \tag{12.42}$$

Including also the $\mathcal{N}(m_i^*, 1)$ density function in the importance sampler $m(\ln \lambda_i^* | \boldsymbol{\Lambda}_{i-1}^*, \phi_i)$ and using the property that the product of normal densities is itself a normal density, a density kernel of $m(\cdot)$ can be written as

$$\begin{aligned} k(\boldsymbol{\Lambda}_i^*, \phi_i) &\propto \exp\left((\phi_{1,i} + m_i^*) \ln \lambda_i^* + \left(\phi_{2,i} - \frac{1}{2} \right) (\ln \lambda_i^*)^2 \right) \\ &= \exp\left(-\frac{1}{2\pi_i^2} (\ln \lambda_i^* - \mu_i)^2 \right) \exp\left(\frac{\mu_i^2}{2\pi_i^2} \right), \end{aligned} \tag{12.43}$$

where

$$\pi_i^2 := (1 - 2\phi_{2,i})^{-1} \tag{12.44}$$

$$\mu_i := (\phi_{1,i} + m_i^*)\, \pi_i^2. \tag{12.45}$$

Then, the integrating constant (12.40) is given by

$$\chi(\boldsymbol{\Lambda}_{i-1}^*, \phi_i) = \exp\left(\frac{\mu_i^2}{2\pi_i^2} - \frac{m_i^{*2}}{2} \right) \tag{12.46}$$

(neglecting the factor $\pi_i \sqrt{2\pi}$ since it depends neither on $\boldsymbol{\Lambda}_{i-1}^*$ nor on ϕ_i).

The choice of the auxiliary parameters is optimized to minimize the MC variance of $\hat{\mathcal{L}}_R(\mathbf{Y}; \boldsymbol{\theta})$. Then, following Richard and Zhang (2007), this problem can be split into n minimization problems of the form

$$\min_{\phi_{i,0},\,\phi_i} \sum_{r=1}^{R} \Big\{ \ln \Big[f(\mathbf{y}_i, \ln \lambda_i^{*(r)} | \mathbf{Y}_{i-1}, \boldsymbol{\Lambda}_{i-1}^{*(r)}, \boldsymbol{\theta}) \chi(\boldsymbol{\Lambda}_i^{*(r)}, \phi_{i+1}) \Big]$$

$$- \phi_{0,i} - \ln k(\boldsymbol{\Lambda}_i^{*(r)}, \phi_i) \Big\}^2 . \tag{12.47}$$

The resulting EIS algorithm involves the following steps:
Step 1: Generate R trajectories $\{\ln \lambda_i^{*(r)}\}_{i=1}^n$ using the direct samplers $\{\mathcal{N}(m_i^*, 1)\}_{i=1}^n$.
Step 2: For each i (from n to 1), estimate by OLS the regression (with R observations) implicit in (12.47), which takes precisely the following form:

$$\delta^{k*} y_i^k \ln \lambda_i^{*(r)} - \left(\lambda_i^{*(r)}\right)^{\delta^{k*}} \sum_{k=1}^{K} \Lambda^{o,k}(t_{i-1}, t_i) + \sum_{k=1}^{K} y_i^k \ln \lambda^{o,k}(t_i; \mathcal{F}_{t_i}^o)$$

$$+ \ln \left(\chi(\boldsymbol{\Lambda}_i^{*(r)}, \phi_{i+1}) \right)$$

$$= \phi_{0,i} + \phi_{1,i} \ln \lambda_i^{*(r)} + \phi_{2,i} (\ln \lambda_i^{*(r)})^2 + u^{(r)}, \; r = 1, \ldots, R, \tag{12.48}$$

(where $u^{(r)}$ is an error term), using $\chi(\boldsymbol{\Lambda}_n^{*(r)}, \phi_{n+1}) = 1$ as initial condition, and then (12.46).
Step 3: Generate R trajectories $\{\ln \lambda_i^{*(r)}\}_{i=1}^n$ using the EIS samplers $\{\mathcal{N}(\mu_i, \pi_i^2)\}_{i=1}^n$ (see (12.44) and (12.45)) to compute $\hat{\mathcal{L}}_R(\mathbf{Y}; \boldsymbol{\theta})$ as defined in (12.38). Per construction of the SCI model, the computation of the terms $\Lambda^{o,k}(t_{i-1}, t_i)$ and $\lambda^{o,k}(t_i; \mathcal{F}_{t_i}^o)$ can be done separately and is done in a step before the EIS algorithm.

The first iteration can be started with a sampler other than the direct one. This is achieved by immediately multiplying the direct sampler by a normal approximation to $\delta^{k*} \ln \lambda_i^* - \lambda_i^{*\delta^{k*}} \sum_{k=1}^K \Lambda^{o,k}(t_{i-1}, t_i)$, using a second order Taylor expansion (TSE) of the argument of the exponential function around $\ln \lambda_i^* = 0$. This yields

$$\delta^{k*} y_i^k \ln \lambda_i^* - \lambda_i^{*\delta^{k*}} \sum_{k=1}^{K} \Lambda^{o,k}(t_{i-1}, t_i)$$

$$\approx \text{constant} + \ln \lambda_i^* - (\ln \lambda_i^*)^2 \sum_{k=1}^{K} \Lambda^{o,k}(t_{i-1}, t_i), \tag{12.49}$$

which implies that $\phi_{1,i} = 1$ and $\phi_{2,i} = -\sum_{k=1}^K \Lambda^{o,k}(t_{i-1}, t_i)$ must be inserted into (12.44) and (12.45) to obtain the moments of the TSE normal importance sampler. In this way, the initial importance sampler takes into account the data and enables one to reduce the number of iterations over the three steps.

The SCI residuals of the kth process are computed on the basis of the trajectories drawn from the sequence of auxiliary samplers characterizing $\mathcal{L}(\mathbf{Y}; \boldsymbol{\theta})$, i.e.,

$$\hat{\varepsilon}_i^{k,(r)} := \int_{t_{i-1}^k}^{t_i^k} \left[\lambda_i^{*(r)}\right]^{\delta^{k*}} \hat{\lambda}^{o,k}(u;\mathcal{F}_u^o)du$$

$$= \sum_{j=N(t_{i-1}^k)}^{N(t_i^k)} \left[\lambda_i^{*(r)}\right]^{\delta^{k*}} \int_{t_j}^{t_{j+1}} \hat{\lambda}^{o,k}(u;\mathcal{F}_u^o)du. \tag{12.50}$$

The residual diagnostics are computed for each of the R sequences separately.[11] Under correct specification, the residuals $\hat{\varepsilon}_i^{k,(r)}$ should be i.i.d. $Exp(1)$. Hence, model evaluations can be done by testing the dynamical and distributional properties of the residual series using the techniques described in Chap. 5.

12.5 SCI Modelling of Multivariate Price Intensities

Bauwens and Hautsch (2006) apply the SCI model to analyze price intensities of several stocks and to test for the presence of a common underlying factor. They use price durations generated from a the stocks AOL, IBM, Coca-Cola, JP Morgan and AT&T, traded at NYSE, during the period from 01/02/2001 to 31/05/2001. The price durations are computed based on multiples of the average size of absolute trade-to-trade midquote changes of the corresponding stock. Using a multiple of 20 (i.e., dp corresponds to 20 times the average absolute trade-to-trade midquote change), yields an aggregation level dp of \$0.225, \$0.463, \$0.086, \$0.196 and \$0.193 for AOL, IBM, Coca-Cola, JP Morgan and AT&T, respectively. The resulting average price durations are within the range of 12 to 20 min.

The multivariate price intensities are modelled using a five-dimensional SCI(1,1) process with a diagonal specification of $\mathbf{B}$ and parameter restrictions $p_r^k = 1$ and $\kappa_r^k = 0 \; \forall, k \neq r$ ruling out interdependencies in the (Burr type) baseline intensities. The seasonality function is assumed to be common for all five processes and is specified as a linear spline function according to (5.33) based on 6 nodes dividing the trading hours from 9:30 to 16:00 into equal-sized time intervals. The models are estimated using the EIS technique discussed in Sect. 12.4.3 based on $R = 50$ Monte Carlo replications, while the efficiency steps in the algorithm are repeated five times. The first iteration starts with the TSE normal importance sampler. The standard errors are computed based on the inverse of the estimated Hessian.

Table 12.8 gives the estimation results reported by Bauwens and Hautsch (2006) of four different five-dimensional SCI specifications with unrestricted innovation impact vectors $\mathbf{A}^k$. Panel A gives the results of a basic ACI specification, whereas Panel B to D account for a joint dynamic factor. It turns out that the autoregressive common component is highly persistent with a^* being close to one. Comparing

[11] Note that it is not reasonable to evaluate the model based on the average trajectory since dispersion and dynamic effects would eliminate each other then and would lead to non-interpretable results.

Table 12.8 ML(-EIS) estimates and corresponding p-values of five-dimensional SCI(1,1) models for the intensities of price durations of the stocks of (1) AOL, (2) IBM, (3) Coca-Cola, (4) JP Morgan, and (5) AT&T. Baseline intensity functions are specified in terms of individual univariate Burr hazard functions. The innovation impact vectors $\mathbf{A}^k$ are fully unrestricted, whereas $\mathbf{B}$ is restricted to be a diagonal matrix. The innovation term is specified according to (12.28). One joint spline function is specified for all processes based on six equally spaced nodes between 9:30 a.m. and 4:00 p.m. Standard errors are based on the inverse of the estimated Hessian. The time series are re-initialized at each trading day

	A		B		C		D	
	est.	p-v.	est.	p-v.	est.	p-v.	est.	p-v.
Baseline intensity parameters								
ω^1	−0.73	0.00	−0.46	0.00	−0.73	0.00	−0.76	0.00
ω^2	−1.10	0.00	−0.81	0.00	−1.11	0.00	−1.13	0.00
ω^3	−1.38	0.00	−1.12	0.00	−1.30	0.00	−1.33	0.00
ω^4	−0.48	0.00	−0.21	0.01	−0.44	0.00	−0.47	0.00
ω^5	−0.74	0.00	−0.47	0.00	−0.70	0.00	−0.73	0.00
p^1	1.29	0.00	1.34	0.00	1.40	0.00	1.39	0.00
p^2	1.25	0.00	1.31	0.00	1.36	0.00	1.36	0.00
p^3	1.31	0.00	1.36	0.00	1.36	0.00	1.37	0.00
p^4	1.29	0.00	1.34	0.00	1.36	0.00	1.36	0.00
p^5	1.33	0.00	1.37	0.00	1.40	0.00	1.41	0.00
κ^1	0.04	0.00	0.03	0.00	0.03	0.00	0.03	0.00
κ^2	0.02	0.00	0.02	0.00	0.02	0.00	0.02	0.00
κ^3	0.02	0.00	0.02	0.00	0.02	0.00	0.02	0.00
κ^4	0.08	0.00	0.06	0.00	0.06	0.00	0.06	0.00
κ^5	0.06	0.00	0.05	0.00	0.05	0.00	0.04	0.00
Innovation parameters								
α_1^1	0.11	0.00	0.12	0.00	0.07	0.00	0.07	0.00
α_2^1	0.04	0.00	−0.00	0.69	0.01	0.27	0.01	0.25
α_3^1	0.05	0.00	−0.00	0.94	0.02	0.03	0.02	0.03
α_4^1	0.05	0.00	0.01	0.36	0.03	0.01	0.03	0.01
α_5^1	0.02	0.12	−0.04	0.00	−0.01	0.38	−0.01	0.36
α_1^2	0.02	0.12	−0.03	0.08	−0.00	0.73	−0.00	0.93
α_2^2	0.13	0.00	0.11	0.00	0.10	0.00	0.10	0.00
α_3^2	0.03	0.01	−0.01	0.27	0.01	0.32	0.01	0.29
α_4^2	0.04	0.00	−0.01	0.36	0.02	0.06	0.02	0.04
α_5^2	0.03	0.01	−0.02	0.14	0.01	0.38	0.01	0.30
α_1^3	0.02	0.10	−0.03	0.10	0.01	0.60	0.00	0.73
α_2^3	0.02	0.12	−0.01	0.32	0.01	0.47	0.01	0.59
α_3^3	0.10	0.00	0.07	0.00	0.09	0.00	0.08	0.00
α_4^3	0.03	0.00	0.01	0.47	0.02	0.03	0.02	0.04
α_5^3	0.04	0.00	0.01	0.43	0.03	0.03	0.02	0.04
α_1^4	0.05	0.00	0.01	0.39	0.02	0.04	0.02	0.05
α_2^4	0.05	0.00	0.01	0.62	0.03	0.01	0.03	0.01
α_3^4	0.02	0.04	−0.03	0.01	0.00	0.76	0.00	0.85

(continued)

Table 12.8 (continued)

	A		B		C		D	
	est.	p-v.	est.	p-v.	est.	p-v.	est.	p-v.
α_4^4	0.10	0.00	0.09	0.00	0.06	0.00	0.06	0.00
α_5^4	0.05	0.00	−0.00	0.88	0.02	0.02	0.02	0.03
α_1^5	0.05	0.00	−0.01	0.60	0.03	0.03	0.03	0.03
α_2^5	0.05	0.00	−0.00	0.79	0.02	0.05	0.02	0.06
α_3^5	0.06	0.00	0.01	0.28	0.04	0.00	0.04	0.00
α_4^5	0.05	0.00	−0.01	0.46	0.03	0.01	0.03	0.01
α_5^5	0.14	0.00	0.09	0.00	0.10	0.00	0.10	0.00
Persistence parameters								
β_{11}	0.98	0.00	0.89	0.00	0.98	0.00	0.98	0.00
β_{22}	0.99	0.00	0.96	0.00	0.98	0.00	0.99	0.00
β_{33}	0.98	0.00	0.98	0.00	0.98	0.00	0.98	0.00
β_{44}	0.98	0.00	0.93	0.00	0.99	0.00	0.99	0.00
β_{55}	0.98	0.00	0.96	0.00	0.98	0.00	0.98	0.00
Seasonality parameters								
s_1	−1.45	0.00	−1.50	0.00	−1.60	0.00	−1.59	0.00
s_2	1.13	0.00	1.13	0.00	1.28	0.00	1.27	0.00
s_3	0.15	0.05	0.20	0.04	0.16	0.12	0.16	0.14
s_4	0.41	0.00	0.35	0.00	0.33	0.00	0.35	0.00
s_5	−0.03	0.74	−0.03	0.67	−0.03	0.80	−0.07	0.55
s_6	0.66	0.00	0.46	0.00	0.52	0.00	0.55	0.00
Latent parameters								
a^*			0.98	0.00				
a_1^*					0.92	0.00	0.92	0.00
a_2^*					0.80	0.00	0.82	0.00
a_3^*					0.64	0.00	0.65	0.00
a_4^*					0.60	0.00	0.57	0.00
σ^*			0.09	0.00				
σ_1^*					0.28	0.00	0.28	0.00
σ_2^*					0.27	0.00	0.27	0.00
σ_3^*					0.16	0.00	0.17	0.00
σ_4^*					0.21	0.00	0.22	0.00
σ_5^*					0.22	0.00	0.24	0.00
ν_1^*							−0.16	0.71
ν_2^*							0.18	0.83
ν_3^*							0.29	0.76
ν_4^*							−0.07	0.95
ν_5^*							−0.83	0.45
ν_6^*							−0.82	0.40
LL	−30,260		−29,562		−29,545		−29,536	
BIC	−30,501		−29,812		−29,828		−29,848	

Table 12.9 Average diagnostics (mean, standard deviation, Ljung–Box statistic), as well as excess dispersion test (see Chap. 5) over all trajectories of SCI residuals $\hat{\Lambda}_i^{k,(r)}$ based on Table 12.8. Significance at the 1%, 5% and 10% levels are denoted by ***, ** and *, respectively

	A	B	C	D	A	B	C	D
	AOL				IBM			
Mean of $\hat{\Lambda}_i$	0.98	0.98	0.98	0.98	0.97	0.96	0.97	0.97
S.D. of $\hat{\Lambda}_i$	1.04	1.04	1.05	1.05	1.04	1.03	1.02	1.03
LB(20) of $\hat{\Lambda}_i$	31.82**	26.13	21.29	21.25	40.17***	34.02***	31.42**	30.78*
Exc. disp.	1.63	1.69*	1.79*	1.76	1.24	1.01	0.82	0.87
	Coca-Cola				JP Morgan			
Mean of $\hat{\Lambda}_i$	0.95	0.95	0.95	0.95	0.98	0.98	0.98	0.98
S.D. of $\hat{\Lambda}_i$	1.00	1.02	1.01	1.01	1.01	1.00	1.01	1.00
LB(20) of $\hat{\Lambda}_i$	40.13***	39.76***	38.49***	37.19**	42.31***	32.07*	35.67**	35.06**
Exc. disp.	0.12	0.54	0.31	0.33	0.37	0.39	0.38	0.37
	AT&T							
Mean of $\hat{\Lambda}_i$	0.97	0.97	0.97	0.97				
S.D. of $\hat{\Lambda}_i$	1.04	1.04	1.05	1.05				
LB(20) of $\hat{\Lambda}_i$	22.14	23.76	17.60	17.51				
Exc. disp.	1.75*	1.62	1.88*	1.88*				

Panels A and B shows that the inclusion of the latent component leads to a strong increase of the model's goodness-of-fit as indicated by the log likelihood and the Bayes Information Criterion. Accounting for a common component induces a reduction in the magnitudes of the persistence parameters β_{kk} with most of the innovation parameters associated with cross-effects, α_k^m, $k \neq m$, becoming insignificant. As indicated by the Ljung–Box statistics (see Table 12.9), the SCI model captures the dynamics in a better way than the pure ACI model. As in Chap. 6 in the context of stochastic MEM processes, these results provide evidence for the existence of a joint factor capturing common underlying dynamics and interdependencies between the individual processes.

As reflected by the parameters σ^* shown in Panel C, the impact of the common component on the individual processes varies to some extent. Moreover, the strength of the serial dependence in the latent component is assumed to depend on the time elapsed since the last price event according to (12.22). The thresholds $\bar{x}_m$ are fixed exogenously and correspond to 50%, 100% and 200% of the average event waiting time in the pooled process. It is shown that the serial dependence in the latent component significantly declines with the length of past durations. Simultaneously, the overall level of serial dependence in the latent factor is reduced whereas the persistence in the observation-driven component is increased. Since a decline of the serial dependence implicitly reduces the unconditional variance of the latent component, we observe a counterbalancing effect by an increase of the scaling parameters σ_k^*. However, as indicated by the BIC, the extra flexibility implied by specification C is not supported by the data.

Panel D gives the estimates of a specification where the dynamics of the observation-driven component are excluded. As now the joint latent component has to capture the dynamics of all individual processes, the parameters a_r^* are driven towards one. This effect is counterbalanced by a significant decline of the variance scaling parameters σ_k^*. It is shown that, according to the BIC, a specification with only one common parameter-driven dynamic but no observation-driven dynamics outperforms the basic ACI model (specification A) in terms of goodness-of-fit. This supports the idea of a common component as a major driving force of the multivariate system. Nevertheless, as in Chap. 7, we observe that one latent factor solely is not sufficient to completely capture the dynamics of the multivariate process.

References

Bauwens L, Giot P, Grammig J, Veredas D (2004) A comparison of financial duration models via density forecasts. Int J Forecast 20:589–609

Bauwens L, Hautsch N (2006) Stochastic conditional intensity processes. J Financ Econom 4: 450–493

Biais B, Hillion P, Spatt C (1995) An empirical analysis of the limit order book and the order flow in the Paris Bourse. J Finance 50:1655–1689

Bowsher CG (2007) Modelling security markets in continuous time: intensity based, multivariate point process models. J Econom 141:876–912

Cox DR, Isham V (1980) Point processes. Chapman and Hall, London

Grandell J (1976) Doubly stochastic poisson processes. Springer, Berlin

Hall AD, Hautsch N (2006) Order aggressiveness and order book dynamics. Empir Econ 30: 973–1005

Hall AD, Hautsch N (2007) Modelling the buy and sell intensity in a limit order book market. J Finan Markets 10:249–286

Hawkes AG (1971) Spectra of some self-exciting and mutually exciting point processes. Biometrika 58:83–90

Parlour CA (1998) Price dynamics in limit order markets. Rev Financ Stud 11(4):789–816

Richard J-F, Zhang W (2007) Efficient high-dimensional importance sampling. J Econom 141:1385–1411

Russell JR (1999) Econometric modeling of multivariate irregularly-spaced high-frequency data. Working Paper, University of Chicago

Chapter 13
Autoregressive Discrete Processes and Quote Dynamics

In this chapter, we discuss dynamic models for discrete-valued data and quote processes. As illustrated in Chap. 4, the time series of the number of events in a given time interval yields a counting process and provides an alternative way to characterize the underlying point process. Section 13.1 presents a class of univariate autoregressive models for count data based on dynamic parameterizations of the conditional mean function in a Poisson distribution. Moreover, we discuss extensions thereof, such as the Negative Binomial distribution and Double Poisson distribution. As illustrated in Sect. 13.2, these approaches have the advantage of being straightforwardly extended to a multivariate framework.

Section 13.3 presents a simple and classical approach to model (mid-)quote and price dynamics in terms of a vector autoregressive framework. Such a setting has been proposed by Hasbrouck (1991) as a reduced-form approach to study price dynamics. It is easily implemented and extended, however, faces the disadvantage of not explicitly accounting for the discreteness of prices and quotes on transaction level. The latter issue is addressed by dynamic models for integer-valued variables. Such frameworks allow to model discrete-valued transaction price changes, bid-ask spreads, indicators of the trading direction (buy vs. sell) or trade sizes which only occur in round lot sizes. Section 13.4 presents an autoregressive conditional multinomial model as proposed by Russell and Engle (2005) while Sect. 13.5 discusses approaches decomposing integer-valued random variables into their directional components (negative, zero, positive) and their magnitudes. These frameworks open up flexible ways to model trade-to-trade price changes. Finally, Sect. 13.6 shows approaches to capture the joint dynamics of both ask and bid quotes. In this context, we discuss a cointegration model for ask and bid quotes as proposed by Engle and Patton (2004) as well as structural models which decompose the bivariate quote process into a common stochastic trend and stationary idiosyncratic components.

N. Hautsch, *Econometrics of Financial High-Frequency Data*,
DOI 10.1007/978-3-642-21925-2_13,

13.1 Univariate Dynamic Count Data Models

13.1.1 *Autoregressive Conditional Poisson Models*

Autoregressive count data models are valuable approaches to model dynamic intensity processes based on aggregated (equi-distant) data. A first textbook treatment of time series of count data is given by Cameron and Trivedi (1998). For a state-of-the-art overview, see Jung and Tremayne (2011). Dynamic count approaches might also be used to model the behavior of positive-valued discrete variables, such as the (absolute) magnitude of trade-to-trade price changes, bid-ask spreads or trade sizes occurring only in round lot sizes.

Let $\{y_i\}$ denote a time series of counts or integer-valued variables. The so-called Autoregressive Conditional Poisson (ACP) model belongs to the class of autoregressive conditional mean models which dynamically parameterize the conditional mean function. It is given by

$$y_i|\mathcal{F}_{i-1} \sim \text{Po}(\lambda_i), \quad \lambda_i = \omega + \sum_{j=1}^{P} \alpha_j y_{i-j} + \sum_{j=1}^{Q} \beta_j \lambda_{i-j}, \tag{13.1}$$

where $\text{Po}(\lambda_i)$ denotes the Poisson probability mass function (p.m.f.) given by

$$\text{Po}(\lambda_i) = \frac{\exp(-\lambda_i)\lambda_i^{y_i}}{y_i!}, \quad y_i = 0, 1, 2 \ldots, \quad \lambda_i > 0. \tag{13.2}$$

The time-varying Poisson intensity, λ_i, equals the conditional mean as well as the conditional variance, i.e.,

$$\mathbb{E}[y_i|\mathcal{F}_{i-1}] = \mathbb{V}[y_i|\mathcal{F}_{i-1}] = \lambda_i. \tag{13.3}$$

This specification has been originally proposed by Rydberg and Shephard (1998, 2003). Extensions of this model are considered by Heinen (2003). Ferland et al. (2006) discuss (13.1) and call it an integer-valued GARCH (INGARCH) process as it can be seen as an integer-valued analogue of a GARCH process. However, though the ACP model captures the dynamics of both the conditional mean *and* the conditional variance, (13.1) is essentially a conditional mean relation linking the conditional mean to past values of y_i and λ_i. In this sense, it is more in the spirit of an ACD specification as discussed in Chap. 5 and is referred in this book to as an Autoregressive Conditional Poisson (ACP) model.

As stressed by Fokianos et al. (2009), (13.1) can be restated in terms of a counting representation. Denote N_i as a Poisson process with unit intensity starting at time point i. Then, (13.1) can be expressed in terms of a sequence of independent Poisson drawings, $\{N_i, i = 1, 2, \ldots\}$. Define $N_i(\lambda_i)$ as the number of events of N_i in the time interval $[0, \lambda_i]$. Then, (13.1) is represented as

$$y_i = N_i(\lambda_i), \quad \lambda_i = \omega + \sum_{j=1}^{P} \alpha_j y_{i-j} + \sum_{j=1}^{Q} \beta_j \lambda_{i-j}. \tag{13.4}$$

Model (13.1) (or, equivalently, (13.4)) belong to the class of observation-driven models as also discussed in Chap. 10. Observation driven models for time series of counts have been studied, among others, by Zeger and Qaquish (1988), Brumback et al. (2000), Davis et al. (2003) and Jung et al. (2006).

Streett (2000), Heinen (2003) and Ferland et al. (2006) derive the stationarity properties of the model. Process (13.1) is stationary if $0 \leq \sum_{j=1}^{P} \alpha_j + \sum_{j=1}^{Q} \beta_j < 1$ with the unconditional mean given by

$$\mathbb{E}[y_i] = \mathbb{E}[\lambda_i] := \mu = \frac{\omega}{1 - \sum_{j=1}^{P} \alpha_j - \sum_{j=1}^{Q} \beta_j}. \tag{13.5}$$

The unconditional covariance function for an ACP(1,1) process is given by

$$\mathrm{Cov}[y_i, y_{i+h}] = \begin{cases} \dfrac{(1 - (\alpha+\beta)^2 + \alpha^2)\mu}{1 - (\alpha+\beta)^2}, & h = 0, \\[2ex] \dfrac{\alpha(1 - \beta(\alpha+\beta))(\alpha+\beta)^{h-1}\mu}{1 - (\alpha+\beta)^2} & h \geq 1. \end{cases} \tag{13.6}$$

Rewriting the unconditional variance as

$$\mathbb{V}[y_i] = \mu \left(1 + \frac{\alpha^2}{1 - (\alpha+\beta)^2}\right), \tag{13.7}$$

it is easily seen that $\mathbb{V}[y_i] > \mathbb{E}[y_i]$ (with equality if $\alpha = 0$). Hence, there is a close link between overdispersion and persistence. In order to ensure the non-negativity of λ_i, the same conditions as for MEM processes apply (see Chap. 5).

Fokianos et al. (2009) consider an exponential version of the ACP model given by

$$\lambda_i = (\omega + \beta \exp(-\gamma \lambda_{i-1}^2))\lambda_{i-1} + \alpha y_{i-1} \tag{13.8}$$

mimicking the structure of the exponential autoregressive model proposed by Haggan and Ozaki (1981). Geometric ergodicity of both specifications (13.1) and (13.8) are proven by Fokianos et al. (2009).

The model is straightforwardly estimated by maximum likelihood. Denote $\boldsymbol{\theta} = (\omega, \alpha_1, \ldots, \alpha_P, \beta_1, \ldots, \beta_Q)$ as the vector of unknown parameters associated with the ACP(P,Q) model. Then, the log likelihood is obtained by (up to a constant)

$$\ln \mathcal{L}(\mathbf{Y}; \boldsymbol{\theta}) = \sum_{i=1}^{n} (y_i \ln \lambda_i(\boldsymbol{\theta}) - \lambda_i(\boldsymbol{\theta})), \tag{13.9}$$

where the score is given by

$$\mathbf{s}(\mathbf{Y};\boldsymbol{\theta}) = \sum_{i=1}^{n} \frac{\partial \ln \mathcal{L}(\mathbf{Y};\boldsymbol{\theta})}{\partial \boldsymbol{\theta}} = \sum_{i=1}^{n} \left(\frac{y_i}{\lambda_i(\boldsymbol{\theta})} - 1 \right) \frac{\partial \lambda_i(\boldsymbol{\theta})}{\partial \boldsymbol{\theta}}, \tag{13.10}$$

and

$$\begin{aligned}
\frac{\partial \lambda_i(\boldsymbol{\theta})}{\partial \boldsymbol{\theta}} &= \left(\frac{\partial \lambda_i(\boldsymbol{\theta})}{\partial \omega}, \frac{\partial \lambda_i(\boldsymbol{\theta})}{\partial \alpha}, \frac{\partial \lambda_i(\boldsymbol{\theta})}{\partial \beta} \right)' \\
\frac{\partial \lambda_i(\boldsymbol{\theta})}{\partial \omega} &= 1 + \beta \frac{\partial \lambda_{i-1}(\boldsymbol{\theta})}{\partial \omega}, \\
\frac{\partial \lambda_i(\boldsymbol{\theta})}{\partial \alpha} &= y_{i-1} + \beta \frac{\partial \lambda_{i-1}(\boldsymbol{\theta})}{\partial \alpha}, \\
\frac{\partial \lambda_i(\boldsymbol{\theta})}{\partial \beta} &= \lambda_{i-1}(\boldsymbol{\theta}) + \beta \frac{\partial \lambda_{i-1}(\boldsymbol{\theta})}{\partial \beta}.
\end{aligned}$$

For the asymptotic properties of the maximum likelihood estimator, see Fokianos et al. (2009).

As the autoregressive count data models discussed above rely on parameterizations of the conditional mean function, various extensions discussed for MEMs in Chap. 6 can be straightforwardly adapted. For instance, a long memory version of the ACP model similar to long memory MEMS discussed in Chap. 6 are proposed by Groß-Klußmann and Hautsch (2011a). In this context, λ_i is parameterized in terms of long memory specifications as discussed in Chap. 6. Adapting, for instance, the specification proposed by Koulikov (2003), a long memory ACP model is given by parameterizing λ_i as

$$\lambda_i = \omega + \alpha(1 - \beta L)^{-1}(1 - L)^{-d}\eta_{i-1} \tag{13.11}$$

with $\eta_i := y_i - \lambda_i$. For more details on long memory specifications, see Chap. 6 and Groß-Klußmann and Hautsch (2011a).

Note that ACP specifications provide only one possibility to model time series of counts. Alternative approaches are, for instance, integer autoregressive models which preserve the discreteness of counts or generalized linear ARMA models. For a survey and a comparison of the different classes of models, see Jung and Tremayne (2011).

13.1.2 Extended ACP Models

A natural extension of the Poisson distribution is the Negative Binomial (NegBin) distribution with parameters r and p and probability mass function given by

$$\begin{aligned}\text{NegBin}(r,p) &= \binom{y_i+r-1}{y_i}(1-p)^r p^{y_i},\\ &= \frac{\Gamma(y_i+r)}{y_i!\Gamma(r)}(1-p)^r p^{y_i}\\ &= \frac{\Gamma(y_i+r)}{\Gamma(y_i)\Gamma(r+1)}(1-p)^r p^{y_i}, \quad y_i = 0,1,2,\ldots,\end{aligned} \tag{13.12}$$

where $r > 0$ and $p \in (0,1)$. The unconditional mean and variance of the NegBin distribution are given by

$$\mathbb{E}[y_i] = r\frac{p}{1-p}, \tag{13.13}$$

$$\mathbb{V}[y_i] = r\frac{p}{(1-p)^2}. \tag{13.14}$$

Hence, the NegBin distribution allows for overdispersion which is a typical feature of most high-frequency time series, see also Chap. 3.

To define an autoregressive conditional NegBin model similar to the ACP model, re-write the distribution in terms of the unconditional mean. Hence, by defining

$$\lambda := \mathbb{E}[y_i] = r\frac{p}{1-p}, \tag{13.15}$$

the p.m.f. can be re-written as

$$\text{NegBin}(r,p) = \frac{\Gamma(y_i+r)}{y_i!\Gamma(r)}\left(\frac{r}{\lambda+r}\right)^r\left(\frac{\lambda}{\lambda+r}\right)^{y_i}, \tag{13.16}$$

with $p = \lambda/(r+\lambda)$ and $\mathbb{V}[y_i] = \lambda + \lambda^2/r$.

Then, an ACP structure is straightforwardly adapted by dynamically parameterizing λ as in (13.1) yielding

$$y_i|\mathcal{F}_{i-1} \sim \text{NegBin}(r,p_i), \quad p_i = \lambda_i/(r+\lambda_i), \tag{13.17}$$

$$\lambda_i = \omega + \sum_{j=1}^{P}\alpha_j y_{i-j} + \sum_{j=1}^{Q}\beta_j\lambda_{i-j}. \tag{13.18}$$

The ACP is nested by letting $r \to \infty$. Such a distributional specification is applied by Rydberg and Shephard (2003) to formulate a discrete generalized linear autoregressive moving average (GLARMA) model to model the size of trade-to-trade price changes.

While the NegBin distribution only allows for overdispersion, even more flexibility is provided by the Double Poisson distribution introduced by Efron (1986). Denote $\text{Po}_y(\lambda)$ as the p.m.f. of the Poisson distribution of random variable y, then the p.m.f. of the Double Poisson distribution of y_i is given by

$$\begin{aligned} \text{DPo}(\lambda,\gamma) &= \gamma^{1/2}\text{Po}_{y_i}(\lambda)^{\gamma}\text{Po}_{\lambda}(\lambda)^{1-\gamma} \\ &= c(\gamma,\lambda)\left(\gamma^{1/2}e^{-\gamma\lambda}\right)\left(\frac{e^{-y_i}y_i^{y_i}}{y_i!}\right)\left(\frac{e\lambda}{y_i}\right)^{\gamma\lambda}, \quad \lambda,\gamma > 0, \end{aligned} \tag{13.19}$$

where $c(\gamma, \lambda)$ denotes a multiplicative constant which ensures that (13.19) is a proper density function which integrates to one. Efron (1986) shows that this constant can be approximated by

$$\frac{1}{c(\gamma,\lambda)} = 1 + \frac{1-\gamma}{12\lambda\gamma}\left(1 + \frac{1}{\lambda\gamma}\right). \tag{13.20}$$

The advantage of the Double Poisson distribution is that it allows for both overdispersion ($\gamma < 1$) and underdispersion ($\gamma > 1$) and nests the Poisson distribution for $\gamma = 1$. As shown by Efron (1986), the mean and variance of the Double Poisson distribution is given by

$$\mathbb{E}[y] = \lambda, \tag{13.21}$$

$$\mathbb{V}[y] \approx \frac{\lambda}{\gamma}, \tag{13.22}$$

where the latter is a close approximation (see Efron 1986, for more details).

Using the Double Poisson distribution to specify a dynamic count data process is proposed by Heinen (2003). He introduces the so-called Autoregressive Conditional Double Poisson (ACDP) model given by

$$y_i|\mathcal{F}_{i-1} \sim \text{DPo}(\lambda_i, \gamma), \tag{13.23}$$

where λ_i is specified as in (13.1). As shown by Heinen (2003), the ACDP model yields unconditional overdispersion, i.e., $\mathbb{V}[y_i] \geq \mathbb{E}[y_i]$ if $\gamma \leq 1$. Empirically, this is observed whenever the overdispersion induced by the estimated autocorrelation is not sufficient to match the overdispersion in the data.

A further advantage of the Double Poisson distribution is that the conditional variance can be specified separately by parameterizing γ. For instance, Heinen (2003) proposes specifying the conditional variance as

$$\mathbb{V}[y_i|\mathcal{F}_{i-1}] := \sigma_i^2 = \lambda_i + \delta\lambda_i^2, \tag{13.24}$$

corresponding to a quadratic relationship between variance and mean. This parameterization requires replacing γ in (13.19) by $\lambda_i/\sigma_i = (1 + \delta\lambda_i)^{-1}$. However, note that this specification generates negative conditional variances whenever $\lambda_i < -\delta^{-1}$. This case might occur in cases of underdispersion when $\delta < 0$.

Alternatively, as proposed Heinen (2003), σ_i^2 might be parameterized in terms of a GARCH specification yielding

Table 13.1 Maximum likelihood estimates of ACP models based on the Poisson, NegBin and Double Poisson distribution. Based on the number of trades per 30-sec intervals for the JPM stock traded at the New York Stock Exchange, June 2009. 17,160 observations

	Poisson		NegBin		Doub Poiss	
par	est	p-value	est	p-value	est	p-value
ω	0.573	0.000	0.438	0.000	0.763	0.000
α	0.207	0.000	0.194	0.000	0.204	0.000
β	0.706	0.000	0.742	0.000	0.687	0.000
$r\|\gamma$			7.243	0.000	0.044	0.000
δ^s	0.152	0.000	0.113	0.000	0.164	0.000
$\delta^s_{s,1}$	0.049	0.000	0.031	0.005	0.058	0.000
$\delta^s_{s,2}$	0.028	0.000	0.019	0.010	0.030	0.000
$\delta^s_{s,3}$	0.004	0.496	-0.001	0.846	0.007	0.232
$\delta^s_{s,4}$	0.005	0.347	0.000	0.967	0.006	0.250
$\delta^s_{c,1}$	0.118	0.000	0.108	0.000	0.130	0.000
$\delta^s_{c,2}$	0.031	0.000	0.029	0.000	0.041	0.000
$\delta^s_{c,3}$	0.030	0.000	0.028	0.000	0.033	0.000
$\delta^s_{c,4}$	0.029	0.000	0.025	0.000	0.031	0.000
LB(20)	278.43		361.88		272.87	

Standard errors are based on QML estimates of the covariance matrix. The distribution parameter is denoted by r for the NegBin distribution and by γ for the Double Poisson distribution. Ljung–Box(20) statistics based on residuals $y_i - \hat{\lambda}_i \exp(\hat{s}_i)$.

$$\sigma_i^2 = \tilde{\omega} + \tilde{\alpha}(y_{i-1} - \lambda_{i-1})^2 + \tilde{\beta}\sigma_{i-1}^2, \tag{13.25}$$

where $\tilde{\omega}$, $\tilde{\alpha}$ and $\tilde{\beta}$ denote parameters. This specification opens up possibilities for various extensions and modifications. Nevertheless, its theoretical properties are still widely unknown.

13.1.3 Empirical Illustrations

Tables 13.1 and 13.2 show the maximum likelihood estimates of ACP models based on the Poisson, NegBin and Double Poisson distribution for the number of trades per 30 s intervals for the JP Morgan and Microsoft stock traded at the NYSE. To account for intraday seasonalities, the conditional mean function multiplicatively interacts with a flexible Fourier form. In this case, the conditional mean is given by $\lambda_i \exp(s_i)$, where s_i is given by (5.34). The lag orders and order of the flexible Fourier form are chosen based on the Bayes Information Criterion.

Clear evidence for overdispersion is found. In both cases of the NegBin distribution as well as Double Poisson distribution, the distributional parameters r and γ, respectively, indicate overdispersion. Moreover, it is shown that the ACP model is not able to fully capture the dynamics in high-frequency trade counts. The Ljung–Box statistics computed based on the residuals $y_i - \hat{\lambda}_i \exp(\hat{s}_i)$ provide

Table 13.2 Maximum likelihood estimates of ACP models based on the Poisson, NegBin and Double Poisson distribution. Based on the number of trades per 30-sec intervals for the MSFT stock traded at the New York Stock Exchange, June 2009. 17,160 observations

	Poisson		NegBin		Doub Poiss	
par	est	p-value	est	p-value	est	p-value
ω	0.540	0.000	0.487	0.000	0.785	0.000
α	0.184	0.000	0.177	0.000	0.183	0.000
β	0.684	0.000	0.709	0.000	0.656	0.000
$r\|\gamma$			4.297	0.000	0.063	0.000
δ^s	0.244	0.000	0.201	0.000	0.235	0.000
$\delta^s_{s,1}$	0.099	0.000	0.083	0.000	0.104	0.000
$\delta^s_{s,2}$	0.076	0.000	0.063	0.000	0.077	0.000
$\delta^s_{s,3}$	0.009	0.245	0.006	0.466	0.014	0.066
$\delta^s_{s,4}$	0.039	0.000	0.035	0.000	0.039	0.000
$\delta^s_{c,1}$	0.137	0.000	0.141	0.000	0.133	0.000
$\delta^s_{c,2}$	0.073	0.000	0.069	0.000	0.080	0.000
$\delta^s_{c,3}$	0.041	0.000	0.037	0.000	0.048	0.000
$\delta^s_{c,4}$	0.035	0.000	0.025	0.000	0.043	0.000
LB(20)	393.45		468.46		518.79	

Standard errors are based on QML estimates of the covariance matrix. The distribution parameter is denoted by r for the NegBin distribution and by γ for the Double Poisson distribution. Ljung–Box(20) statistics based on residuals $y_i - \hat{\lambda}_i \exp(\hat{s}_i)$.

evidence for remaining serial dependence. The same is true even if higher order lags are included. However, given Ljung–Box statistics of 30-sec trade counts of approximately 1e6 and 7e5 for JPM and MSFT, respectively, the model implied reductions of Ljung–Box statistics are quite satisfying. Further improvements could be achieved by long memory ACP models, see Groß-Klußmann and Hautsch (2011a).

13.2 Multivariate ACP Models

As proposed by Heinen and Rengifo (2007), an autoregressive count data model can be extended to a multivariate setting by modelling the conditional mean in terms of VAR(MA)-type system. Assume a K-dimensional count data process $\{y_i^j\}$, $j = 1, \ldots, K$, with the marginal distributions following a Double Poisson distribution,

$$y_i^j | \mathcal{F}_{i-1} \sim DPo(\lambda_i^j, \gamma^j), \quad j = 1, \ldots, K, \tag{13.26}$$

with $\mathbb{E}[N_i^j | \mathcal{F}_{i-1}] = \lambda_i^j$ and $\mathbb{V}[N_i^j | \mathcal{F}_{i-1}] = \lambda_i^j / \gamma^j$. Denote $\mathbf{y}_i := (y_i^1, \ldots, y_i^K)'$ and $\boldsymbol{\lambda}_i := (\lambda_i^1, \ldots, \lambda_i^K)'$. Then, multivariate dynamics are captured by a VARMA process,

$$\boldsymbol{\lambda}_i = \boldsymbol{\omega} + \sum_{j=1}^{P} \mathbf{A}_j \mathbf{y}_{i-j} + \sum_{j=1}^{Q} \mathbf{B}_j \boldsymbol{\lambda}_{i-j}, \tag{13.27}$$

which is stationary if the eigenvalues of $(\mathbf{I} - \mathbf{A} - \mathbf{B})$ lie within the unit circle.

To capture contemporaneous dependencies, Heinen and Rengifo (2007) propose using a copula function. Let $F(y^1, \ldots, y^K)$ denote a continuous K-variate cumulative distribution function with univariate marginal c.d.f.'s $F_j(y^j)$, $j = 1 \ldots, K$. As shown by Sklar (1959), there exists a function C – a so-called copula – mapping $[0, 1]^K$ into $[0, 1]$ such that

$$F(y^1, \ldots, y^K) = C(F_1(y^1), \ldots, F_K(y^K)). \tag{13.28}$$

The joint p.d.f. is given by the product of the marginals and the copula density,

$$\frac{\partial F(y^1, \ldots, y^K)}{\partial y^1 \ldots \partial y^K} = \prod_{j=1}^{K} f_j(y^j) \frac{\partial C(F_1(y^1), \ldots, F_K(y^K))}{\partial F_1(y^1) \ldots \partial F_K(y^K)}. \tag{13.29}$$

Correspondingly, the copula of a multivariate distribution with $U[0, 1]$ margins is given by

$$C(z^1, \ldots, z^K) = F(F_1^{-1}(z^1), \ldots, F_K^{-1}(z^K)), \tag{13.30}$$

where $z^k := F_k(y^k)$ for $k = 1, \ldots, K$.

Heinen and Rengifo (2007) propose using a Gaussian copula given by

$$C(z^1, \ldots, z^K; \boldsymbol{\Sigma}) = \boldsymbol{\Phi}^K(\Phi^{-1}(z^1), \ldots, \boldsymbol{\Phi}^{-1}(z^K); \boldsymbol{\Sigma}), \tag{13.31}$$

where $\boldsymbol{\Phi}^K$ is the K-dimensional c.d.f. of the standard normal distribution and $\boldsymbol{\Phi}^{-1}$ is the inverse of the standard univariate normal distribution function. The corresponding copula density is given by

$$c(z^1, \ldots, z^K; \boldsymbol{\Sigma}) = |\boldsymbol{\Sigma}|^{-1/2} \exp\left(\frac{1}{2}(\mathbf{q}'(\mathbf{I}_K - \boldsymbol{\Sigma}^{-1})\mathbf{q})\right) \tag{13.32}$$

with $\mathbf{q} := (q^1, \ldots, q^K)'$ and $q^k := \boldsymbol{\Phi}^{-1}(z^k)$, $k = 1, \ldots, K$. If the variables $y^1, \ldots, y^k$ are mutually independent, the copula density equals one and $\boldsymbol{\Sigma}$ equals the identity matrix $\mathbf{I}_K$.

As shown by Sklar (1959), C is unique as long as the marginal distributions are continuous. This is, however, violated in case of counting variables. In this case, the probability integral transformation theorem stating that if Y is a continuous variable with c.d.f. F, then $Z = F(Y) \sim U[0, 1]$, does not hold. To overcome this problem, Heinen and Rengifo (2007) propose applying the continued extension argument by Denuit and Lambert (2005). The major principle is to continuously extend a discrete random variable Y by a $U[0,1]$ distributed random variable U (independent of Y) yielding

$$Y^* = Y + (U - 1). \tag{13.33}$$

Then, with $[Y]$ denoting the integer part of Y and defining $f_Y(y) := \mathbb{P}r[Y = y]$, it is easily shown that

$$\begin{aligned} z^* :&= F^*(y^*) = F^*(y + (u-1)) = F([y^*]) + f_{[Y^*]+1}([y^*]+1)u \\ &= F(y-1) + f_Y(y)u \sim U[0,1]. \end{aligned} \tag{13.34}$$

Applying this principle to multivariate count data with Double Poisson marginals yields the joint c.d.f.

$$f(y_i^1, \ldots, y_i^K; \boldsymbol{\theta}, \boldsymbol{\Sigma}) = \prod_{j=1}^{K} f_{DPo}(y_i^j, \lambda_i^j, \gamma^j) c(\mathbf{q}_i; \boldsymbol{\Sigma}), \tag{13.35}$$

where $\boldsymbol{\theta}:=(\boldsymbol{\omega}, \text{vec}(\mathbf{A}), \text{vec}(\mathbf{B}))$, $f_{DPo}(y_i^j, \lambda_i^j, \gamma^j)$ denotes the Double Poisson p.d.f. and $c(\mathbf{q}_i, \boldsymbol{\Sigma})$ denotes the Gaussian copula density with $\mathbf{q}_i:=(\boldsymbol{\Phi}^{-1}(z_i^1), \ldots, \boldsymbol{\Phi}^{-1}(z_i^K))'$. The variables z_i^j are the probability integral transform of the continued extension of y_i^j and are given by

$$z_i^j = F_{ij}^*(y_i^{j*}) = F_{ij}(y_i^j - 1) + f_{ij}(y_i^j)u_i^j, \tag{13.36}$$

where F_{ij}^* is the c.d.f. of the continued extension and F_{ij} and f_{ij} denote the c.d.f. and p.d.f. of the Double Poisson distribution. Correspondingly, the log likelihood contribution of the ith observation is given by

$$\ln \mathcal{L}(\mathbf{Y}; \boldsymbol{\theta}) = \sum_{j=1}^{K} \ln(f_{DPo}(y_i^j, \lambda_i^j, \gamma^j) + \ln c(\mathbf{q}_i; \boldsymbol{\Sigma}). \tag{13.37}$$

Heinen and Rengifo (2007) suggest estimating the model in two steps, where in the first step the individual marginal models are estimated using a Double Poisson distribution. In the second step, the parameters of the copula are estimated given the estimates of $\boldsymbol{\theta}$.

13.3 A Simple Model for Transaction Price Dynamics

A seminal paper on the modelling of trade-to-trade quote processes is Hasbrouck (1991). Define p_i as the mid-quote or alternatively transaction price at trade i and define y_i^b as an indicator variable which equals 1 for buyer-initiated trades and -1

for seller-initiated trades. Then, Hasbrouck (1991) proposes a VAR(∞) model to study the effects of trade-related information on mid-quotes:

$$\Delta mq_i = \sum_{j=1}^{\infty} a_j \Delta mq_{i-j} + \sum_{j=1}^{\infty} b_j y_{i-j}^b + \varepsilon_{1,i} \tag{13.38}$$

$$y_i^b = \sum_{j=1}^{\infty} c_j \Delta mq_{i-j} + \sum_{j=1}^{\infty} d_j y_{i-j}^b + \varepsilon_{2,i}, \tag{13.39}$$

where $\Delta mq_i := mq_i - mq_{i-1}$ represents the revision in the midquote, a_j, b_j, c_j and d_j are coefficients and $\varepsilon_{1,i}$ and $\varepsilon_{2,i}$ are two white noise processes. To estimate the model by least squares, the lag order of the process is often truncated at a certain number of lags (5). Then, the model can be seen as an approximation to a VAR(∞) process. However, an obviously more parsimonious alternative is to extend the model by a moving average component.

Moreover, as the VAR specification does not account for the discreteness of quote changes and y_i^b is an indicator variable, OLS estimation is obviously not efficient as due to the discrete nature of y_i^b, model errors are subject to heteroscedasticity. Therefore, robust standard errors should be used.

Hasbrouck's (1991) VAR model can be seen as a reduced form approach to model quote dynamics. Its major advantage is its flexibility to be easily extended in various ways. For instance, Dufour and Engle (2000) propose to extend the Hasbrouck (1991) model by accounting for the time between trades. They propose to replace the parameter b_j in equation (13.38) by

$$b_j = \gamma_j^{\Delta} + \sum_{k=1}^{K} \lambda_{k,j}^{\Delta} D_{k,i-j} + \delta_j^{\Delta} \ln x_{j-1},$$

where $D_{k,i-j}$ is a set of time-of-day dummy variables, x_j denotes the trade duration between t_{j-1} and t_j and γ_j^{Δ}, $\lambda_{k,j}^{\Delta}$, and δ_j^{Δ} are coefficients. Then, the δ coefficients capture the influence of previous trade durations on the price impact of a trade. Accordingly, the λ's allow to capture time-of-day effects in the price impact of a trade.

Analogously, the coefficient d_j in (13.39) is modified to

$$d_j = \gamma_j^{x} + \sum_{k=1}^{K} \lambda_{k,j}^{x} D_{k,i-j} + \delta_j^{x} \ln x_{j-1}.$$

Here, the parameters $\lambda_{k,j}^{x}$ and δ_j^{x} capture the impact of time-of-day-effects and past trade durations on the autocorrelation of signed trades. Dufour and Engle (2000) show that the time between trades has a significant impact on the price impact of a trade. In particular, as the duration between consecutive trades decreases, the price impact of trades increases, the speed of price adjustment to trade-related information increases, and the positive autocorrelation of signed trades increases.

Finally, the VAR approach has the advantage of being easily extended to higher dimensions, for instance, to include also trade sizes, bid-ask spreads or depth. Recall from Chap. 7 that a Vector MEM process can be re-written in terms of a VARMA model. Consequently, a VAR(P) model can be re-written in terms of an MEM(P,0) whose parameters can be consistently (though not efficiently) estimated by least squares. Such an approach is advantageous whenever datasets and/or dimensions are huge making numerical optimizations (as, e.g., required for maximum likelihood estimation) tedious. In such a situation, VAR estimates serve at least as best linear predictors.[1] However, as stressed above, when the persistence of the underlying data is strong, the lag order Q of the VAR process has to be chosen high requiring to estimate a high number of parameters. This obviously limits the model's usefulness in such a situation. A further limitation of applications of VAR models to discrete-valued variables is that implied predictions are difficult to interpret and to use as they do not account for the discreteness of the underlying data.

13.4 Autoregressive Conditional Multinomial Models

As illustrated in Chap. 3, trade-to-trade quote changes, price changes or bid-ask spreads can be very discrete, particularly if the underlying liquidity of the asset is high. In such situations, VAR or MEM type approaches still yield consistent estimates of the underlying dynamics but can be quite inefficient. Moreover, as soon as we are interested in predictions of integer-valued variables, the discreteness of the underlying variable has to be explicitly taken into account.[2] This calls for dynamic models for integer-valued random variables. In this section, we discuss the Autoregressive Conditional Multinomial model proposed by Russell and Engle (2005) to model dynamic multinomial distributed random variables.

Consider an integer-valued process $\{y_i\}$ with $y_i \in \{1, \ldots, K\}$ defining a multinomial random variable reflecting K states. A typical example is the trade-to-trade price change which moves only in ticks and is very discrete (see Chap. 3). Capturing the discreteness of trade-to-trade price changes motivated early work on high-frequency data employing microeconometric concepts to model categorical data. A seminal paper in this line of research is Hausman et al. (1992) applying an ordered probit model to model trade-to-trade price changes. Bollerslev and Melvin (1994) employ a similar approach to model the bid-ask spread at foreign exchange markets. While these models straightforwardly allow to account for

[1] See, for instance, Groß-Klußmann and Hautsch (2011b) for an application to estimate high-frequency market responses to publications of automated news feeds.

[2] For an analysis of the effects of neglected discreteness, see Harris (1990), Gottlieb and Kalay (1985) or Ball (1988).

conditioning information by including appropriate regressors, the dynamic nature of the underlying process is not easily captured. This is due to the fact that in ordered response models, the underlying (continuous) process is unobservable but is only categorically observed whenever it crosses an underlying threshold. However, capturing dynamics in latent processes inevitably results in integrals of the dimension of the underlying sample size, see, e.g., the discussion of a stochastic MEM in Chap. 6. A possibility to overcome the difficulties implied by a dynamic latent factor model is to rely on an observation-driven approach as discussed in Chap. 10. Actually, the autoregressive conditional proportional hazard model presented in Chap. 10 relies on an observation-driven dynamic extension of an underlying ordered response model. This approach could be easily adapted to an ordered probit model in the spirit of Hausman et al. (1992).

An alternative idea, however, is pursued by Russell and Engle (2005) by dynamically extending a multinomial distribution. The major principle underlying the so-called *Autoregressive Conditional Multinomial* (ACM) model is to capture the dynamic behavior of y_i by dynamically parameterizing the conditional probability to observe the state y_i. Denote $\tilde{\mathbf{d}}_i$ as the $K \times 1$ vector indicating the realization of y_i. In particular, $\tilde{\mathbf{d}}_i$ is the jth column of the $K \times K$ identity matrix if $Y_i = j$. Let $\tilde{\pi}_i$ denote the $K \times 1$ vector of conditional probabilities associated with the states, i.e., the jth element of $\tilde{\pi}_i$ corresponds to the probability that the jth element of $\tilde{\mathbf{d}}_i$ takes the value one. Then, the conditional distribution of $\tilde{\mathbf{d}}_i$ is completely characterized by $\tilde{\pi}_i$.

Russell and Engle (2005) assume that $\tilde{\pi}_i$ is driven by a first-order Markov chain

$$\tilde{\pi}_i = \mathbf{P}\tilde{\mathbf{d}}_{i-1}, \tag{13.40}$$

where $\mathbf{P} = \{P_{i,j}\}_{i,j=1,\ldots,K}$ is a $K \times K$ transition matrix with elements $P_{i,j} = \mathbb{P}r[Y_i = y_i | Y_{i-1} = j]$. The fundamental principle behind the ACM model is to dynamically parameterize the elements of $\mathbf{P}$. To automatically guarantee that all elements of $\mathbf{P}$ are non-negative and all columns sum to unity, Russell and Engle (2005) propose dynamically parameterizing the log probability ratios (log odds ratios). Let π_i and $\mathbf{d}_i$ denote the $(K-1) \times 1$ vectors consisting of the first $K-1$ elements of $\tilde{\pi}_i$ and $\tilde{\mathbf{d}}_i$,

$$\pi_i := (\tilde{\pi}_{i,1}, \ldots, \tilde{\pi}_{i,K-1})' := (\pi_{i,1}, \ldots, \pi_{i,K-1})' \tag{13.41}$$

and

$$\mathbf{d}_i := (\tilde{d}_{i1}, \ldots, \tilde{d}_{i,K-1})' := (d_{i1}, \ldots, d_{i,K-1})'. \tag{13.42}$$

Furthermore, define $\mathbf{P}^*$ as the $(K-1) \times (K-1)$ matrix of log ratios of transition probabilities with mjth element given by $P^*_{mj} := \ln(P_{mj}/P_{Kj})$. Hence $\mathbf{P}^*$ is given by

$$\mathbf{P}^* = \begin{bmatrix} \ln(P_{11}/P_{K1}) & \cdots & \ln(P_{1,K-1}/P_{K,K-1}) \\ \vdots & \vdots & \vdots \\ \ln(P_{K-1,1}/P_{K1}) & \cdots & \ln(P_{K-1,K-1}/P_{K,K-1}) \end{bmatrix}_{(K-1)\times(K-1)} . \quad (13.43)$$

Then, the log odds ratio of state m relative to state K with $m < K$ is given by

$$\begin{aligned} \ln(\tilde{\pi}_{im}/\tilde{\pi}_{iK}) &= \ln\left(\sum_{j=1}^{K} P_{mj}\tilde{d}_{i-1,j}\right) - \ln\left(\sum_{j=1}^{K} P_{Kj}\tilde{d}_{i-1,j}\right) \\ &= \sum_{j=1}^{K} \ln(P_{mj}/P_{Kj})\tilde{d}_{i-1,j} \\ &= \sum_{j=1}^{K-1} \ln(P_{mj}/P_{Kj})\tilde{d}_{i-1,j} + \ln(P_{mK}/P_{KK})\tilde{d}_{i-1,K} \\ &= \sum_{j=1}^{K-1} (\ln(P_{mj}/P_{Kj}) - \ln(P_{mK}/P_{KK}))\tilde{d}_{i-1,j} + \ln(P_{mK}/P_{KK}) \\ &:= \sum_{j=1}^{K-1} P^*_{mj} d_{i-1,j} + c_m. \end{aligned}$$

Define $\mathbf{c} := (c_1, \ldots, c_{k-1})'$. Then, Russell and Engle (2005) suggest to parameterize $\mathbf{P}^*\mathbf{d}_i + \mathbf{c}$ in terms of the inverse logistic function

$$\mathbf{P}^*\mathbf{d}_i + \mathbf{c} = \ln(\boldsymbol{\pi}_i/(1 - \boldsymbol{\iota}'\boldsymbol{\pi})) := h(\boldsymbol{\pi}_i), \quad (13.44)$$

where $\boldsymbol{\iota}$ is a conforming vector of ones. The conditional state probabilities are easily recovered from the logistic transformation

$$\frac{\boldsymbol{\pi}_i}{1 - \boldsymbol{\iota}'\boldsymbol{\pi}_i} = \exp[\mathbf{P}^*\mathbf{d}_{i-1} + \mathbf{c}] \quad (13.45)$$

yielding

$$\boldsymbol{\pi}_i = \frac{\exp[\mathbf{P}^*\mathbf{d}_{i-1} + \mathbf{c}]}{1 + \boldsymbol{\iota}'\exp[\mathbf{P}^*\mathbf{d}_{i-1} + \mathbf{c}]} \quad (13.46)$$

and thus

$$P_{mn} = \frac{\exp[P^*_{mn} + c_m]}{1 + \sum_{j=1}^{k-1} \exp[P^*_{jn} + c_j]}. \quad (13.47)$$

The ACM(P,Q) model is then given by

$$h(\boldsymbol{\pi}_i) = \sum_{j=1}^{P} \mathbf{A}_j \boldsymbol{\xi}_{i-j} + \sum_{j=1}^{Q} \mathbf{B}_j h(\boldsymbol{\pi}_{i-j}) + \mathbf{z}'_{i-1}\boldsymbol{\gamma}, \tag{13.48}$$

where $\mathbf{A}_j$ and $\mathbf{B}_j$ denote $(K-1) \times (K-1)$ parameter matrices and $\mathbf{z}_i$ is a vector of explanatory variables with corresponding parameter vector $\boldsymbol{\gamma}$. To allow for an intercept, the first element of $\mathbf{z}_i$ is assumed to equal one. Moreover, $\boldsymbol{\xi}_i$ with $\boldsymbol{\xi}_i := (\xi_{1i}, \ldots, \xi_{K-1,i})'$ is a vector of innovations with elements which can be chosen as

$$\xi_{ji} := d_{ji} - \pi_{ji} \tag{13.49}$$

or

$$\xi_{ji} = \frac{d_{ji} - \pi_{ji}}{\sqrt{\pi_{ji}(1-\pi_{ji})}}. \tag{13.50}$$

Both specifications ensure that ξ_{ji} are martingale differences. Specification (13.50) is suggested by Rydberg and Shephard (2003) and ensures a variance of one.

Due to the linear structure of the ACM model, the choice of the base state is arbitrary. The first $(K-1)$ conditional probabilities are easily recovered from

$$\boldsymbol{\pi}_i = \exp\left[\sum_{j=1}^{P} \mathbf{A}_j(\mathbf{d}_{i-j} - \boldsymbol{\pi}_{i-j}) + \sum_{j=1}^{Q} \mathbf{B}_j h(\boldsymbol{\pi}_{i-j}) + \mathbf{z}'_{i-1}\boldsymbol{\gamma}\right]$$
$$\times \left(1 + \boldsymbol{\iota}' \exp\left[\sum_{j=1}^{P} \mathbf{A}_j(\mathbf{d}_{i-j} - \boldsymbol{\pi}_{i-j}) + \sum_{j=1}^{Q} \mathbf{B}_j h(\boldsymbol{\pi}_{i-j}) + \mathbf{z}'_{i-1}\boldsymbol{\gamma}\right]\right)^{-1},$$

where the Kth probability is given by the restriction $\boldsymbol{\iota}'\boldsymbol{\pi} = 1$.

As $\boldsymbol{\xi}_i$ is a martingale difference sequence, the stationarity conditions solely depend on the matrix $\mathbf{B}$. In particular, if the eigenvalues of $\mathbf{B}$ are distinct and lie inside the unit circle, we can rewrite the ACM(1,1) model with $z_i = 1$ as

$$h(\boldsymbol{\pi}_i) = \sum_{j=1}^{\infty} \mathbf{P}\boldsymbol{\Lambda}^{j-1}\mathbf{P}^{-1}\mathbf{A}(\mathbf{d}_{i-j} - \boldsymbol{\pi}_{i-j}) + (\mathbf{I} - \mathbf{B})^{-1}\boldsymbol{\gamma}, \tag{13.51}$$

where $\mathbf{B} = \mathbf{P}\boldsymbol{\Lambda}\mathbf{P}^{-1}$ and $\boldsymbol{\Lambda}$ is the diagonal matrix with the eigenvalues of $\mathbf{B}$ along the diagonal and $\mathbf{I}$ denotes a unity matrix. Russell and Engle (2005) prove the positiveness of $\boldsymbol{\pi}_i$ whenever the process is stationary.

The log likelihood function is given by

$$\ln \mathcal{L}(\mathbf{Y};\boldsymbol{\theta}) = \sum_{i=1}^{N}\sum_{j=1}^{K} \tilde{d}_{ij} \ln(\tilde{\pi}_{ij}) = \sum_{i=1}^{N} \tilde{\mathbf{d}}_i' \ln(\tilde{\boldsymbol{\pi}}_i), \tag{13.52}$$

where model diagnostics can be performed based on the vector of errors $\mathbf{d}_i - \hat{\boldsymbol{\pi}}_i$, which should form a martingale difference sequence under correct dynamic specification.

As the ACM model resembles the structure of a MEM, many of the extensions for MEM specifications discussed in Chap. 6 can be straightforwardly adopted. Russell and Engle (2005) combine the ACM model with an ACD model to jointly model the time between trades and the resulting price movement. Based on IBM transaction data they show that the ACM-ACD model successfully captures the dynamic properties of the underlying process.

13.5 Autoregressive Models for Integer-Valued Variables

An alternative way to model dynamic integer-valued processes is to extend a count data distribution to cover not only positive-valued but also negative-valued outcomes and to capture zero outcomes in form of a hurdle approach. In this context, a so-called *integer count hurdle* (ICH) model is proposed by Liesenfeld et al. (2006) and is used to model discrete trade-to-trade price changes. Let Y_i denote an integer valued random variable with outcomes $y_i \in \mathbb{Z}$. The main principle is to decompose the process into three components determining the sign of realizations (positive, negative, zero) as well as the size thereof given the realization is nonzero. Formally, the conditional p.d.f. of Y_i given $\mathcal{F}_{i-1}$ can be decomposed as

$$\mathbb{P}r[Y_i = y_i|\mathcal{F}_{i-1}] = \begin{cases} \mathbb{P}r[Y_i < 0|\mathcal{F}_{i-1}]\mathbb{P}r[Y_i = y_i|Y_i < 0, \mathcal{F}_{i-1}] & \text{if } y_i < 0, \\ \mathbb{P}r[Y_i = 0|\mathcal{F}_{i-1}] & \text{if } y_i = 0, \\ \mathbb{P}r[Y_i > 0|\mathcal{F}_{i-1}]\mathbb{P}r[Y_i = y_i|Y_i > 0, \mathcal{F}_{i-1}] & \text{if } y_i > 0. \end{cases} \tag{13.53}$$

Hence, the processes $\mathbb{P}r[Y_i = y_i|Y_i > 0, \mathcal{F}_{i-1}]$ and $\mathbb{P}r[Y_i = y_i|Y_i < 0, \mathcal{F}_{i-1}]$ represent the size of y_i conditional on its direction and are defined over the set of strictly positive or negative integers, respectively. Accordingly, $\mathbb{P}r[Y_i < 0|\mathcal{F}_{i-1}]$, $\mathbb{P}r[Y_i > 0|\mathcal{F}_{i-1}]$ and $\mathbb{P}r[Y_i = 0|\mathcal{F}_{i-1}]$ drive the direction of y_i.

Rydberg and Shephard (2003) propose to parameterize $\mathbb{P}r[Y_i = y_i|Y_i > 0, \mathcal{F}_{i-1}]$ and $\mathbb{P}r[Y_i = y_i|Y_i < 0, \mathcal{F}_{i-1}]$ by simply specifying the p.d.f. of $Y_i - 1$ given $Y_i > 0$ using a standard count data distribution, such as a Poisson or NegBin distribution.

Alternatively, building on the work by Mullahy (1986), Liesenfeld et al. (2006) propose using the p.d.f. of a standard count data distribution which is truncated at zero. Correspondingly, $\mathbb{P}r[Y_i = y_i | Y_i > 0, \mathcal{F}_{i-1}]$ and $\mathbb{P}r[Y_i = y_i | Y_i < 0, \mathcal{F}_{i-1}]$ are given by

$$\mathbb{P}r[Y_i = y_i | Y_i > 0, \mathcal{F}_{i-1}] := p^+(y_i | \mathcal{F}_{i-1}) = \frac{f^+(y_i | \mathcal{F}_{i-1})}{1 - f^+(0 | \mathcal{F}_{i-1})}, \tag{13.54}$$

$$\mathbb{P}r[Y_i = y_i | Y_i < 0, \mathcal{F}_{i-1}] := p^-(y_i | \mathcal{F}_{i-1}) = \frac{f^-(-y_i | \mathcal{F}_{i-1})}{1 - f^-(0 | \mathcal{F}_{i-1})}, \tag{13.55}$$

where $p^+()$ and $p^-()$ denote the p.d.f. of standard count data distributions for positive and negative realizations of y_i, respectively.

If one assumes that the truncated p.d.f.'s $p^+()$ and $p^-()$ stem from the *same* distribution, it is sufficient to parameterize a (conditional) p.d.f. of *absolute* realizations $S_i := |Y_i|$,

$$\mathbb{P}r[S_i = s_i | S_i > 0, D_i, \mathcal{F}_{i-1}] := p(s_i | D_i, \mathcal{F}_{i-1}) \tag{13.56}$$

with

$$D_i := \begin{cases} -1 & \text{if } Y_i < 0, \\ 0 & \text{if } Y_i = 0, \\ 1 & \text{if } Y_i > 0, \end{cases}$$

yielding to the p.d.f. of y_i given by

$$\begin{aligned} \mathbb{P}r[Y_i = y_i | \mathcal{F}_{i-1}] = \mathbb{P}r[Y_i < 0 | \mathcal{F}_{i-1}]^{\delta_i^-} \mathbb{P}r[Y_i = 0 | \mathcal{F}_{i-1}]^{\delta_i^0} \mathbb{P}r[Y_i > 0 | \mathcal{F}_{i-1}]^{\delta_i^+} \\ \times [p(s_i | D_i, \mathcal{F}_{i-1})]^{(1-\delta_i^0)}, \end{aligned} \tag{13.57}$$

where $\delta_i^- := \mathbb{1}_{\{Y_i < 0\}}$, $\delta_i^0 := \mathbb{1}_{\{Y_i = 0\}}$ and $\delta_i^+ := \mathbb{1}_{\{Y_i > 0\}}$.

Then, the resulting log likelihood is given by

$$\mathcal{L}(\mathbf{Y}; \boldsymbol{\theta}) = \sum_{i=1}^n \ln \mathbb{P}r[Y_i = y_i | \mathcal{F}_{i-1}] = \sum_{i=1}^n \mathcal{L}_{i,1} + \sum_{i=1}^n \mathcal{L}_{2,i}, \tag{13.58}$$

where

$$\sum_{i=1}^n \mathcal{L}_{1,i} := \delta_i^- \ln \mathbb{P}r[Y_i < 0 | \mathcal{F}_{i-1}] + \delta_i^0 \ln \mathbb{P}r[Y_i = 0 | \mathcal{F}_{i-1}] + \delta_i^+ \ln \mathbb{P}r[Y_i > 0 | \mathcal{F}_{i-1}],$$

$$\sum_{i=1}^n \mathcal{L}_{1,2} := (1 - \delta_i^0) \ln p(s_i | D_i, \mathcal{F}_{i-1}).$$

As long as there are no parametric restrictions on $\mathcal{L}_{1,i}$ and $\mathcal{L}_{2,i}$, both components can be maximized separately.

Liesenfeld et al. (2006) propose modelling the dynamics of the sign of Y_i, represented by D_i, in terms of an ACM specification as discussed in Sect. 13.4. Denote $\pi_{ji} := \mathbb{P}r[D_i = j|\mathcal{F}_{i-1}]$ with $j \in \{-1, 0, 1\}$ and the bi-variate vector $\boldsymbol{\pi}_i := (\pi_{-1i}/\pi_{0i}, \pi_{1i}/\pi_{0i})'$. Then, the inverse logistic transformation of $\boldsymbol{\pi}_i$, $h(\boldsymbol{\pi}_i)$, with $h(\boldsymbol{\pi}_i) = \ln(\boldsymbol{\pi}_i/(1-\sum_{j=-1}^{1}\pi_{ji})$ is modelled according to a bi-variate version of (13.48), i.e.,

$$h(\boldsymbol{\pi}_i) = \sum_{j=1}^{P} \mathbf{A}_j \xi_{i-j} + \sum_{j=1}^{Q} \mathbf{B}_j h(\boldsymbol{\pi}_{i-j}) + \mathbf{z}_{i-1}' \gamma^{\pi}, \tag{13.59}$$

where $\mathbf{z}_i$ are (weakly exogenous) explanatory variables with parameter vector γ^{π} and ξ_i is given by (13.49) or (13.50) with

$$d_i := (d_{-1i}, d_{1i}) = \begin{cases} (1,0)' & \text{if } Y_i < 0, \\ (0,0)' & \text{if } Y_i = 0, \\ (0,1)' & \text{if } Y_i > 0. \end{cases}$$

Then, π_{ji} is recovered by

$$\pi_{ji} = \frac{\exp(h(\pi_{ji}))}{1 + \sum_{j=-1}^{1} \exp(h(\pi_{ji}))}. \tag{13.60}$$

To evaluate the model, Liesenfeld et al. (2006) propose using the standardized residuals

$$\mathbf{e}_i := (e_{-1i}, e_{1i})' = \hat{\mathbb{V}}[d_i|\mathcal{F}_{i-1}]^{-1/2}(d_i - \hat{\mathbb{E}}[d_i|\mathcal{F}_{i-1}]) \tag{13.61}$$

where $\hat{\mathbb{V}}[d_i|\mathcal{F}_{i-1}]^{-1/2}$ is the inverse of the Cholesky factor of $\hat{\mathbb{V}}[d_i|\mathcal{F}_{i-1}]$. Under correct specification, the series $\{\mathbf{e}_i\}$ are uncorrelated with mean zero and identity covariance matrix.

Liesenfeld et al. (2006) propose the dynamics of the size S_i to be modelled by generalized linear autoregressive moving average (GLARMA) model based on a conditional NegBin distribution which is truncated at zero with p.m.f.

$$\begin{aligned} p(s_i|D_i, \mathcal{F}_{i-1}) &= p(y_i|D_i, \mathcal{F}_{i-1})/(1 - p(0|D_i, \mathcal{F}_{i-1})) \\ &= \frac{\Gamma(s_i + r)}{\Gamma(r)\Gamma(s_i + 1)} \left(\left[\frac{r + \lambda_i}{r} \right]^r - 1 \right)^{-1} \left(\frac{\lambda_i}{\lambda_i + r} \right)^{s_i}, \end{aligned} \tag{13.62}$$

with $s_i = 1, 2, \ldots$, $p(y_i|D_i, \mathcal{F}_{i-1})$ denoting the p.m.f. of a NegBin(r, p_i) distribution with $p_i = \lambda_i/(r + \lambda_i)$ as given in Sect. 13.1.2.

Its conditional moments are given by

$$\mathbb{E}[S_i | S_i > 0, D_i, \mathcal{F}_{i-1}] := \lambda_i^+ = \frac{\lambda_i}{1-\vartheta_i},$$

$$\mathbb{V}[S_i | S_i > 0, D_i, \mathcal{F}_{i-1}] := \sigma_i^{2+} = \frac{\lambda_i}{1-\vartheta_i} - \frac{\lambda_i^2}{(1-\vartheta_i)^2}\left(\vartheta_i - \frac{1-\vartheta_i}{r}\right),$$

where $\vartheta_i = [r/(r+\lambda_i)]^r$. Then, $\ln \lambda_i$ is parameterized in terms of a GLARMA structure as proposed by Rydberg and Shephard (2003),

$$\ln \lambda_i = \omega + \sum_{j=1}^{P} \tilde{\mathbf{A}}_j \varepsilon_{i-j} + \sum_{j=1}^{Q} \tilde{\mathbf{B}}_j \ln \lambda_{i-j} + \mathbf{z}_{i-1}' \gamma^\lambda, \tag{13.63}$$

where $\varepsilon_i := (S_i - \lambda_i^+)/\sigma_i^+$ which is evaluated based on the residuals $\tilde{e}_i := (S_i - \hat{\lambda}_i^+)/\hat{\sigma}_i^+$.

The integer count hurdle model is a modification of the decomposition model by Rydberg and Shephard (2003) who decompose Y_i into three components

$$Y_i = A_i S_i D_i, \tag{13.64}$$

where $A_i := \mathbb{1}_{\{Y_i \neq 0\}}$, D_i denoting the sign of Y_i given $A_i \neq 0$ and S_i defined as above. Rydberg and Shephard (2003) model the dynamics of A_i and D_i using an autologistic model, formulated for A_i by

$$\mathbb{P}r[A_i = 1 | \mathcal{F}_{i-1}] := p(\theta_i^A), \tag{13.65}$$

$$p(\theta_i^A) = \frac{\exp(\theta_i^A)}{1+\exp(\theta_i^A)}, \tag{13.66}$$

$$\theta_i^A = \sum_{j=1}^{P} \alpha_j A_{i-j} + \mathbf{z}_{i-1}' \gamma. \tag{13.67}$$

Rydberg and Shephard (2003) and Liesenfeld et al. (2006) apply these models to model the dynamics of transaction price changes. Alternative applications might be the modelling of trade directions or the sum of signed trades in fixed intervals. An extension of the integer count hurdle model to a multivariate setting is proposed by Bien et al. (2011). The authors develop a model for the conditional inflated multivariate density of integer count variables where contemporaneous dependencies are captured by a copula. The model is applied to the modelling of the conditional bivariate density of high-frequency ask and bid quote changes.

Finally, note that the models discussed in this section can be seen as reduced-form approaches to capture the dynamics of transaction prices and quotes. More structural approaches in line with classical market microstructure theory are discussed in Sect. 13.6.2.

13.6 Modelling Ask and Bid Quote Dynamics

13.6.1 Cointegration Models for Ask and Bid Quotes

Modelling mid-quotes as, e.g., in Hasbrouck's (1991) model (see Sect. 13.3), allows to study underlying price evolutions but does not capture possible asymmetries in ask and bid quote dynamics. To obtain a deeper understanding of quote dynamics and their implications for bid-ask spread and mid-quote movements, Engle and Patton (2004) propose modelling the processes of log bid prices ($\ln b_i$) and log ask prices ($\ln a_i$) as a VAR system. As quote processes follow stochastic trends (i.e., are integrated of the order one), the corresponding model should be set up in most generality in terms of a cointegrated VAR(P) model which is given in vector error correction (VEC) form for $\Delta y_i := y_i - y_{i-1}$,

$$\Delta \mathbf{y}_i = \boldsymbol{\mu} + \boldsymbol{\alpha}\boldsymbol{\beta}'\mathbf{y}_{i-1} + \sum_{j=1}^{P} \Gamma_j \Delta \mathbf{y}_{i-j} + \boldsymbol{\varepsilon}_i, \tag{13.68}$$

where $\mathbf{y}_i := [\ln a_i, \ln b_i]'$, $\boldsymbol{\mu}$ is a 2×1 vector of constants, $\boldsymbol{\alpha}$ and $\boldsymbol{\beta}$ denote the 2×1 loading and cointegrating vectors, and $\boldsymbol{\Gamma}_j$, $j = 1, \ldots, P-1$, is a 2×2 parameter matrix. The vector of noise terms $\boldsymbol{\varepsilon}_i$ is assumed to be serially uncorrelated with zero mean and covariance $\boldsymbol{\Sigma}$.

The cointegration vector $\boldsymbol{\beta}$ can be estimated using Johansen's (1991) maximum likelihood approach. In practice, however, estimates of $\boldsymbol{\beta}$ are very close to $\hat{\boldsymbol{\beta}} \approx (1, -1)'$ revealing the bid-ask spread as natural cointegration relationship. Consequently, Engle and Patton (2004) fix $\boldsymbol{\alpha}\boldsymbol{\beta}' y_{i-1}$ in (13.68) to the log bid-ask spread, i.e., $\boldsymbol{\alpha}\boldsymbol{\beta}' y_{i-1} = s_i$ with $s_i := \ln a_i - \ln b_i$. This leads to a straightforward vector error correction model for the series of log ask returns and log bid returns $\Delta \ln a_i := \ln a_i - \ln a_{i-1}$ and $\Delta \ln b_i := \ln b_i - \ln b_{i-1}$ of the form

$$\begin{pmatrix} \Delta \ln a_i \\ \Delta \ln b_i \end{pmatrix} = \begin{pmatrix} \gamma_0 \\ \delta_0 \end{pmatrix} + \begin{pmatrix} \gamma_1 & \gamma_2 \\ \delta_1 & \delta_2 \end{pmatrix} \cdot \begin{pmatrix} \Delta \ln a_{i-1} \\ \Delta \ln b_{i-1} \end{pmatrix} + \begin{pmatrix} \gamma_3 \\ \delta_3 \end{pmatrix} \cdot s_{i-1} + \begin{pmatrix} \gamma_4 \\ \delta_4 \end{pmatrix} \cdot \mathbf{z}_{i-1} + \begin{pmatrix} \varepsilon_i^a \\ \varepsilon_i^b \end{pmatrix}, \tag{13.69}$$

where $\mathbf{z}_i$ denotes a vector of additional explanatory variables and ε_i^a and ε_i^b denote white noise error terms. The model can be straightforwardly estimated by ordinary least squares.

The advantage of this specification is that it leads to a corresponding model for the difference in the log-spread $\Delta s_i := s_i - s_{i-1}$, and the log difference in the mid-quote, $\Delta \ln mq_i := \ln mq_i - \ln mq_{i-1}$, where $\ln mq_i := 0.5(\ln a_i + \ln b_i)$. In particular, multiplying the model by the matrix

$$\begin{pmatrix} 1 & -1 \\ 0.5 & 0.5 \end{pmatrix}$$

yields

$$\begin{pmatrix} \Delta s_i \\ \Delta \ln mq_i \end{pmatrix} = \begin{pmatrix} \gamma_0 - \delta_0 \\ 0.5(\gamma_0 + \delta_0) \end{pmatrix} + \begin{pmatrix} \gamma_1 - \delta_1 & \gamma_2 - \delta_2 \\ 0.5(\gamma_1 + \delta_1) & 0.5(\gamma_2 + \delta_2) \end{pmatrix} \cdot \begin{pmatrix} \Delta \ln p_{i-1}^a \\ \Delta \ln p_{i-1}^b \end{pmatrix}$$
$$+ \begin{pmatrix} \gamma_3 - \delta_3 \\ 0.5(\gamma_3 + \delta_3) \end{pmatrix} \cdot s_{i-1} + \begin{pmatrix} \gamma_4 - \delta_4 \\ 0.5(\gamma_4 + \delta_4) \end{pmatrix} \cdot \mathbf{z}_{i-1}$$
$$+ \begin{pmatrix} \varepsilon_i^a - \varepsilon_i^b \\ 0.5(\varepsilon_i^a + \varepsilon_i^b) \end{pmatrix}.$$

Alternative manipulations of the above expression yield

$$\begin{pmatrix} s_i \\ \Delta \ln mq_i \end{pmatrix} = \begin{pmatrix} \gamma_0 - \delta_0 \\ 0.5(\gamma_0 + \delta_0) \end{pmatrix} + \begin{pmatrix} \gamma_4 - \delta_4 \\ 0.5(\gamma_4 + \delta_4) \end{pmatrix} \cdot \mathbf{z}_{i-1}$$
$$= \begin{pmatrix} 1 + 0.5(\gamma_1 - \delta_1 - \gamma_2 + \delta_2) + \gamma_3 - \delta_3 & \gamma_1 - \delta_1 + \gamma_2 - \delta_2 \\ 0.25(\gamma_1 + \delta_1 - \gamma_2 - \delta_2) + 0.5(\gamma_3 + \delta_3) & 0.5(\gamma_1 + \delta_1 + \gamma_2 + \delta_2) \end{pmatrix}$$
$$\cdot \begin{pmatrix} s_{i-1} \\ \Delta \ln mq_{i-1} \end{pmatrix} + \begin{pmatrix} -0.5(\gamma_1 - \delta_1 - \gamma_2 + \delta_2) & 0 \\ -0.25(\gamma_1 + \delta_1 - \gamma_2 - \delta_2) & 0 \end{pmatrix} \cdot \begin{pmatrix} s_{i-2} \\ \Delta \ln q_{i-2} \end{pmatrix}$$
$$+ \begin{pmatrix} \varepsilon_i^a - \varepsilon_i^b \\ 0.5(\varepsilon_i^a + \varepsilon_i^b) \end{pmatrix}.$$

Hence, the VEC(1) model is "rotated" into a VAR(2) model for the log spread and the log-difference of the midquote. Similar approaches and modifications of this framework are considered by Lo and Sapp (2006) and Escribano and Pascual (2006).

Hautsch and Huang (2009) extend this model to include not only ask and bid quotes but also several levels of order book depth on both sides of the market. They illustrate that this approach serves as a convenient but powerful way to model limit order book dynamics. For a deeper discussion of this framework in a limit order book context, see Chap. 9.

Finally, note that this model captures the joint dynamics of ask and bid quotes but does not capture the quote's discreteness. A model capturing the dynamics of *discrete* ask and bid quotes is proposed by Hasbrouck (1996). This approach builds on the idea of decomposing ask and bid quotes into a common underlying random walk component (capturing the "efficient price") and market side specific market-making costs inducing a bid-ask spread. Such a framework will be discussed in more detail in the following section.

13.6.2 Decomposing Quote Dynamics

While the dynamic models for integer-valued variables discussed in Sects. 13.4 and 13.5 can be seen as reduced-form type approaches, classical market microstructure approaches aim for connecting quote dynamics to the dynamics of an underlying ("efficient") price process. A typical framework, as used, e.g., by Glosten and Harris (1988), Hasbrouck (1996) or Madhavan et al. (1997) builds on the assumption of an underlying random walk process for the unobservable efficient ("true", fundamental) price m_i. Accordingly, m_i, a_i and b_i are given by

$$m_i = m_{i-1} + \varepsilon_i, \tag{13.70}$$

$$a_i = m_i + c, \tag{13.71}$$

$$b_i = m_i - c, \tag{13.72}$$

where c denotes the market-making costs driving the bid-ask spread. This framework can be used as starting point for capturing quote dynamics as well as discreteness. To capture the latter, Hasbrouck (1996) adds the equation

$$p_i = \begin{cases} \text{round}(b_i) & \text{if } y_i^b = 0, \\ \text{round}(a_i) & \text{if } y_i^b = 1, \end{cases}$$

where p_i denotes the resulting transaction price and y_i^b is a trading indicator taking the value one (zero) for a sell (buy). Moreover, by modifying (13.71) and (13.72) by

$$b_i = \text{Floor}(m_i - b), \tag{13.73}$$

$$a_i = \text{Ceiling}(m_i - a), \tag{13.74}$$

Hasbrouck (1996) captures market side specific transaction cost components a and b. For more extensions and modifications of this framework, see Hasbrouck (2007).

Pascual and Veredas (2010) propose a generalization of the framework outlined above by allowing the trading costs to be time-varying and all processes to be conditionally heteroscedastic. The model is in the spirit of Hasbrouck (1993) and Madhavan et al. (1997) and decomposes quotes into a common stochastic trend and transitory noise components:

$$\begin{pmatrix} a_i \\ b_i \end{pmatrix} = \beta \begin{pmatrix} 1 \\ -1 \end{pmatrix} + \begin{pmatrix} 1 & 0 & 1 \\ 0 & 1 & 1 \end{pmatrix} \begin{pmatrix} S_{a,i} \\ S_{b,i} \\ m_i \end{pmatrix}, \tag{13.75}$$

where β captures the average half bid-ask spread and $S_{a,i} := a_i - \beta$ and $S_{b,i} := \beta_i - b$ denote the transitory (transaction cost induced) components of the corresponding quotes. Hence, the vector $(1,-1)'$ correspond to the (a priori fixed) cointegration vector as in Engle and Patton (2004), see Sect. 13.6.1.

Pascual and Veredas (2010) propose modelling the dynamics of the three components m_i, $S_{a,i}$ and $S_{b,i}$ as

$$\begin{pmatrix} S_{a,i} \\ S_{b,i} \\ m_i \end{pmatrix} = \begin{pmatrix} \phi_a & \phi_{ab} & 0 \\ \phi_{ba} & \phi_b & 0 \\ 0 & 0 & 1 \end{pmatrix} \begin{pmatrix} S_{a,i-1} \\ S_{b,i-1} \\ m_{i-1} \end{pmatrix} + \begin{pmatrix} \varepsilon_{a,i} \\ \varepsilon_{b,i} \\ \varepsilon_{m,i} \end{pmatrix} \tag{13.76}$$

with

$$\begin{pmatrix} \varepsilon_{a,i} \\ \varepsilon_{b,i} \\ \varepsilon_{m,i} \end{pmatrix} \sim \mathcal{N}\left(\begin{pmatrix} 0 \\ 0 \\ 0 \end{pmatrix}, \boldsymbol{\Sigma}_i \right),$$

and

$$\boldsymbol{\Sigma}_i = \begin{pmatrix} \sigma^2_{a,i} & 0 & 0 \\ 0 & \sigma^2_{b,i} & 0 \\ 0 & 0 & \sigma^2_{m,i} \end{pmatrix}.$$

This system allows for dynamic interdependencies between the transaction cost components $S_{a,i}$ and $S_{b,i}$. Conversely, m_i is driven by a random walk process which does not interfere with $S_{a,i}$ and $S_{b,i}$. To capture conditional heteroscedasticity, the conditional variances might be dependent on intraday seasonality functions and/or GARCH processes. For instance, Pascual and Veredas (2010) allow the long run variance $\sigma^2_{m,i}$ to follow an EGARCH specification according to Nelson (1991), whereas the short-run variance components $\sigma^2_{a,i}$ and $\sigma^2_{b,i}$ vary only according to intraday periodicities. In addition, conditional variances as well as transaction cost components $S_{a,i}$ and $S_{b,i}$ might depend on additional regressors. Pascual and Veredas (2010) illustrate how to estimate the model using the Kalman filter and apply it to 5-min limit order book data of the Spanish Stock Exchange.

Zhang et al. (2008) propose a similar decomposition inducing an asymmetric rounding mechanism generating discrete bid and ask quotes from a latent continuous process and estimate the model by Markov chain Monte Carlo methods. Hautsch et al. (2011) adapt and extend the model by Pascual and Veredas (2010) to model the ask and bid *return* processes and corresponding volatilities in periods of news announcements. This setup is presented in Chap. 8.

References

Ball C (1988) Estimation bias induced by discrete security prices. J Finance 43:841–865

Bien K, Nolte I, Pohlmeier W (2011) An inflated multivariate integer count hurdle model: an application to bid and ask quote dynamics. J Appl Econom 26:669–707

Bollerslev T, Melvin M (1994) Bid-ask spreads and volatility in the foreign exchange market. J Int Econ 36:355–372

Brumback BA, Ryan LM, Schwartz JD, Neas LM, Stark PC, Burge HA (2000) Transitional regression models with application to environmental time series. J Am Stat Assoc 85:16–27

Cameron AC, Trivedi PK (1998) Regression analysis of count data. Cambridge University Press, Cambridge

Davis RA, Dunsmuir WTM, Streett SB (2003) Observation-driven models for Poisson counts. Biometrika 90:777–790

Denuit M, Lambert P (2005) Constraints on concordance measures in bivariate discrete data. J Multivar Anal 93:40–57

Dufour A, Engle RF (2000) The ACD model: predictability of the time between consecutive trades. Working Paper, ISMA Centre, University of Reading

Efron B (1986) Double exponential families and their use in generalized linear regression. J Am Stat Assoc 81:709–721

Engle RF, Patton A (2004) Impact of trades in an error-correction model of quote prices. J Financ Markets 7:1–25

Escribano A, Pascual R (2006) Asymmetries in bid and ask responses to innovations in the trading process. Empir Econ 30(4):913–946

Ferland R, Latour A, Oraichi D (2006) Integer-valued GARCH processes. J Time Series Anal 27:923–942

Fokianos K, Rahbek A, Tjostheim D (2009) Poisson autoregression. J Am Stat Assoc 104: 1430–1439

Glosten LR, Harris LE (1988) Estimating the components of the bid/ask spread. J Finan Econ 21:123–142

Gottlieb G, Kalay A (1985) Implications of the discreteness of observed stock prices. J Finance 40:135–154

Groß-Klußmann A, Hautsch N (2011a) Predicting bid-ask spreads using long memory autoregressive conditional poisson models. Working Paper, Humboldt-Universität zu Berlin

Groß-Klußmann A, Hautsch N (2011b) When machines read the news: using automated text analytics to quantify high frequency news-implied market reactions. J Empir Financ 18:321–340

Haggan V, Ozaki T (1981) Modelling nonlinear random vibrations using an amplitude-dependent autoregressive time series model. Biometrika 68:189–196

Harris L (1990) Estimation of stock variances and serial covariances from discrete observations. J Financ QuantAnal 25:291–306

Hasbrouck J (1991) Measuring the information content of stock trades. J Finance 46:179–207

Hasbrouck J (1993) Assessing the quality of a security market: a new approach to transaction costs measurement. Rev Finan Stud 6(1):191–212

Hasbrouck J (1996) The dynamics of discrete bid and ask quotes. J Finance 6:2109–2142

Hasbrouck J (2007) Empirical market microstructure. Oxford University Press, Oxford

Hausman JA, Lo AW, MacKinlay AC (1992) An ordered probit analysis of transaction stock prices. J Finan Econ 31:319–379

Hautsch N, Hess D, Veredas D (2011) The impact of macroeconomic news on quote adjustments, noise, and informational volatility. J Bank Finance 35:2733–2746

Hautsch N, Huang R (2009) The market impact of a limit order. Discussion Paper 2009/23, Collaborative Research Center 649 "Economic Risk", Humboldt-Universität zu Berlin

Heinen A (2003) Modeling time series count data: an autoregressive conditional Poisson model. Discussion paper, Université Catholique de Louvain

Heinen A, Rengifo E (2007) Multivariate autoregressive modelling of time series count data using copulas. Empir Financ 14:564–583
Johansen S (1991) Estimation and hypothesis testing of cointegration vectors in Gaussian vector autoregressive models. Econometrica 59:1551–1580
Jung RC, Kukuk M, Liesenfeld R (2006) Time series of count data: modeling, estimation and diagnostics. Comput Stat Data Anal 51:2350–2364
Jung RC, Tremayne AR (2011) Useful models for time series of counts or simply wrong ones? Adv Stat Anal 95:59–91
Koulikov D (2003) Modeling sequences of long memory non-negative covariance stationary random variables. Discussion Paper 156, CAF
Liesenfeld R, Nolte I, Pohlmeier W (2006) Modelling financial transaction price movements: a dynamic integer count model. Empir Econ 30:795–825
Lo I, Sapp S (2006) A structural error-correction model of best prices and depths in the foreign exchange limit order market. Working paper, Bank of Canada
Madhavan A, Richardson M, Roomans M (1997) Why do security prices changes? a transaction-level analysis of NYSE stocks. Rev Financ Stud 10(4):1035–1064
Mullahy Y (1986) Specification and testing of some modified count data models. J Econom 33:341–365
Nelson D (1991) Conditional heteroskedasticity in asset returns: a new approach. J Econom 43:227–251
Pascual R, Veredas D (2010) Does the open limit order book matter in explaining long run volatility? J Financ Econom 8(1):57–87
Russell JR, Engle RF (2005) A discrete-state continuous-time model of financial transactions prices and times: the autoregressive conditional multinomial-autoregressive conditional duration model. J Bus Econ Stat 23:166–180
Rydberg TH, Shephard N (1998) Bin models for trade-by-trade data: modelling the number of trades in a fixed interval of time. Working Paper, Nuffield College, Oxford
Rydberg TH, Shephard N (2003) Dynamics of trade-by-trade price movements: decomposition and models. J Financ Econom 1:2–25
Sklar A (1959) Fonctions de répartitions à n dimensions et leurs marges. Public Institute of Statistics of the University of Paris 8:229–231
Streett S (2000) Some observation driven models for time series of counts. Ph.D. thesis, Colorado State University
Zeger SL, Qaquish B (1988) Markov regression models for time series: a quasi-likelihood approach. Biometrics 44:1019–1031
Zhang MY, Russell JR, Tsay RS (2008) Determinants of bid and ask quotes and implications for the cost of trading. J Empir Financ 15(4):656–678

Appendix A
Important Distributions for Positive-Valued Data

In the following, the gamma function $\Gamma(m)$ is defined as

$$\Gamma(m) := \int_0^\infty x^{m-1} \exp(-x) dx$$

for $m > 0$.

Poisson Distribution

Acronym: $x \sim Po(\lambda)$.

Probability mass function:

$$P(X = x) = \frac{\exp(-\lambda)\lambda^x}{x!}, \qquad x = 0, 1, 2, \ldots, \lambda > 0.$$

Mean:

$$\mathbb{E}[x] = \lambda.$$

Variance:

$$\mathbb{V}[x] = \lambda.$$

Negative Binomial Distribution

Acronym: $x \sim NegBin(r, p)$.

Probability mass function:

N. Hautsch, *Econometrics of Financial High-Frequency Data*,
DOI 10.1007/978-3-642-21925-2,

$$P(X = x) = \binom{x + r - 1}{x} (1 - p)^r p^x,$$
$$= \frac{\Gamma(x + r)}{x!\Gamma(r)} (1 - p)^r p^x$$
$$= \frac{\Gamma(x + r)}{\Gamma(x)\Gamma(r + 1)} (1 - p)^r p^x, \quad x = 0, 1, 2, \ldots, r > 0, p \in (0, 1).$$

Mean:

$$\mathbb{E}[x] = \frac{pr}{1 - p}.$$

Variance:

$$\mathbb{V}[x] = \frac{pr}{(1 - p)^2}.$$

Remark: If $r \to \infty$, then $x \to Po(\lambda)$.

Log-Normal Distribution

Acronym: $x \sim LN(\mu, \sigma^2)$.
Probability density function:

$$f(x) = \frac{1}{x\sigma\sqrt{2\pi}} \exp\left(-\frac{1}{2\sigma^2}(\ln(x) - \mu)^2\right), \qquad x > 0.$$

Cumulative density function:

$$F(x) = \Phi\left(\frac{\ln(x) - \mu}{\sigma}\right),$$

where $\Phi(\cdot)$ denotes the c.d.f. of the standard normal distribution.
Uncentered moments:

$$\mathbb{E}[x^s] = \exp\left(s\mu + \frac{\sigma^2 s^2}{2}\right).$$

Variance:

$$\mathbb{V}[x] = \exp\left(2\mu + \sigma^2\right)\left(\exp(\sigma^2) - 1\right).$$

Remark: If $x \sim LN(\mu, \sigma^2)$, then $\ln(x) \sim N(\mu, \sigma^2)$.

Exponential Distribution

Acronym: $x \sim Exp(\lambda)$.
Probability density function:

$$f(x) = \frac{1}{\lambda}\exp\left(-\frac{x}{\lambda}\right), \qquad \lambda > 0.$$

Cumulative density function:

$$F(x) = 1 - \exp\left(-\frac{x}{\lambda}\right).$$

Uncentered moments:

$$\mathbb{E}[x^s] = \lambda^s \Gamma(1+s).$$

Variance:

$$\mathbb{V}[x] = \lambda^2.$$

Hazard function:

$$\tilde{\lambda}(x) = 1/\lambda.$$

Gamma Distribution

Acronym: $x \sim \mathcal{G}(\lambda, m)$.
Probability density function:

$$f(x) = \frac{x^{m-1}\exp(-x/\lambda)}{\lambda^m \Gamma(m)}, \qquad x > 0, \lambda > 0, m > 0.$$

Uncentered moments:

$$\mathbb{E}[x^s] = \frac{\lambda^s \Gamma(m+s)}{\Gamma(m)}, \qquad m + s > 0.$$

Variance:

$$\mathbb{V}[x] = m\lambda^2.$$

Remark: If $x \sim \mathcal{G}(\lambda, m)$, then $x/\lambda \sim \mathcal{G}(1, m)$. A $\mathcal{G}(\lambda, 1)$ distribution is equivalent to an *Exp*(λ) distribution.

Weibull Distribution

Acronym: $x \sim W(\lambda, a)$.

Probability density function:

$$f(x) = a\lambda^{-a}x^{a-1}\exp\left[-\left(\frac{x}{\lambda}\right)^a\right], \qquad x > 0, \lambda > 0, a > 0.$$

Cumulative density function:

$$F(x) = 1 - \exp\left[-\left(\frac{x}{\lambda}\right)^a\right].$$

Uncentered moments:

$$\mathbb{E}[x^s] = \lambda^s \Gamma(1 + s/a).$$

Variance:

$$\mathbb{V}[x] = \lambda^2\left[\Gamma(1 + 2/a) - \Gamma(1 + 1/a)^2\right].$$

Hazard function:

$$\tilde{\lambda}(x) = \frac{a}{\lambda}\left(\frac{x}{\lambda}\right)^{a-1}.$$

Remark: A $W(\lambda, 1)$ distribution is equivalent to an *Exp*(λ) distribution. A Weibull distribution with $\lambda = 1$ is called standard Weibull distribution.

Generalized Gamma Distribution

Acronym: $x \sim \mathcal{GG}(\lambda, a, m)$.

Probability density function:

$$f(x) = \frac{a}{\lambda^{am}\Gamma(m)}x^{ma-1}\exp\left[-\left(\frac{x}{\lambda}\right)^a\right], \qquad x > 0, \lambda > 0, a > 0, m > 0.$$

Uncentered moments:

$$\mathbb{E}[x^s] = \lambda^s\frac{\Gamma(m + s/a)}{\Gamma(m)}.$$

Variance:

$$\mathbb{V}[x] = \lambda^2 \left[\frac{\Gamma(m + 2/a)}{\Gamma(m)} - \left(\frac{\Gamma(m + 1/a)}{\Gamma(m)} \right)^2 \right].$$

Remark: A $\mathcal{GG}(\lambda, 1, m)$ distribution is equivalent to a $\mathcal{G}(\lambda, m)$ distribution. A $\mathcal{GG}(\lambda, a, 1)$ distribution is equivalent to a $W(\lambda, a)$ distribution.

Generalized F Distribution

Acronym: $x \sim GF(\lambda, a, m, \eta)$.

Probability density function:

$$f(x) = \frac{a x^{am-1} [\eta + (x/\lambda)^a]^{(-\eta-m)} \eta^\eta}{\lambda^{am} \mathcal{B}(m, \eta)}, \qquad x > 0, \lambda > 0, a > 0, m > 0, \eta > 0.$$

where $\mathcal{B}(\cdot)$ describes the complete beta function with $\mathcal{B}(m, \eta) = \frac{\Gamma(m)\Gamma(\eta)}{\Gamma(m+\eta)}$.

Uncentered moments:

$$\mathbb{E}[x^s] = \lambda^s \eta^{s/a} \frac{\Gamma(m + s/a)\Gamma(\eta - s/a)}{\Gamma(m)\Gamma(\eta)}, \qquad s < a\eta.$$

Remark: For $\eta \to \infty$, the $GF(\lambda, a, m, \eta)$ distribution converges to a $\mathcal{GG}(\lambda, a, m)$ distribution. A $GF(\lambda, a, 1, 1)$ distribution is equivalent to a log-logistic distribution.

Burr Distribution

(according to Lancaster 1997)

Acronym: $x \sim Burr(\lambda, a, \eta)$.

Probability density function:

$$f(x) = \frac{a}{\lambda} \left(\frac{x}{\lambda} \right)^{a-1} \left[1 + \eta \left(\frac{x}{\lambda} \right)^a \right]^{-(1+\eta^{-1})}, \qquad x > 0, \lambda > 0, a > 0, \eta > 0.$$

Cumulative density function:

$$F(x) = 1 - (1 + \eta \lambda^{-a} x^a)^{-1/\eta}.$$

Uncentered moments:

$$\mathbb{E}[x^s] = \lambda^s \frac{\Gamma(1 + s/a)\Gamma(\eta^{-1} - s/a)}{\eta^{1+s/a}\Gamma(1 + \eta^{-1})}, \qquad s < a/\eta.$$

Hazard function:

$$\tilde{\lambda}(x) = \frac{a}{\lambda}\left(\frac{x}{\lambda}\right)^{a-1}(1 + \eta\lambda^{-a}x^{a})^{-1}.$$

Remark: A *Burr*$(\lambda, a, 1)$ distribution is equivalent to a log-logistic distribution. For $\eta \to 0$, the *Burr*(λ, a, η) distribution converges to a $W(\lambda, a)$ distribution.

Extreme Value Type I Distribution

(Gumbel (minimum) distribution)

Acronym: $x \sim EV(\lambda, m)$.

Probability density function:

$$f(x) = \frac{1}{\lambda}\exp\left(\frac{x-m}{\lambda} - \exp\left(\frac{x-m}{\lambda}\right)\right).$$

Cumulative density function:

$$F(x) = \exp\left(-\exp\left(\frac{x-m}{\lambda}\right)\right).$$

Mean:

$$\mathbb{E}[x] = m + \lambda \cdot 0.5772.$$

Variance:

$$\mathbb{V}[x] = \frac{\lambda^2\pi^2}{6}.$$

Hazard function:

$$\tilde{\lambda}(x) = \exp\left(\frac{x-m}{\lambda}\right).$$

Remark: An $EV(1, 0)$ distribution is called standard extreme value type I distribution (standard Gumbel (minimum) distribution).

Burr Type II Distribution

(according to Johnston et al. 1994)

Acronym: $x \sim BurrII(\eta)$.

Probability density function:

$$f(x) = \frac{\eta \exp(-x)}{[1 + \exp(-x)]^{\eta+1}}, \qquad \eta > 0.$$

Cumulative density function:

$$F(x) = [1 + \exp(-x)]^{-\eta}.$$

Remark: For $\eta \rightarrow \infty$, the $BurrII(\eta)$ distribution converges to an (type I) $EV(1,0)$ distribution. A $BurrII(1)$ distribution is equivalent to a logistic distribution.

Pareto Distribution

Acronym: $x \sim P(\nu, m)$.

Probability density function:

$$f(x) = \frac{\nu m^{\nu}}{x^{\nu+1}}, \qquad x > m > 0.$$

Cumulative density function:

$$F(x) = 1 - \left(\frac{m}{x}\right)^{\nu}.$$

Uncentered moments:

$$\mathbb{E}[x^s] = \frac{\nu m^s}{\nu - s}, \qquad \nu > s.$$

Variance:

$$\mathbb{V}[x] = \nu m^2 \left(\frac{1}{\nu - 2} - \frac{\nu}{(\nu - 1)^2}\right).$$

Hazard function:

$$\tilde{\lambda}(x) = \frac{\nu}{x}.$$

Index

N. Hautsch, *Econometrics of Financial High-Frequency Data*,
DOI 10.1007/978-3-642-21925-2, © Springer-Verlag Berlin Heidelberg 2012